The Convergent Evolution of Agriculture in Humans and Insects

Vienna Series in Theoretical Biology

Gerd B. Müller, editor-in-chief

Thomas Pradeu and Katrin Schäfer, associate editors

The Evolution of Cognition, edited by Cecilia Heyes and Ludwig Huber, 2000

Origination of Organismal Form, edited by Gerd B. Müller and Stuart A. Newman, 2003

Environment, Development, and Evolution, edited by Brian K. Hall, Roy D. Pearson, and Gerd B. Müller, 2004

Evolution of Communication Systems, edited by D. Kimbrough Oller and Ulrike Griebel, 2004

Modularity: Understanding the Development and Evolution of Natural Complex Systems, edited by Werner Callebaut and Diego Rasskin-Gutman, 2005

Compositional Evolution: The Impact of Sex, Symbiosis, and Modularity on the Gradualist Framework of Evolution, by Richard A. Watson, 2006

Biological Emergences: Evolution by Natural Experiment, by Robert G. B. Reid, 2007

Modeling Biology: Structure, Behaviors, Evolution, edited by Manfred D. Laubichler and Gerd B. Müller, 2007

Evolution of Communicative Flexibility, edited by Kimbrough D. Oller and Ulrike Griebel, 2008

Functions in Biological and Artificial Worlds, edited by Ulrich Krohs and Peter Kroes, 2009

Cognitive Biology, edited by Luca Tommasi, Mary A. Peterson, and Lynn Nadel, 2009

Innovation in Cultural Systems, edited by Michael J. O'Brien and Stephen J. Shennan, 2009

The Major Transitions in Evolution Revisited, edited by Brett Calcott and Kim Sterelny, 2011

Transformations of Lamarckism, edited by Snait B. Gissis and Eva Jablonka, 2011

Convergent Evolution: Limited Forms Most Beautiful, by George McGhee, 2011

From Groups to Individuals, edited by Frédéric Bouchard and Philippe Huneman, 2013

Developing Scaffolds in Evolution, Culture, and Cognition, edited by Linnda R. Caporael, James Griesemer, and William C. Wimsatt, 2013

Multicellularity: Origins and Evolution, edited by Karl J. Niklas and Stuart A. Newman, 2016

Vivarium: Experimental, Quantitative, and Theoretical Biology at Vienna's Biologische Versuchsanstalt, edited by Gerd B. Müller, 2017

Landscapes of Collectivity in the Life Sciences, edited by Snait B. Gissis, Ehud Lamm, and Ayelet Shavit, 2017

Rethinking Human Evolution, edited by Jeffrey H. Schwartz, 2018

Convergent Evolution in Stone-Tool Technology, edited by Michael J. O'Brien, Briggs Buchanan, and Metin I. Erin, 2018

Evolutionary Causation: Biological and Philosophical Reflections, edited by Tobias Uller and Kevin N. LaLand, 2019

Convergent Evolution on Earth: Lessons for the Search for Extraterrestrial Life, by George McGhee, 2019

Contingency and Convergence: Toward a Cosmic Biology of Body and Mind, by Russell Powell, 2020

Rethinking Cancer: A New Understanding for the Post-Genomics Era, edited by Bernhard Strauss, Marta Bertolaso, Ingemar Ernberg, and Mina J. Bissell, 2021

Levels of Organization in the Biological Sciences, edited by Daniel S. Brooks, James DiFrisco, and William C. Wimsatt, 2021

The Convergent Evolution of Agriculture in Humans and Insects

Edited by Ted R. Schultz, Richard Gawne, and Peter N. Peregrine

The MIT Press
Cambridge, Massachusetts
London, England

This book was set in Times New Roman by Westchester Publishing Services. Printed and bound in the United States of America. The MIT Press would like to thank the anonymous peer reviewers who provided comments on drafts of this book. The generous work of academic experts is essential for establishing the authority and quality of our publications. We acknowledge with gratitude the contributions of these otherwise uncredited readers.

Library of Congress Cataloging-in-Publication Data

Names: Schultz, Ted R., editor. | Gawne, Richard, editor. | Peregrine, Peter N. (Peter Neal), 1963– editor.
Title: The convergent evolution of agriculture in humans and insects / Ted R. Schultz, Richard Gawne, Peter N. Peregrine.
Other titles: Vienna series in theoretical biology.
Description: Cambridge, Massachusetts : The MIT Press, [2021] | Series: Vienna series in theoretical biology | Includes bibliographical references and index.
Identifiers: LCCN 2021000486 | ISBN 9780262543200 (paperback)
Subjects: LCSH: Agriculture—Origin. | Domestication. | Insect rearing. | Beneficial insects—Evolution. | Crops—Evolution.
Classification: LCC GN799.A4 C668 2021 | DDC 630.9—dc23
LC record available at https://lccn.loc.gov/2021000486

10 9 8 7 6 5 4 3 2 1

Contents

Series Foreword ix
Gerd B. Müller, Thomas Pradeu, and Katrin Schäfer

Introduction 1
Ted R. Schultz, Richard Gawne, and Peter N. Peregrine

I **COMPARATIVE ANALYSES OF HUMAN AND NONHUMAN AGRICULTURE**

1 **Convergent Evolution of Agriculture in Bilaterian Animals: An Adaptive Landscape Perspective** 7
George R. McGhee

2 **The Convergent Evolution of Agriculture: A Systematic Comparative Analysis** 31
Peter N. Peregrine

II **CONFLICT AND COOPERATION IN HUMAN AND NONHUMAN AGRICULTURE**

3 **If Group Selection Is Weak, What Can Agriculture Learn from Fungus-Farming Insects?** 49
R. Ford Denison

4 **The Sociobiology of Domestication** 61
Duur K. Aanen and Niels P. R. Anten

5 **Lifetime Commitment between Social Insect Families and Their Fungal Cultivars Complicates Comparisons with Human Farming** 73
Jacobus J. Boomsma

III **THE DIVERSITY OF INSECT AGRICULTURE**

6 **Fungus-Growing Termites: An Eco-Evolutionary Perspective** 89
Judith Korb

7 **Mycangia Define the Diverse Ambrosia Beetle–Fungus Symbioses** 105
Chase G. Mayers, Thomas C. Harrington, and Peter H. W. Biedermann

8 **Agricultural and Proto-Agricultural Symbioses in Ants** 143
Ana Ješovnik and Ted R. Schultz

9 **Plant Farming by Ants: Convergence and Divergence in the Evolution of Agriculture** 161
Guillaume Chomicki

IV **PATTERNS OF CONVERGENCE IN AGRICULTURALISTS, DOMESTICATES, AND PARASITES**

10 **Coevolution in the Arable Battlefield: Pathways to Crop Domestication, Cultural Practices, and Parasitic Domesticoids** 177
Dorian Q. Fuller and Tim Denham

11 **Convergent Adaptation and Specialization of Eukaryotic Pathogens across Agricultural Systems** 209
Nicole M. Gerardo

12 **Evaluating Potential Proximate and Ultimate Causes of Phenotypic Change in the Human Skeleton over the Agricultural Transition** 225
Lumila P. Menéndez and Laura T. Buck

13 **Hammond's Law: A Mechanism Governing the Development and Evolution of Form in Domesticated Organisms** 257
Richard Gawne and Kenneth Z. McKenna

14 **The Convergent Evolution of Agriculture in Humans and Fungus-Farming Ants** 281
Ted R. Schultz

Contributors 315
Index 317

Series Foreword

Biology is a leading science in this century. As in all other sciences, progress in biology depends on the interrelations between empirical research, theory building, modeling, and societal context. But whereas molecular and experimental biology have evolved dramatically in recent years, generating a flood of highly detailed data, the integration of these results into useful theoretical frameworks has lagged behind. Driven largely by pragmatic and technical considerations, research in biology continues to be less guided by theory than seems indicated. By promoting the formulation and discussion of new theoretical concepts in the biosciences, this series intends to help fill important gaps in our understanding of some of the major open questions of biology, such as the origin and organization of organismal form, the relationship between development and evolution, and the biological bases of cognition and mind. Theoretical biology has important roots in the experimental tradition of early-twentieth-century Vienna. Paul Weiss and Ludwig von Bertalanffy were among the first to use the term *theoretical biology* in its modern sense. In their understanding the subject was not limited to mathematical formalization, as is often the case today, but extended to the conceptual foundations of biology. It is this commitment to a comprehensive and cross-disciplinary integration of theoretical concepts that the Vienna Series intends to emphasize. Today, theoretical biology has genetic, developmental, and evolutionary components, the central connective themes in modern biology, but it also includes relevant aspects of computational or systems biology and extends to the naturalistic philosophy of sciences. The Vienna Series grew out of theory-oriented workshops organized by the KLI, an international institute for the advanced study of natural complex systems. The KLI fosters research projects, workshops, book projects, and the journal *Biological Theory*, all devoted to aspects of theoretical biology, with an emphasis on—but not restriction to—integrating the developmental, evolutionary, and cognitive sciences. The series editors welcome suggestions for book projects in these domains.

Gerd B. Müller, Thomas Pradeu, and Katrin Schäfer

Introduction

Ted R. Schultz, Richard Gawne, and Peter N. Peregrine

Agriculture originated at least 14 times—and perhaps as many as 23 times—in human history (chapter 10, this volume). Agriculture also originated a minimum of 20 times in nonhuman animals, including at least 15 times in insects (chapters 1, 6, 7, and 8, this volume). In humans, agriculture has dramatically increased the numbers of both farmers and domesticates as compared to their nonfarming, undomesticated close relatives. For example, the biomass of extant humans and their mammalian and avian livestock outweighs that of all wild mammals and birds by over 17 times (Bar-On, Phillips, and Milo 2018). Insect agriculturalists likewise dominate the biosphere. The colonies of fungus-farming termites and leaf-cutting ants are highly eusocial "superorganisms" consisting of many millions of individuals divided into physically differentiated castes, enabling highly refined division of labor that has been called "civilization by instinct" (Hölldobler and Wilson 2010). The multiple origins of agriculture in both humans and nonhumans represent remarkable, repeated cases of convergent evolution that deserve to be carefully examined in a comparative context. Moreover, the demonstrable success of human and insect agriculturalists raises the question of why, in the history of life on Earth, agriculture has only arisen dozens of times instead of hundreds or thousands of times. In other words, why is agriculture so rare in animals?

To explore such questions, two previous working groups of anthropologists, archaeologists, and entomologists were convened at the Santa Fe Institute in 2014 (August 28–29) and 2016 (April 10–12). These working groups concluded that the convergent evolution of agriculture may be explained, at least in part, by energetic benefits that include increased buffering against environmental variability due to dependable food resources and reduction of time spent foraging, especially for wild sources of protein for which competition from other species is fierce. The groups also identified a set of specific preconditions that likely must be present for agriculture to evolve, including (1) future farmers must be generalized foragers that create central places for food storage, distribution, and consumption; (2) future farmers must interact with a plant or animal species that is genetically or behaviorally pre-adapted for domestication; (3) future farmers must be highly social and capable of communicating in a manner that spans generations and allows for social learning or conditioning; (4) favorable mutations or innovations must be replicated rapidly across generations; and (5) the climate must be relatively stable. These preconditions may be quite rare, which thus explains the relative rarity of agriculture. Although these conclusions were intriguing, they were judged to be preliminary, and so, on June 13–16, 2019, the 38th Altenberg Workshop in Theoretical Biology was convened at the Konrad Lorenz Institute in Klosterneuburg, Austria, in order to more thoroughly explore the topic of "The Convergent Evolution of Agriculture in Humans and Insects."

This volume contains the products of that workshop. In part I, two different analyses codify and quantify the traits held in common across human and animal agriculturalists, one more broadly, spanning the entire animal kingdom (chapter 1), and one more narrowly, focusing on multiple human groups and a subset of insect agriculturalists (chapter 2). In part II, various authors examine both human and insect agriculture from a group-versus-individual-selection perspective, reasoning that it is unlikely that natural selection will generate an evolutionarily stable agricultural mutualism when farmers and cultivars—as well as individual cultivars—have conflicting interests. These chapters provide insights about the limited conditions under which we might expect agriculture to evolve and, thus, why it may be rare. The chapters in part III summarize the biology and evolution of the major groups of insect agriculturalists, including aspects of their biologies that have never before been reviewed. In part IV, chapters 10 and 12 explore different aspects of convergent evolution globally across human agricultural systems; chapter 11 compares fungal parasites of ant agriculture with fungal parasites of human agriculture; chapter 13 explores a general developmental mechanism for generating morphological form that provides insights into the rapid evolution of human and possibly nonhuman plant and animal domesticates; and chapter 14 provides an in-depth comparison of the evolution of agriculture in humans and in a single, relatively well-studied insect group, fungus-farming ants.

The evolution of agriculture is studied by scholars working in many different fields. For the most part, research on this topic has been conducted in relative isolation, discipline by discipline. The isolation of students of human and nonhuman agriculture, respectively, is particularly dramatic. To date, there has been remarkably little contact between researchers in these fields, and as a result many opportunities for fruitful interactions have been missed. The aim of this volume is to bring researchers working on human and insect agriculture together for the first time, with the goal of promoting an interactive dialogue that has the potential to bring about new empirical and theoretical advances. Ultimately, we hope that this book will serve as a foundation for future studies of agriculture from a comparative point of view.

Definitions

Throughout this book, the authors use the terms "cultivation," "domestication," and "agriculture." Unless otherwise stated in a specific instance, these terms are defined as follows:

cultivation: Targeted intervention in the life cycle of a particular species in order to promote its growth, including practices such as tillage, planting, and harvesting.

domestication: Genetic modification of one species (the domesticate) by another (the farmer) in ways that benefit the farmer but that would reduce the fitness of the domesticate in its original niche.

agriculture: Cultivation on a large scale in which the farmer has become obligately dependent on the cultivated species for nutrition.

Workshops

2014 Santa Fe Institute Working Group: Jennifer A. Dunne, David C. Krakauer, George R. McGhee, Ulrich G. Mueller, Peter N. Peregrine, Ted R. Schultz, Bruce D. Smith, Melinda A. Zeder

2016 Santa Fe Institute Working Group: Duur K. Aanen, Richard Gawne, Marcus J. Hamilton, Jiri Hulcr, George R. McGhee, Ulrich G. Mueller, Peter N. Peregrine, Ted R. Schultz

Participants in the 38th Altenberg Workshop in Theoretical Biology, "The Convergent Evolution of Agriculture in Humans and Insects," June 13–16, 2019, Konrad Lorenz Institute, Klosterneuburg, Austria. (1) Ted R. Schultz, (2) R. Ford Denison, (3) Judith Korb, (4) Dorian Q. Fuller, (5) Peter H.W. Biedermann, (6) Rick Gawne, (7) Peter N. Peregrine, (8) Hanna Kokko, (9) Jacobus J. Boomsma, (10) Susan Milius, (11) Duur K. Aanen, (12) Ana Ješovnik, (13) Nicole Gerardo, (14) George R. McGhee. Not pictured: Jiri Hulcr.

2019 KLI Altenberg Workshop: Duur K. Aanen, Peter H.W. Biedermann, Jacobus J. (Koos) Boomsma, R. Ford Denison, Dorian Q. Fuller, Richard Gawne, Nicole M. Gerado, Jiri Hulcr, Ana Jesovnik, Hanna Kokko, Judith Korb, George R. McGhee, Susan Milius, Dolores R. Piperno, Peter N. Peregrine, Ted R. Schultz

Acknowledgments

The editors would like to express their gratitude to the staff of the Konrad Lorenz Institute for hosting the workshop that led to this book. We are particularly grateful to KLI President Gerd B. Müller for his commitment to our cross-disciplinary topic; to Scientific Director Guido Caniglia for facilitating our intellectual interactions; to Executive Manager Isabella Sarto-Jackson for her help with program scheduling and management; and to Coordinator Eva Lackner for handling all of the complex logistics associated with our travel, lodging, and activities. We add special thanks to Mustafa Mujkic, Vahida Mujkic, and Viktoria Zsanko for preparing the conference facilities. The editors would like to thank all of the workshop attendees for their willingness to risk participating in a workshop devoted to such an unusual subject. In retrospect, the rare confluence of individuals from such diverse disciplines proved to be exceptionally thought-provoking and inspiring. In addition to the memorable intellectual stimulation of the workshop presentations and discussions, we recall many fascinating and productive conversations over lunches, dinners, and drinks. Finally, we thank the contributing authors for making this edited volume unique and, we hope, synthetic.

References

Bar-On, Y. M., R. Phillips, and R. Milo. 2018. "The Biomass Distribution on Earth." *Proceedings of the National Academy of Sciences* 115(25): 6506. doi:10.1073/pnas.1711842115.

Hölldobler, B., and E. O. Wilson. 2010. *The Leafcutter Ants: Civilization by Instinct.* New York: W. W. Norton.

I COMPARATIVE ANALYSES OF HUMAN AND NONHUMAN AGRICULTURE

1 Convergent Evolution of Agriculture in Bilaterian Animals: An Adaptive Landscape Perspective

George R. McGhee

In all cases of two very distinct species furnished with apparently the same anomalous organ, it should be observed that, although the general appearance and function of the organ may be the same, yet some fundamental difference can generally be detected. I am inclined to believe that in nearly the same way as two men have sometimes independently hit on the very same invention, so natural selection, working for the good of each being and taking advantage of analogous variations, has sometimes modified in very nearly the same manner two parts in two organic beings, which owe but little of their structure in common to inheritance from the same ancestor.
—Charles Darwin, *On the Origin of Species* (1859, 193–194)

In 1859 Darwin faced the dilemma of dealing with the problem of convergence in biological evolution—that is, of accounting for morphological traits and instinctive behaviors that were "almost identically the same in animals so remote in the scale of nature, that we cannot account for their similarity by inheritance from a common parent, and must therefore believe that they have been acquired by independent acts of natural selection" (Darwin 1859, 235–236). In an attempt to explain to his readers how natural selection might produce convergent biological evolution, Darwin (1859, 193–194) called upon the human analogy that "two men have sometimes independently hit on the very same invention"—a phenomenon he was very well aware of as the theory of natural selection itself had been independently formulated by both Darwin and Wallace, just as the logic of the calculus had been independently formulated by both Newton and Leibniz, and so on throughout human history.

Lycett (2011, 157) has pointed out that in 1859, when Darwin used the analogy of two men independently creating the very same invention, he was de facto giving an "example of convergence within the realm of human technology." The analysis of convergent evolution in a very specific form of human technology—the creation of similar stone tools in different human cultures—was the subject of the 33rd Altenberg Workshop in Theoretical Biology at the Konrad Lorenz Institute for Evolution and Cognition Research (O'Brien, Buchanan, and Eren 2018).

The 38th Altenberg Workshop in Theoretical Biology at the KLI continues that research focus by examining the convergent evolution of agriculture—except this time the focus is not only on the convergent evolution of agricultural techniques in different human cultures, but also on the convergent evolution of agricultural techniques in different clades of animals that are not human. Agriculture has evolved, independently and convergently, in nine phylogenetic lineages of bilaterian animals: actinopterygian vertebrates, sarcopterygian vertebrates, polychaete annelids, patellogastropod molluscs, neogastropod molluscs, littorinoid molluscs, polyneopteran insects, hymenopteran insects, and coleopteran insects (table 1.1). The analysis of that agricultural convergence from an adaptive landscape perspective is the subject of this chapter.

Table 1.1
Phylogeny of the convergent evolution of agriculture in bilaterian animals

Metazoa
– Ctenophora
– †Archaeocyatha
– Porifera
– Parahoxozoa
– – Placozoa
– – Eumetazoa
– – – Cnidaria
– – – BILATERIA (agriculturalist lineages 1–9)
– – – – Unnamed clade
– – – – – Xenoturbellida
– – – – – Acoelomorpha
– – – – – – Nemertodermatida
– – – – – – Acoela
– – – – Nephrozoa
– – – – – DEUTEROSTOMIA (agriculturalist lineages 1–2)
– – – – – – Ambulacraria
– – – – – – – Echinodermata
– – – – – – – Hemichordata
– – – – – – CHORDATA
– – – – – – – Vertebrata
– – – – – – – – Gnathostomata
– – – – – – – – – Osteichthyes
– – – – – – – – – – ACTINOPTERYGII (agriculturalist lineage 1)
– – – – – – – – – – – Neopterygii
– – – – – – – – – – – – Teleostei
– – – – – – – – – – – – – Euteleostei
– – – – – – – – – – – – – – Eurypterygii
– – – – – – – – – – – – – – – Ctenosquamata
– – – – – – – – – – – – – – – – Acanthomorpha
– – – – – – – – – – – – – – – – – Acanthopterygii
– – – – – – – – – – – – – – – – – – Percomorpha
– – – – – – – – – – – – – – – – – – – Ovalentaria
– Pomacentridae
– Pomacentrinae (*Stegastes nigricans*) **lineage 1**
– – – – – – – – – – SARCOPTERYGII (agriculturalist lineage 2)
– – – – – – – – – – – Tetrapodomorpha
– – – – – – – – – – – – Tetrapoda
– – – – – – – – – – – – – Reptiliomorpha

Table 1.1
(continued)

– – – – – – – – – – – – – – Amniota
– – – – – – – – – – – – – – – Synapsida
– – – – – – – – – – – – – – – – Therapsida
– – – – – – – – – – – – – – – – – Cynodontia
– – – – – – – – – – – – – – – – – – Mammalia
– – – – – – – – – – – – – – – – – – – Theria
– Eutheria
– Boreoeutheria
– Euarchontoglires
– Primates
– Anthropoidea
– Catarrhini
– Hominoidea
– Hominidae
– Homininae
– Hominini (*Homo sapiens*) **lineage 2**
– – – – – PROTOSTOMIA (agriculturalist lineages 3–9)
– – – – – – SPIRALIA (agriculturalist lineages 3–6)
– – – – – – – Platytrochozoa
– – – – – – – – Lophotrochozoa
– – – – – – – – – ANNELIDA (agriculturalist lineage 3)
– – – – – – – – – – Polychaeta
– – – – – – – – – – – Aciculata
– – – – – – – – – – – – Phyllodocida
– – – – – – – – – – – – – Nereididae (*Platynereis dumerilii*) **lineage 3**
– – – – – – – – – MOLLUSCA (agriculturalist lineages 4–6)
– – – – – – – – – – Eumollusca
– – – – – – – – – – – Ganglioneura
– – – – – – – – – – – – Visceroconcha
– – – – – – – – – – – – – Gastropoda
– – – – – – – – – – – – – – Patellogastropoda
– – – – – – – – – – – – – – – Patelloidea
– – – – – – – – – – – – – – – – Scutellastridae (*Scutellastra longicosta*) **lineage 4**
– – – – – – – – – – – – – – Caenogastropoda
– – – – – – – – – – – – – – – Hypsogastropoda
– – – – – – – – – – – – – – – – Neogastropoda
– – – – – – – – – – – – – – – – – Buccinoidea
– – – – – – – – – – – – – – – – – – Nassariidae (*Bullia digitalis*) **lineage 5**
– – – – – – – – – – – – – – – – Littorinimorpha
– – – – – – – – – – – – – – – – – Littorinoidea
– – – – – – – – – – – – – – – – – – Littorinidae (*Littoraria irrorata*) **lineage 6**
– – – – – – ECDYSOZOA (agriculturalist lineages 7–9)
– – – – – – – Panarthropoda
– – – – – – – – Arthropoda
– – – – – – – – – Euarthropoda

(*continued*)

Table 1.1
(continued)

– – – – – – – – – – Mandibulata
– – – – – – – – – – – Pancrustacea
– – – – – – – – – – – – Hexapoda
– – – – – – – – – – – – – Insecta
– – – – – – – – – – – – – – POLYNEOPTERA
– – – – – – – – – – – – – – – Isoptera
– – – – – – – – – – – – – – – – Termitidae
– – – – – – – – – – – – – – – – – Macrotermitinae (*Odontotermes formosanus*) **lineage 7**
– – – – – – – – – – – – – – HOLOMETABOLA
– – – – – – – – – – – – – – – HYMENOPTERA
– – – – – – – – – – – – – – – – Formicidae
– – – – – – – – – – – – – – – – – Myrmicinae
– – – – – – – – – – – – – – – – – – Attini
– – – – – – – – – – – – – – – – – – – Attina
– Neoattina (*Atta texana*) **lineage 8**
– – – – – – – – – – – – – – – COLEOPTERA
– – – – – – – – – – – – – – – – Curculionidae
– – – – – – – – – – – – – – – – – Scolytinae
– – – – – – – – – – – – – – – – – – Xyleborini
– – – – – – – – – – – – – – – – – – – *Ambrosiodmus/Ambrosiophilus* clade
– (*Ambrosiophilus atratus*) **lineage 9**

Sources: Phylogeny modified from Nielsen (2012), Near et al. (2013), Bourguignon et al. (2014), Misof et al. (2014), Struck et al. (2014), Benton (2015), Li et al. (2015), and Branstetter et al. (2017).

Note: Major clades containing agriculturalist animals are given in capital letters. Extinct taxa are marked with a dagger "†."

The Phenomenon of Convergent Evolution in Agricultural Behaviors

A list of the farming species and their agricultural crops is given in table 1.2. Five of the farming lineages cultivate plants, and four cultivate fungi. Of the plant crops, four lineages farm marine algae, and one lineage farms land plants. Of the fungal crops, three lineages farm land fungi, and one lineage farms marine fungi.

A list of 12 agricultural traits is here used to create a comparative measure of the agricultural complexity found in the nine independent lineages of agriculturalist animals, modified from the list of agricultural traits found in insects and humans compiled by Mueller et al. (2005). These traits are: (A) preparing the substrate for growing crops, (B) planting the crops, (C) usage of naturally occurring organic fertilizers for the crops, (D) usage of artificially produced chemical fertilizers for the crops, (E) protecting the crops from raiders, (F) weeding invasive species out of the crops, (G) using chemicals for pest control of parasites or diseases in the crops, (H) using living microbes for pest control of parasites or diseases in the crops, (I) development of a one-way symbiosis between the farming species and its crop (either the crop is obligately dependent on the farmer, or the farmer is obligately dependent on the crop), (J) development of a two-way mutualism between the farming species and its crop (both farmer and crop are obligately dependent

Table 1.2
Convergent agricultural lineages, their farming animals, and their crops

Lineage	Farmer	Crop
1	Damselfish (*Stegastes nigricans*)	Photosynthetic marine red alga (*Polysiphonia* sp. 1)
2	Human primate (*Homo sapiens*)	Photosynthetic terrestrial plants (numerous species)
3	Polychaete annelid worm (*Platynereis dumerilii*)	Photosynthetic marine algae (several species)
4	Limpet snail (*Scutellastra longicosta*)	Photosynthetic marine brown alga (*Ralfsia verrucosa*)
5	Whelk snail (*Bullia digitalis*)	Photosynthetic marine green alga (*Eugomontia sacculata*)
6	Periwinkle snail (*Littoraria irrorata*)	Saprophytic intertidal fungi (species of *Phaerosphaeria* and *Mycosphaerella*)
7	Macrotermitine termite (*Odontotermes formosanus*)	Saprophytic terrestrial fungi (species of *Termitomyces*)
8	Leaf-cutter ant (*Atta texana*)	Saprophytic terrestrial fungus (*Leucoagaricus gonglyophorus*)
9	Ambrosia beetle (*Ambrosiophilus atratus*)	Saprophytic terrestrial fungus (*Flavodon ambrosius*)

on one another), (K) the usage of artificial selection to improve the crops, and (L) the usage of genetic engineering to improve the crops.

The distribution of these 12 agricultural traits across the nine lineages of agriculturalist animals is given in table 1.3. The metric of "agricultural complexity" is here defined as the number of agricultural traits found in a given lineage of agriculturalist animals. The agricultural-complexity level achieved by each of the nine agriculturalist lineages is given in the "sum of convergent agricultural behaviors" row at the bottom of table 1.3. In this summation a two-way obligate mutualism between the farmer and the crop (trait J) is counted as being the sum of two one-way obligate symbioses (trait I); that is, trait I (Y) + trait I (Y) = trait J (2Y).

Two of the agriculturalist lineages are found in the clade of the deuterostomes—the damselfishes and humans (table 1.1). Humans have complex agricultures that can exhibit nine of the 12 agricultural traits chosen here to measure differential agricultural complexity (table 1.3). In contrast to some other agriculturalist animals, humans and their crops have a generally facultative relationship with one another rather than an obligate one, and humans do not use (as yet) microbial pest-control methods (table 1.3; see also the discussion in Mueller et al. 2005). Human agriculture is itself convergent, in that several populations of humans made the transition from hunter-gatherers to agriculturalists independently of each other around the world about 10,000 years ago (Gupta 2004; chapter 10, this volume). Of the other agriculturalist animals only the clades of the ambrosia beetles and damselfishes have exhibited numerous independent convergences in agricultural evolution: 11 independent lineages of ambrosia beetles have arisen in the past 50 million years (Jordal and Cognato 2012; Li et al. 2015; chapter 7, this volume), and, in the damselfishes, "multiple independent transitions to algaculture have occurred" (Aanen 2010, 3).

Table 1.3
Convergent evolution of specific agricultural traits in nine lineages of bilaterian animals

	Phylogenetic lineage								
Behavior	1	2	3	4	5	6	7	8	9
A. Substrate preparation for crops	Y	Y	Y	Y	Y	Y	Y	Y	Y
B. Planting the crops	n	Y	Y	n	n	Y	Y	Y	Y
C. Fertilizing the crops: organics	n	Y	Y	Y	n	Y	Y	Y	n
D. Fertilizing the crops: chemicals	n	Y	n	n	n	n	n	n	n
E. Protecting the crops from raiders	Y	Y	n	Y	Y	n	Y	Y	Y
F. Weeding invasive species out of crops	Y	Y	n	Y	Y	n	Y	Y	Y
G. Pest control: chemicals	n	Y	n	n	n	n	n	Y	n
H. Pest control: microbes	n	n	n	n	n	n	n	Y	Y
I. Symbiosis: one-way obligate	n	n	n	Y	n	Y	n	n	n
J. Mutualism: two-way obligate	2Y	n	n	n	n	n	2Y	2Y	2Y
K. Artificial selection to improve crops	n	Y	n	n	n	n	Y	Y	n
L. Genetic engineering to improve crops	n	Y	n	n	n	n	n	n	n
Sum of convergent agricultural behaviors	5	9	3	5	3	4	8	10	7

Notes: Y = yes, present; n = no, absent. Phylogenetic lineages: 1 = damselfish, 2 = human primate, 3 = polychaete annelid worm, 4 = limpet snail, 5 = whelk snail, 6 = periwinkle snail, 7 = macrotermitine termite, 8 = leaf-cutter ant, 9 = ambrosia beetle.

In contrast to humans, the damselfishes have simpler agricultures that exhibit only five of the 12 agricultural traits (table 1.3). Hata and Kato (2006, 159) have observed that species of the damselfishes maintain "dense stands of filamentous algae as algal farms." In particular, the "territorial damselfish, *Stegastes nigricans*, maintains algal farms by excluding invading herbivores and weeding unpalatable algae from its territories" (Hata and Umezawa 2011, 809); it also carefully grazes its crops to stimulate the algae to remain in the rapid-growth, early-succession phase (Hato, Watanabe, and Kato 2010; Aanen 2010). Hata and Kato (2006) have further noted that the mutualism between the damselfish *Stegastes nigricans* and the red alga *Polysiphonia* sp. 1 is "the first record of an obligate plant-herbivore cultivation mutualism in a marine ecosystem" (see also Hata and Kato 2004; Hata, Watanabe, and Kato 2010; Hata and Umezawa 2011). Aanen (2010, 1, 3) proposed that this "fascinating agricultural mutualism" was started by the damselfish's "weeding or consumption of all the algae in a territory, and that the fast-growing algae remained, and started to specialize on this new niche. With their behavior, the fish shows some parallels with cattle, which by grazing create and improve their own niche—grassland. The difference is that the fish seem to remove unpalatable algae without consuming these, and that they are territorial."

Hata, Watanabe, and Kato (2010, 2), however, have noted that "this fish neither sows nor transplants the algae." Further, *Stegastes nigricans* has not been observed to fertilize its crops or to use any type of pest control against crop parasites or diseases, and the damselfish does not use artificial selection or genetic engineering to improve its crops, unlike humans.

The other seven agriculturalist animals are all protostomes. Four of the agriculturalist lineages are found in the clade of the spiralian protostomes and all are marine—the polychaete annelid worms, the limpet snails, the whelk snails, and the periwinkle snails (table 1.1). Of these, the limpet snails have the most complex agricultures, consisting of five of the 12 agricultural traits; thus, this gastropod mollusc has an agricultural complexity rivaling that of an actinopterygian chordate (tables 1.1 and 1.3). Species of the scutellastrid limpets cultivate *Ralfsia verrucosa* algae in two types of gardens: periphery gardens (*Scutellastra cochlear*, *S. flexuosa*, *S. mexicana*, and others) and patch gardens (*Scutellastra longicosta*, *S. laticostra*, *S. chapmani*, and others; Lindberg 2007). Periphery-gardening limpets cultivate algae in a zone around the periphery of a permanent home base, in the center of which the snail resides. These limpets prepare their periphery substrates by rasping the coral surface, "and the alga within the garden area is restricted to the interstices of the rasped surface" (Lindberg 2007, 222). The more mobile patch-gardening limpets cultivate algae in larger patches over which the snails graze.

Both types of gardening limpets fertilize their gardens, but in a different manner: by the release of nitrogenous excretions (ammonium and urea in their urine and feces) in the periphery gardeners, and by the spreading of nutrient-rich mucus in mucus trails from their feet by the patch gardeners (Lindberg 2007). Field measurements have shown that the periphery-gardening species *Scutellastra cochlear* fertilizes its crop with "nitrogenous excretions released along the shell edge" (Lindberg 2007, 229). Plagányi and Branch (2000, 113) have demonstrated that the amount of ammonium (NH_4) excreted by *Scutellastra cochlear* on its gardens "supplied approximately 30% of the algal garden's daily nitrogen growth requirements" and that these limpets "enhance the productivity of their algal gardens through the regeneration of limiting nutrients."

The patch-gardening snail *Scutellastra longicosta* has been demonstrated to weed its gardens—it "removes algal species other than *Ralfsia* from its territory"—and its weeding activity is "important in preventing the competitive exclusion of *Ralfsia* by foliose algae," particularly the sea lettuce *Ulva* (McQuaid and Froneman 1993, 132). The scutellastrid limpets also defend their crops from raiding herbivores, as *Scutellastra longicosta* was observed to have "actively excluded vagrant grazers, preventing overgrazing" of the crop (McQuaid and Froneman 1993, 128). Likewise, "by keeping the site free of other grazers, *Patella* [= *Scutellastra*] *longicosta* ensures the survival of *Ralfsia* plants" (McQuaid and Froneman 1993, 132).

In their study of the "mutualism between the territorial intertidal limpet *Patella* [= *Scutellastra*] *longicosta* and crustose alga *Ralfsia verrucosa*," McQuaid and Froneman (1993, 128) noted that "adult *Patella* [= *Scutellastra*] are found exclusively in association with *Ralfsia*, on which they feed, while *Ralfsia* occurs primarily, but not exclusively, in *Patella* [= *Scutellastra*] territories." That is, the farmer species is obligately dependent on its particular crop but the crop species, although benefiting from the activities of the farmer species, is not obligately dependent upon it (table 1.3).

As with the agriculturalist damselfishes, the agriculturalist limpets have not been observed to plant their algal crops or to use any type of pest control against crop parasites or diseases, and they are not known to use artificial selection or genetic engineering to improve their crops. Although they do use natural, organic fertilizers, they do not use chemical ones as the agriculturalist humans do.

The remaining three agriculturalist spiralian-protostome lineages (table 1.1) all have even simpler agricultures, consisting of only three or four of the 12 agricultural traits (table 1.3).

The nereidid polychaete annelid *Platynereis dumerilii* uses its own self-grown mucus-tube habitat as the substratum for its algal garden, according to Levinton (1982, 284), who has noted that some species of nereidids "attach pieces of sea lettuce (*Ulva*) to their tubes and maintain algal gardens." Gambi et al. (2000) have further described the gardening of filamentous algae by the nereidid *Platynereis dumerilii* on leaves of the Mediterranean seagrass *Posidonia oceanica*, whereby the worms feed on erect filaments of algae, particularly the large macroalgae *Cystoseria*. They have noted that the worms "live inside semi-permanent mucous tubes, that are generally attached to macroalgal thalli. . . . *P. dumerilii* generally feeds close to the tube entrance, to which worms attach small pieces of algae" (Gambi et al. 2000, 190). At the tube entrance, these pieces of algae are also in close proximity to the nitrogenous urine and feces of the polychaete and thus may be fertilized, even if inadvertently.

Woodin (1977) has argued that the gardening behavior of *Platynereis dumerilii* is an adaptation to reduce the risk of predation. Gambi et al. (2000, 199) agree with this hypothesis, noting that "*P. dumerilii* feeding requires the partial or total exit of the worm from the tube" and that during such "searching for food, the worms are more vulnerable to potential predators." Maintaining a garden of crop algae on the worm's tube itself, near its opening, is clearly advantageous to the worm.

The nassariid whelk snail *Bullia digitalis* also has an agricultural technique that uses only three of the 12 agricultural traits, but they are not the same three traits as those used by the polychaete worms (table 1.3). Like *Platynereis dumerilii*, *Bullia digitalis* uses its own, self-grown calcareous-shell habitat as the substratum for its algal garden. Da Silva and Brown (1984, 64) have noted that the snail "frequently has an algal growth on the upper surface of its shell and especially on the last whorl," and that "only a single species of alga is present and that in all cases it appears to have been cropped almost down to the shell." Harris et al. (1986) further note that the garden consists of chlorophyte algae that bore into the shell material of the snail, that the snail periodically grazes on filamentous strands of the algae that protrude from its shell, and that the most commonly seen alga appears to be *Eugomontia sacculata*.

Unlike *Platynereis dumerilii*, *Bullia digitalis* has not been observed to plant the alga on its shell, but the presence of "only a single species of alga" (Da Silva and Brown 1984, 64) argues for weeding behavior by the snail to prevent other algal species from overgrowing its crop, and obviously the snail does not allow other species to graze its shell, thus defending its crop. Because the snail is mobile in high-energy sandy beaches (Harris et al. 1986), it is unlikely that it fertilizes its crop with nitrogenous excretions as the agriculturalist limpets do, since these excretions would be quickly washed away rather than being absorbed by the alga. The snail appears to use its algal garden as a supplementary source of food to augment its normal scavenging mode of feeding.

Last, the littorinid periwinkle snail *Littoraria irrorata* also has a simpler agricultural technique consisting of only four agricultural traits (table 1.3), but of overwhelming interest is that this periwinkle snail farms saprophytic fungi rather than photosynthetic algae (table 1.2). Thus, in its crop choice this agriculturalist marine species is convergent with the land-dwelling insect agriculturalists, all of which cultivate fungi instead of plants (table 1.2). As summarized by Silliman and Newell (2003, 15643), field observations confirm a "farming mutualism between *Littoraria* and intertidal fungi. Snails graze live grass primarily not to feed but to prepare substrate for fungal growth and consume invasive fungi. . . .

These results provide a case of fungal farming in the marine environment and outside the class Insecta."

Littoraria irrorata prepares the substrate for its garden by actively producing longitudinal wounds with its radula on the leaves of the salt marsh cordgrass *Spartina alternifora*; these wounds are then colonized by species of the ascomycete fungi *Phaerosphaeria* and *Mycosphaerella* (Silliman and Newell 2003; Sieg et al. 2013). Silliman and Newell (2003, 15643) have argued that "(*i*) *Littoraria* promote fungal growth on live *Spartina* plants through their grazing activities and direct application of fecal pellets and that this growth promotion has a positive effect on snail growth; and (*ii*) fungi benefit from snail wound-grazing by gaining access to nutritious and relatively defenseless inner plant tissues and by receiving supplements (potentially nutrients and/or propagules) from snail feces." Further, Sieg et al. (2013, 122) have observed that *Littoraria irrorata* snails "consume fungal hyphae during garden maintenance"; they have also been seen engaging in "selectively depositing hyphae-laden feces within wounded plant tissues to facilitate fungal establishment and growth" and actively engaging in the "establishment of new fungal cultures on plants by defecating in wounded plant tissues" (Sieg et al. 2013, 128).

Silliman and Newell (2003, 15647) have further argued that *Littoraria irrorata* "snails can grow only on fungus-colonized *Spartina* or pure mycelium and not on sterile leaves or ones colonized by bacteria. These growth-study results indicate that *Littoraria* obligately employs fungus-promoting feeding strategies . . . and consumption of facilitated invasive fungi." Thus, the farmer is obligately dependent upon the presence of the fungal crop, but the fungal crop, although benefiting from the activities of the farmer species, is not obligately dependent upon it (table 1.3). Sieg et al. (2013, 122) have further noted that this "snail-fungal mutualism increases exposure of the foundation species *Spartina altinoflora* to infection, whereas fungal farming on other local marsh plants is rarely observed. . . . We propose that the weak chemical defenses produced by *S. alterniflora* make it a preferred plant for fungal farming, whereas less abundant species that invest in potential chemical defenses against grazers and fungi are not as frequently colonized, consumed, or subjected to fungal farming by *L. irrorata*."

The last three of the agriculturalist lineages are found in the clade of the ecdysozoan protostomes and all are land dwellers—the macrotermitine termites, the leaf-cutter ants, and the *Ambrosiodmus*/*Ambrosiophilus*-clade ambrosia beetles (table 1.1). All three ecdysozoan agriculturalist animals farm fungi rather than plants, and all three have a two-way obligate relationship between the farming animals and their crops (tables 1.2 and 1.3).

The macrotermitine termites have an agricultural technique that uses eight of the 12 agricultural traits and is almost as complex as human agriculture (table 1.3). Unlike the human-plant facultative agricultural mutualism, the relationship between the macrotermitine termites and their crops, species of *Termitomyces* fungi, is one of obligate mutualism: "termites depend on the fungi for food, and the fungi depend on the termites for growth and protection" (Aanen 2006, 209; see also Aanen et al. 2002; Mueller et al. 2005; Aanen and Eggleton 2017; chapter 6, this volume). Also unlike humans, the termites are not known to use any type of pest control against crop parasites or diseases (Mueller et al. 2005).

In addition to substrate preparation, crop planting, crop fertilization with organics, and protection and weeding of crops (Mueller et al. 2005), Aanen (2006, 209) also observed that "the termites 'artificially' select for high nodule production" to improve their *Termitomyces* crops and that "genetic screening of *Termitomyces* strain diversity happens in at least some of the genera either directly through active selection of symbionts or indirectly through inter-strain competition for comb space."

The leaf-cutter ants have an agricultural technique that uses ten of the 12 agricultural traits—an agriculture that is more complex than human agriculture in this analysis (table 1.3). Like humans, the ants use "chemical herbicides to combat pests," but, unlike humans (as yet), the ants also use "disease-suppressant microbes for biological pest control" (Mueller et al. 2005, 565; chapter 11, this volume), in particular an "antibiotic produced by the *Pseudonocardia* bacterial symbiont" that is used against fungal parasites (Schultz and Brady 2008). Also unlike the human-plant agricultural mutualism, the mutualism between the leaf-cutter ants and their fungal crops is obligate: all attine ants obligately depend on the cultivation of fungus gardens for food, and, whereas the fungal cultivars of the non-leaf-cutting "lower" attine ants are facultative symbionts, ants in the leaf-cutter genera *Atta* and *Acromyrmex* cultivate "higher attine" fungi (principally *Leucoagaricus gongylophorus*) that are incapable of living separately from their ant farmers (Schultz and Brady 2008; Branstetter et al. 2017; chapters 8 and 14, this volume).

Like termites, ants practice artificial selection through the selection of variant fungal symbionts (Mueller et al. 2005). Thus leaf-cutter ants are known to utilize all of the agricultural techniques used by humans in this analysis except two: the usage of artificially produced chemical fertilizers and genetic engineering to improve their crops. However, it is possible that the ants may even use genetic engineering—through the introduction of viruses and horizontal-gene transfer into their crops—but this remains to be proved (T. R. Schultz, personal communication, 2014).

Last, of the 11, independently evolved, fungus-farming lineages of ambrosia beetles, the *Ambrosiodmus/Ambrosiophilus*-clade ambrosia beetles utilize the most complex agricultural techniques (Jordal and Cognato 2012; Li et al. 2015). Their agricultures use seven of the 12 agricultural traits (table 1.3), although the ten other convergent agriculturalist lineages of ambrosia beetles may use fewer than these seven traits and have simpler agricultures (Farrell et al. 2001; Mueller et al. 2005). Like the fungus-farming termites and ants, the *Ambrosiodmus/Ambrosiophilus*-clade ambrosia beetles are obligate mutualists with their fungal crop species *Flavodon ambrosius* (Li et al. 2015; Kasson et al. 2016; Simmons et al. 2016).

The *Ambrosiodmus/Ambrosiophilus*-clade ambrosia beetles possess the general agricultural traits of substrate preparation, crop planting, crop protection, and crop weeding (Mueller et al. 2005; Li et al. 2015; Kasson et al. 2016). Like the ants, they use microbes for biological pest control (U. G. Mueller, personal communication, 2014), but they are not known to fertilize their crops (table 1.3). They also are not known to practice artificial selection to improve their crops, but some are known to practice another type of crop choice: "Several Asian species within the genus *Ambriosiophilus* engage in another inter-specific interaction—fungus stealing (mycocleptism; Hulcr and Cognato 2010). Instead of making their galleries in uninhabited wood, these parasitic *Ambrosiophilus* species search for galleries established by the much larger ambrosia beetles in the genus *Beaverium* and excavate their galleries immediately next to the existing tunnels. The fungus established by *Beaverium* spp. therefore immediately grows in the gallery of *Ambriosiophilus* spp." (Kasson et al. 2016, 94).

Potential Causes of Convergent Agricultural Evolution

Why does convergent biological evolution occur? Convergence arises because the possible evolutionary pathways available to life are not endless, but *limited* (McGhee 2011, 2015). If the number of possible evolutionary pathways were infinite, then each species on Earth would be morphologically different from every other species, and each species would have its own unique ecological role or niche. Such an Earth does not exist. Instead, repeated evolutionary convergences on similar morphologies, niches, molecules, and even mental states is the norm for life on Earth (McGhee 2011).

Evolutionary limits are the product of functional constraints, developmental constraints, and the two processes acting in concert. Functional constraints are imposed by the laws of physics, chemistry, and geometry and are *extrinsic* to the organisms affected by those constraints (McGhee 2007). Convergence results from the fact that there are limited numbers of ways to solve a functional problem within the boundaries imposed by these extrinsic constraints. In contrast to functional constraints, developmental constraints are *intrinsic* and are imposed by the biology and phylogeny of specific organisms (McGhee 2007). First, the developmental pathways that are available to specific organisms are limited by what has been variously called "phylogenetic legacy," "phylogenetic inertia," or "phylogenetic constraint" (see discussion in McGhee 2007). Second, even given a certain phylogenetically available repertoire of traits, the processes of developmental bias, such as the "repeated, differential re-use of developmental modules, which enables novel phenotypes to arise by developmental rearrangements of ancestral elements, as in the parallel evolution of animal eyes" (Laland et al. 2015, 3), may make the development of some of those traits more probable than others. Third, "phenotypic variation can be channeled and directed towards functional types by the process of development"; this phenomenon is known as "facilitated variation" and can "sometimes elicit substantial, non-random, well-integrated and apparently adaptive innovations in the phenotype" (Laland et al. 2015, 3).

Is convergent agricultural evolution the product of functional constraints, developmental constraints, or both? The functional constraint hypothesis proposes that there exists a limited number of ways to successfully cultivate crops and that convergent agricultural techniques must necessarily result because independent lineages of animals have a limited number of farming options to discover. The developmental constraint hypothesis proposes that there exists a limited repertoire of shared agricultural behaviors that are inherited within a phylogenetic lineage and that convergent agricultural techniques must necessarily result because members of those phylogenetic lineages of animals have a limited repertoire of behavioral options to utilize in farming.

A phylogenetic classification of the nine lineages of agriculturalist bilaterian animals is given in table 1.1. As discussed in the previous section of the chapter, agriculture has evolved independently in both the clades of the deuterostome and protostome bilaterians. The molecular analyses of Dos Reis et al. (2015) indicate that the deuterostome-protostome split occurred by 578 million years ago (Mya) in the Ediacaran period of the Neoproterozoic (Gradstein et al. 2012), and possibly as early as 653 Ma in the older Cryogenian period. Thus, if developmental constraint is responsible for the convergent evolution of agriculture in the nine agriculturalist lineages of animals, then those constraints would have had to be present in animals that existed before the deuterostome-protostome divergence in animal evolution. Metazoan nervous systems evolved independently and convergently in the ctenophores and eumetazoans before the deuterostome-protostome divergence (table 1.1; Dunn et al. 2008; Ryan et al. 2013; Hejnol 2014; Moroz et al. 2014; see discussion in McGhee 2019). The eumetazoan nervous system is surely a prerequisite for the later evolution of agricultural behaviors in the bilaterian animals, but possession of that nervous system does not constrain these animals to any particular behavior in terms of farming crops. Rather, convergence in agricultural behaviors appears to be the result of functional constraint.

The ecdysozoan macrotermitine termites, leaf-cutter ants, and ambrosia beetles do not farm fungi because their ecdysozoan-style development constrains them to that crop; instead they farm fungi because fungi do not require light in order to grow—thus fungi can be cultivated in dark tunnels within wood and underground in soil. Deuterostomes are not developmentally constrained to farm plants—plants are also farmed by annelids and gastropods (table 1.2), animals with protostome developmental systems (table 1.1). Animals in these different clades simply farm plants because light is available in the habitats in which they live, and plants require light in order to grow. Damselfishes, annelid worms, and gastropods have no manipulative appendages, yet this developmental constraint does not prevent them from farming.

Convergence in the agricultural methods used by animals for the farming of plants and fungi (table 1.3) appears to be the result of similar functional constraints imposed by these two very different types of crop organisms. Although plants require light and fungi do not, both crop types grow better when their growth substratum is prepared for them, when they are fertilized, when they are weeded, and when they are protected from raiders and parasites. The hypothesis of the primacy of functional constraint in the convergent evolution of agriculture will be explored in more detail in the next section of this chapter.

Adaptive Landscapes: A Spatial Approach to Evolutionary Analysis

The concept of the adaptive landscape was first proposed by Wright (1932) and has become a "standard imagination prosthesis for evolutionary theorists. It has proven its worth in literally thousands of applications, including many outside evolutionary theory" (Dennett 1996: 190; see discussion in McGhee 2007). It is usually portrayed as a three-dimensional grid surface, where the two dimensions of the *x-y* plane are morphological or ecological traits and the vertical *z*-dimension is the degree of adaptation of permutations of those traits (figure 1.1). Regions in the landscape where trait permutations have a high degree of adaptation are called *adaptive peaks*, and regions of trait permutations with low degrees of adaptation are called *adaptive valleys*. The geometric arrangement of the adaptive peaks within the landscape thus represents, in a spatial fashion, the different possible ways of life available to organisms. The spatial distribution of the adaptive valleys and plains represents trait permutations that range from nonfunctional to lethal in nature. The adaptive landscape can also be portrayed as a two-dimensional contour map, where the third dimension of the degree of adaptation is given by the contour values within the map (figure 1.1).

A basic rule of modeling evolution in an adaptive landscape is that natural selection will operate to move a population up the slope of an adaptive peak, from lower degrees of adaptation to higher degrees of adaptation. Thus, using the adaptive landscape concept,

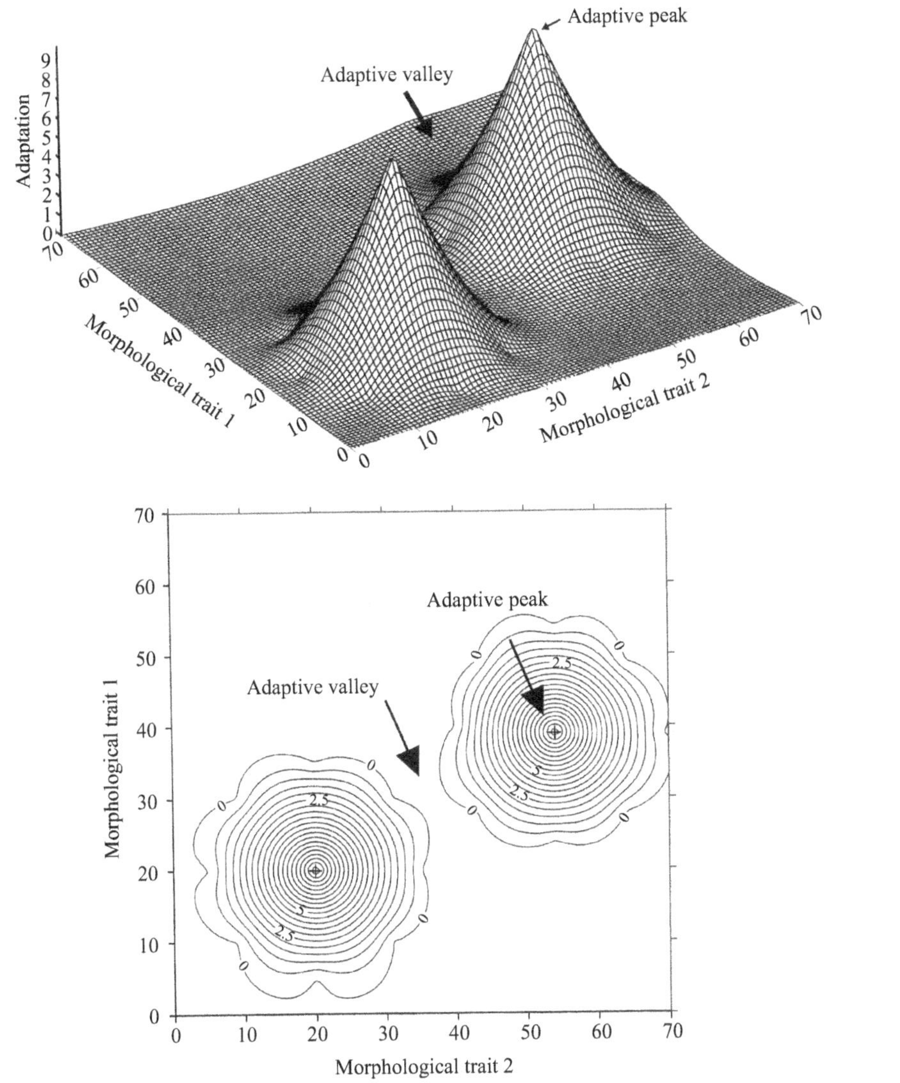

Figure 1.1
A hypothetical adaptive landscape, portrayed as a three-dimensional grid at the top of the figure and a two-dimensional contour map at the bottom. Topographic highs represent adaptive morphologies that function well in natural environments (and therefore are selected for), while topographic lows represent nonadaptive morphologies that function poorly in natural environments (and therefore are selected against). From McGhee (2007).

evolution via the process of natural selection can be spatially visualized as a journey across adaptive hills and valleys, mountains and ravines (McGhee 2007).

The actual utilization of the adaptive landscape concept in the analysis of evolution requires a metric for measuring the different degrees of adaptation of different morphological traits. One commonly used metric is *efficiency*—for example, the filtration efficiencies of different colony geometries in the analysis of evolution in the filter-feeding bryozoans, or the swimming efficiencies of different shell forms in the analysis of evolution in swimming ammonoid cephalopods (McGhee 2007, 2015).

In the analysis of the convergent evolution of agriculture an obvious analog would be a metric of agricultural efficiency, perhaps some measure of the cost to the farmer in producing a crop in contrast to the profit the farmer receives on selling the crop. I am not an anthropologist or economist, but it seems to me that measures of agricultural efficiency would be difficult to obtain even for different human agricultures and nearly impossible to obtain from nonhuman agricultures. Thus, for the current analysis I am using the metric of agricultural complexity (table 1.3) as a proxy for the degree of adaptation of different animal agricultural systems. The assumption here is that the more complex agricultural systems are more highly adapted to the efficient cultivation of crops, and that the less complex agricultural systems are less so.

Figure 1.2 gives a contour map of the agricultural complexities of the nine phylogenetic lineages of agriculturalist animals with respect to the two crop types farmed by these animals, plants or fungi. Two main landscape peaks are apparent, with similar peak heights and slopes. The plant-crop landscape has a maximum agricultural-complexity peak of 9, with human agriculture at the top of the peak and damselfish and polychaete agricultures on the lower slopes of the peak (figure 1.2). A second and smaller plant-crop peak on the slope from the main plant-crop peak has limpet agriculture at the top, with a peak complexity of 5, and whelk agriculture lower on the slope of the overall peak.

The fungus-crop peak has a maximum agricultural-complexity peak of 10, with leaf-cutter ant agriculture at the top of the peak and termite, beetle, and periwinkle agricultures located on progressively lower positions on the slope of the peak. At first glance at the top of the fungus-crop peak it would appear that the ecdysozoan protostomes, the insects, are the exclusive cultivators of fungus crops and that the deuterstome and spiralian protostome animals are all cultivators of plant crops. This is not true, however, as located farther down from the top of the fungus-crop peak, at complexity level 4, the periwinkle agriculturalists also cultivate fungi, and they are snails, spiralian protostomes, not insects (figure 1.2).

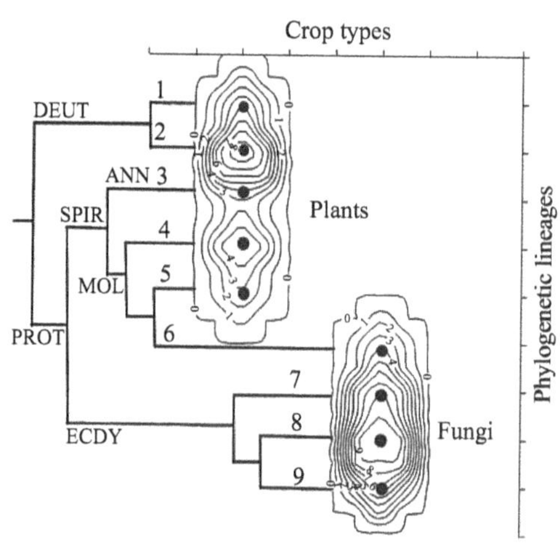

Figure 1.2
Contour map of the distribution of agricultural-complexity levels found in the nine agriculturalist animal lineages that cultivate either plants or fungi. The phylogenetic relationship of the nine lineages is given in the cladogram at the left of the figure; where DEUT = deuterostomes, PROT = protostomes, SPIR = spiralians, ECDY = ecdysozoans, ANN = annelids, and MOL = molluscs. The identities of the phylogenetic lineages are: 1 = damselfish, 2 = human primate, 3 = polychaete annelid worm, 4 = limpet snail, 5 = whelk snail, 6 = periwinkle snail, 7 = macrotermitine termite, 8 = leaf-cutter ant, and 9 = ambrosia beetle.

The two main peaks present in figure 1.2 are solely the result of the two crop types farmed by agriculturalist animals, plants or fungi. The height of the peaks, and their slopes, are very similar, just as the range of agricultural complexities exhibited by plant farmers and fungus farmers are similar (table 1.3). This similarity supports the hypothesis that there exists a single adaptive peak, a best way to conduct agriculture, that applies both to the farming of plants and the farming of fungi.

In contrast to figure 1.2, figure 1.3 gives a contour map of the agricultural complexities of the nine agriculturalist lineages with respect to the two habitat types in which the farms of these animals are located, marine or land. Two main landscape peaks are apparent, but they have noticeably different peak heights and slopes. The overall marine-habitat landscape has a maximum peak height of 5 but possesses two peaks of that height, one with damselfish agriculture at the top and one with limpet agriculture at the top. On the slopes of this second peak, the periwinkle, whelk, and polychaete agricultures are progressively located on lower levels.

The land-habitat peak has a maximum agricultural complexity of 10 and is occupied by the ants, with human, termite, and beetle agriculturalists progressively located lower on the slopes of the land-habitat peak. Note the large "phylogenetic gap" between the deuterostome human agriculturalists and ecdysozoan-protostome insect agriculturalists (figure 1.3). All of the intervening spiralian-protostome agriculturalists are marine farmers, as are the deuterostome damselfishes. In addition to the height of the land-habitat peak, which is twice that of the marine farmers, note the steepness of the slope. This is due to the fact that all of the farmers on land have agricultures of high complexity, ranging from a maximum of level 10 to a minimum of level 7, below which the slope sharply drops off to the base of the land-habitat peak (figure 1.3).

In contrast to figure 1.2, figure 1.3 could support the hypothesis that two agricultural adaptive peaks exist—a land-agriculture peak and a marine-agricultural peak—where the

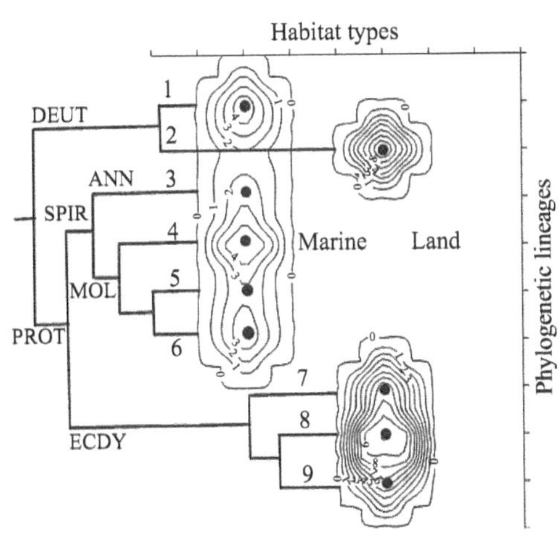

Figure 1.3
Contour map of the distribution of agricultural-complexity levels found in the nine agriculturalist animal lineages that farm either in marine habitats or on land. The phylogenetic relationship of the nine lineages is given in the cladogram at the left of the figure; where DEUT = deuterostomes, PROT = protostomes, SPIR = spiralians, ECDY = ecdysozoans, ANN = annelids, and MOL = molluscs. The identities of the phylogenetic lineages are: 1 = damselfish, 2 = human primate, 3 = polychaete annelid worm, 4 = limpet snail, 5 = whelk snail, 6 = periwinkle snail, 7 = macrotermitine termite, 8 = leaf-cutter ant, and 9 = ambrosia beetle.

height of the land-agricultural peak is twice that of the marine-agricultural peak (figure 1.4). This two-peak hypothesis would imply that animals that live on land are capable of better agricultural efficiencies and higher degrees of adaptation than marine animals (figure 1.5), or that land-dwelling animals must develop more efficient agricultures in order to survive whereas marine agriculturalists can survive with simpler agricultural techniques.

In a summary contrasting fungal farming by land-dwelling and marine animals, Silliman and Newell (2003, 15646–15647) have argued that "evolutionary biologists have recently suggested that fungus-growing animals, like human agriculturalists (Diamond 1999), use a range of cultivation strategies, varying from 'low-' to 'high-level food production'(Schultz et al. 2005). . . . In terrestrial ecosystems, instances of high-level fungal cultivation have been extensively documented," but examples "of low-level fungal production have not been experimentally demonstrated (Schultz et al. 2005), although evolution of this strategy could be common, given its relative engineering simplicity." The agricultural behaviors of substrate preparation, weeding, fertilizing, and defense of the crop from other herbivores have been independently evolved in marine animal clades, though not all of these traits may be present in a given single lineage (table 1.3). Silliman and Newell (2003, 15647) argue that "if there is an important message to be learned from *Littoraria*'s distinct lower-level fungal production strategy, it may be that evolutionary success of fungal farmers may not depend on intricate pest management and inoculation techniques as long as cultivated fungi naturally occur and are successful even without farmer's care (i.e., fungi are effective dispersers and have strong pathogen and competitor resistance)." That is, it may be that marine farms do not need to be as complex as agriculture on dry land. Other traits that marine agricultural systems have

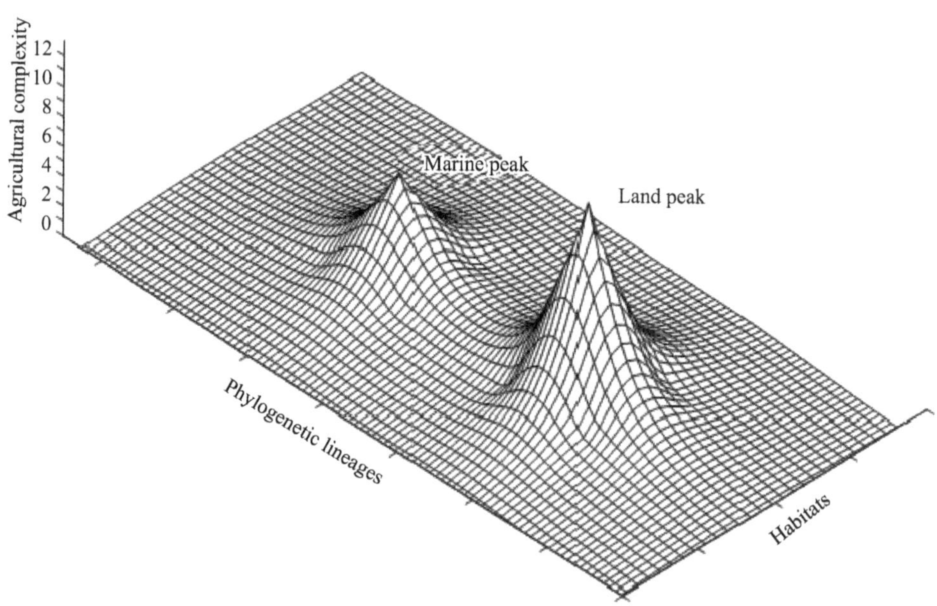

Figure 1.4
Three-dimensional grid representation of the two adaptive peak hypothesis, where a marine agriculture adaptive peak exists with a maximum agricultural-complexity peak (level 6) that is half the magnitude of that of a land agricultural adaptive peak (level 12).

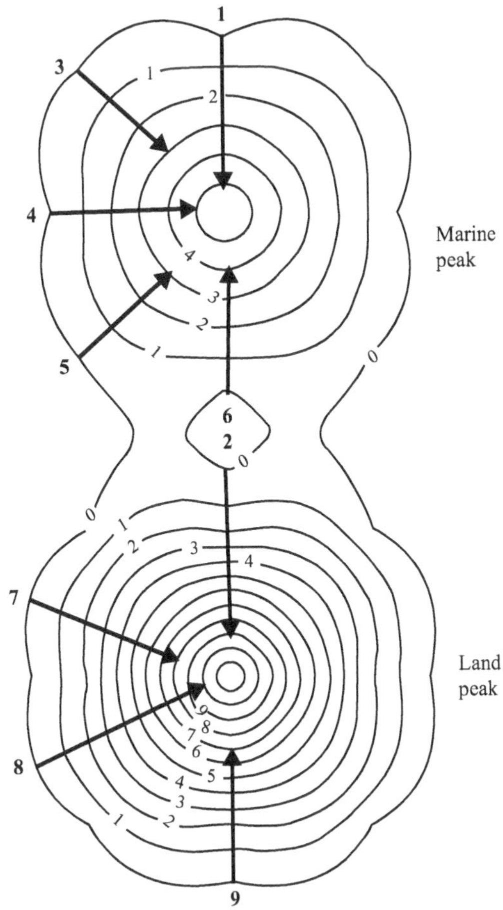

Figure 1.5
The evolution of five agriculturalist animal lineages on a hypothesized marine adaptive peak (top of figure) and the evolution of four agriculturalist animal lineages on a hypothesized land adaptive peak (bottom of figure). Vectors illustrate the maximum agricultural-complexity levels achieved by each lineage on the respective adaptive peak. The identities of the phylogenetic lineages are: 1 = damselfish, 2 = human primate, 3 = polychaete annelid worm, 4 = limpet snail, 5 = whelk snail, 6 = periwinkle snail, 7 = macrotermitine termite, 8 = leaf-cutter ant, and 9 = ambrosia beetle.

in common are the sizes of both the farms and the farmer populations. Marine farms are small in areal extent compared to the huge farms of ants, termites, and some human agricultures. The population sizes of marine farming species are small in contrast to the huge population sizes of ant, termite, and human farming species.

Nevertheless, from the perspective of theoretical morphology, the demonstrable convergent evolution of even simple agricultural behaviors in clades of animals in radically different environmental conditions—in the oceans and on dry land—can be argued as evidence that there is a limited number of ways to successfully cultivate crops, and that convergent behavior must necessarily result because independent lineages of animals have a limited number of farming options to discover—whether in the oceans or on dry land.

Thus the alternative to the two-adaptive-peak agricultural hypothesis (figures 1.4 and 1.5) is the single-adaptive-peak hypothesis (figure 1.6). In figure 1.6 the hypothesis is that

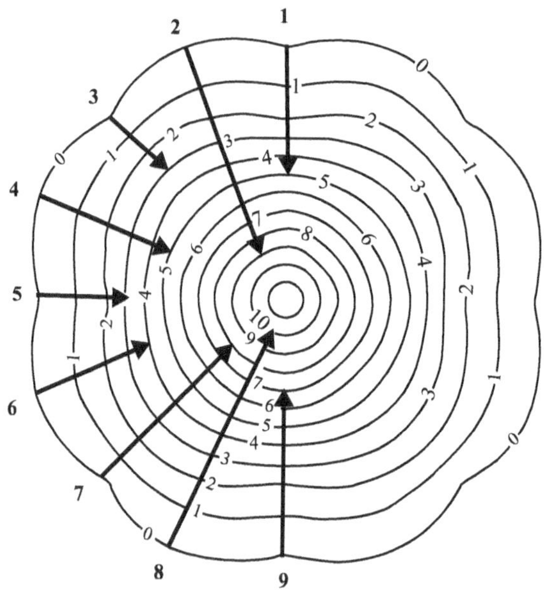

Figure 1.6
The evolution of nine agriculturalist animal lineages on a hypothesized single adaptive peak of agriculture, representing a limited number of ways to successfully cultivate crops that are found in both marine and land habitats. Vectors illustrate the maximum agricultural-complexity levels achieved by each lineage on the single adaptive peak. The identities of the phylogenetic lineages are: 1 = damselfish, 2 = human primate, 3 = polychaete annelid worm, 4 = limpet snail, 5 = whelk snail, 6 = periwinkle snail, 7 = macrotermitine termite, 8 = leaf-cutter ant, and 9 = ambrosia beetle.

there exists a limited number of ways to do agriculture and that these constitute the complexity levels of a single adaptive peak—regardless of whether the crop type is plants or fungi, and whether the habitat is marine or terrestrial. Independent agricultural-animal clades differ in the number of those limited ways that they have discovered or have been able to master. Even the agricultural-animal clades that have managed to reach the higher levels of the adaptive peaks—ants, humans, and termites—initially started out at lower levels of the peak in the evolution of their agricultural systems (Schultz et al. 2005; Schultz and Brady 2008; Branstetter et al. 2017; Gupta 2004; Lieberman 2018; Aanen and Eggleton 2017; chapter 14, this volume).

Types of Agricultural Convergence

There exist three pathways to convergent evolution that can be formalized by using a unified terminology of convergence (McGhee et al. 2018; see discussion in McGhee 2019, 21–31). In that terminology the *phenomenon* of convergent evolution is a *set*, and the *types* of convergent evolution are *subsets* of that phenomenon. In this unified terminology the *phenomenon* of convergence is the *suffix* of the term, and the *type* of convergent evolution is the *prefix* of the term. The unified convergence terminology is outlined in table 1.4. The set of the convergence phenomenon contains all of the same or very similar traits that have evolved independently in different lineages of organisms. This set contains three subsets that specify

Table 1.4
The types of convergent evolution

SET:
CONVERGENT EVOLUTION: The evolution of the same or very similar traits independently in different lineages of organisms.
SUBSETS:
1. Allo-convergent evolution: The independent evolution of the same or very similar new trait from different precursor traits in different lineages. 2. Iso-convergent evolution: The independent evolution of the same or very similar new trait from the same precursor trait in different lineages. 3. Retro-convergent evolution: The independent re-evolution of the same or very similar trait to an ancestral trait in different lineages.
SUBSETS:
3a. Retro-alloconvergent evolution: The independent re-evolution of the same or very similar trait to an ancestral trait from different precursor traits in different lineages. 3b. Retro-isoconvergent evolution: The independent re-evolution of the same or very similar trait to an ancestral trait from the same precursor trait in different lineages.

Sources: McGhee et al. (2018), McGhee (2019).

the three pathways by which evolution may produce convergence: allo-convergent evolution, iso-convergent evolution, and retro-convergent evolution (table 1.4).

Allo-convergent evolution is the independent evolution of the same or very similar new trait from *different precursor traits* in different lineages (the prefix "allo" means "different"). Iso-convergent evolution is the independent evolution of the same or very similar new trait from the *same precursor trait* in different lineages (the prefix "iso" means "same"). Retro-convergent evolution is the independent *re-evolution* of the same or very similar trait to an ancestral trait in different lineages (the prefix "retro" means "reverse").

The general subset of retro-convergent evolution contains two specific subsets that further differentiate the type of retro-convergence that may occur: retro-alloconvergent evolution and retro-isoconvergent evolution (table 1.4). Retro-alloconvergent evolution is the independent *re-evolution* of the same or very similar trait to an ancestral trait from *different precursor traits* in different lineages. Retro-isoconvergent evolution is the independent *re-evolution* of the same or very similar trait to an ancestral trait from the *same precursor trait* in different lineages.

The convergent evolution of agricultural traits by the nine lineages of bilaterian animals (tables 1.1 and 1.3) is a mixture of both allo-convergence and iso-convergence. Allo-convergence is evidenced by the evolution of the same agricultural behaviors from three different precursor feeding types: hunter-gatherers, herbivorous grazers, and scavengers (table 1.5). The fact that animals independently evolved the same agricultural behaviors (table 1.3) while starting from three different ancestral modes of feeding (table 1.5) can be argued as evidence that there exists a limited number of ways to successfully cultivate crops regardless of the ancestral feeding behaviors of the agriculturalist lineages, and as evidence that a single adaptive peak for agriculture exists (figure 1.6).

Iso-convergence is evidenced by the independent evolution of the same agricultural behaviors (table 1.3) by three different lineages of ancestral hunter-gatherers in parallel—the

Table 1.5
Pre-agricultural feeding methods in nine lineages of agriculturalist bilaterian animals

Farmer	Clade	Predominant pre-agricultural feeding method within clade
1	Damselfishes	hunter-gatherers (carnivorous-herbivorous)
2	Humans	hunter-gatherers (carnivorous-herbivorous)
3	Polychaetes	scavengers (omnivorous)
4	Limpets	grazers (herbivorous)
5	Whelks	scavengers (carrion-feeders)
6	Periwinkles	scavengers (herbivorous)
7	Termites	grazers (herbivorous)
8	Ants	hunter-gatherers (carnivorous-herbivorous)
9	Beetles	grazers (herbivorous)

Sources: 1 = Aanen (2010), Hata and Umezawa (2011); 2 = Gupta (2004), Lieberman (2018); 3 = Gambi et al. (2000); 4 = Lindberg (2007); 5 = Da Silva and Brown (1984), Harris et al. (1986); 6 = Sieg et al. (2013); 7 = Aanen (2010), Aanen and Eggleton (2017), D. Aanen (personal communication 2018); 8 = Mueller et al. (2001), Branstetter et al. (2017); 9 = Farrell et al. (2001), Jordal and Cognato (2012), Li et al. (2015), J. Hulcr (personal communication 2018).

damselfishes, humans, and ants (table 1.5). Likewise, the same agricultural behaviors were iso-convergently evolved by three different lineages of herbivorous grazers in parallel—the limpets, termites, and ambrosia beetles—and three different lineages of scavengers in parallel—the polychaete worms, whelks, and periwinkles (table 1.5).

Moreover, *within* some of these iso-convergent bilaterian lineages the iso-convergent evolution of agricultural behaviors has independently occurred *repeatedly*—a phenomenon known as *iterative evolution* (Bayer and McGhee 1984; see discussion in McGhee 2007). Thus Aanen (2010, 3) notes that within the damselfish lineage "multiple independent transitions to algaculture have occurred" from ancestral fishes that fed on both animals and plants (Kato and Umezawa 2011). Likewise, multiple groups of humans independently transitioned to agriculture from ancestral hunter-gatherers about 10,000 years ago (Gupta 2004; Lieberman 2018; chapter 10, this volume), and 11 independent lineages of ambrosia beetles iteratively evolved agriculture in the past 50 million years (Jordal and Cognato 2012; Li et al. 2015; chapter 7, this volume). In contrast, the convergent evolution of agricultural behaviors has occurred only once in the lineages of the *Platynereis* polychaete worms (Gambi et al. 2000), scutellastrid limpets (Lindberg 2007), *Bullia* nassariid whelks (Da Silva and Brown 1984; Harris et al. 1986), *Littoraria* periwinkles (Silliman and Newell 2003; Sieg et al. 2013), macrotermitine termites (Mueller et al. 2005; Bourguignon et al. 2014; chapter 6, this volume), and attine ants (Mueller et al. 2005; Branstetter et al. 2017; chapters 8 and 14, this volume). However, it should be noted that Silliman and Newell (2003, 15647) have posed the question "Is fungal farming common but overlooked?" in marine environments and have argued that "given the relative engineering simplicity of low-level fungal production, the benefits of having predictable food supply . . . [suggest that] fungal farming on live, senescing, and/or dead plants may be more geographically and phylogenetically widespread then presently envisioned, especially in systems where fungal spores are abundant, grazers can manipulate fungus-growing media, and fungus is a major diet component of consumers."

Last, retro-convergence has also occurred in the history of the evolution of agriculture, where an agriculturalist species reverts back to a prior ancestral mode of feeding that is nonagricultural. Thus *Homo sapiens*, a species that has iteratively evolved agriculture in the past, also has populations that have abandoned agriculture and gone back to being hunter-gatherers (Oota et al. 2005). However, they are not primarily hunter-gatherers in the sense that their hunter-gatherer mode of feeding is secondary, a retro-convergence on an ancestral mode of feeding (table 1.5).

Some of the agriculturalist ambrosia beetles "have lost their capacity to culture their own fungal gardens and depend on mycoclepty, or fungus theft" (Hulcr and Cognato 2010, 3). This phenomenon does not meet the strict definition of retro-convergence (table 1.4) because a feeding mode involving fungal-theft scavenging is not the ancestral feeding mode of the agriculturalist ambrosia beetles, which was grazing on dead phloem tissues in plants by ancestral bark beetles (Farrell et al. 2001; J. Hulcr, personal communication, 2018). Farrell et al. (2001, p. 2011) further state that the ambrosia beetle's agricultural "habits have evolved repeatedly and are unreversed." Still, this phenomenon represents another interesting example of the loss of agriculture in an agriculturalist lineage.

Conclusions

Agriculture has evolved independently and convergently in nine phylogenetic lineages of bilaterian animals: actinopterygian vertebrates, sarcopterygian vertebrates, polychaete annelids, patellogastropod molluscs, neogastropod molluscs, littorinoid molluscs, polyneopteran insects, hymenopteran insects, and coleopteran insects. Using a measure of agricultural complexity, it can be demonstrated that the most complex agricultural systems exist in terrestrial animals and that lower-complexity agricultural systems exist in marine animals, regardless of whether the farmed crops are plants or fungi. Convergence in the agricultural methods used by animals for the farming of plants and fungi appears to be the result of similar functional constraints imposed by these two very different types of crop organisms, and not by developmental constraints within the clades of the agriculturalist animals. Allo-convergence in the evolution of agriculture is evidenced by the independent appearance of the same agricultural behaviors in the nine lineages of bilaterian animals while starting from three different, nonagricultural, ancestral types of feeders—hunter-gatherers, herbivorous grazers, and scavengers. Once acquired, agricultural behaviors have rarely been lost.

From an adaptive landscape perspective, there appears to exist a limited number of ways to do agriculture that constitute the levels of a single adaptive peak—regardless of whether the crop type is plants or fungi and whether the habitat is marine or terrestrial. Independent agricultural-animal clades differ in the number of those limited ways that they have discovered or been able to master.

Acknowledgments

I thank the organizers, Peter Peregrine, Ted Schultz, and Rick Gawne, for inviting me to the Altenberg Workshop on the convergent evolution of agriculture, and I thank the Konrad

Lorenz Institute for Evolution and Cognition Research for its present and past support of my research on the phenomenon of convergent evolution.

References

Aanen, D. K. 2006. "As You Reap, So Shall You Sow: Coupling of Harvesting and Inoculating Stabilizes the Mutualism between Termites and Fungi." *Biological Letters* 2:209–212.

Aanen, D. K. 2010. "As You Weed, So Shall You Reap: On the Origin of Algaculture in Damselfish." *BMC Biology 2010* 8:81. http://www.biomedcentral.com/1741-7007/8/81.

Aanen, D. K., and P. Eggleton. 2017. "Symbiogenesis: Beyond the Endosymbiosis Theory?" *Journal of Theoretical Biology* 434:99–103.

Aanen, D., P. Eggleton, C. Rouland-Lefèvre, T. Guldberg-Frøslev, S. Rosendahl, et al. 2002. "The Evolution of Fungus-32 Growing Termites and Their Mutualistic Fungal Symbionts." *Proceedings of the National Academy of Sciences* 99:14887–14892.

Bayer, U., and G. R. McGhee. 1984. "Iterative Evolution of Middle Jurassic Ammonite Faunas." *Lethaia* 17:1–16.

Benton, M. J. 2015. *Vertebrate Palaeontology*, 4th ed. Chichester, UK: Wiley-Blackwell.

Bourguignon, T., N. Lo, S. L. Cameron, J. Šobotnik, Y. Hayashi, et al. 2014. "The Evolutionary History of Termites as Inferred from 66 Mitochondrial Genomes." *Molecular Biology and Evolution* 32:406–421.

Branstetter, M. G., A. Ješovnik, J. Sosa-Calvo, M. W. Lloyd, B. C. Faircloth, et al. 2017. "Dry Habitats Were Crucibles of Domestication in the Evolution of Agriculture in Ants." *Proceedings of the Royal Society*, B284:20170095. http://dx.doi.org/10.1098/rspb.2017.0095.

Darwin, C. 1859. *On the Origin of Species by Means of Natural Selection; or the Preservation of Favoured Races in the Struggle for Life*. London: Murray.

Da Silva, F. M., and A. C. Brown. 1984. "The Gardens of the Sandy-Beach Whelk *Bullia digitalis* (Dillwyn)." *Journal of Molluscan Studies* 50:64–65.

Dennett, D. C. 1996. *Darwin's Dangerous Idea*. New York: Simon and Schuster.

Diamond, J. 1999. *Guns, Germs, and Steel*. New York: Norton.

Dos Reis, M., Y. Thawornwattana, K. Angelis, M. J. Telford, P. C. J. Donoghue, et al. 2015. "Uncertainty in the Timing of Origin of Animals and the Limits of Precision in Molecular Timescales." *Current Biology* 25:2939–2950.

Dunn, C. W., A. Hejnol, D. Q. Matus, K. Pang, K., W. E. Browne, et al. 2008. "Broad Phylogenetic Sampling Improves Resolution of the Animal Tree of Life." *Nature* 452:745–749.

Farrell, B. D., A. S. Sequeira, B. C. O'Meara, B. B. Normark, J. H. Chung, and B. H. Jordal. 2001. "The Evolution of Agriculture in Beetles (Curculionidae: Scolytinae and Platypodinae)." *Evolution* 55:2011–2027.

Gambi, M. C., V. Zupo, M. C. Buia, and L. Mazella. 2000. "Feeding Ecology of *Platynereis dumerilii* (Audouin & Milne-Edwards) in the Seagrass *Posidonia oceanica* System: The Role of the Epiphytic Flora (Polychaeta, Nereididae)." *Ophelia* 53:189–202.

Gradstein, F. M., J. G. Ogg, M. Schmitz, and G. Ogg. 2012. *The Geologic Time Scale 2012*. Amsterdam, The Netherlands: Elsevier B. V.

Gupta, A. K. 2004. "Origin of Agriculture and Domestication of Plants and Animals Linked to Early Holocene Climate Amelioration." *Current Science* 87:54–59.

Harris, S. A., F. M. da Silva, J. J. Bolton, and A. C. Brown. 1986. "Algal Gardens and Herbivory in a Scavenging Sandy-Beach Nassariid Whelk." *Malacologia* 27: 299–305.

Hata, K., and M. Kato. 2004. "Monoculture and Mixed-Species Algal Farms on a Coral Reef Are Maintained through Intensive and Extensive Management by Damselfishes." *Journal of Experimental Marine Biology and Ecology* 313:285–296.

Hata, K., and M. Kato. 2006. "A Novel Obligate Cultivation Mutualism between Damselfish and *Polysiphonia* Algae." *Biological Letters* 2:593–596.

Hata, K., and Y. Umezawa. 2011. "Food Habits of the Farmer Damselfish *Stegastes nigricans* Inferred by Stomach Content, Stable Isotope, and Fatty Acid Composition Analyses." *Ecological Research* 26:809–818.

Hata, K., K. Watanabe, and M. Kato. 2010. "Geographic Variation in the Damselfish–Red Alga Cultivation Mutualism in the Indo-West Pacific." *BMC Evolutionary Biology* 10:185. http://www.biomedcentral.com/1471-2148/10/185.

Hejnol, A. 2014. "Excitation over Jelly Nerves." *Nature* 510:38–39.

Hulcr, J., and A. I. Cognato. 2010. "Repeated Evolution of Crop Theft in Fungus-Farming Ambrosia Beetles." *Evolution* 64:3205–3212.

Jordal, B. H., and A. I. Cognato. 2012. "Molecular Phylogeny of Bark and Ambrosia Beetles Reveals Multiple Origins of Fungus Farming during Periods of Global Warming." *BMC Evolutionary Biology* 12:133. http://www.biomedcentral.com/1471-2148/12/133.

Kasson, M. T., K. L. Wickert, C. M. Stauder, A. M. Macias, M. C. Berger, et al. 2016. "Mutualism with Aggressive Wood-Degrading *Flavon ambrosius* (Polyporales) Facilitates Niche Expansion and Communal Social Structure in *Ambrosiophilus* Ambrosia Beetles." *Fungal Ecology* 23:86–96.

Laland, K. N., T. Uller, M. W. Feldman, K. Sterelny, G. B. Müller, et al. 2015. "The Extended Evolutionary Synthesis: Its Structure, Assumptions and Predictions." *Proceedings of the Royal Society* B282:20151019. https://doi.org/10.1098/rspb.2015.1019.

Levinton, J. S. 1982. *Marine Ecology*. Englewood Cliffs, NJ: Prentice-Hall.

Li, Y., D. R. Simmons, C. C. Bateman, D. P. G. Short, M. T. Kasson, et al. 2015. "New Fungus-Insect Symbiosis: Culturing, Molecular, and Histological Methods Determine Saprophytic Polyporales Mutualists of *Ambrosiodmus* Ambrosia Beetles." *PLoS ONE* 10(9): e0137689. https://doi.org/10.1371/journal.pone.0137689.

Lindberg, D. R. 2007. "Reproduction, Ecology and Evolution of the Indo-Pacific Limpet *Scutellastra flexuosa*." *Bulletin of Marine Science* 81:219–234.

Lieberman, D. E. 2018. "Being a Carnivorous Hominin in the Lower Paleolithic: A Biological Perspective on Convergence and Stasis." In *Convergent Evolution in Stone-Tool Technology*, edited by M. J. O'Brien, B. Buchanan, and M. I. Eren, 91–105. Cambridge, MA: MIT Press.

Lycett, S. J. 2011. "'Most Beautiful and Most Wonderful': Those Endless Stone Tool Forms." *Journal of Evolutionary Psychology* (Budapest) 9:143–171.

McGhee, G. R. 2007. *The Geometry of Evolution: Adaptive Landscapes and Theoretical Morphospaces*. Cambridge: Cambridge University Press.

McGhee, G. R. 2011. *Convergent Evolution: Limited Forms Most Beautiful*. Cambridge, MA: MIT Press.

McGhee, G. R. 2015. "Limits in the Evolution of Biological Form: A Theoretical Morphologic Perspective." *Interface Focus* 5:1–6. https://doi.org/10.1098/rsfs.2015.0034.

McGhee, G. R. 2019. *Convergent Evolution on Earth: Lessons for the Search for Extraterrestrial Life*. Cambridge, MA: MIT Press.

McGhee, G. R., I. Hue, J. Dardaillon, and P. Pontarotti. 2018. "A Proposed Terminology of Convergent Evolution." In *Origin and Evolution of Biodiversity*, edited by P. Pontarotti, 331–340. Berlin: Springer International Publishing AG.

McQuaid, C. D., and P. W. Froneman. 1993. "Mutualism between the Territorial Intertidal Limpet *Patella longicosta* and the Crustose Alga *Ralfsia verrucosa*." *Oecologia* 96:128–133.

Misof, B., S. Liu, K. Meusemann, R. S. Peters, A. Donath, C. Mayer, et al. 2014. "Phylogenomics Resolves the Timing and Pattern of Insect Evolution." *Science* 346:763–767.

Moroz, L. L., K. M. Kocot, M. R. Citarella, S. Dosung, T. P. Norekian, et al. 2014. "The Ctenophore Genome and the Evolutionary Origins of Neural Systems." *Nature* 510:109–114.

Mueller, U. G., N. M. Gerardo, D. K. Aanen, D. L. Six, and T. R. Schultz. 2005. "The Evolution of Agriculture in Insects." *Annual Review of Ecology, Evolution and Systematics*. 36:563–595.

Mueller, U. G., T. R. Schultz, C. R. Currie, R. M. M. Adams, and D. Malloch. 2001. "The Origin of the Attine Ant-Fungus Mutualism." *Quarterly Review of Biology* 76:169–197.

Near, T. J., A. Dornburg, R. I. Eytan, B. P. Keck, W. L. Smith, et al. 2013. "Phylogeny and Tempo of Diversification in the Superradiation of Spiny-Rayed Fishes." *Proceedings of the National Academy of Sciences* 110:12738–12743.

Nielsen, C. 2012. *Animal Evolution: Interrelationships of the Living Phyla*. Oxford: Oxford University Press.

O'Brien, M. J., B. Buchanan, and M. I. Eren, eds. 2018. *Convergent Evolution in Stone-Tool Technology*. Cambridge, MA: MIT Press.

Oota, H., B. Pakendorf, G. Weiss, A. von Haeseler, S. Pookajorn, et al. 2005. "Recent Origin and Cultural Reversion of a Hunter-Gatherer Group." *PLoS Biology* 3(3): e71. doi:10.1371/journal.pbio.0030071.

Plagányi, E. E., and G. M. Branch. 2000. "Does the Limpet *Patella cochlear* Fertilize Its Own Algal Garden?" *Marine Ecology Progress Series* 194:113–122.

Ryan, J. R., K. Pang, C. E. Schnitzler, A.-D. Nguyen, R. T. Moreland, et al. 2013. "The Genome of the Ctenophore *Mnemiopsis leidyi* and Its Implications for Cell Type Evolution." *Science* 342:1242592. doi:10.1126/science.1242592.

Schultz, T. R., and S. G. Brady. 2008. "Major Evolutionary Transitions in Ant Agriculture." *Proceedings of the National Academy of Sciences* 105:5435–5440.

Schultz, T. R., U. G. Mueller, C. R. Currie, and S. A. Rehner. 2005. "Reciprocal Illumination: A Comparison of Agriculture in Humans and Ants." In *Ecological and Evolutionary Advances in Insect-Fungal Associations*, edited by F. Vega and B. Meredith, 149–190. Oxford: Oxford University Press.

Sieg, R. D., K. Wolfe, D. Willey, V. Ortiz-Santiago, and J. Kubanek. 2013. "Chemical Defenses against Herbivores and Fungi Limit Establishment of Fungal Farms on Salt Marsh Angiosperms." *Journal of Experimental Marine Biology and Ecology* 446:122–130.

Silliman, B. R., and S. Y. Newell. 2003. "Fungal Farming in a Snail." *Proceedings of the National Academy of Sciences* 100:15643–15648.

Simmons, D. R., Y. Li, C. C. Bateman, and J. Hulcr. 2016. "*Flavodon ambrosius* sp. nov., a Basidiomycetous Mycosymbiont of *Ambrosiodmus* Ambrosia Beetles." *Mycotaxon* 131:277–285.

Struck, T. H., A. R. Wey-Fabrizius, A. Golombek, L. Herig, A. Weigert, et al. 2014. "Platyzoan Paraphyly Based on Phylogenomic Data Supports a Noncoelomate Ancestry of Spiralia." *Molecular Biology and Evolution* 31:1833–1849.

Woodin, S. A. 1977. "Algal 'Gardening' Behaviour by Nereid Polychaetes: Effects on Soft-Bottom Community Structure." *Marine Biology* 44:39–42.

Wright, S. 1932. "The Roles of Mutation, Inbreeding, Crossbreeding and Selection in Evolution." *Proceedings of the Sixth International Congress of Genetics* 1:356–366.

2 The Convergent Evolution of Agriculture: A Systematic Comparative Analysis

Peter N. Peregrine

A central question of this volume is the extent to which human and insect agriculture is an example of convergent evolution. To help explore that question, I conducted a comparative study of human and insect agricultural systems to see if there are empirical similarities and differences that might indicate a convergent evolutionary process. Data for this study were collected through a questionnaire consisting of 60 dichotomous or ordinal variables completed by four experts on specific insect species or human cultural groups (different cultural groups were considered different "species" for the purposes of this study). Seven indices were created from these data and used to test six hypotheses concerning similarities and differences between human and insect agricultural systems and the impact of agriculture on human and insect societies. I found that while humans and insects differed in terms of agricultural practices, there were remarkable similarities in the impact of agriculture on human and insect societies. I suggest this has important implications for understanding the evolution of agriculture as a convergent process.

Human Agriculture

In this section I provide a brief introduction to agriculture as it is practiced by humans and the impacts that have been hypothesized to have stemmed from it. I define *agriculture* following the other authors in the volume, as economic reliance on domesticated plants and animals. *Domesticated* refers to genetic modification of one species by another in a way that appears to benefit the domesticating species but reduces the fitness of the domesticated species in the wild.

Agriculture evolved independently at least seven times among humans (Bellwood 2005), but the processes through which domestication and agriculture evolved appear to differ, sometimes dramatically, in each area (see figure 2.1). For example, in the Fertile Crescent (the general area of the Zagros Mountains and Tigris/Euphrates Valley in modern Lebanon, Turkey, Syria, and Iraq), agriculture appears to have evolved as foragers harvesting grasses artificially selected for desired characteristics, including a larger seed that did not fall off the stalk while harvesting and with a husk that was easy to remove. This led to grasses that could not readily reproduce in the wild, but could be readily harvested and

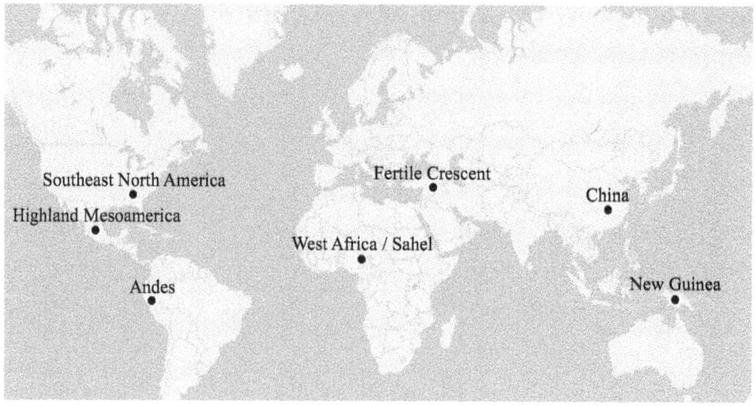

Figure 2.1
Seven locations where human agriculture originated.

sown by humans (Zeder 2011). In contrast, the bottle gourd was domesticated (or further domesticated—there is controversy over the bottle gourd's origins) in Mesoamerica not as a food source but as a utensil for carrying liquids, and the process of domestication did not strictly consist of the selection for desired traits in the bottle gourd but also of geographical expansion as humans transplanted bottle gourds to locations where they did not typically grow (Asch and Hart 2004).

The evolution of agriculture among humans is considered one of the key events in human history, one that globally transformed human societies (see, e.g., Morgan 1877; Childe 1936; Barker 2006). While the causes of this "Neolithic revolution" have been debated for decades (see Bellwood 2005, 19–25, for a quick overview of these debates), the consequences have been less so. Most scholars agree that the transition to agriculture is associated with (and perhaps causally so) increases in population, sedentism, social complexity, and the elaboration of material culture (see, e.g., Diamond 2002; Jobling, Hurles, and Tyler-Smith 2004, 305–306; Dow and Reed 2015). In addition, many scholars believe that agriculture led to a decline in health and to increased conflict between groups (e.g., Larsen 2018).

In North America, for example, the adoption of maize-based agriculture led to a dramatic increase in dental caries (cavities) and diseases related to poor nutrition such as porodic hyperostosis ("spongy" cranial bones) and enamel hypoplasia (linear growth irregularities on the teeth) (Cohen and Crane-Kramer 2007). In locations such as the central Mississippi River valley large agricultural communities emerged with formal political leaders who orchestrated large social and religious events such as feasting and calendrical ceremonies. Long-distance trade in both exotic raw materials and manufactured goods intensified, leading to the creation of trade entrepots across midcontinental North America. Conflict between local agricultural groups increased, and political-economic systems began to regularly collapse and rebuild, probably due to both external and internal conflict often associated with crop failure (Milner 2004, 165–168).

Thus the evolution of agriculture among humans has long been understood to have led to profound changes in demography, health and nutrition, and social organization. To what extent is this true for the agricultural insects?

A Comparative Perspective on the Evolution of Agriculture

In this section I outline the materials and methods used here to better understand the impact of agriculture on both human and insect societies. I assume that the reader can obtain an understanding of the variety of insect agriculture from other chapters in this volume, so I will not cover that here. I will, however, attempt to clearly distinguish the systematic comparative method used here from the comparative examples provided by many of this volume's authors.

While most research on the evolution and impact of agriculture has been in the form of case studies focused on individual sites or regions (e.g., Cowan and Watson 2006; Price and Bar-Yosef 2011; Barker and Goucher 2015) or, in the case of insects, of species (see, e.g., chapter 9, this volume), this study takes a different approach. Here I perform a systematic comparative study of 29 human and insect agricultural cases. The underlying logic of systematic comparison is to look across cases to identify broad similarities and differences using statistical analyses. If statistically robust associations can be identified from a wide range of cases, then there is no a priori reason to assume that they are not generalizable to all agricultural organisms. An additional strength of this type of comparison is that it allows one to test whether or not an assumed predictive condition is empirically associated with assumed effects across a wide range of cases. Finally, such comparison allows patterns of traits and relationships between variables to be described probabilistically, which in turn provides the potential to test the statistical significance (or to evaluate, from a Bayesian perspective, the likelihood) of identified relationships (Ember and Ember 2009).

A relatively large and well-defined sample is required to perform systematic comparative analyses of this type. The sample must provide examples from the entire range of expected variation, and the cases must be, as much as possible, independent of one another in order to avoid autocorrelation. Here I employ a random sample of 15 human cases from the eHRAF Collection of Archaeology (ehrafarchaeology.yale.edu) and a sample of 14 insect cases (three ant, seven termite, and four beetle) suggested by other authors in this volume. These cases are listed in table 2.1.

The human cases, which I have employed in previous research (Peregrine, Ember, and Ember 2004, 2007), represent a sample of evolutionary sequences from three centers of independent agricultural evolution: Mesopotamia, Highland Mesoamerica, and the Central Andes. Being sequences, these cases allow for both synchronic (assuming all cases exist at the same time) and diachronic (looking over time within each sequence) analysis. The insect cases were selected by the individual experts who coded them (Ted Schultz for ants, Duur Aanen for termites, and Jiri Hulcr for beetles—an effort for which I thank them heartily). These cases were, like the human cases, intended to be proxies for evolutionary sequences (e.g., from lower attines to leaf-cutter ants), but after long discussion it was decided that the lack of certainty about insect evolution makes the identification of evolutionary "sequences" for termites and beetles impossible, so no diachronic analyses were performed.

In addition to a carefully selected sample, systematic comparison also requires clearly defined and reliably coded variables. Both the insect and human cases were coded on a total of 60 dichotomous or ordinal variables as presented in table 2.2. Evaluation of reliability showed that the variables are all low inference and reliability is very good (over

Table 2.1
Cases used in the analyses

Highland Mesoamerican Archaic
Highland Mesoamerican Early Preclassic
Highland Mesoamerican Late Preclassic
Central Mexico Classic
Central Mexico Postclassic
Epipaleolithic
Aceramic Neolithic
Ceramic Neolithic
Ubaid
Late Chalcolithic Mesopotamia
Highland Andean Archaic
Highland Andean Formative
Chavin
Andean Regional Development
Huari
Ants lower attine
Ants higher attine
Ants leaf-cutter
Termites clade I (Acantho)
Termites clade II (Odonto)
Termites clade III (Macro)
Termites clade IIIa (Macro b)
Termites clade V (Synacantho)
Termites clade V (Ancistro)
Termites clade VI (Micro)
Beetles (Ambrosiophilus)
Beetles (Xylosandrus)
Beetles (Trypodendron)
Beetles (Xyleborus)

90% inter-coder agreement). For analysis the variables were standardized and combined into the seven indices shown in table 2.3. Alpha values for each index, also given in table 2.3, indicate that all indices are reliable (alpha values approaching 1.0 indicate high scale reliability; see DeVellis 2012).

Analyses were performed using SPSS (IBM Corporation 2019) and the nonparametric statistics options, as at least one of the variables in all of the analyses was ordinal. Mann-Whitney U tests were used for comparison of insects versus humans, and Kruskall-Wallace H tests were used to compare cases with different degrees of dependence on agriculture. Tau-b correlations were also computed for these comparisons to examine possible linear associations. Bayesian analyses were also performed using the SPSS Bayesian statistics options, but there were no major differences between the nonparametric findings and Bayesian estimates, so only the nonparametric tests are discussed here because they are much simpler to interpret.

Table 2.2
Codebook

I. "Agricultural" practice variables
VI.1 Selecting substrate (universal)
1 = low specificity
2 = moderate specificity
3 = high specificity
9 = missing
VI.2 Internal sustainability (harvested domesticates provide source for next crop cycle)
1 = 0 to 33%
2 = 34 to 67%
3 = 68 to 100%
9 = missing
VI.3 Planting crops
1 = low investment
2 = moderate investment
3 = high investment
9 = missing
VI.4 Preparing substrate
1 = low investment
2 = moderate investment
3 = high investment
9 = missing
VI.5 Dimensions of substrate
1 = 2d
2 = 3d
9 = missing
VI.6 Temporal variation in cultivation
1 = discrete (seasonal/crop rotation/fallowing)
2 = continuous
9 = missing
VI.7 Diversity of domesticates (at a single location/within a single group)
1 = single domesticate
2 = two or three domesticates
3 = four or more domesticates
9 = missing
VI.8 Monitoring crops for disease or thieves/predators
1 = 0 to 33% of the time
2 = 34 to 67% of the time
3 = 68 to 100% of the time
9 = missing
VI.9 "Weeding": Physical removal of invasive pests/predators
1 = 0 to 33% of pests removed
2 = 34 to 67% of pests removed
3 = 68 to 100% of pests removed
9 = missing

(*continued*)

Table 2.2
(continued)

VI.10 Engineering for optimal growth condition (climate control, watering, etc.)
- 1 = low investment
- 2 = moderate investment
- 3 = high investment
- 9 = missing

VI.13 Fertilizing: Organic
- 1 = 0 to 33% of crops treated
- 2 = 34 to 67% of crops treated
- 3 = 68 to 100% of crops treated
- 9 = missing

VI.15 Reproductive isolation from free-living populations (reproductive barriers)
- 1 = low isolation
- 2 = moderate isolation
- 3 = high isolation
- 9 = missing

VI.16 Controlling breeding partners (controlling recombination and sexual selection)
- 1 = low control
- 2 = moderate control
- 3 = extensive control
- 9 = missing

VI.17 Artificial selection for domesticate improvement
- 1 = no selection performed
- 2 = selection done, but less than annually
- 3 = selection common (annually or more frequent)
- 9 = missing

II. Agriculture process variables

VII.1 Degree of dependence on domesticated resources (estimated through caloric intake or productive effort)
- 1 = 0 to 33% of crops
- 2 = 34 to 67% of crops
- 3 = 68 to 100% of crops
- 9 = missing

VII.2 Sociality
- 1 = asocial/solitary
- 2 = ultrasocial/communal
- 3 = eusocial
- 9 = missing

VII.3 Task specialization
- 1 = no agricultural task specialists
- 2 = one or two specialists
- 3 = three or more specialists
- 9 = missing

VII.4 Use of extrasomatic technology
- 1 = absent
- 2 = present
- 9 = missing

Table 2.2
(continued)

VII.6 Information transmission
- 1 = genetic
- 2 = developmental
- 3 = traditional
- 9 = missing

VII.7 Storage of domesticates
- 1 = absent
- 2 = seasonal but less than a year
- 3 = more than a year
- 9 = missing

III. Uses of domesticates variables

VIII.1 Subsistence foods
- 1 = absent
- 2 = present
- 9 = missing

VIII.2 Secondary foods
- 1 = absent
- 2 = present
- 9 = missing

VIII.3 “Drug” foods
- 1 = absent
- 2 = present
- 9 = missing

VIII.4 Raw materials
- 1 = absent
- 2 = present
- 9 = missing

VIII.5 Utensils
- 1 = absent
- 2 = present
- 9 = missing

VIII.6 Labor
- 1 = absent
- 2 = present
- 9 = missing

IV. Biological impacts of agriculture variables

VIV.1 Population density
- 1 = declines
- 2 = stable
- 3 = increases
- 9 = missing

VIV.2 Community size
- 1 = declines
- 2 = stable

(*continued*)

Table 2.2
(continued)

3 = increases
9 = missing
VIV.3 Number of communities
1 = declines
2 = stable
3 = increases
9 = missing
VIV.4 Catchment area
1 = declines
2 = stable
3 = increases
9 = missing
VIV.6 Longevity
1 = declines
2 = stable
3 = increases
9 = missing
VIV.8 Sex ratios
1 = declines
2 = stable
3 = increases
9 = missing
VIV.11 Age of reproduction
1 = declines
2 = stable
3 = increases
9 = missing
VIV.13 Pathogen load
1 = declines
2 = stable
3 = increases
9 = missing
VIV.14 Nutrition
1 = declines
2 = stable
3 = increases
9 = missing
VIV.16 "Wear and tear"
1 = declines
2 = stable
3 = increases
9 = missing
VIV.19 Ecological diversity
1 = declines
2 = stable
3 = increases
9 = missing

Table 2.2
(continued)

VIV.20 Ecological assemblage 1 = declines 2 = stable 3 = increases 9 = missing
V. "Sociocultural" impacts of agriculture variables
VV.1 Sedentarism 1 = declines 2 = stable 3 = increases 9 = missing VV.2 Intra-community communication/coordination 1 = declines 2 = stable 3 = increases 9 = missing VV.3 Inter-community communication/coordination 1 = declines 2 = stable 3 = increases 9 = missing VV.4 Intra-community territoriality/ownership 1 = declines 2 = stable 3 = increases 9 = missing VV.5 Inter-community territoriality/ownership 1 = declines 2 = stable 3 = increases 9 = missing VV.6 Intra-community violence 1 = declines 2 = stable 3 = increases 9 = missing VV.7 Inter-community violence 1 = declines 2 = stable 3 = increases 9 = missing VV.8 Intra-community exchange/transmission/diffusion 1 = declines 2 = stable 3 = increases 9 = missing

(continued)

Table 2.2
(continued)

VV.9 Inter-community exchange/transmission/diffusion
- 1 = declines
- 2 = stable
- 3 = increases
- 9 = missing

VV.10 Kinship structure
- 1 = declines
- 2 = stable
- 3 = increases
- 9 = missing

VV.11 Size of kin group
- 1 = declines
- 2 = stable
- 3 = increases
- 9 = missing

VV.12 Access to and control of resources
- 1 = declines
- 2 = stable
- 3 = increases
- 9 = missing

VV.13 Access to and control of reproduction (social and physical)
- 1 = declines
- 2 = stable
- 3 = increases
- 9 = missing

VV.14 Access to leadership
- 1 = declines
- 2 = stable
- 3 = increases
- 9 = missing

VV.15 Differential survivorship
- 1 = declines
- 2 = stable
- 3 = increases
- 9 = missing

VV.18 Diversity of tasks
- 1 = declines
- 2 = stable
- 3 = increases
- 9 = missing

VV.19 Specialization of tasks
- 1 = declines
- 2 = stable
- 3 = increases
- 9 = missing

Table 2.2
(continued)

VV.21 Genetic social control mechanisms
1 = declines
2 = stable
3 = increases
9 = missing
VV.23 Authoritarian social control mechanisms
1 = declines
2 = stable
3 = increases
9 = missing
VV.28 Pheromonal communication
1 = declines
2 = stable
3 = increases
9 = missing
VV.30 Acoustic communication
1 = declines
2 = stable
3 = increases
9 = missing
VV.31 Visual communication
1 = declines
2 = stable
3 = increases
9 = missing

Table 2.3
Indices used in the analysis

Index name	Mean of the listed standardized variables	Cronbach's alpha
Agricultural Practices	VII.1, VII.2, VII.3, VII.4, VI.5, VI.6, VI.7, VI.8, VI.9, VI.10, VI.13, VI.15, VI.16	0.908
Agricultural Processes	VII.1, VII.3, VII.4, VII.7	0.831
Agricultural Product Uses	VIII.1, VIII.2, VIII.3, VIII.4, VIII.5, VIII.6	0.899
Demographics	VIV.1, VIV.2, VIV.6, VIV.8, VIV.11	0.907
Community Relations	VV.2, VV.3, VV.4, VV.5, VV.6, VV.7, VV.8, VV.9	0.837
Task Specialization	VV.18, VV.19	0.948
Social Control	VV.20, VV.22, VV.23, VV.24, VV.26	0.749

Results

In this section I present the results of the systematic comparisons. Some of them are not surprising, but others indicate important similarities among all of the cases that, I will argue in the conclusions, indicate convergent evolution. The analyses discussed in this section are all presented in table 2.4, which also gives the six specific hypotheses tested here. These hypotheses were developed during two working group meetings on agriculture in humans and insects held at the Santa Fe Institute in 2014 and 2016 (see Introduction, this volume).

The analysis confirms the first hypothesis, that insects are generally obligate agriculturalists while humans are not. This finding is not surprising, but does raise a general conceptual issue with the analyses of hypotheses 4 to 6. Since the insects are obligate agriculturalists, it is reasonable to ask how one can consider variation in dependence on agriculture as an independent variable, as there is no variation among the insects coded for this project. The answer is that humans and insects are considered equivalent cases here. Future scholars might add nonagricultural insect cases to the dataset to see if the results change, but doing so is not necessary to evaluate hypotheses 4 to 6. For those hypotheses cases are treated as equivalent, just as are environmental context, language group, political organization, or any other variable that might be considered important in the evaluation of a different hypothesis. Here, all that is examined is the impact of domestication on the cases, whether or not they are insects or humans.

The analysis indicates that insects and humans differ in the range of domesticates employed, as expected under hypothesis 2. Specifically, humans use a wider range of domesticates. In Highland Mexico, for example, the three primary crops used were from completely different orders: *Poales* (corn), *Fabales* (beans), and *Curcurbitales* (squash) (Bellwood 2005). In contrast, all fungus-farming ants cultivate a single order of fungus, *Agaricales* (see chapter 8, this volume). This illustrates the related fact that not only do insects tend to be obligate agriculturalists, but they also tend to focus on a specific domesticated species. The differences in the range of domesticates used is mirrored by great differences in the management of domesticated crops.

Humans also employ a far greater diversity of crop management methods than do insects, as expected under hypothesis 3. Insects employ very specific management techniques for creating and maintaining substrates, planting crops, controlling weeds and pests, and the like. Insects also tend to have "enclosed" agricultural systems that are isolated from outside environmental fluctuations, pests, weeds, and so on (see, e.g., the discussion in chapter 6, this volume). The most "open" and diverse insect management system occurs among the ambrosia beetles, which cultivate symbiont fungi in tunnels that they bore into dead or stressed trees (Hulcr and Stelinski 2017; chapter 7, this volume). In contrast, because of the diversity of plants humans cultivate and because humans practice agriculture in a wide range of environments, humans require flexible means of management that are adapted to specific plants in specific environments. Nowhere is this clearer than in Peru where specific varieties of potatoes were developed for individual temperature, sunlight, and altitude conditions—perhaps as many as 4,000 different varieties (Brush et al. 1995).

Turning to the impact of agriculture, the analysis indicates that greater dependence on domesticates produces demographic changes, but these are not directly associated with an

Table 2.4
Results of hypothesis tests

H_1 Insects are generally obligate agriculturalists while humans are not.

Independent variable = species
Dependent variable = VII.1 Degree of dependence on domesticates
Mann-Whitney U = 10 (p < .000)
Mean rank: humans = 8.67; insects = 19.50

H_2 Insects and humans differ in the range of domesticates employed.

Independent variable = species
Dependent variables = Ag. Uses Index
Mann-Whitney U = 10 (p < .000)
Mean rank: humans = 17.33; insects = 6.50

H_3 Insects and humans differ in the management of domesticated crops.

Independent variable = species
Dependent variables = Ag. Process Index; Ag. Practice Index
Ag. Proc. Index: Mann-Whitney U = 22 (p < .003)
Mean Rank: Humans = 9.47; Insects = 18.30
Ag. Prac. Index: Mann-Whitney U = 19 (p < .003)
Mean rank: humans = 8.86; insects = 17.60

H_4 Greater dependence on domesticates produces demographic changes.

Independent variable = VII.1 Degree of dependence on domesticates
Dependent variables = Demographics Index
Tau-b = .342 (p < .043)
Kruskal-Wallace H = 4.72 (p < .047, 1-tailed)

Mean rank:	0 to 33%	= 7.20
	34 to 67%	= 12.75
	68 to 100%	= 15.58

H_5 Greater dependence on domesticates produces greater task specialization.

Independent variable = VII.1 Degree of dependence on domesticates
Dependent variables = Task Specialization Index
Tau-b = .54 (p < .005)
Kruskal-Wallace H = 9.82 (p < .007)

Mean rank:	0 to 33%	= 6.10
	34 to 67%	= 13.25
	68 to 100%	= 15.71

H_6 Greater dependence on domesticates produces greater social control.

Independent variable = VII.1 Degree of dependence on domesticates
Dependent variables = Social Control Index
Tau-b = .62 (p < .006)
Kruskal-Wallace H = 4.72 (p < .016)

Mean Rank:	0 to 33%	= 3.40
	34 to 67%	= 10.06
	68 to 100%	= 11.25

increase in population or community size, primarily because there is more variation in the population density and community size of groups with a low dependence on agriculture. Thus the underlying notion in hypothesis 4, that population increases with greater dependence on domesticates, is not supported. Looking more broadly at demographics, however, there is an overall increase in population size, fertility, and longevity in more agricultural groups. For example, Ješovnik and Schultz (chapter 8, this volume) describe a colony of *Atta* leaf-cutter ants as being the "equivalent of a large vertebrate herbivore in terms of biomass, life span, and consumption." A massive transformation of demographics is also seen in humans, whose agricultural communities tend to be larger and to have higher rates of juvenile mortality than nonagricultural ones (Cohen and Crane-Kramer 2007).

As hypothesis 5 suggests, greater dependence on domesticates produces greater task specialization. This relationship makes sense, as the cultivation of domesticates requires individuals to bring select skills and expertise to the process. Among humans these skills often relate to technology, from the creation of tools (e.g., metalwork), to the extraction of raw materials to produce those tools (e.g., mining), and to the movement and marketing of both raw materials and finished products. Agriculture among humans also tends to require specialization in such skills and knowledge that provide continued access to agricultural lands—for example, establishing legal ownership and transfer, enforcing rules of ownership and use, and protecting lands from outsiders. Among insects specialization often takes the form of behavioral variation, with individuals spending their lives gathering raw materials (e.g., leaves), "weeding" crops, or harvesting. Indeed, specialization among insect farming lineages is often produced through specific phenotypic variations (see, e.g., chapters 6, 7, 8, and 14, this volume).

Finally, and as hypothesized, greater dependence on domesticates produces greater degrees of social control. This makes sense for humans, given the need for specialization and protection of agricultural lands. Laws must be followed to ensure access to land and resources and to maintain open markets for the exchange of raw materials, finished goods, and agricultural products. Enforcement of some kind must also be present or those laws are meaningless. The relationship between social control and agriculture is less clear among insects, as they are often eusocial, and thus control of behavior is intrinsic to the group (Wilson 1971). Among insects, this social control takes the form of eusociality in general and phenotypic specialization specifically. Aanen and Anten (chapter 4, this volume) discuss this in terms of kin selection and the social evolution of cooperative behavior, specifically in gut microbiota where nonreproductive symbionts can only increase fitness by maximizing host fitness.

The systematic comparative analysis thus demonstrates that while humans and insects differ in terms of agricultural practices, the evolution of agriculture produces similar changes in social organization. These findings lead to the conclusions I present in the next section.

The Convergent Evolution of Agriculture in Humans and Insects

Many of the chapters in this volume highlight variation in both human and insect agriculture, identifying both unique and shared features among them. These chapters frequently take a comparative approach, either within species or between them, but none takes the systematic comparative approach employed here. The approach here provides a quantitative and statistical perspective on the similarities and differences between human and insect agriculture. The major differences, and those pointed out in many of this volume's chapters, are in the variety of domesticates that are relied upon and the degree to which that reliance is obligate. It is clear that humans have far greater diversity in the domesticates employed than any of the insects and are far less dependent upon those domesticates for survival. Those are interesting findings, but, to me, not as interesting as the ones concerning social similarities and differences.

It is surprising that, despite the diversity in domesticates, the diversity in agricultural practices, and the diversity in species considered in this volume, agriculture appears to have similar social impacts in most of the cases discussed. In the analyses here those similarities are quantified to show empirically that dependence on agriculture is directly associated with greater degrees of task specialization and social control. Agricultural societies are more "politically" and "economically" centralized in both insects and humans. Although not considered here, other chapters (e.g., chapters 3, 4, and 5, this volume) make the case that agricultural societies are more cooperative or integrated than nonagricultural ones.

Clearly agriculture is a successful adaptation. The agricultural insects dominate their environments (e.g., chapters 6, 8, and 14, this volume), and I suggest that it is heightened sociality and social control that provide the ability to do this (chapters 4 and 14, this volume). A dependence on agriculture makes cooperation and specialization necessary (chapters 3 and 8, this volume), but this also means that social control mechanisms to maintain cooperation and division of labor are necessary. What agricultural systems create, then, is a complex, centralized society that can use its coordinated strength to outcompete other organisms in its environment. This is precisely why human agriculture has long been seen as a precursor of states and empires, and it is why agricultural insects can come to dominate their environments.

I argue that these similarities in social impacts across agricultural species points to convergent evolution—not convergence that is rooted in a particular means of getting food, but rather one that is rooted in a food production system that promotes highly coordinated and cooperative behavior. There are other ways of accomplishing this, and many of the social insects have evolved some form of tight coordination and cooperation. But agriculture takes this a step further, and this is where agriculture itself becomes important. Not only do agricultural societies demand coordination and cooperation, but they also provide a more reliable food source than foraging and thus allow for demographic changes that foster overall fitness. Simply put, agricultural societies can outcompete nonagricultural ones in the same environment. They can support more individuals than foraging, they offer ready mechanisms for defense, and they create a context where coordination provides a better survival option to would-be defectors than self-interested behavior.

If agriculture provides all these evolutionary benefits, then one may reasonably ask why agriculture is not more common. Here McGhee (chapter 1, this volume) provides excellent guidance. He points out that agricultural behavior occurs not only in humans and insects, but also among fishes and mollusks, so that it is more widespread among species than one might perhaps think. But McGhee also demonstrates that there are significant functional constraints surrounding the evolution of agriculture so that the conditions allowing agriculture to evolve may be fairly rare. Given the apparent range of organisms that have been found to be agricultural and the strict functional constraints limiting the ability for organisms to evolve agricultural systems, it is perhaps not surprising that, though a widespread and successful adaptation, agriculture is relatively rare.

Returning to the central question of this volume, whether or not agriculture is an example of convergent evolution, I suggest the systematic comparative analysis presented here provides an obvious answer: yes.

References

Asch, David, and John P. Hart. 2004. "Crop Domestication in Prehistoric Eastern North America." In *Encyclopedia of Plant and Crop Science*, edited by R. Goodman, 314–319. Boca Raton, FL: CRC Press.

Barker, Graeme. 2006. *The Agricultural Revolution in Prehistory*. New York: Oxford University Press.

Barker, Graeme, and Candice Goucher, eds. 2015. *The Cambridge World History: Volume 2, A World with Agriculture, 12000 BCE–500 CE*. Cambridge: Cambridge University Press.

Bellwood, Peter. 2005. *First Farmers: The Origins of Agricultural Societies*. Oxford: Blackwell.

Brush, Stephen, R. Kesseli, R. Ortega, P. Cisneros, K. Zimmerer, and C. Quiros. 1995. "Potato Diversity in the Andean Center of Crop Domestication." *Conservation Biology* 9(5): 1189–1198.

Childe, V. Gordon. 1936. *Man Makes Himself*. New York: Oxford University Press.

Cohen, Mark N., and Gillian M. M. Crane-Kramer, eds. 2007. *Ancient Health: Skeletal Indicators of Agricultural and Economic Intensification*. Gainesville: University of Florida Press.

Cowan, C. Wesley, and Patty Jo Watson, eds. 2006. *The Origins of Agriculture: An International Perspective*. Tuscaloosa: University of Alabama Press.

DeVellis, R. F. 2012. *Scale Development*. Los Angeles: Sage.

Diamond, Jared. 2002. "Evolution, Consequences and Future of Plant and Animal Domestication." *Nature* 418(August 9): 700–707.

Dow, Gregory, and Clyde Reed. 2015. "The Origins of Sedentism: Climate, Population, and Technology." *Journal of Economic Behavior and Organization* 119(November): 56–71.

Ember, Carol R., and Melvin Ember. 2009. *Cross-Cultural Research Methods*, 2nd ed. Lanham, MD: AltaMira.

Hulcr, Jiri, and Lukasz Stelinski. 2017. "The Ambrosia Symbiosis: From Evolutionary Ecology to Practical Management." *Annual Review of Entomology* 62:285–303.

IBM Corporation. 2019. *IBM SPSS Statistics for Windows, Version 26.0*. Armonk, NY: IBM Corp.

Jobling, M. A., M. E. Hurles, and C. Tyler-Smith. 2004. *Human Evolutionary Genetics*. New York: Garland Science.

Larsen, Clark S. 2018. "The Bioarchaeology of Health Crisis: Infectious Disease in the Past." *Annual Review of Anthropology* 47:295–313.

Milner, George. 2004. *The Moundbuilders: Ancient Peoples of Eastern North America*. London: Thames and Hudson.

Morgan, Lewis Henry. 1877. *Ancient Society*. New York: Henry Holt.

Peregrine, Peter N., Carol R. Ember, and Melvin Ember. 2004. "Universal Patterns in Cultural Evolution: An Empirical Analysis Using Guttman Scaling." *American Anthropologist* 106(1): 145–149.

Peregrine, Peter N., Carol R. Ember, and Melvin Ember. 2007. "Modeling State Origins Using Cross-Cultural Data." *Cross-Cultural Research* 41(1): 1–12.

Price, T. Douglas, and Ofer Bar-Yosef. 2011. "The Origins of Agriculture: New Data, New Ideas." *Current Anthropology* 52(S4): S163–S174.

Wilson, Edward O. 1971 *The Insect Societies*. Cambridge, MA: Belknap.

Zeder, Melinda. 2011. "The Origins of Agriculture in the Near East." *Current Anthropology* 52(S4): S221–S235.

II CONFLICT AND COOPERATION IN HUMAN AND NONHUMAN AGRICULTURE

3 If Group Selection Is Weak, What Can Agriculture Learn from Fungus-Farming Insects?

R. Ford Denison

The theme of this volume echoes the proposal by Mueller et al. (2005, 564) that "it may be fruitful to examine the short-term and long-term solutions that have evolved convergently in insect agriculture for possible application to human agriculture." Although they cited Denison, Kiers, and West (2003) in support of this idea, our 2003 review did not actually mention insect agriculture explicitly. However, like Darwin, we contrasted improvement of crop wild ancestors by natural selection, over millions of years, with "man's feeble efforts" (Darwin 1859, 61) to make further improvements. Could the long-term improvement of insect agriculture by natural selection offer useful ideas for human agriculture? Possibly, but we need to consider what natural selection can improve and what it cannot.

We argued that natural selection is unlikely to have missed simple improvements that would have consistently improved individual plant fitness. However, in contrast to individual plants, we argued that "there is no reason to expect the structure of natural ecosystems . . . to be a reliable blueprint for agricultural ecosystems" (Denison, Kiers, and West et al. 2003, 146). The fungal gardens of insects may be intermediate between individuals and ecosystems as potential targets of natural selection (Denison 2012). So, how likely is it that "solutions that have evolved convergently in insect agriculture" will be useful models for human agriculture?

To answer this key question, I first explain why natural selection has not consistently improved the overall organization of natural ecosystems—of things like species composition or spatial and temporal patterns. Then, I consider the extent to which these arguments apply to the organization of the fungal gardens of insects. Finally, I consider a specific aspect of insect agriculture: Namely, whether selection imposed by fungus-farming insects on their fungal cultivars or on other beneficial microbes are plausible contributors to their success.

If we measure the sustainability of agricultural practices by how long they can persist, then the survival of traditional rice-growing practices in Asia for over 4,000 years is certainly impressive (King 1911). But cattle have been raised for milk for at least 9,000 years (Evershed et al. 2008), and fungus-farming ants have been growing fungi for food for about 60 million years (Branstetter et al. 2017). Few natural ecosystems have persisted with little change for more than a few thousand years, given ice ages and human impacts on ecosystems worldwide.

Even if an ecosystem has persisted for only a few thousand years, we can say that its sustainability has been tested over time. But maybe a different version of that ecosystem (one with a different ratio of legumes to grasses, say, or more-uniform spatial distribution of trees) would not only have persisted, but would also have been more productive or more resilient under drought. For an entity to *improve* over time requires something beyond mere persistence. Competitive selection among alternatives is a well-understood mechanism for such improvements. For example, direct or indirect competition among individuals improves traits that benefit individuals, although this may not always benefit the species, community, or ecosystem (Denison 2012).

My doubts about modeling agriculture on natural ecosystems contrasts with claims that natural ecosystems are "the most appropriate standard available to sustainable agriculture" (Jackson and Piper 1989, 1592). I agree that, if we completely understood what features or processes make a particular natural ecosystem productive, sustainable, and so on, then it might make sense to copy those features or processes. This understanding-based approach is probably what Mueller et al. (2005) implied by "*possible* application to human agriculture." But if natural ecosystems are "beyond complete human comprehension" (Jackson and Piper 1989, 1591), then we risk copying ecosystem features or processes that undermine rather than support desired outcomes such as productivity and sustainability. Copying without understanding was the basis of South Pacific "cargo cults" and what Richard Feynman called "cargo-cult science" (Feynman 1985, 340), such as emphasizing fancy equipment over disproof of hypotheses. We therefore rejected "mindless mimicry" of natural ecosystems by human agriculture (Denison, Kiers, and West 2003, 146). Instead, we suggested that natural ecosystems could offer both "models to imitate" and "mistakes to avoid" (Denison, Kiers, and West 2003, 161). Is this also true of insect agriculture?

If we can be reasonably confident that something has been consistently improved by natural selection over millions of years, is it a good model for copying by humans, even if without complete understanding? Maybe. Natural selection may at least approach a local optimum, superior to any minor variants. Although there is no guarantee that it will find a global optimum, humans may not be able to do that either. For example, consider current attempts to dramatically improve crop photosynthesis. The C4 photosynthetic pathway has evolved repeatedly (Kellogg 1999), so it is reasonable to assume that it has some advantages over the older C3 pathway. Current attempts to transform C3 rice into a C4 crop may not succeed, but this might be worth trying even if we knew nothing about the mechanism of C4 photosynthesis and its potential benefits. But if the nearest natural ecosystem has three layers of leaves or a 70:30 ratio of grasses to legumes, should human agriculture copy those features, even without a thorough understanding of their consequences? Probably not. Similarly, before we knew about antagonism among fungal cultivars (Poulsen and Boomsma 2005), the success of insects' fungal monocultures over millions of years might have led us to overlook potential benefits of more diverse cropping systems.

Natural Selection Has Not Consistently Improved Ecosystem Organization

Our rejection of mindless mimicry of natural ecosystems was based on two hypotheses. First, natural selection is the only process that consistently improves biological entities over millennia. To avoid semantic arguments, we can limit this to improvement by criteria relevant to agriculture. Second, natural selection has much less power to improve ecosystems than it does to improve genes and the traits of individuals (Dawkins 1976).

First, consider the relative power of natural selection to improve genes, genomes, individuals, groups, communities, and ecosystems. Some genomes enhance individual fitness (survival and reproduction) more than others. But high-fitness genomes are not inherited intact by offspring, at least in species with sexual recombination. So selection acts on genes, not genomes. Usually, the most successful alleles are those that enhance organismal fitness, but this is not always true. For example, natural selection can favor male-sterility alleles of cytoplasmic genes in plants (Dominguez 1995). This is because eliminating production of pollen, which does not carry cytoplasmic genes, frees resources for seeds, which do carry these genes. Cytoplasmic genes can therefore have a conflict of interest with nuclear genes, which are transmitted in both seeds and pollen. As expected, natural selection favors alleles of nuclear genes that restore male fertility (Dominguez 1995). Similarly, the fungal cultivars of ants are transmitted only by females, so the fungi could benefit from female-biasing the sex ratio of the ants (Mueller 2002).

Bacterial genomes may be passed more or less intact to offspring, albeit with some possibilities for horizontal gene transfer and recombination. But conflicts of interest among bacteria are common. To what extent do these conflicts undermine the potential benefits of symbiotic bacteria to plant hosts, or the potential benefits of bacteria to insects and their fungal cultivars? People may argue about whether a given microbial symbiont is "cheating" its host (Jones et al. 2015; Gano-Cohen et al. 2019), but what would be a "fair" exchange rate of nitrogen or phosphorus for plant carbohydrates? It may be simpler to think of a bacterial strain as a "free-rider" if it invests less than other strains in maintaining the health of a shared host plant or insect colony upon which they all depend (Kiers and Denison 2008).

Mathematical modeling showed that such cheating of fellow symbionts is often favored by natural selection, especially as the number of strains per host increases (West et al. 2002). For a legume-rhizobia symbiosis with a realistic number of rhizobial strains per plant, the model predicted that strains investing nothing in nitrogen fixation would outcompete those that invest anything. This would change, however, if legumes impose "sanctions" that reduce the relative fitness of rhizobia in root nodules that fix less nitrogen. Host-imposed sanctions against less beneficial bacteria were assumed not to "reform" the behavior of cheaters, but only to decrease their reproduction, making them less common in subsequent generations. Sanctions could result from hosts attacking less beneficial rhizobia or simply from allocating more resources to more beneficial root nodules. Sanctions against less beneficial rhizobia were subsequently reported in soybeans, where decreases in nodule oxygen permeability are apparently involved (Kiers et al. 2003), and in several other legume species (Simms et al. 2006; Oono, Anderson, and Denison 2011; Regus et al. 2017). Sanctions by *Medicago truncatula* against less beneficial mycorrhizal fungi have also been reported (Kiers et al. 2011), as have sanctions by figs against less beneficial fig wasps (Jander and Herre 2016). Sanctions complicate any definition of "cheater" based on actual symbiont fitness, because "cheating" (behavior that *would have* increased a symbiont's relative fitness, at the expense of the host, *in the absence of sanctions*) could result in lower *actual* fitness in a host that does impose sanctions.

Throughout the biological world, the evolutionary persistence of cooperation depends on mechanisms that limit the fitness payoff to cheaters. Individuals have evolved mechanisms such as nuclear genes that restore male fertility (Dominguez 1995) to reduce the negative effects of conflicts of interest among genes. Some (most?) species that host symbionts have evolved mechanisms (e.g., sanctions) to reduce the negative effects of conflicts of interest among symbionts (Kiers et al. 2011; Oono, Anderson, and Denison 2011; Jander and Herre 2016). These mechanisms enhance individual host fitness in the species that evolved them. Furthermore, the mechanisms are inherited accurately through DNA.

At the ecosystem level, in contrast, there are no documented mechanisms to limit individual behaviors that undermine ecosystem stability or productivity. In contrast to individuals, ecosystems do not inherit species composition or spatial patterns from other ecosystems with anything like the accuracy of DNA-based inheritance. Furthermore, communities and ecosystems do not compete as unified entities the way individuals do. This is because, unlike genes in individuals, species in ecosystems can disperse independently.

In communities and ecosystems, we find many conflicts of interest, which natural selection among individuals has little or no power to resolve. For example, the wild ancestors of our crops inherited traits that enhanced individual plant fitness in past environments, but many of those traits are counterproductive from a community perspective (Denison 2012). Although a plant community with only short plants would waste fewer resources on stems, leaving more for seeds, natural selection favors taller plants that outcompete shorter neighbors for light. Similarly, solar tracking by upper leaves increases shading of lower leaves enough that it can decrease total photosynthesis (Denison, Fedders, and Harter 2010), but natural selection has kept solar tracking, perhaps because many shadows fall on competing neighbors. Natural selection can lead to wasteful overinvestment in roots that "steal" soil water from neighbors, increasing their own productivity but decreasing that of their neighbors more (Zhang, Sun, and Jiang 1999). Individual Irish elk bulls with larger antlers than their challengers won more mates, even as the large antlers were contributing to species extinction (Moen, Pastor, and Cohen 1999). Because natural selection is driven mainly by competition among individuals, it does not always promote the collective interests of each species, much less those of the entire community.

Quantitative comparisons of natural versus human-managed ecosystems, including wild rice or reindeer managed with minimal external inputs, support the conclusion that humans can sometimes design agricultural ecosystems that outperform natural ones, particularly in their ability to export human-edible food (Denison 2012). For example, some reindeer herders enhance productivity by maintaining a female-biased sex ratio (Muuttoranta and Mäki-Tanila 2012), in contrast to the natural 50:50 ratio. They presumably thought of this independently, rather than copying fungi that manipulate insect sex ratios. Similarly, deploying crop diversity over time, via complex crop rotations, may control pests and pathogens better than mimicking natural ecosystems that have similar mixtures of plants in successive years (Xu 2011).

Has Group Selection or Kin Selection Improved Insect Agriculture?

To what extent do the arguments against mindless mimicry extend from ecosystems to insect farms? In contrast to ecosystems, colonies of ants and termites do compete. Furthermore, the behavioral traits of workers in successful colonies are inherited via the queen's DNA. To what extent have the genetically controlled agricultural practices of fungus-farming insects been optimized by group (or kin) selection, driven by competition among colonies?

Recent modeling (Gardner and Grafen 2009, 666) confirms that "between-group selection can lead to group adaptation, but only . . . where within-group selection can be considered to have negligible impact upon phenotypic evolution." With respect to social insects—the term "eusocial" is now used very differently than when it was coined (Batra 1995)—Gardner and Grafen argue that worker policing of reproduction by other workers may meet this criterion, whereas worker sterility per se is neither necessary nor sufficient. Although within-colony conflict may still occur over issues such as queen formation (Hughes and Boomsma 2008) or the sex ratio of offspring (Mehdiabadi, Reeve, and Mueller 2003), it seems unlikely that within-colony conflicts over agricultural practices would undermine selection among different colonies (competing directly or indirectly for territory or resources) for more effective practices.

But can selection among different colonies also maintain colony-benefiting activities of fungal cultivars and other beneficial microbes, when those activities have a metabolic or other cost to the microbes? That seems much less likely. Even if microbes reproduce clonally, a clone will inevitably include less beneficial mutants. The power of within-group selection to undermine cooperation increases rapidly with group size. Mathematical modeling suggests that group size of less than 25 unrelated individuals is needed for natural selection among groups to maintain within-group cooperation (Levin and Kilmer 1974).

Consider bacteria that make a chemical that kills the weed fungus *Escovopsis* (Currie et al. 1999). Each cell that makes the chemical must pay some metabolic cost not paid by mutant strains that do not make that chemical. Unless bacterial cells whose secretions kill *Escovopsis* somehow enhance their own reproduction, relative to mutants that make fewer or different antifungals, within-colony selection will favor the mutants. Given the very large number of bacterial individuals in a colony and the large number of bacterial generations over the life of a colony, it seems very unlikely that natural selection among different colonies is sufficient to outweigh within-colony selection favoring cheating bacteria.

Consistent with this hypothesis, *Pseudonocardia*, which lives on ants and can suppress *Escovopsis* (Currie et al. 1999), can also "kill or strongly suppress ant-cultivated fungi" (Sen et al. 2009, 17807). Suggested alternative roles for these microbes could include "protection of ants or sanitation of the nest" (Sen et al. 2009, 17809). However, free-rider mutants that fail to provide these functions would still have a reproductive advantage, if selection occurs only at the level of the nest. Similarly, within-ant selection for free-riders could override among-ant selection for bacteria whose secretions protect ants from pathogens. An alternative, individual-based hypothesis is that production of antifungal compounds confers a fitness benefit to bacteria (an individual cell or a cluster of clonemates) that outweighs its metabolic cost. A plausible individual-cell benefit to *Pseudonocardia* from producing antifungal compounds involves interactions with (competition with or predation on) black yeasts with which they share the bodies of fungus-growing ants. These yeasts compete with and suppress the growth of *Pseudonocardia* (Little and Currie 2008), so there could be an individual benefit to the bacteria from making antifungals that suppress nearby yeast cells. Suppression of *Escovopsis* by the same antifungals would then be a beneficial side effect that the ants could exploit, albeit with some possible risk to their fungal cultivar.

Based on the above, it seems unlikely that selection among colonies can override within-colony selection for bacterial free-riders that invest less in suppressing pest fungi. Given reasonable mutation rates and short bacterial generation times, bottlenecks during colony formation would provide only short-term relief. Similar arguments could apply to the fungal cultivars: although among-colony selection would favor more beneficial fungal genotypes, within-colony selection might favor fungal traits that enhance fungal fitness at the expense of the insects. Two examples of possible fungal "cheating" traits are (1) manipulation of insect sex ratio, mentioned as a hypothetical possibility above, and (2) sexual reproduction and dispersal of fungi from the colony (Mueller 2002).

If the fungi produced hormones that increased the frequency of female insects, that could reduce fitness of the insects, but it would benefit the fungi. This is because only females transmit fungi to new colonies (Mueller 2002). However, it is not clear that fungal "individuals"—a complex concept in fungi, not explored here—would benefit, relative to fungal free-riders that skip the metabolic cost of making the hormone. Human pathogens offer an analogous case, where the progress of infection depends on the production of metabolically expensive molecules by the pathogen. In such cases, virulent populations can be invaded by less virulent free-riders (Köhler, Buckling, and van Delden 2009). Similarly, individual-versus-community conflicts among fungi provide an alternative explanation for their apparent failure actually to manipulate insect sex ratio (Mueller, 2002). Sexual reproduction by fungi, in contrast, might offer an individual benefit. It is therefore not surprising that there are numerous reports of fungal fruiting structures that "are actively destroyed by ants" (Mueller 2002, S90) although there may be cases in which sexual reproduction by the fungi benefits ants. Self-dispersal could also offer individual fitness benefits to fungi, but it seems to be rare (Mueller 2002).

How Effective Is Selection among Microbes Imposed by Fungus-Growing Insects?

I have argued above that among-colony selection is too weak to prevent the evolution of harmful free-riding in the fungal cultivars of insects and the bacteria that potentially protect those cultivars. But among-colony selection may be fairly effective in favoring the evolution of insect behaviors that benefit the whole colony. High within-colony relatedness, especially ancestrally (Boomsma et al. 2011), and limited reproductive opportunities for workers should minimize, although not eliminate (Mehdiabadi, Reeve, and Mueller 2003; Hughes and Boomsma 2008), conflicts of interest that might undermine efficient agricultural practices. One such practice could be for insects to impose selection on their fungal crops, on bacteria that kill *Escovopsis*, or perhaps even on "integrated crop-microbe consortia" (Mueller et al. 2005, 563). Where among-colony selection fails, with respect to beneficial microbes, can insect-imposed selection fill the gap?

An insect can presumably select some volume of material—a "patch" of fungal garden—to be transplanted, discarded, or perhaps treated with antifungal secretions (Fernández-Marín et al. 2006, 2009). But any evolutionary improvement in crop fungi or beneficial bacteria by such processes depends on (1) the ability of insects to identify "good" and "bad" patches and (2) within-patch genetic diversity being low relative to diversity among different patches. These two requirements will be discussed in order.

Insects imposing selection on fungal cultivars seems roughly analogous to what human plant breeders do, but the vigor of individual plants may influence some plant breeders more than it should. Insects may be less likely to make this mistake, since individual bacterial cells are so small. True, some of the target traits for which plant breeders select can be evaluated easily in individual plants. Disease resistance and flavor are two examples. But I have argued that most increases in yield potential—the yield of groups of plants, given good control of pests and disease—have required accepting tradeoffs between individual plant fitness and whole-crop yield (Denison 2015). In a historical example, Peter Jennings (1964) targeted shorter stature and more erect leaves in designing the first Green Revolution rice (figure 3.1), reversing some effects of past natural selection and sacrificing competitiveness (Jennings and de Jesus 1968). Both traits were among those proposed by Colin Donald for a more cooperative "ideotype" of wheat (Donald 1968). In practice, selection for cooperative traits has often come from selecting among plots containing many plants of a single genotype, without much consideration of tradeoffs. By selecting among field plots for yield, plant breeders increased leaf angle and reduced the excessive size of tassels (the male flowers of maize) without having either as a specific objective (Duvick and Cassman 1999), although this took 60 years. Further exploitation

Figure 3.1
Traits that enhance individual fitness in mixtures, such as greater height in plants or decreased investment in "public goods" by beneficial microbes, tend to undermine community productivity. For example, IR8 rice yielded more grain than older varieties because it invested less in trying to grow above its neighbors. Photo by R. Ford Denison.

of tradeoffs between individual competitiveness and yield remains a promising route for genetic improvement of crop plants (Denison 2015; Anten and Vermeulen 2016; Weiner 2019). It is unclear whether tradeoff-blind selection among plots of genetically uniform plants (current practice in late stages of crop improvement) or a tradeoff-based approach (like that of Jennings or Donald) will work best. In rice, Yuan et al. (2011, 168) found that selection based on yield increased yield more than "trait-based selection," but their unexplained target traits were inconsistent with a tradeoff-based approach. In particular, their target height for "trait-based" breeding was greater than a tradeoff-based approach would suggest (Donald 1968; Denison 2012).

Returning to insects, they might be good at identifying which patches produce the most edible fungi. As for bacteria, even if they could identify patches with bigger individual bacterial cells, it seems unlikely that they would favor those patches under the faulty assumption that a community of such cells will control *Escovopsis* better.

So we turn to the second criterion: low within-patch diversity. This requirement might be met for crop fungi. However, bacterial diversity is probably much greater, due to population size, even if we consider only a single strain and its mutants. Therefore, the ability of insects to improve "microbial consortia" by selecting among garden patches would seem to be severely constrained by the "effective deme sizes of less than 25 and usually closer to 10" criterion of Levin and Kilmer (1974, 544).

Despite this theory, some published experimental results have appeared to show beneficial microbial consortia evolving under human-imposed selection (Swenson, Wilson, and Elias 2000; Panke-Buisse et al. 2014). If these results are valid, or if we obtain clear evidence for beneficial microbial consortia from insect-imposed group selection, evolutionary theory might need to be revised. Swenson, Wilson, and Elias (2000, 9112) noted that, under current theory, their "artificial ecosystem selection" experiments (selecting soils for better or worse plant growth) "should not have worked." Unfortunately, their experiment did not really show whether their version of group selection actually worked. The main problem was pseudo-replication. In each generation, a single good- or poor-growth composite soil sample (pooled from several plants each with good or poor growth) was used to inoculate replicate test plants. Thus, there was replication to show whether the two soil inocula had consistent effects on plant growth, but no replication of selection regime. Even ignoring this problem, their results cast doubt on a consistent effect of selection. Both selection regimes showed seemingly random but parallel changes in plant growth—changing growing conditions?—for ten generations. In generation 10, plants from both selection regimes had much less growth than in any previous generation. Then growth improved in both, but more so in the plants inoculated with soil from good-growth plants. Given divergence, from whatever cause, there is a 50% probability that the good-growth treatment would be superior, by chance.

A similar experiment, selecting for soil-microbe effects on flowering date in *Arabidopsis* (Panke-Buisse et al., 2014), had the same pseudo-replication problem. If we treat each of these publications as a replicate supporting the same general conclusion, ignoring the risk of publication bias, we still have only N=2. So, someone needs to repeat these soil experiments with true replication. Standard practice in experimental-evolution experiments is to maintain multiple independent lineages for each selection regime (Garland and Rose 2009), as we recently did in selecting for multicellularity (Ratcliff et al. 2012). For selection among microbiomes, Mueller and Sachs (2015) recommend a minimum of six independent lines for each selection regime, noting that all six would have to show the same directional response for statistical significance in a sign test.

Like the soil-selection experiments, the convergent evolution of similar agricultural practices in ants and termites has a sample size of only two. The multiple evolutionary origins of agriculture in ambrosia beetles offer more statistical power, but perhaps less convergence (chapter 7, this volume).

In summary, the hypothesis that insect-imposed selection among garden patches can effectively favor more beneficial strains seems somewhat plausible for fungal cultivars, assuming low within-patch diversity, but not for beneficial bacteria. For example, I would expect bacteria to evolve to make whatever antifungal compounds maximize their own ratio of benefits to costs (including inclusive-fitness effects on nearby clonemates), whatever the effect on *Escovopsis* and the fungal cultivar. However, if someone shows that insects really are imposing effective selection for microbial consortia that benefit the whole colony, beyond what can be explained by microbial inclusive fitness, that would be an exciting result, requiring changes to our current understanding of natural selection. This result would also suggest that it might be possible for our crops to impose beneficial selection on entire microbial communities around their roots, rather than only on the intimate partners (rhizobia and mycorrhizal fungi) that are subject to host sanctions (Kiers and Denison 2008). The main point of this chapter is that I do not expect this to happen.

References

Anten, N. P. R., and P. J. Vermeulen. 2016. "Tragedies and Crops: Understanding Natural Selection to Improve Cropping Systems." *Trends in Ecology and Evolution* 31:429–439.

Batra, S. W. T. 1995. "The Evolution of 'Eusocial' and the Origin of 'Pollen Bees.'" *Maryland Naturalist* 39:1–4.

Boomsma, J. J., M. Beekman, C. K. Cornwallis, A. S. Griffin, L. Holman, et al. 2011. "Only Full-Sibling Families Evolved Eusociality." *Nature* 471:E4.

Branstetter M. G., A. Ješovnik, J. Sosa-Calvo, M. W. Lloyd, B. C. Faircloth, et al. 2017. "Dry Habitats Were Crucibles of Domestication in the Evolution of Agriculture in Ants." *Proceedings of the Royal Society B—Biological Sciences* 284:20170095.

Currie, C. R., J. A. Scott, R. C. Summerbell, and D. Malloch. 1999. "Fungus-Growing Ants Use Antibiotic-Producing Bacteria to Control Garden Parasites." *Nature* 398:701–704.

Darwin, C. 1859. *On the Origin of Species by Means of Natural Selection, or the Preservation of Favoured Races in the Struggle for Life*. London: John Murray.

Dawkins, R. 1976. *The Selfish Gene*. Oxford: Oxford University Press.

Denison, R. F. 2012. *Darwinian Agriculture: How Understanding Evolution Can Improve Agriculture*. Princeton: Princeton University Press.

Denison, R. F. 2015. "Evolutionary Tradeoffs as Opportunities to Improve Yield Potential." *Field Crops Research* 182:3–8.

Denison, R. F., J. M. Fedders, and B. L. Harter. 2010. "Individual Fitness Versus Whole-Crop Photosynthesis: Solar Tracking Tradeoffs in Alfalfa." *Evolutionary Applications* 3:466–472.

Denison, R. F., E. T. Kiers, and S. A. West. 2003. "Darwinian Agriculture: When Can Humans Find Solutions beyond the Reach of Natural Selection?" *Quarterly Review of Biology* 78:145–168.

Dominguez, C. A. 1995. "Genetic Conflicts of Interest in Plants." *Trends in Ecology and Evolution* 10:412–416.

Donald, C. M. 1968. "The Breeding of Crop Ideotypes." *Euphytica* 17:385–403.

Duvick, D. N., and K. G. Cassman. 1999. "Post–GreenRevolution Trends in Yield Potential of Temperate Maize in the North-Central United States." *Crop Science* 39:1622–1630.

Evershed, Richard P., Sebastian Payne, Andrew G. Sherratt, Mark S. Copley, Jennifer Coolidge, et al. 2008. "Earliest Date for Milk Use in the Near East and Southeastern Europe Linked to Cattle Herding." *Nature* 455:528–531.

Fernández-Marín, Hermógenes, Jess K. Zimmerman, Stephen A. Rehner, and William T. Wcislo. 2006. "Active Use of the Metapleural Glands by Ants in Controlling Fungal Infection." *Proceedings of the Royal Society B—Biological Sciences* 273:1689–1695.

Fernández-Marín, Hermógenes, Jess K. Zimmerman, David R. Nash, Jacobus J. Boomsma, and William T. Wcislo. 2009. "Reduced Biological Control and Enhanced Chemical Pest Management in the Evolution of Fungus Farming in Ants." *Proceedings of the Royal Society B—Biological Sciences* 276:2263–2269.

Feynman, R. 1985. *Surely You're Joking, Mr. Feynman!* New York: Norton.

Gano-Cohen, K. A., C. E. Wendlandt, K. Al Moussawi, P. J. Stokes, K.W. Quides, et al. 2019. "Interspecific Conflict and the Evolution of Ineffective Rhizobia." *Ecology Letters* 22:914–924.

Gardner, A., and A. Grafen. 2009. "Capturing the Superorganism: A Formal Theory of Group Adaptation." *Journal of Evolutionary Biology* 22:659–671.

Garland, T., and M. R. Rose. 2009. *Experimental Evolution: Concepts, Methods, and Applications of Selection Experiments.* Berkeley: University of California Press.

Hughes, W. O. H., and J. J. Boomsma. 2008. "Genetic Royal Cheats in Leaf-Cutting Ant Societies." *Proceedings of the National Academy of Sciences* 105:5150–5153.

Jackson, W., and J. Piper. 1989. "The Necessary Marriage between Ecology and Agriculture." *Ecology* 70:1591–1593.

Jander, K. C., and E. A. Herre. 2016. "Host Sanctions in Panamanian *Ficus* Are Likely Based on Selective Resource Allocation." *American Journal of Botany* 103:1753–1762.

Jennings, P. R. 1964. "Plant Type as a Rice Breeding Objective." *Crop Science* 4:13–15.

Jennings, P. R., and J. de Jesus. 1968. "Studies on Competition in Rice. I. Competition in Mixtures of Varieties." *Evolution* 22:119–124.

Jones, E. I., M. E. Afkhami, E. Akçay, J. L. Bronstein, R. Bshary, et al. 2015. "Cheaters Must Prosper: Reconciling Theoretical and Empirical Perspectives on Cheating in Mutualism." *Ecology Letters* 18:1270–1284.

Kellogg, E. A. 1999. "Phylogenetic Aspects of the Evolution of C4 Photosynthesis." In *C4 Plant Biology*, edited by R. F. Sage and R. K. Monson, 411–444. San Diego: Academic Press.

Kiers, E.T., and R. F. Denison. 2008. "Sanctions, Cooperation, and the Stability of Plant-Rhizosphere Mutualisms." *Annual Review of Ecology, Evolution, and Systematics* 39:215–236.

Kiers, E.T., M. Duhamel, Y. Yugandgar, J. A. Mensah, O. Franken, et al. 2011. "Reciprocal Rewards Stabilize Cooperation in the Mycorrhizal Symbiosis." *Science* 333:880–882.

Kiers, E. T., R. A. Rousseau, S. A. West, and R. F. Denison. 2003. "Host Sanctions and the Legume-Rhizobium Mutualism." *Nature* 425:78–81.

King, F. 1911. *Farmers of Forty Centuries: Or, Permanent Agriculture in China, Korea and Japan.* Emmaus, PA: Rodale Press.

Köhler, Thilo, Angus Buckling, and Christian van Delden. 2009. "Cooperation and Virulence of Clinical *Pseudomonas aeruginosa* Populations." *Proceedings of the National Academy of Sciences* 106:6339–6344.

Levin, B. R., and W. L. Kilmer. 1974. "Interdemic Selection and the Evolution of Altruism: A Computer Simulation Study." *Evolution* 28:527–545.

Little, A. E. F., and C. R. Currie. 2008. "Black Yeast Symbionts Compromise the Efficiency of Antibiotic Defenses in Fungus-Growing Ants." *Ecology* 89:1216–1222.

Mehdiabadi, N. J., H. K. Reeve, and U. G. Mueller. 2003. "Queens versus Workers: Sex-Ratio Conflict in Eusocial *Hymenoptera*." *Trends in Ecology and Evolution* 18:88–93.

Moen, R. A., J. Pastor, and Y. Cohen. 1999. "Antler Growth and Extinction of Irish Elk." *Evolutionary Ecology Research* 1:235–249.

Mueller, U. G. 2002. "Ant versus Fungus Versus Mutualism: Ant-Cultivar Conflict and the Deconstruction of the Attine Ant-Fungus Mutualism." *American Naturalist* 160:S67–S98.

Mueller, U. G., N. M. Gerardo, D. K. Aanen, D. L. Six, and T. R. Schultz. 2005. "The Evolution of Agriculture in Insects." *Annual Review of Ecology, Evolution, and Systematics* 36:563–595.

Mueller, U. G., and J. L. Sachs. 2015. "Engineering Microbiomes to Improve Plant and Animal Health." *Trends in Microbiology* 23:606–617.

Muuttoranta, K., and A. Mäki-Tanila. 2012. "Regional Differences in Reindeer Herding Operations in Finland." *Acta Agriculturae Scandinavica, Section A* 62:142–152.

Oono, R., C. G. Anderson, and R. F. Denison. 2011. "Failure to fix Nitrogen By Non-Reproductive Symbiotic Rhizobia Triggers Host Sanctions That Reduce Fitness of Their Reproductive Clonemates." *Proceedings of the Royal Society of London B—Biological Sciences* 278:2698–2703.

Panke-Buisse, Kevin, A. C. Poole, J. K. Goodrich, R. E. Ley, and J. Kao-Kniffin. 2014. "Selection on Soil Microbiomes Reveals Reproducible Impacts on Plant Function." *ISME Journal* 9:980–989.

Poulsen, M., and J. J. Boomsma. 2005. "Mutualistic Fungi Control Crop Diversity in Fungus-Growing Ants." *Science* 307:741–744.

Ratcliff, W. C., R. F. Denison, M. Borrello, and M. Travisano. 2012. "Experimental evolution of multicellularity." *Proceedings of the National Academy of Sciences* 109:5: 1595–1600.

Regus, J. U., K. W. Quides, M. R. O'Neill, R. Suzuki, E. A. Savory, et al. 2017. "Cell Autonomous Sanctions in Legumes Target Ineffective Rhizobia in Nodules with Mixed Infections." *American Journal of Botany* 104:1299–1312.

Sen, R., H. D. Ishak, D. Estrada, S. E. Dowd, E. Hong, et al. 2009. "Generalized Antifungal Activity and 454-Screening of *Pseudonocardia* and *Amycolatopsis* Bacteria in Nests of Fungus-Growing Ants." *Proceedings of the National Academy of Sciences* 106:17805–17810.

Simms, E. L., D. L. Taylor, J. Povich, R. P. Shefferson, J. L. Sachs, et al. 2006. "An Empirical Test of Partner Choice Mechanisms in a Wild Legume-Rhizobium Interaction." *Proceedings of the Royal Society of London B—Biological Sciences* 273:77–81.

Swenson, W., D. S. Wilson, and R. Elias. 2000. "Artificial Ecosystem Selection." *Proceedings of the National Academy of Sciences* 97:9110–9114.

Weiner, J. 2019. "Looking in the Wrong Direction for Higher-Yielding Crop Genotypes." *Trends in Plant Science* 24:927–933.

West, S. A., E. T. Kiers, E. L. Simms, and R.F. Denison. 2002. "Sanctions and Mutualism Stability: Why Do Rhizobia Fix Nitrogen?" *Proceedings of the Royal Society of London B—Biological Sciences* 269:685–694.

Xu, X. M. 2011. "A Simulation Study on Managing Plant Diseases by Systematically Altering Spatial Positions of Cultivar Mixture Components between Seasons." *Plant Pathology* 60:857–865.

Yuan, W., S. Peng, C. Cao, P. Virk, D. Xing, et al. 2011. "Agronomic Performance of Rice Breeding Lines Selected Based on Plant Traits or Grain Yield." *Field Crops Research* 121:168–174.

Zhang, D. Y., G. J. Sun, and X. H. Jiang. 1999. "Donald's Ideotype and Growth Redundancy: A Game Theoretical Analysis." *Field Crops Research* 61:179–187.

4 The Sociobiology of Domestication

Duur K. Aanen and Niels P. R. Anten

Farming, whereby one organism cultivates a population of another organism, is usually associated with tending of livestock and crop plants by humans. However, similar behavior exists in other organisms, and those examples predate human agriculture by millions of years. The best-known examples of nonhuman farming are the fungus-farming insects (Mueller et al. 2005). In this chapter we use a broader concept of farming, including mutualistic host-symbiont relationships in general, and ask how useful this concept is for analyzing their evolution.

The host in a mutualistic host-symbiont interaction has the same interest as a farmer who cultivates a field of rice or breeds a herd of cattle—namely, maximizing the productivity of its domesticates. Simply put, farmers can be considered as hosts and their crops and livestock as symbionts. Collective performance of the symbionts usually depends on cooperation (Frank 1996). However, within a group of symbionts a "tragedy of the commons" (Hardin 1968) may occur: individuals within the herd may increase their relative fitness within the group at the cost of collective herd productivity (see chapter 3, this volume). This is a fundamental problem of social evolution in a wide variety of biological (and socioeconomic) interactions, ranging from the eukaryotic cell with thousands of mitochondrial genomes, to natural populations of plants or animals, to the farmer with thousands of livestock. Because of the difference in numbers between partners, *domestication*, defined as the genetic modification of one species by another in ways that benefit the modifying species but that would reduce the fitness of the modified species in its original niche, is essentially a process of social evolution. How is lower-level competition within the herd prevented from disrupting collective performance, and thus herd productivity?

First, we review the data from a variety of natural host-symbiont interactions and identify key mechanisms that hosts use to enforce sociality in their livestock. Second, we consider human agriculture and husbandry and identify parallels with nonhuman host-symbiont interactions. We also identify examples of conscious or unconscious group selection of domesticated species by farmers and of unwanted consequences of selection at the level of the individual. Third, we identify opportunities for group selection. While the conditions for selection for cooperation in nature are highly specific and rare, humans can create those conditions in their selection programs and be inspired by natural examples of host-symbiont interactions. Explicitly considering social effects in artificial selection is a relatively unexplored and potentially extremely important category of improvement of crops and livestock, not only for higher yields but also for greater product quality and, in husbandry, animal welfare.

Is the Distinction between a "Host" and a "Symbiont" Useful?

Many symbioses are asymmetrical, consisting of a large host and smaller symbiotic organisms (Frank 1997). The host can be a multicellular individual with gut bacteria; a eukaryotic cell with intracellular symbionts; a farmer with crops; or a colony itself, such as a colony of social insects that cultivate fungal symbionts. There are various possible criteria for distinguishing the host from the symbionts (table 4.1). In this chapter, we argue that a useful criterion is based on a difference in numbers. Almost invariably, a mutualistic entity consists of a single host individual or an "organism" at a different level in the biological hierarchy, and a group of symbionts. We define "organism" as a relatively conflict-free entity in the biological hierarchy (Queller and Strassmann 2009); for example, an organism can be a single individual or a colony of individuals founded by a singly mated ant queen or a monogamous pair of termite reproductives. Distinguishing host and symbiont based on the number criterion allows us to explicitly consider the significance of cooperation between members of one partner species, the symbionts, and how the other partner, the host, can influence that.

Mechanisms Whereby a Host Maximizes "Symbiont Productivity"

Identified mechanisms whereby a host maximizes "symbiont productivity" can be placed in three main categories (table 4.2): (1) mechanisms that facilitate kin selection; (2) mechanisms that reduce correlation between competitive success of symbionts and reproductive success; and (3) mechanisms that favor competitiveness. All three are based on the assumption that the host organism profits from the collective performance of the symbionts, but they differ in the assumptions they make about the relationship between collective performance of the symbionts and their competitive traits. The first two assume that individual competitiveness of symbionts negatively correlates with collective performance, while the third is the exceptional case in that the host has an interest in competitive traits.

Table 4.1
Differences between host and symbiont

Differences in "power" (Frank 1997). The host is often envisaged as being the partner in control, although this is partially due to an anthropocentric perspective.
Differences in the generation times, with the symbionts generally having multiple generations per host generation, leading to a difference in evolutionary rate. In animal husbandry, the difference in generation time has further dramatically increased due to modern techniques such as embryo transplantation, artificial insemination, and cloning.
(Potential) differences in evolutionary rate. A difference in the generation times potentially results in a higher rate of evolution of the symbionts (but see Bergstrom and Lachmann 2003).
Differences in being external (the host) or internal—that is to say protected from the external environment (the symbionts) or otherwise compartmentalized (Law and Lewis 1983).
Differences in the frequency of sex, with symbionts typically having reduced frequency of sex (Law and Lewis 1983). The examples discussed by Law and Lewis (1983), such as between algae and fungi in lichens and between arbuscular mycorrhizal fungi and plants, provide support for this idea.

Table 4.2
The three main categories of mechanisms whereby a host maximizes "symbiont productivity"

1. Facilitating kin selection		
Indirect	Control symbiont reproduction (vertical uniparental transmission, with bottleneck)	Fungus-farming ants, eukaryotic cells (chloroplasts, mitochondria), some fungus-farming termites
Direct	Control symbiont genetic variation ("weeding")	Fungus-farming ants (Howe, Schiøtt, and Boomsma 2019), fungus-farming termites (Aanen et al. 2009), fig trees for their mutualistic wasps (Frank 1996; Herre 1985)
2. Reducing correlation between symbiont competitive abilities and reproductive success		
Indirect	Early split reproductive and "somatic" symbionts	"Germline" bacteria that remain quiescent until new eggs are formed, and "somatic bacteria" that provide benefits in sucking lice (Buchner 1965); some rhizobia (Denison 2012); modern human agriculture (germline provided by seed companies or breeding companies)
	Randomize chances to be transmitted analogously to meiosis	*Not known to us*
Direct	Sanctioning "bad" symbionts	*Rhizobia* and legumes (Kiers et al. 2003); arbuscular mycorrhiza (Kiers et al. 2011)
	Choice of "good" symbionts.	Squids and luminescent bacteria (Koropatnick et al. 2004; Sachs et al. 2004); artificial selection in human agriculture
3. Favoring competitiveness		
	When the competitive trait is the trait exploited by the host, also known as *desirable tragedies* (Anten and Vermeulen 2016)	Bacteria that produce antibiotics (Scheuring and Yu, 2012); in human agriculture, e.g., production of timber or large/abundant flowers, breeding of fighting animals

Below we will explain these categories and provide examples. In table 4.2, an overview of the three categories and subcategories and examples of each are provided.

Facilitating Kin Selection

The first means for improving symbiont productivity is by facilitating kin selection. Hamilton's (1964a, 1964b) rule ($rB–C>0$) has three parameters: the extra number of offspring in social partners due to a social behavior (the benefit, B), the reduced number of offspring of the individual performing the behavior (the cost, C), and the relatedness (r) between the individual performing and the individual receiving the social behavior. Increasing relatedness among symbionts will increase the indirect fitness component of inclusive fitness (chapter 5, this volume). In many symbioses, symbiont relatedness is an indirect consequence of symbiont transmission. For example, many endosymbionts are transmitted vertically and uniparentally, usually via the egg cell. Uniparental transmission is also usually associated with a bottleneck, so that individuals start with limited genetic variation. For mitochondrial evolution, it has been suggested that uniparental transmission is an important means to reduce the scope for competitive traits that have "virulent" consequences for host fitness (Cosmides and Tooby 1981; Frank 1996). While vertical uniparental transmission is most strict in endosymbioses, this transmission mode also occurs in some other host-symbiont interactions such as the fungus-farming ants and two groups of fungus-growing termites. Of note is that vertical transmission in the two independent groups of fungus-growing termites is uniparental in both cases, but by different sexes (Johnson et al. 1981; Nobre et al. 2011). In all studied species of the genus *Microtermes*, female alates transmit the fungus to newly founded colonies, but in the species *Macrotermes bellicosus*, males do (Nobre et al. 2011).

Symbiont relatedness can also be manipulated directly by the host. For example, colonies of most species of fungus-growing termites do not regularly have vertical uniparental transmission, except for the two groups just mentioned (Korb and Aanen 2003). Instead, most species rely on horizontal symbiont acquisition via sexual spores in the environment, produced by mushrooms from termite colonies. Even though this would give ample opportunity for mixed cultures, all colonies examined so far are associated with a single fungal clone (Aanen et al. 2009). It has been shown that termites achieve a monoculture even when started from a mixture of strains, because the most common strain achieves more cooperative interactions and produces a disproportionate number of asexual spores that are used to inoculate new fragments of the fungus garden (Aanen et al. 2009; Bastiaans et al. 2015). Also, in the fungus-growing ants, where vertical uniparental transmission is the default transmission mode, frequent horizontal transmission occurs (Howe, Schiøtt, and Boomsma 2019), and mechanisms exist whereby symbiont diversity is reduced (Poulsen and Boomsma 2005). Another example wherein a host enforces high symbiont relatedness are leguminous plants, which carry nitrogen-fixing rhizobial bacteria in root nodules. Each nodule is colonized by a single bacterium, so that relatedness among bacteria of a single nodule is maximal. This organization of genetic variation among groups of bacteria allows plants to effectively discriminate nodules and to sanction unproductive rhizobial strains (Kiers et al. 2003; discussed further below). Finally, fig trees have an interesting pollination mutualism with fig wasps, in which fig wasps lay eggs in fig flowers, and the emerging fig wasps disperse pollen to other plants. Since mating between newly emerged wasps occurs inside the fig, and since males die upon mating, males do not disperse pollen. Each male consumes a seed, however, and thus represents a virulent trait from the fig tree's perspective. Since there is local mate competition, fig wasps adjust the sex ratio depending on the number of foundresses. With a single foundress the sex ratio is female biased, while the proportion of males increases with additional foundresses (Herre 1985). It is thus in the interest of a fig tree to minimize the number of foundresses per flower, and it has been demonstrated that a fig inflorescence is sealed off within 24 hours after colonization, thus limiting the chance of additional foundresses (Frank 1996; Herre 1985).

Steve Frank (1996) has pointed out that there is an important distinction between the direct effect of increased symbiont relatedness on collective performance and the indirect effect due to kin selection facilitated by high symbiont relatedness. An important question is whether symbiont mixing has an immediately negative effect on collective performance—that is, whether competitiveness is induced upon mixing, or whether it is an evolved obligate characteristic because of an evolutionary history of individual selection. For fungi associated with fungus-growing social insects, it has been found that the effect of symbiont mixing is induced upon mixing (Aanen et al. 2009; Poulsen and Boomsma 2005). It has also been shown that there is a direct benefit of "monoculture" of mitochondria for mice

(Sharpley et al. 2012). The same question can be asked for plant communities in human agriculture. Are competitive traits constitutive or induced upon recognition of unrelated individuals, either of the same species or of different species?

Reducing Correlation between Symbiont Competitive Abilities and Reproductive Success

Next to kin selection, repression of competition is the second main cornerstone for the evolution of cooperation (Frank 2003). When opportunities for competition against neighbors are limited within groups, individuals can increase their own success only by enhancing the efficiency and productivity of their group. Therefore, characters that repress competition within groups promote cooperation and enhance group success (Frank 2003). Repression of competition can indirectly lead to more cooperative traits by favoring selection of cooperative individuals or directly, for example, by partner choice or sanctioning. An example of the indirect selection for more cooperative traits is an early split between reproductive and "somatic" symbionts. A nice example is provided by a particular group of gut symbionts of sucking lice. Those symbionts are divided at an early developmental stage into "germline" bacteria, which remain quiescent until new eggs are formed, and "somatic bacteria," which provide nutritional benefits to the lice (Buchner 1965; see figure 4.1). Since the somatic bacteria are an evolutionary dead end, their only way to increase

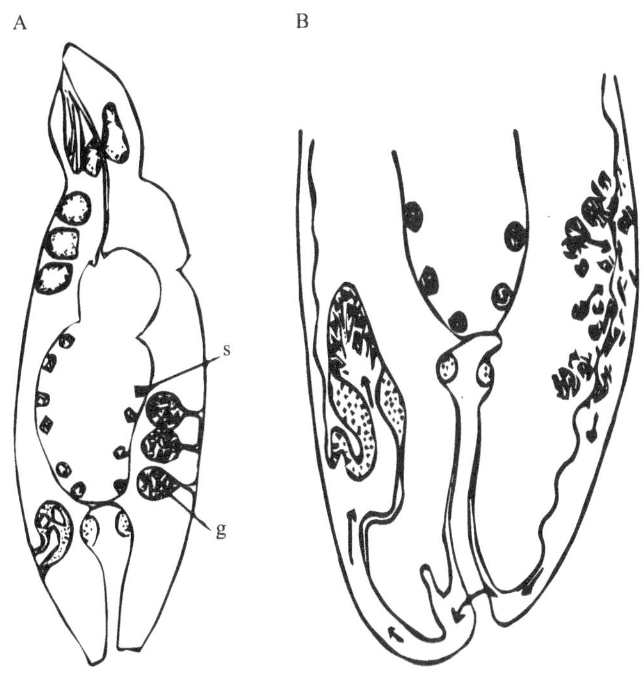

Figure 4.1
The early sequestration of symbionts into germ and soma in the sucking louse (*Haematopinus suis*; redrawn based on Buchner 1965). (A) The symbionts are separated at an early developmental stage into a somatic (*s*) and a germline (*g*) population. The somatic population is contained in storage organs lining the gut, and the germ-line population is temporarily stored in special organs. (B) Germ-line storage organs degenerate and symbionts are released and flow toward the developing ovaries. The direction of movement is indicated by the arrows.

in fitness is indirectly, by maximizing host benefit, which increases the chance of survival of their genetically related germline symbionts. This example is analogous to the early soma-germline sequestration between cells found in metazoa. There is a nice analogy here with modern agriculture: the seeds of most crops are produced separately from those used in production fields by specialized seed companies. Similarly, in many livestock species, individuals are produced by specialized breeding companies.

An example of direct selection of cooperative individuals is the sanctioning of "bad" symbionts, or the choosing of "good" symbionts. For example, for the earlier mentioned leguminous plants, which live in symbiosis with nitrogen-fixing rhizobial bacteria, it has been found that nonproductive strains can be sanctioned by providing them with less carbohydrates relative to productive strains (Kiers et al. 2003). A similar mechanism has been found in arbuscular mycorrhiza (Kiers et al. 2011). Partner choice has been found in squids and their luminescent symbiotic bacteria (Koropatnick et al. 2004; Sachs et al. 2004). Those bacteria are housed in a special light organ, which probably functions as a camouflage organ that disguises them from predators that swim below the squid. Symbionts are acquired horizontally, not only when squids emerge from the egg, but also continuously throughout their lives. Each day 95% of the symbiotic bacteria are expelled and renewed by horizontal acquisition from the environment. A mechanism has been hypothesized for how the host can choose specific partners based on their luciferase activity. The high concentration of peroxidase produced in crypts of the squid light organs are poisonous (Visick and McFall-Ngai 2000). This poison may act specifically against nonluminous strains. Because the functioning bacterial luciferase has a higher binding affinity for oxygen than for the peroxidases, luminous strains may escape the effects of the deadly poison (Visick and McFall-Ngai 2000). There is a direct analogy here with modern human agriculture: humans artificially select and have selected their various crops and pets. The question is whether the selection criteria can be improved if group productivity is explicitly taken into account.

Favoring Competitiveness

The above sections show that competition between symbionts often negatively affects performance (e.g., yield) at the group level. An exception to the general rule that competitiveness conflicts with host interest is when the competitive trait is exploited by the host. An example is bacteria that produce antibiotics, which have been found in several insects (Currie et al. 1999; Kaltenpoth et al. 2005; Scheuring and Yu 2012; Visser et al. 2012). More generally, if a host has the ability to recognize desirable characteristics and to select for them, competitive traits can be the target of selection. For example, within eukaryotic cells, mitochondria are tested for functionality on an individual basis by alternation between fusion and fission (Kowald and Kirkwood 2011).

Analogies with Human Agriculture

Competitive Traits in Plants: Examples of and Opportunities for Group Selection

From an agricultural perspective, maximum population-level performance is generally the objective. That is, farmers tend to want to maximize the yield or resource–use efficiency of a crop field or herd of farm animals. If maximization of functions such as seed production or resource-use efficiency is the objective, then naturally evolved competitive traits resulting in tragedies of the commons (TOCs) are undesirable. Here we discuss examples of such traits, with regard to plants and their relation to crops and with regard to animal breeding. We then discuss how lessons for breeding can be drawn from knowledge of the existence of TOC phenomena.

The most-studied competitive trait in plants is plant height. Height growth involves costs. This is first because taller structures require stronger support structures (McMahon 1973), and taller plants thus have a higher fractional allocation of resources to stems and branches and a smaller allocation to leaves (Stutzel, Charles-Edwards, and Beech 1988). Second, in trees there are hydraulic limitations to height caused by increased resistance in longer transport vessels and increased gravitational potential opposing the ascent of water. Consequently, leaf stomatal conductance and photosynthesis become constrained in tall trees (Ryan, Phillips, and Bond 2006). Thus, a population of relatively short plants has more energy to invest in fitness-enhancing functions such as seed production. However, such vegetation would not be evolutionarily stable as it could be invaded by mutant plants that are taller and thus shade their neighbors.

Many other similarly competitive traits with the potential to produce TOCs exist in plants. For instance, it is known that plants may produce extra roots in the presence of non-self neighbors at the expense of shoot and seed production (Gersani et al. 2001; but see Chen et al. 2015). Leaf area indices of both crop and natural vegetation stands have been found to be larger than values that would maximize whole-canopy carbon gain, because populations with optimal leaf areas for maximal community performance are prone to invasion by plants with larger leaf areas that capture more light (Anten and During 2011). Horizontal leaf angles may have evolved similarly. Vertical leaves favor a better light distribution in the canopy and may thus favor whole-stand canopy photosynthesis, but horizontal leaves are more efficient in competition (Hikosaka and Hirose 1997). The traits mentioned so far relate to physical resource acquisition, but reproductive traits may also have evolved in a similar way. For instance, plants invest energy in fruits or flowers to attract dispersers or pollinators, and there is probably an optimal size of these organs that maximizes visibility of these structures per unit of invested energy. However, if one individual produces more or bigger fruits or flowers, it may become more attractive to dispersers/pollinators and thus attract a bigger share of them (Anten and Vermeulen 2016).

Competitive Traits in Animals: Examples of and Opportunities for Group Selection

In animal breeding there are opportunities for maximizing group performance by breeding for indirect genetic effects (Bijma and Wade 2008). Bill Muir (1996) was the first to state in the literature that selection schemes should therefore take the social aspect into account. Classical, individual-level selection for survival paradoxically led to a decrease in survival because the individuals surviving were the most competitive, that is to say aggressive, ones. Selecting at the level of the family group, in contrast, favored less aggressive, more social individuals. In this way, a fast response to selection against mortality and increased longevity was achieved (Muir 1996). Similarly, in other domesticated animals increases in yield may be achieved by selecting for more social behaviors. For example, a regular practice in dairy farming is the removal of the horns of cattle to prevent them from harmful fighting. Horns are an individually selected competitive trait that decreases group productivity, and hornless breeds have been selected to increase group productivity. Likewise, in poultry, the beaks of chicken are trimmed. Recent work by Piter Bijma (Bijma and Wade 2008) and others (Muir 1996; Denison 2012) shows that selection for sociality can work.

What Can Crop Breeding Learn from Other Host-Symbiont Interactions?

Given the aforementioned conflict between individual-based selection in many natural settings and the inherent focus on group performance in agriculture, we may reasonably ask what steps have (possibly unwittingly) been made in selecting for increased group performance and what lessons can be learned for the future.

One of the great advances in crop breeding has been the green revolution, which involved strong reductions in plant height in several of our major crops (Khush 1999). The main advantage derived was reduced investment in stem tissue, which allows greater investment in seeds and a smaller chance of lodging (plants buckling, usually due to wind forces or under their own weight). In many fruit crops, grafting is an interesting technique whereby shoots of less competitive and highly productive genotypes can be connected to stem and root systems of more competitive genotypes (Bulley et al. 2005). But probably a broader set of traits has been similarly optimized in crop breeding. When plants were grown at high population density, yields were twice as high in maize varieties released in the 1990s as in varieties released in the 1930s, but there was no difference at low density. These patterns were associated with trait values that run counter to those that produce a TOC: more vertical leaves, smaller investment in male flowers, and slower leaf senescence in more recent varieties (Duvick and Cassman 1999). In another study higher yields in cotton were associated with smaller leaf area (Lu et al. 1994). It has also been documented that modern high-yield varieties produce fewer roots or roots with less competitive architectures than older varieties (York et al. 2015). Together, these findings indicate that crop breeding has, albeit unwittingly, been successful in selecting for traits that run counter to the expression of a TOC and that this development has contributed to significant increases in crop production.

This raises the question of whether further improvements in breeding along these lines are possible, as advocated by several researchers (e.g., Weiner et al. 2010; Denison 2011, 2012). Root traits may provide an interesting avenue. While, as noted, there is evidence that root traits in modern varieties might be more cooperative, their invisibility and difficulty to measure mean that they have been rather understudied in crop breeding. Similarly, the physiological regulation of trait expression and potential interactions between the regulation of different traits have rarely been explicitly considered as breeding targets. An interesting phenomenon in this respect is the apparent interaction that exists between competitive responses to the presence of neighbor plants and expression of defense against pests. Plants respond to light cues reflected from neighbor plants (i.e., light with relatively less red and more far-red) through a set of responses including stem elongation and reduced branching, a syndrome known as *shade avoidance* (Pierik and de Wit 2014). These same light cues have been shown to also suppress expression of defense (Ballare 2014). This has been suggested to reflect a TOC, the idea being that a certain defense expression favors the community as it may deter pests at relatively low costs. A mutant that suppresses defense may, however, benefit as it foregoes these costs while still benefiting from the pest deterrence provided by its neighbors (Ballare 2014; but see, e.g., Douma et al. 2019). If the expression of this TOC really takes place under farmer field conditions, which still needs to be shown, it would entail a highly interesting breeding target.

Clearly, there are various potential cases for which crop performance could be further improved by selecting for more cooperative traits. The question is to what extent can this be achieved in phenotypic selection methods. This is best achieved through some sort of group selection by growing monocrops of genetic variants and evaluating their performance at the stand level. But it is important to note that in order to maximize the chance of finding the variation that can be used to select for improved plant performance over a range of conditions, one would want to test as many genotypes in as many conditions as possible in the realistic conditions and high densities used by farmers. However, because of limits on space, time, and money, there is a limit to the number of plants one can accurately score in terms of phenotype (Araus and Cairns 2014). The simple solution is to grow only a few individuals per genotype in mixtures. But this enhances the chance of selecting for competitive rather than cooperative traits (Bos 2008), which as noted can be counterproductive to trying to increase population-level performance. This may hold especially for cross-pollinating species, in which maintenance of genetic variation is essential for preventing inbreeding depression, and for evolutionary breeding practices, in which the seed ratio of the next year depends on the yields of the varieties that are grown mixed in genetically diverse stands (Doring et al. 2011). Selection can also directly occur at the individual level through mass selection (Yabe et al. 2013a, 2013b), in which competition between potential targets can be even stronger when these best-performing individuals are grown in higher frequencies at high densities.

Thus, individual selection allows one to test more genotypes under more conditions while group selection allows one to select for less competitive traits. Hybrids of the two could be used whereby, for example, mass selection is applied to the first (few) generation(s) following a cross, and then the most promising varieties are tested in mono-variety groups (kin selection) (Murphy et al. 2017). The balance between the two selection levels (i.e., how much individual versus how much kin) depends on where one expects the greatest yield gains to be achieved: through exploring the overall genetic variation or through directed selection for cooperative traits.

Conclusion and Outlook

An often overlooked characteristic of mutualistic host-symbiont associations is a difference in numbers: in many cases a single host is associated with a group of symbionts. According to this criterion, agriculture is a host-symbiont association in that the farmer generally cultivates fields of crops and herds of livestock. Because of the difference in numbers between partners, domestication is essentially a process of social evolution. Assuming that the symbionts collectively provide a benefit to their host and that resources used for individual competition would divert resources from that collective benefit, an important question is: How can symbionts become and remain social, or behave socially, so that they collectively provide a benefit to the host, and how can the host influence symbiont sociality, either via genetic selection or phenotypic induction? We considered the two main categories of explanation for social evolution, kin selection and repression of competition, which are not mutually exclusive. Examples of the first category are enforcement of monoculture cultivation by fungus-cultivating insects and selection for social chickens in human agriculture. An example of the second category is the sequestration of "reproductive symbionts" from "worker symbionts," which has been found in sucking louse gut symbionts. Of note is that in modern human agriculture a similar sequestration of "germline" and "soma" is found, with the germline being provided by specialized seed companies for crops or breeding companies for livestock. An exception to the above-sketched general pattern of increasing sociality of symbionts are mutualisms wherein the benefit of symbionts is based on competitive traits. An example of this category is antibiotic production, which can be selected by interference competition. Another example is a preference for fast-growing crops as found in the mutualism between damselfish and algae (Aanen 2010). In this mutualism, the preferred algae are those that are fast growing and that do not invest in persistent cell walls. In the nonmutualistic situation, these algae are found in the early stages of succession. By their continual weeding, damselfish continuously recreate these early succession conditions. Host interference is a major mechanism to facilitate sociality in a world that is otherwise dominated by individual-level selection and numerous TOCs. This insight can be used more explicitly in agriculture, both in plant and animal breeding, than it is in current practice. There are major unexplored opportunities for selecting at the level of the group in order to improve collective performance of domesticated plants and animals, which usually is correlated with total harvestable yield.

References

Aanen, D. K. 2010. "As You Weed, So Shall You Reap: On the Origin of Algaculture in Damselfish." *BMC Biology* 8:81–84.

Aanen, D. K., H. H. D. Licht, A. J. M. Debets, N. A. G. Kerstes, R. F. Hoekstra, and J. J. Boomsma. 2009. "High Symbiont Relatedness Stabilizes Mutualistic Cooperation in Fungus-Growing Termites." *Science* 326:1103–1106. doi:10.1126/science.1173462.

Anten, N. P. R., and H. J. During. 2011. "Is Analysing the Nitrogen Use at the Plant Canopy Level a Matter of Choosing the Right Optimization Criterion?" *Oecologia* 167:293–303. doi:10.1007/s00442-011-2011-3.

Anten, N. P. R., and P. J. Vermeulen. 2016. "Tragedies and Crops: Understanding Natural Selection to Improve Cropping Systems." *Trends in Ecology & Evolution* 31:429–439. doi:10.1016/j.tree.2016.02.010.

Araus, J. L., and J. E. Cairns. 2014. "Field High-Throughput Phenotyping: The New Crop Breeding Frontier." *Trends in Plant Science* 19:52–61. doi:10.1016/j.tplants.2013.09.008.

Ballare, C. L. 2014. Light Regulation of Plant Defense." *Annual Review of Plant Biology* 65:335–363.

Bastiaans, E., A. J. M. Debets, and D. K. Aanen. 2015. "Experimental Demonstration of the Benefits of Somatic Fusion and the Consequences for Allorecognition." *Evolution* 69:1091–1099. doi:10.1111/evo.12626.

Bergstrom, C. T., and M. Lachmann. 2003. "The Red King Effect: When the Slowest Runner Wins the Coevolutionary Race." *Proceedings of the National Academy of Sciences* 100:593–598. doi:10.1073/pnas.0134966100.

Bijma, P., and M. J. Wade. 2008. "The Joint Effects of Kin, Multilevel Selection and Indirect Genetic Effects on Response to Genetic Selection." *Journal of Evolutionary Biology* 21:1175–1188. doi:10.1111/j.1420–9101.2008.01550.x.

Bos, I., and P. Caligari. 2008. *Selection Methods in Plant Breeding*, 2nd ed. Dordrecht, The Netherlands: Springer.

Buchner, P. 1965. *Endosymbiosis of Animals with Plant Microorganisms*. New York: Interscience Publishers.

Bulley, S. M., F. M. Wilson, P. Hedden, A. L. Phillips, S. J. Croker, and D. J. James. 2005. "Modification of Gibberellin Biosynthesis in the Grafted Apple Scion Allows Control of Tree Height Independent of the Rootstock." *Plant Biotechnology Journal* 3:215–223. doi:10.1111/j.1467–7652.2005.00119.x.

Chen, B. J. W., H. J. During, P. J. Vermeulen, H. de Kroon, H. Poorter, and N. P. R. Anten. 2015. "Corrections for Rooting Volume and Plant Size Reveal Negative Effects of Neighbour Presence on Root Allocation in Pea." *Functional Ecology* 29:1383–1391. doi:10.1111/1365–2435.12450.

Cosmides L. M., and J. Tooby. 1981. "Cytoplasmic Inheritance and Intragenomic Conflict." *Journal of Theoretical Biology* 89:83–129. doi:10.1016/0022–5193(81)90181–8.

Currie, C. R., J. A. Scott, R. C. Summerbell, and D. Malloch. 1999. "Fungus-Growing Ants Use Antibiotic-Producing Bacteria to Control Garden Parasites." *Nature* 398:701–704. doi:10.1038/19519.

Denison, R. F. 2011. "Past Evolutionary Tradeoffs Represent Opportunities for Crop Genetic Improvement and Increased Human Lifespan." *Evolutionary Applications* 4:216–224. doi:10.1111/j.1752–4571.2010.00158.x.

Denison, R. F. 2012. *Darwinian Agriculture. How Understanding Evolution Can Improve Agriculture.* Princeton: Princeton University Press.

Doring, T. F., S. Knapp, G. Kovacs, K. Murphy, and M. S. Wolfe. 2011. "Evolutionary Plant Breeding in Cereals—Into a New Era." *Sustainability* 3:1944–1971. doi:10.3390/su3101944.

Douma, J. C., J. de Vries, E. H. Poelman, M. Dicke, N. P. R. Anten, and J. B. Evers. 2019. "Ecological Significance of Light Quality in Optimizing Plant Defence." *Plant Cell and Environment* 42:1065–1077. doi:10.1111/pce.13524.

Duvick, D. N., and K. G. Cassman. 1999. "Post–Green Revolution Trends in Yield Potential of Temperate Maize in the North-Central United States." *Crop Science* 39:1622–1630. doi:10.2135/cropsci1999.3961622x.

Frank, S. A. 1996. "Host-Symbiont Conflict over the Mixing of Symbiotic Lineages." *Proceedings of the Royal Society of London B* 263:339–344.

Frank, S. A. 1997. "Models of Symbiosis." *American Naturalist* 150:S80–S99.

Frank, S. A. 2003. "Perspective: Repression of Competition and the Evolution of Cooperation." *Evolution* 57:693–705.

Gersani, M., J. S. Brown, E. E. O'Brien, G. M. Maina, and Z. Abramsky. 2001. "Tragedy of the Commons as a Result of Root Competition." *Journal of Ecology* 89:660–669. doi:10.1046/j.0022–0477.2001.00609.x.

Hamilton, W. D. 1964a. "Genetical Evolution of Social Behaviour 1." *Journal of Theoretical Biology* 7:1–16. doi.org/10.1016/0022-5193(64)90038-4.

Hamilton, W. D. 1964b. "Genetical Evolution of Social Behaviour 2." *Journal of Theoretical Biology* 7: 17–52. doi:10.1016/0022–5193(64)90039–6.

Hardin, G. 1968. "Tragedy of the Commons." *Science* 162:1243–1248.

Herre, E. A. 1985. "Sex-Ratio Adjustment in Fig Wasps." *Science* 228:896–898. doi:10.1126/science.228.4701.896.

Hikosaka, K., and T. Hirose. 1997. "Leaf Angle as a Strategy for Light Competition: Optimal and Evolutionarily Stable Light-Extinction Coefficient within a Leaf Canopy." *Ecoscience* 4:501–507. doi:10.1080/11956860.1997.11682429.

Howe, J., M. Schiøtt, and J. J. Boomsma. 2019. "Horizontal Partner Exchange Does Not Preclude Stable Mutualism in Fungus-Growing Ants." *Behavioral Ecology* 30:372–382. doi:10.1093/beheco/ary176.

Johnson, R. A., R. J. Thomas, T. G. Wood, and M. J. Swift. 1981. "The Inoculation of the Fungus Comb in Newly Founded Colonies of Some Species of the Macrotermitinae (Isoptera) from Nigeria." *Journal of Natural History* 15:751–756.

Kaltenpoth, M., W. Gottler, G. Herzner, and E. Strohm. 2005. "Symbiotic Bacteria Protect Wasp Larvae from Fungal Infestation." *Current Biology* 15:475–479. doi:10.1016/j.cub.2004.12.084.

Khush, G. S. 1999. "Green Revolution: Preparing for the 21st Century." *Genome* 42:646–655. doi:10.1139/gen-42-4-646.

Kiers, E., M. Duhamel, Y. Beesetty, J. Mensah, O. Franken, et al. 2011. "Reciprocal Rewards Stabilize Cooperation in the Mycorrhizal Symbiosis." *Science* 333:880–882.

Kiers, E. T., R. A. Rousseau, S. A. West, and R. F. Denison. 2003. "Host Sanctions and the Legume-Rhizobium Mutualism." *Nature* 425:78.

Korb, J., and D. K. Aanen. 2003. "The Evolution of Uniparental Transmission of Fungal Symbionts in Fungus-Growing Termites (Macrotermitinae)." *Behavioral Ecology and Sociobiology* 53:65–71.

Koropatnick, T. A., J. T. Engle, M. A. Apicella, E. V. Stabb, W. Goldman, et al. 2004. "Microbial Factor-Mediated Development in a Host-Bacterial Mutualism." *Science* 306:1186–1188. doi:10.1126/science.1102218.

Kowald, A., and T. B. L. Kirkwood. 2011. "Evolution of the Mitochondrial Fusion-Fission Cycle and Its Role in Aging." *Proceedings of the National Academy of Sciences* 108:10237–10242. doi:10.1073/pnas.1101604108.

Law, R., and D. H. Lewis. 1983. "Biotic Environments and the Maintenance of Sex—Some Evidence from Mutualistic Symbioses." *Biological Journal of the Linnean Society* 20:249–276. doi:10.1111/j.1095–8312.1983.tb01876.x.

Lu, Z. M., J. W. Radin, E. L. Turcotte, R. Percy, and E. Zeiger. 1994. "High Yields in Advanced Lines of Pima Cotton Are Associated with Higher Stomatal Conductance, Reduced Leaf-Area and Lower Leaf Temperature." *Physiologia Plantarum* 92:266–272. doi:10.1111/j.1399–3054.1994.tb05336.x.

McMahon, T. 1973. "Size and Shape in Biology." *Science* 179:1201–1204. doi:10.1126/science.179.4079.1201.

Mueller, U. G., N. M. Gerardo, D. K. Aanen, D. L. Six, and T. R. Schultz. 2005. "The Evolution of Agriculture in Insects." *Annual Review of Ecology Evolution and Systematics* 36:563–595.

Muir, W. M. 1996. "Group Selection for Adaptation to Multiple-Hen Cages: Selection Program and Direct Responses." *Poultry Science* 75:447–458.

Murphy, G.P., C. J. Swanton, R. C. Van Acker, and S. A. Dudley. 2017. "Kin Recognition, Multi-Level Selection and Altruism in Crop Sustainability." *Journal of Ecology* 105:930–934.

Nobre, T., C. Fernandes, J. J. Boomsma, J. Korb, and D. K. Aanen. 2011. "Farming Termites Determine the Genetic Population Structure of *Termitomyces* Fungal Symbionts." *Molecular Ecology* 20:2023–2033. doi:10.1111/j.1365–294X.2011.05064.x.

Pierik, R., and M. de Wit. 2014. "Shade Avoidance: Phytochrome Signalling and Other Aboveground Neighbour Detection Cues." *Journal of Experimental Botany* 65:2815–2824. doi:10.1093/jxb/ert389.

Poulsen, M., and J. J. Boomsma. 2005. "Mutualistic Fungi Control Crop Diversity in Fungus-Growing Ants." *Science* 307:741–744. doi:10.1126/science.1106688.

Queller, D. C., and J. E. Strassmann. 2009. "Beyond Society: The Evolution of Organismality." *Philosophical Transactions of the Royal Society B—Biological Sciences* 364:3143–3155. doi:10.1098/rstb.2009.0095.

Ryan, M. G., N. Phillips, and B. J. Bond. 2006. "The Hydraulic Limitation Hypothesis Revisited." *Plant Cell and Environment* 29:367–381. doi:10.1111/j.1365–3040.2005.01478.x.

Sachs, J. L., U. G. Mueller, T. P. Wilcox, and J. J. Bull. 2004. "The Evolution of Cooperation." *Quarterly Review of Biology* 79:135–160.

Scheuring, I., and D. W. Yu. 2012. "How to Assemble a Beneficial Microbiome in Three Easy Steps." *Ecology Letters* 15:1300–1307. doi:10.1111/j.1461–0248.2012.01853.x.

Sharpley, M. S., C. Marciniak, K. Eckel-Mahan, M. McManus, M. Crimi, et al. 2012. "Heteroplasmy of Mouse mtDNA Is Genetically Unstable and Results in Altered Behavior and Cognition." *Cell* 151:333–343. doi:10.1016/j.cell.2012.09.004.

Stutzel, H., D. A. Charles-Edwards, and D. F. Beech. 1988. "A Model of the Partitioning of New Above-Ground Dry-Matter." *Annals of Botany* 61:481–487. doi:10.1093/oxfordjournals.aob.a087579.

Visick, K. L., and M. J. McFall-Ngai. 2000. "An Exclusive Contract: Specificity in the *Vibrio fischeri Euprymna scolopes* Partnership." *Journal of Bacteriology* 182:1779–1787. doi:10.1128/jb.182.7.1779–1787.2000.

Visser, A. A., T. Nobre, C. R. Currie, D. K. Aanen, and M. Poulsen. 2012. "Exploring the Potential for Actinobacteria as Defensive Symbionts in Fungus-Growing Termites." *Microbial Ecology* 63:975–985. doi:10.1007/s00248-011-9987-4.

Weiner, J., S. B. Andersen, W. K. M. Wille, H. W. Griepentrog, and J. M. Olsen. 2010. "Evolutionary Agroecology: The Potential for Cooperative, High Density, Weed-Suppressing Cereals." *Evolutionary Applications* 3: 473–479. doi:10.1111/j.1752–4571.2010.00144.x.

Yabe, S., R. Ohsawa, and H. Iwata. 2013b. "Potential of Genomic Selection for Mass Selection Breeding in Annual Allogamous Crops." *Crop Science* 53:95–105. doi:10.2135/cropsci2012.03.0167.

York, L. M., T. Galindo-Castaneda, J. R. Schussler, and J. P. Lynch. 2015. "Evolution of US Maize (*Zea mays* L.) Root Architectural and Anatomical Phenes over the Past 100 Years Corresponds to Increased Tolerance of Nitrogen Stress." *Journal of Experimental Botany* 66:2347–2358. doi:10.1093/jxb/erv074.

5 Lifetime Commitment between Social Insect Families and Their Fungal Cultivars Complicates Comparisons with Human Farming

Jacobus J. Boomsma

Farming is always a form of mutualistic symbiosis in which farmers' efforts are rewarded by edible produce, and crops benefit because farmers propagate cultivar genes to future generations. As Darwin (1859) realized, human farmers practice artificial selection, both directly by manipulating differential reproduction of cultivars and indirectly by propagating close kin of preferred individual crop plants or livestock. Over time, human farmers adapted to the nutritional landscapes created by their farming practices, but only as a secondary population-level response. Rapid cultural evolution implied active nutritional niche construction (Scott-Phillips et al. 2014) and the risk of subsequent mismatch when the rate of cultural innovation exceeded the pace by which natural selection could respond with novel genetic adaptations (Tishkoff et al. 2006; Lucock et al. 2014). This human scenario differs fundamentally from the social insect fungus-farming symbioses that are often highlighted for their analogies with human farming. In contrast to humans, the fungus-growing ants and termites co-adapted with their crops in a direct pairwise manner. This is because each social insect farming family commits to a single cultivar for life (Poulsen and Boomsma 2005: Aanen et al. 2009) so that natural selection directly targets the joint reproductive success of each symbiotic colony as a whole (figure 5.1).

This chapter explores the implications of always living in lifetime matrimony with a cultivar and develops a set of analogies with mating system evolution in the social insects. These explorations show that social insect farms are bilaterally symmetrical forms of exploitation for mutual benefit, while human farming symbioses are unilateral forms of exploitation once crop-domestication has been achieved. The niche construction metaphor is useless in the fungus-farming social insects because there is no constructing agent that asymmetrically creates conditions for a partner species. Instead, each matrimonial partnership exploits, and copes with, a joint environmental niche that is fixed for the lifetime of a colony. That difference does not necessarily make farming-practice analogies between social insects and humans invalid, but it changes the causal perspective of such comparisons.

A Hamiltonian Gene's Eye Perspective on Cooperation and Conflict in Mating and Farming

I began studying Panamanian fungus-growing ants in the early 1990s, driven by a natural history interest in the manifestations of social cooperation and conflict. My initial questions

Figure 5.1
Lifetime commitment in sex and symbiosis. (A) A queen of *Atta colombica* from Panama sitting on one of the fungus gardens in a laboratory colony while being guarded by a retinue of minor workers, medium workers, and a large soldier. The queen is > 20 mm long, and her abdominal gaster contains a sperm storage organ of ca. 5 mm diameter with several hundred million stored spermatozoa that will allow her to fertilize up to a hundred million eggs during her maximally achievable life span of 20–30 years. (B) A one-year-old field colony of *A. colombica* exposed from the side with the queen on top, showing the tight fit of the excavated cavity around the colony's fungus garden, on the surface of which small newly added leaf fragments are visible. A fully mature *Atta* colony may contain hundreds of similarly sized fungus gardens, all belonging to the same cultivar clone. (C) A freshly excavated and opened royal chamber of South African *Macrotermes natalensis* showing the queen with her hugely swollen abdomen, her monogamous mate (the colony's king) just above her, and a retinue of guarding adult soldiers and immature (white) nymphs. The queen and king that founded the colony years earlier are recognizable by their blackish thoraces on which the remnants of their shed wings—used during the dispersal flight that brought them together—are still visible. At that point in time they had the same body size so the dark segments on the queen's abdomen illustrate how much her ovaries, needed to lay thousands of eggs each day, have grown. In the image she is ca. 10 cm long. These *Macrotermes* colonies are believed to have a maximal life span of several decades, similar to *Atta* colonies. (D) Fungus-garden cavities of *M. natalensis* showing the older (brown) fungus comb, recently added (grey) fungus comb, and the small white fungal nodules growing from the oldest comb (bottom), consisting mainly of asexual spores that the termites eat so that they are deposited with their primary feces to form the new inoculated (grey) comb material. (E) A medium-sized colony of *M. natalensis* near Pretoria, South Africa. Photos: David Nash (A), Michael Poulsen (D), and the author (B, C, E).

did not relate to farming but they shaped many of the hypotheses that my colleagues and I came to test in later years, because sexual conflicts between females and males and reproductive conflicts between hosts and symbionts turned out to have intriguing parallels (Frank 1996; Poulsen and Boomsma 2005; Boomsma 2007). Given our pioneering studies of the use of genetic markers for mapping pedigrees and mating systems (Van der Have, Boomsma, and Menken 1988), I found the leaf-cutting ants to be intriguing because queens of *Atta* fungus-farming ants had been observed to always mate with multiple males (Hamilton 1964). In the 1980s, many considered multiple insemination of queens to be an explanatory challenge to sterile-worker altruism, but not Hamilton, who wrote that "if the trend to multiple insemination occurs after the firm establishment of the worker caste, its threat to colonial discipline is a rather remote one" (Hamilton 1964, 33–34).

By the end of the 1990s, we had shown that exclusive single-queen mating is ancestral across the fungus-farming ant genera of Panama and that only the crown-group of *Acromyrmex* and *Atta* leaf-cutting ants always has multiply-inseminated queens (Villesen et al. 2002). It later appeared that Hamilton's prediction implied the deeper principle that strict lifetime monogamy could be conjectured to have been a necessary condition for the evolution of permanently unmated nursing castes in general (Boomsma 2007). This idea was confirmed by a large-scale comparative analysis across the ants, bees, and wasps (Hughes et al. 2008), and it also appeared to be valid in the termites (Boomsma 2009; Hartke and Baer 2011; Boomsma 2013). A total lack of ancestral promiscuity thus appeared to explain the evolutionary origins of colonial superorganismality as defined by Wheeler (1911), with pre-imaginal caste differentiation as the sole criterion for having reached that irreversibly enhanced state of organizational complexity. Because strict monogamy maximizes relatedness between siblings, this was a gratifying confirmation of inclusive fitness theory, but it also raised irresistible new questions about the expression and regulation of sperm competition conflicts when obligate multiple queen-mating evolved as a secondary superorganismal elaboration, as had happened multiple times in the ants, corbiculate bees, and vespine wasps (Boomsma, Kronauer, and Pedersen 2009). In following Hamilton's lead in never believing one had understood cooperation without having stared the possible corrupting forces in the face, it seemed obvious to ask why sperm competition had not rendered multiply-inseminated queens infertile.

The queens of ants, social bees, and social wasps are known to never re-mate later in life (Boomsma, Baer, and Heinze 2005), so it seemed reasonable to imagine that multiple ejaculates entering a queen's sperm storage organ on the same afternoon would compete for later paternity. Strife for a higher share of sperm storage space could thus become a tragedy of the commons when ejaculates would continue to kill each other's sperm, leaving a queen without options to fertilize eggs well before her colony would be large enough to reproduce. We therefore hypothesized that leaf-cutting ant queens must have evolved molecular mechanisms to regulate what seemed to be a serious sexual conflict, and we realized that this conflict was mostly a female challenge to manage "endosymbiotic" sperm for maximal overall quality and diversity (Boomsma and Ratnieks 1996; Boomsma, Baer, and Heinze 2005). Over the years we could show that these expectations were correct. Unrelated ejaculates of phylogenetically basal attine ants with obligate single queen-mating peacefully coexisted on microscope slides, consistent with there never having been selection for aggressive sperm competition traits. However, the same *in vitro* confrontations between ejaculates of polyandrous leaf-cutting ants (and honeybees) gave rapid sperm mortality. And, as predicted, adding secretion from queen sperm storage organs terminated mutual sperm killing, at least in *Atta* leaf-cutting ants (Den Boer et al. 2010) (figure 5.1A, B).

We recently characterized the (rather few) proteins involved in this conflict and in its ultimate resolution and showed that they likely evolve rapidly as expected in antagonistic arms races (Dosselli et al. 2019). The bottom line of this study was that, in spite of damaging potential conflict, multiple insemination appears to be evolutionarily stable because queens have the agency to remain in control over a genetically variable set of unruly sperm "symbionts." Ultimately, both males and females came to benefit from the advantages of more genetically diverse colonies (Hughes and Boomsma 2004, 2006) after the potentially devastating tragedy of the stored-sperm commons had been tamed. We thus inferred that for millions of years any new male mutant producing more competitive ejaculates had likely been countered by an effective female response before it could drive populations extinct. It then seemed logical as well that a crucial condition for queens having gained the upper hand must have been that the stored-sperm commons remained closed for newly arriving ejaculates: ancestral lineages with monogamous, full-sibling colonies could thus give rise only to derived lineages with chimeric (multiple patriline) colonies when queens retained their ancestral habit of never re-mating later in life (Boomsma, Baer, and Heinze; Boomsma 2007). What had seemed to be an idiosyncratic peculiarity of social insect mating systems now appeared to make beautiful sense. Lifetime commitment between the sexes—ancestrally between a single queen and a single male and later by a single queen and a specific set of males—had to be a closed game to keep the reproductive interests of the sexes aligned, albeit with the stored ejaculate symbionts in a powerless role devoid of any independent agency.

Very similar Hamiltonian predictions of cooperation and conflict turned out to apply to interspecific host-symbiont interactions, such as those between farming ant families and their fungal cultivars, as was consistent with theory by Frank (1996). Such interactions should be stable if they were as exclusive as the lifetime association between a mitochondrial clone and two sexual gametes in a zygote. However, if symbionts were acquired promiscuously, conflict between genetically different lineages would be expected to threaten symbiotic stability. This is because all garden cultivars need to grow, both in their own interest and to be worth their keep as mutualists, so that multiple cultivars coexisting in the same host-provided niche would inevitably compete for farmer attention. Strife between symbiont lineages would then reduce their overall crop service to the farming colony, in a way that is analogous to sperm competition compromising male fertilization services to queens and their colonies unless inter-ejaculate conflicts are suppressed. Frank's logic thus implied the expectation that farming ants had evolved mechanisms to avoid mixing symbionts in chimeric fungus gardens and that no worker should thus be allowed to introduce a novel unrelated cultivar—for example a fragment from a vacated garden of a newly founded colony nearby.

In this case the conceptual logic again delivered. We could show that *Acromyrmex* colonies always have a single clonal fungus garden, that the fungal cultivars express incompatibility reactions when confronted with mycelium from other fungus gardens, that worker fecal fluid contains nourishing proteins from the resident cultivar that cannot be smoothly absorbed by unrelated cultivars, and that foraging workers are reluctant to adopt alien cultivar fragments (Bot, Rehner, and Boomsma 2001; Poulsen and Boomsma 2005; Ivens et al. 2009; Kooij, Poulsen, et al. 2015). As it turned out, there is lifetime commitment throughout, not only between a queen and the single or multiple male(s) that inseminate her, but also between every ant family and its clonal fungal cultivar (figure 5.1A, B). It appeared that once colonies have workers, a cultivar can neither be replaced nor be secondarily mixed with other cultivars under natural conditions (Poulsen and Boomsma 2005; Howe et al. 2019). Thus, potential conflicts between cultivar symbionts and farming hosts were either avoided or regulated, but, as in sperm storage after queen-mating, at the expense of parties losing all options for later promiscuous exchange. It seemed likely, in fact, that the very origin of ant fungus farming had been conditional on the enforcement of lifetime commitment between a core family of ant farmers and a single cultivar in each colony.

Comparing Exclusively Committed Insect Farming with Promiscuous Human Farming

Conceptually, the findings summarized above cannot be reconciled with the practices of human farming. We would not dream of being dependent on a single clonal cultivar for life, and human farmers are not lifetime unmated altruists comparable to ant workers. There are surely many analogies between farming practices across the human and social insect domains (chapter 14, this volume), but it makes a difference whether we evaluate the deeper roots of unilateral human management to maximize yield or the sustainability of a co-adapted mutualism over evolutionary time. Because the farming ants have practiced the mutual co-adaptation model during millions of years of relentless natural selection on joint performance, they often surpass us in specific efficiency targets. Not only did ants in general evolve sperm banks at ambient temperature that last a queen's potential life span of two to three decades (Den Boer et al. 2009), but they also somehow prevented the evolution of resistance by specialized *Escovopsis* garden pathogens against biocontrol compounds obtained from Actinobacteria that they rear on their cuticles (De Man et al. 2016; Holmes et al. 2016; Heine et al. 2018) (chapter 11, this volume). Recent work has further indicated that the fungus-growing termites are equally efficient in keeping their colonies as free from pathogens as the leaf-cutting ants appear to be (Otani et al. 2019; see also figure 5.1C, D, E). Relative to the extreme specialization of social insect farmers, human farmers are jacks of all trades in their interactions with domesticated crops, and we remain extremely vulnerable to endemic and epidemic diseases of our cultivars.

Given at least a partial understanding of the power of superorganismal farming symbioses, it seems also important to note that the impressive achievements of fungus-farming ants and termites came with an important tradeoff that should make us very reluctant to adopt them as role models. To phrase it provocatively, admiring any superorganismal colonies for their productivity benchmarks is analogous to singing the praise of Nazi Germany because the Third Reich was superbly efficient in building motorways. Defined as they are by stringent and unconditional commitments for life, superorganismal colonies practice what we would conceive in the human domain as *Blut und Boden* principles from the cradle to the grave. As already noted by Wheeler (1911), all individual agency in the sense of independent decision-making has disappeared out of their family lives no matter if one is a queen, a worker, or a soldier (Boomsma 2013; Boomsma and Gawne 2018). Lifetime unmated workers are as committed to their family as somatic cells are to a metazoan body. The main difference is that every somatic cell analog has a brain and six legs (figure 5.1).

It is a truism that the two sexes need each other to reproduce and that the same applies to the farming families of ants and their cultivar clones. However, the notion that none of them can pass on any genes to future generations without a thriving mutualistic partner undermines anthropomorphic extrapolations of human farmer agency to farming insects. A narrative based on fungal cultivars having enslaved an addicted ant family is equally credible. The same is true for the fungus-growing termites. Here, the founding pair establishes a colony and produces its first cohort of workers, who begin to forage, digest, and defecate while creating a fungus comb where sexual *Termitomyces* spores germinate because they were mixed with the termite excrements (Aanen et al. 2009; Nobre et al. 2011). Different *Termitomyces* haplotypes then merge into dikaryons and compete for being propagated asexually so that invariably only a single clone remains. This requires only that the farming termites continue to harvest and eat garden nodules with asexual spores and to produce new inoculated comb material (figure 5.1D). That process ultimately produces an optimal match between a single cultivar clone and a termite family, but it is then equally reasonable to maintain that the garden symbionts domesticated the farming termites as it is to emphasize that the termites domesticated *Termitomyces* (Aanen et al. 2009; chapter 4, this volume). Symbiotic farming agency is evenly split between the insect partner and the cultivar, and none of them can opt out to join a neighboring colony once their host-symbiont marriage has been consummated (see below for this not yet being the case during colony foundation). Strict lifetime commitments thus unambiguously ensure cooperative harmony via monogamous sex and exclusive symbiotic partnership, and they secure a joint interest in suppressing any secondary events of sexual or symbiotic promiscuity.

It seems obvious to consider the fungus-growing termite symbiosis as symmetrically mutualistic (Aanen et al. 2002; Korb and Aanen 2003; Nobre et al. 2011), but fungus-growing ants have always been considered to have unilateral farmer agency because dispersing queens vertically transmit a cultivar clone to the next generation of colonies. However, we actually do not know how vertical the colony-founding matchmaking between a farmer ant queen and a cultivar actually is. Our recent work has challenged the general vertical transmission belief by showing that founding queens of two Panamanian *Acromyrmex* species often acquire their cultivar from nests of other founding queens (Howe, Schiøtt, and Boomsma 2019). This is possible because these queens need to forage from open burrows so they frequently succumb to predation, while other queens lose their incipient garden because of mold or fungus-feeding predators and start looking for a replacement garden. Molecular marker data from natural colony-founding aggregations of *Acromyrmex* made us conjecture that horizontal cultivar acquisition may well be common throughout all attine ant lineages except for *Atta*, which is the only genus that evolved claustral colony founding and sufficient queen fat reserves to rear a first garden and worker cohort in complete seclusion and without foraging.

The study by Howe, Schiøtt, and Boomsma (2019) revealed an unexpected parallel between the fungus-growing ants and the fungus-growing termites where horizontal (*h*) cultivar acquisition is phylogenetically basal and where two species/genus-level lineages secondarily evolved vertical (*v*) cultivar acquisition (Aanen et al. 2002). The difference in cultivar transmission between the independently evolved farming systems of ants and termites now appears to be a matter of degree rather than of principle. A colony-founding *Acromyrmex* queen can principally acquire a single cultivar at the time in some *v*, *h*, *h*, . . . sequence, and she would then commit for life to the last of these before her first workers hatch. Only when no accidents happen will she remain with her original (*v*) cultivar (Poulsen et al. 2009). The fungus-growing termites acquire multiple haplospores from the leaf litter that the first workers bring in, then passively stage competition between the emerging dikaryons in their first garden substrate, and finally let positive frequency-dependent selection sort out which single diploid clone will obtain the monopoly of termite care for the lifetime of the colony (Aanen et al. 2009). Foraging for cultivars—creating *h*, *h*, . . . sequences—thus occurs in both symbioses. The same strategy must also have applied to early human farmers when they tested plants for their suitability as crops (Smith 1998). The difference is that our own culturally evolving ancestors could easily retain their unilateral exploitative agency as farmers—mixing and replacing cultivars as they saw fit—while the insect farming symbioses could remain stable over evolutionary time only by exclusive mutual commitment and symmetrically shared agency in every pair of farming family and cultivar crop.

Rethinking the Natural History of Insect Fungus Farming

The logic outlined above implies that there is host-cultivar recombination across the generations in both ant and termite fungus-farming colonies. Recombination between the fungus-growing termites and their *Termitomyces* cultivars is unconstrained, but two limitations apply in the fungus-growing ants. First, vertical cultivar transmission will often result in vertical cultivar acquisition—all Howe, Schiøtt, and Boomsma (2019) showed is that we cannot take this for granted even in ecological time. There remains, therefore, a significant difference of degree between conditional swapping of cultivars in the fungus-growing ants and the rampant and essentially random host-symbiont recombination across the generations in most fungus-growing termites. This is related to almost all fungus-farming termite genera having retained sexual reproduction of cultivars via spore-producing mushrooms (chapter 6, this volume), while the evolutionarily derived attine cultivars may well be completely asexual and thus unable to recombine genotypes within the fungal cultivar populations (Nobre et al. 2011; Shik et al. 2020). Interestingly, the cultivars of *Acromyrmex* and *Atta* became polynucleate chimeras (Kooij, Aanen, et al. 2015) around the same time that the farming ants evolved chimeric families through multiple queen-insemination (Nygaard et al. 2016). Both parties in the leaf-cutting ant symbioses can therefore effectively express genes from a polyploid set of chromosomes. It is important to stress that this secondarily evolved bilateral form of partner chimerism between the ants and their crops does not challenge the lifetime commitment principle—it just implies that more partner genomes became involved on both sides.

A consequence of stringent "matrimonial" commitments in sex and symbiosis is that there are—at each moment in time and in each population—an equal number of mature termite farmer families alive as there are *Termitomyces* fungal cultivar clones. This also applies throughout the attine ants and their cultivars, although only approximately so in the mutually chimeric colonies of the leaf-cutting ants (Kooij, Aanen, et al. 2015). The same is true for the number of males and females alive in all committed royal pairs of termites and in all attine genera with singly inseminated queens, as long as one acknowledges that single colony fathers remain functionally alive when they are represented by stored sperm (figure 5.1A, B). As I have argued elsewhere (Boomsma 2009, 2013), lifetime commitment is a very powerful mechanism for enforcing evolutionarily stable altruism within species, because promiscuity never intervenes and the offspring produced by both partners always have maximal possible relatedness to each other. Family life of such stringent coherence provides the highest possible incentive for older siblings to raise younger siblings rather than their own offspring. That same logic should also tie different species together in obligate mutualisms when their partnerships remain "monogamously" exclusive (Wyatt, West, and Gardner 2013). In all these cases, partner commitment is so strong that one dies with the only sexual or mutualistic partner(s) that one will ever have (figure 5.1A–D). The contrast with human mate choice and our interactions with domesticated crops and livestock could hardly be more striking.

A corollary of my "vertical transmission unless . . ." interpretation of attine symbiotic commitment is that vertical (parent-offspring) cultivar transmission may have become obligate only in claustral colony-founding *Atta* (e.g., Baer, Armitage, and Boomsma 2006) and in *Macrotermes bellicosus* and the genus *Microtermes* as far as the fungus-growing termites go (Aanen et al. 2002). That would make strict vertical cultivar acquisition a consistently derived evolutionary trait, emerging either from ancestries whose horizontal cultivar transmission was merely rare (*Acromyrmex* and likely all other phylogenetically more basal genera of attine ants) or omnipresent (all other fungus-growing termites). Such a perspective would make fungal cultivar transmission in *Atta* comparable with the secondary loss of sex and meiosis in multicellular organisms, because host-symbiont recombination becomes impossible or significantly compromised when vertical transmission becomes obligate, just as intragenomic recombination ceases when parthenogenesis is no longer alternated with bouts of sexual reproduction. The extent to which such a transition to pure vertical transmission has increased genetic load in fungal cultivars (i.e., reduced average fitness due to accumulating deleterious mutations) is an empirical question, because rare sex and recombination can be disproportionally effective in preventing this problem (D'Souza and Michiels 2010). The degree of year-to-year horizontal cultivar transmission now documented for *Acromyrmex* (Howe, Schiøtt, and Boomsma 2019) is likely to be sufficient to avoid cultivar symbiont deterioration by mutation accumulation (also known as Muller's ratchet (Kondrashov 1994)). This is not because the cultivars fail to accumulate deleterious mutations, but because host-symbiont recombination is frequent enough and subsequent colony-level selection is strong enough to remove cultivars of compromised quality. Even populations of *Atta* cultivars may be protected against Muller's Ratchet, perhaps not so much by rare host-symbiont recombination when the nest cavities of founding colonies might sometimes meet each other underground, but because *Atta* and *Acromyrmex* have been assumed to rear the same species of *Leucoagaricus gongylophorus*. Although high resolution genetic markers showed that the symbionts of these two ant genera in Panama belong to distinct sympatric populations (Kooij, Poulsen, et al. 2015), it might be that founding *Atta* colonies occasionally acquire a fitter *Acromyrmex* cultivar.

My main argument so far has been that horizontal cultivar acquisition is possible whenever founding queens (ants) or first worker cohorts (termites) need to forage, because that will inevitably imply exposure to, respectively, unrelated incipient gardens or sexual cultivar spores. Evolved defenses against chimerism should then imply that secondary cultivar introductions will not be able to replace established gardens (Poulsen and Boomsma 2005; Aanen et al. 2009), but that new cultivars will have a chance of successful take-over in the colony founding phase (Aanen et al. 2009; Poulsen et al. 2009) before matrimonial permanency is established. In this context it is interesting to note that a third group of highly diverse small-scale fungus farmers, the ambrosia beetles, are permanently confined to their nest galleries, which are excavated almost invariably in dead or dying wood. The complete absence of foraging outside the gallery, both by helping offspring and by nest-founding parents, might imply that ambrosia beetles may be the only clade of fungus farmers in which vertical cultivar transmission is both ancestral and has remained obligate ever since (Kirkendall, Biedermann, and Jordal 2015). This is consistent with these beetles being the only fungus-farming lineage where specialized mycangia organs for cultivar co-dispersal are always found (Mayers et al. 2015; Hulcr and Stelinski 2017; Smith et al. 2018; chapter 7, this volume). Even here, however, foundresses may sometimes tunnel next to each other in the same piece of wood so closely that gallery systems could secondarily merge and exchange cultivars.

Retrospectively, it probably makes good sense that cultivar transmission in fungus-farming insects is almost never completely set in stone. Farming mutualisms are all ecto-symbioses (Law 1985) (figure 5.1), where partners can survive without each other's company for at least short periods. This is the key factor enabling rare but statistically predictable cultivar swapping in ecological time. In contrast, microbial endosymbionts have such faithful transmission that punctuation of vertical inheritance by horizontal swaps is detectable only on an evolutionary time scale (McCutcheon, Boyd, and Dale 2019). Transfers at the latter time scale have also been documented in attine ants (Green, Adams, and Mueller 2002: Mehdiabadi et al. 2012). However, ecological and evolutionary time scales are different by orders of magnitude and should not be confused because faithful vertical transmission over evolutionary time implies that we do not expect defensive traits against chimerism to have evolved. That reasoning is again analogous to mixed ejaculates from species with exclusively singly inseminated queens not expressing hostile sperm competition, whereas ejaculates of males from species with multiply inseminated queens do (Den Boer et al. 2010; Dosselli et al. 2019). Only chimeric encounters on an ecological time scale will establish natural selection for defense, enabled by random mutations that can be coopted for expressing mutually hostile phenotypes, be it among ejaculates or fungal cultivars. Vertically transmitted ectosymbionts are thus expected to have defensive traits against competing strains that offer similar services to hosts, while vertically transmitted endosymbionts are likely to be devoid of independent competitive agency. However, the evolution of defensive traits against chimerism will proceed only when there are no other a-priori mechanisms that remove selection for such hostile traits. It took us a while to understand this logic. We initially looked for incompatibility reactions between *Termitomyces* clones similar to those expressed between *Acromyrmex* cultivars, until we realized that positive frequency-dependent propagation of asexual spores within the farming termite colonies selects for rapid cultivar growth and maximal production of edible nodules with asexual spores, rather than for active direct defenses against competing strains (Aanen and Boomsma 2006; Aanen et al. 2009). For these aspects of farming the termites are just passive spectators and defecators of inoculated comb (figure 5.1D).

Finally, it is worth noting that insect farmers differ fundamentally from human farmers because they cultivate heterotrophic crops that need organic substrate rather than autotrophic cultivars that need light and inorganic nutrients. It is interesting that farming of photosynthesizing crops is rare outside humans (Aanen 2010; Wcislo 2020). This is true even among the ants, where neither devil's gardens in the Amazon rain forest (Frederickson and Gordon 2007) nor *Philidris* ants farming *Squamellaria* epiphytes on some of the Fijian islands (Chomicki, Kadereit et al. 2020; chapter 9, this volume) have realized adaptive radiations remotely comparable in diversity to the fungus-farming ants, termites, and ambrosia beetles. A difference of this kind is perhaps not surprising, because farming mutualisms with autotrophic crop symbionts depend on the plant partners providing both above-ground housing (domatia) and nourishment (floral nectaries) to ant colonies. These are then traded against anti-herbivory services or delivery of manure in the form of inorganic waste. If there is competition in these systems, it is for domatia space between alternative ant colonies, which implies that the farming analogy becomes rather indirect and that mutualistic agency rests primarily with the plants, not with the ants (Ward and Branstetter 2017; Marting et al. 2018). In contrast, fungal agriculture is almost invariably based on the farming insects excavating nest cavity "domatia" in abiotic (i.e., not co-adapting) space (figure 5.1B), in which fungal cultivars either have an unchallenged clonal monopoly of farmer care (ambrosia beetles) or gain such monopoly after outcompeting other cultivar strains that are either primarily acquired (in termites) or secondarily invading (in ants).

Conclusions

We are intuitively inclined to think of farmers as subjects and crops as objects because "agency" in the sense of who or what domesticated and maintains what or whom appears to be fundamentally asymmetric. However, it is not without conceptual problems to extrapolate this anthropomorphic narrative to insect farmers. Human farms promiscuously combine and exchange crops; they hedge their bets to secure profit even when some crops fail. In contrast, fungus-farming symbioses of ants and termites lack promiscuity: their families consist of lifetime-committed sets of parents and lifetime unmated altruistic sibling workers who are jointly committed for life to rear a single cultivar clone and make the best of that assignment in terms of reproductive success for both parties. Farmer agency seems indisputable when cultivars are exclusively vertically propagated, but strict mutual dependence and horizontal cultivar acquisition makes that notion ambiguous—with equal right one could then say that cultivars acquire a lifetime devoted nursing insect family to pass on fungal genes to future generations. Recent work shows that it has not been sufficiently appreciated that vertical transmission does not preclude horizontal cultivar acquisition as long as founding queens or first worker cohorts actively or passively forage for alternative cultivars before resident crops acquire enough agency to fight for their monopoly of farmer attention (Nobre et al. 2011; Howe, Schiøtt, and Boomsma 2019). Only the fungus-growing ambrosia beetles never forage outside their burrows, so these beetles may be the most convincing example of vertical cultivar acquisition being both ancestral and universally extant.

Using a conceptual framework of mutualism theory that acknowledges potential conflicts between farming hosts and cultivar ectosymbionts, I have argued that the dynamics of social insect fungus farming are fundamentally different from human farming practices, in spite of interesting analogies in the optimal management of human and insect farms. With both sexual- and host-symbiont partners being committed for life, there are uncanny parallels between the ways in which sexual conflict and host-symbiont conflict were resolved in order to align reproductive interests—perspectives that are irrelevant in human farming. Rather than comparing insect fungus farming with endosymbiotic domestication processes reminiscent of how mitochondria became enslaved by eukaryote cells, it seems more appropriate to consider nest cavities with fungus gardens as artifactual farmer-built "domatia," in which closely related competing cultivar strains always exclude each other, driven by the necessity to obtain a monopoly of care by a farmer family whose fitness is also maximized by rearing only a single clonal cultivar. No such considerations ever apply to culturally evolved human farming practices, which has left our cultivars powerless because they cannot enforce exclusive farmer commitments.

Throughout this edited volume, *domestication* is defined as the genetic modification of one species by another in ways that benefit the modifying species (the farmer) but that would have reduced the fitness of the modified species (the domesticate) in its original niche. This definition, in spite of its apparent generality, implicitly assumes that agency is fundamentally asymmetric, and I have argued that it therefore only partly applies to fungus-farming insects. In these insect farming symbioses it is equally legitimate to consider the insects as domesticates of the crop fungi as the other way around because neither of the parties can still survive in their ancestral nonfarming niche. In my elaboration of the tensions emanating from this mismatch, I have aimed to achieve three objectives: (1) to establish the notion that we should be careful when using analogies between human and insect farming—however fascinating they are—in such broad-brush narrative strokes that the fundamental difference between promiscuous culturally driven artificial selection and naturally selected one-to-one co-adaptation becomes opaque; (2) to evaluate the striking analogies between lifetime commitments in sex and symbiosis that characterize insect fungus farmers; and (3) to illustrate the usefulness of evolutionary theory for a parallel understanding of conflict regulation in mating and farming—an approach that generates both testable predictions and the perspective of a more encompassing understanding of cooperation and conflict.

Acknowledgments

The Smithsonian Tropical Research Institute in Panama has facilitated my research since the early 1990s. Discussions with and comments by Ted Schultz, Rick Gawne, Duur Aanen, Ford Denison, and Peter Biedermann, both during and after the KLI Altenberg Workshop, were important for writing the final version of this chapter. Financial support by KLI for attending the workshop has been highly appreciated.

References

Aanen, D. K. 2010. "As You Weed, So Shall You Reap: On the Origin of Algaculture in Damselfish." *BMC Biology* 8:81.

Aanen, D. K., and J. J. Boomsma. 2006. "The Evolutionary Origin and Maintenance of the Mutualistic Symbiosis between Termites and Fungi." In *Insect Symbiosis*, edited by K. Bourtzis and T. Miller, 2:79–95. Boca Raton, FL: CRC Press.

Aanen, D. K., P. Eggleton, C. Rouland-Lefevre, T. Guldberg-Froslev, S. Rosendahl, et al. 2002. "The Evolution of Fungus-Growing Termites and Their Mutualistic Fungal Symbionts." *Proceedings of the National Academy of Sciences* 99(23): 14887–14892.

Aanen, D. K., H. H. D. Licht, A. J. M. Debets, N. A. G. Kerstes, R. F. Hoekstra, et al. 2009. "High Symbiont Relatedness Stabilizes Mutualistic Cooperation in Fungus-Growing Termites." *Science* 326(5956): 1103–1106.

Baer, B., S. A. O. Armitage, and J. J. Boomsma. 2006. "Sperm Storage Induces an Immunity Cost in Ants." *Nature* 441(7095): 872–875.

Boomsma, J. J. 2007. "Kin Selection Versus Sexual Selection: Why the Ends Do Not Meet." *Current Biology* 17(16): R673–R683.

Boomsma, J. J. 2009. "Lifetime Monogamy and the Evolution of Eusociality." *Philosophical Transactions of the Royal Society B—Biological Sciences* 364(1533): 3191–3207.

Boomsma, J. J. 2013. "Beyond Promiscuity: Mate-Choice Commitments in Social Breeding." *Philosophical Transactions of the Royal Society B—Biological Sciences* 368:20120050.

Boomsma, J. J., B. Baer, and J. Heinze. 2005. "The Evolution of Male Traits in Social Insects." *Annual Review of Entomology* 50:395–420.

Boomsma, J. J., and R. Gawne. 2018. "Superorganismality and Caste Differentiation as Points of No Return: How the Major Evolutionary Transitions Were Lost in Translation." *Biological Reviews* 93:28–54.

Boomsma, J. J., D. J. C. Kronauer, and J. S. Pedersen. 2009. "The Evolution of Social Insect Mating Systems." In *Organization of Insect Societies: From Genome to Sociocomplexity*, edited by J. Gadau and J. Fewell, 3–25. Cambridge, MA: Harvard University Press.

Boomsma, J. J., and F. L. W. Ratnieks. 1996. "Paternity in Eusocial Hymenoptera." *Philosophical Transactions of the Royal Society B—Biological Sciences* 351(1342): 947–975.

Bot, A. N. M., S. A. Rehner, and J. J. Boomsma. 2001. "Partial Incompatibility between Ants and Symbiotic Fungi in Two Sympatric Species of *Acromyrmex* Leaf-Cutting Ants." *Evolution* 55(10): 1980–1991.

Chomicki, G., G. Kadereit, S. S. Renner, and E. T. Kiers. 2020. "Tradeoffs in the Evolution of Plant Farming by Ants." *Proceedings of the National Academy of Sciences* 117:2535–2543.

D'Souza, T. G., and N. K. Michiels. 2010. "The Costs and Benefits of Occasional Sex: Theoretical Predictions and a Case Study." *Journal of Heredity* 101:S34–S41.

Darwin, C. 1859. *On the Origin of Species*. London: John Murray.

De Man, T. J. B., J. E. Stajich, C. P. Kubicek, C. Teiling, K. Chenthamara, et al. 2016. "Small Genome of the Fungus *Escovopsis weberi*, a Specialized Disease Agent of Ant Agriculture." *Proceedings of the National Academy of Sciences* 113:3567–3572.

Den Boer, S. P. A., B. Baer, and J. J. Boomsma. 2010. "Seminal Fluid Mediates Ejaculate Competition in Social Insects." *Science* 327(5972): 1506–1509.

Den Boer, S. P. A., B. Baer, S. Dreier, S. Aron, D. R. Nash, et al. 2009. "Prudent Sperm Use by Leaf-Cutter Ant Queens." *Proceedings of the Royal Society B—Biological Sciences* 276(1675): 3945–3953.

Dosselli, R., J. Grassl, S. P. A. Den Boer, M. Kratz, J. M. Moran, et al.. 2019. "Protein-Level Interactions as Mediators of Sexual Conflict in Ants." *Molecular & Cellular Proteomics* 18:S34–S45.

Frank, S. A. 1996. "Host-Symbiont Conflict over the Mixing of Symbiotic Lineages." *Proceedings of the Royal Society B—Biological Sciences* 263(1368): 339–344.

Frederickson, M. E., and D. Gordon. 2007. "The Devil to Pay: The Cost of Mutualism with *Myrmelachista schumanni* Ants in 'Devil's Gardens' Is Increased Herbivory on *Duroia hirsuta* Trees." *Proceedings of the Royalty Society B—Biological Sciences* 274:1117–1123.

Green, A. M., R. M. M. Adams, and U. G. Mueller. 2002. "Extensive Exchange of Fungal Cultivars between Sympatric Species of Fungus-Growing Ants." *Molecular Ecology* 11:191–195.

Hamilton, W. D. 1964. "Genetical Evolution of Social Behaviour 2." *Journal of Theoretical Biology* 7(1): 17–52.

Hartke, T. R., and B. Baer. 2011. "The Mating Biology of Termites: A Comparative Review." *Animal Behaviour* 82:927–936.

Heine, D., N. A. Holmes, S. F. Worsley, A. C. A. Santos, T. M. Innocent, et al. 2018. "Chemical Warfare between Leafcutter Ant Symbionts and a Co-Evolved Pathogen." *Nature Communications* 9:2208.

Holmes, N. A., T. M. Innocent, D. Heine, M. Al Bassam, S. F. Worsley, et al. 2016. "Genome Analysis of Two *Pseudonocardia* Phylotypes Associated with *Acromyrmex* Leafcutter Ants Reveals Their Biosynthetic Potential." *Frontiers in Microbiology* 7:2073.

Howe, J., M. Schiøtt, and J. J. Boomsma. 2019. "Horizontal Partner Exchange Does Not Preclude Stable Mutualism in Fungus-Growing Ants." *Behavioral Ecology* 30:372–382.

Hughes, W. O. H., and J. J. Boomsma. 2004. "Genetic Diversity and Disease Resistance in Leaf-Cutting Ant Societies." *Evolution* 58(6): 1251–1260.

Hughes, W. O. H., and J. J. Boomsma. 2006. "Does Genetic Diversity Hinder Parasite Evolution in Social Insect Colonies?" *Journal of Evolutionary Biology* 19(1): 132–143.

Hughes, W. O. H., B. P. Oldroyd, M. Beekman, and F. L. W. Ratnieks. 2008. "Ancestral Monogamy Shows Kin Selection Is Key to the Evolution of Eusociality." *Science* 320(5880): 1213–1216.

Hulcr, J., and L. L. Stelinski. 2017. "The Ambrosia Symbiosis: From Evolutionary Ecology to Practical Management." *Annual Review of Entomology* 62:285–303.

Ivens, A. B. F., D. R. Nash, M. Poulsen, and J. J. Boomsma. 2009. "Caste-Specific Symbiont Policing by Workers of *Acromyrmex* Fungus-Growing Ants." *Behavioral Ecology* 20(2): 378–384.

Kirkendall, L. R., P. H. W. Biedermann, and B. H. Jordal. 2015. "Evolution and Diversity of Bark and Ambrosia Beetles." In *Bark Beetles: Biology and Ecology of Native and Invasive Species*, edited by F. E. Vega and R. W. Hofstetter, 85–156. Cambridge, MA: Academic Press.

Kondrashov, A. S. 1994. "Muller's Ratchet under Epistatic Selection." *Genetics* 136:1469–1473.

Kooij, P. W., D. K. Aanen, M. Schiøtt, and J. J. Boomsma. 2015. "Evolutionarily Advanced Ant Farmers Rear Polyploid Fungal Crops." *Journal of Evolutionary Biology* 28:1911–1924.

Kooij, P. W., M. Poulsen, M. Schiøtt, and J. J. Boomsma. 2015. "Somatic Incompatibility and Genetic Structure of Fungal Crops in Sympatric *Atta colombica* and *Acromyrmex echinatior* Leaf-Cutting Ants." *Fungal Ecology* 18:10–17.

Korb, J., and D. K. Aanen. 2003. "The Evolution of Uniparental Transmission of Fungal Symbionts in Fungus-Growing Termites (Macrotermitinae)." *Behavioral Ecology and Sociobiology* 53:65–71.

Law, R. 1985. "Evolution in a Mutualistic Environment." In *The Biology of Mutualism: Ecology and Evolution*, edited by D. H. Boucher, 145–170. London: Croom Helm.

Lucock, M. D., C. E. Martin, Z. R. Yates, and M. Veysey. 2014. "Diet and Our Genetic Legacy in the Recent Anthropocene: A Darwinian Perspective to Nutritional Health." *Journal of Evidence-Based Complementary and Alternative Medicine* 19:68–83.

Marting, P. R., N. M. Kallman, W. T. Wcislo, and S. C. Pratt. 2018. "Ant-Plant Sociometry in the *Azteca-Cecropia* Mutualism." *Scientific Reports* 8:17968.

Mayers, C. G., D. L. McNew, T. C. Harrington, R. A. Roeper, S. W. Fraedrich, et al. 2015. "Three Genera in the Ceratocystidaceae Are the Respective Symbionts of Three Independent Lineages of Ambrosia Beetles with Large, Complex Mycangia." *Fungal Biology* 119:1075–1092.

McCutcheon, J. P., B. M. Boyd, and C. Dale. 2019. "The Life of an Insect Endosymbiont from the Cradle to the Grave." *Current Biology* 29:R485–R495.

Mehdiabadi, N. J., U. G. Mueller, S. G. Brady, A. G. Himler, and T. R. Schultz. 2012. "Symbiont Fidelity and the Origin of Species in Fungus-Growing Ants." *Nature Communications* 3:840.

Nobre, T., C. Fernandes, J. J. Boomsma, J. Korb, and D. K. Aanen. 2011. "Farming Termites Determine the Genetic Population Structure of Termitomyces Fungal Symbionts." *Molecular Ecology* 20(9): 2023–2033.

Nygaard, S., H. Hu, C. Li, M. Schiøtt, Z. Chen, et al. 2016. "Reciprocal Genomic Evolution in the Ant-Fungus Agricultural Symbiosis." *Nature Communications* 7:12233.

Otani, S., V. L. Challinor, N. B. Kreuzenbeck, S. Kildgaard, S. K. Christensen, et al. . 2019. "Disease-Free Monoculture Farming by Fungus-Growing Termites." *Scientific Reports* 9:8819.

Poulsen, M., and J. J. Boomsma. 2005. "Mutualistic Fungi Control Crop Diversity in Fungus-Growing Ants." *Science* 307(5710): 741–744.

Poulsen, M., H. Fernandez-Marin, C. R. Currie, and J. J. Boomsma. 2009. "Ephemeral Windows of Opportunity for Horizontal Transmission of Fungal Symbionts in Leaf-Cutting Ants." *Evolution* 63(9): 2235–2247.

Scott-Phillips, T. C., K. N. Laland, D. M. Shuker, T. E. Dickins, and S. A. West. 2014. "The Niche Construction Perspective: A Critical Appraisal." *Evolution* 68:1231–1243.

Shik, J. Z., P. W. Kooij, D. A. Donoso, J. C. Santos, E. B. Gomez, et al. 2020. "Nutritional Niches Reveal Fundamental Domestication Tradeoffs in Fungus-Farming Ants." *Nature Ecology and Evolution* 5:122–134.

Smith, B. D. 1998. *The Emergence of Agriculture*. New York: Scientific American Library.

Smith, S. M., D. S. Kent, J. J. Boomsma, and A. J. Stow. 2018. "Monogamous Sperm Storage and Permanent Worker Sterility in a Long-Lived Ambrosia Beetle." *Nature Ecology & Evolution* 2:1009–1018.

Tishkoff, S. A., F. A. Reed, A. Ranciaro, B. F. Voight, C. C. Babbitt, et al. 2006. "Convergent Adaptation of Human Lactase Persistence in Africa and Europe." *Nature* 39:31–40.

Van der Have, T. M., J. J. Boomsma, and S. B. J. Menken. 1988. "Sex-Investment Ratios and Relatedness in the Monogynous Ant *Lasius niger* (L.)." *Evolution* 42(1): 160–172.

Villesen, P., T. Murakami, T. R. Schultz, and J. J. Boomsma. 2002. "Identifying the Transition between Single and Multiple Mating of Queens in Fungus-Growing Ants." *Proceedings of the Royal Society B—Biological Sciences* 269(1500): 1541–1548.

Ward, P. S., and M. G. Branstetter. 2017. "The Acacia Ants Revisited: Convergent Evolution and Biogeographic Context in an Iconic Ant/Plant Mutualism." *Proceedings of the Royal Society B—Biological Sciences* 284:20162569.

Wcislo, W. T. 2020. "Fijian Farming Ants Resolve the Guns-or-Butter Dilemma for Their Crop Plants." *Proceedings of the National Academy of Sciences* 117:3357–3359.

Wheeler, W. M. 1911. "The Ant-Colony as an Organism." *Journal of Morphology* 22:307–325.

Wyatt, G. A. K., S. A. West, and A. Gardner. 2013. "Can Natural Selection Favour Altruism between Species?" *Journal of Evolutionary Biology* 26:1854–1865.

III THE DIVERSITY OF INSECT AGRICULTURE

6 Fungus-Growing Termites: An Eco-Evolutionary Perspective

Judith Korb

Fungus-growing termites (Macrotermitinae) dominate African savannahs (Bignell and Eggleton 2000; Hausberger et al. 2011). This ecological success has been attributed to their obligate ectosymbiosis with fungi of the genus *Termitomyces,* which they cultivate inside their subterranean or mound nests (e.g., Bignell and Eggleton 2000; da Costa et al. 2019a). The termites forage for dead plant material. After a first gut passage during which simple sugars are extracted by the termites, the resulting primary feces is used to build so-called fungal combs on which *Termitomyces* develops (Badertscher et al. 1983; Leuthold, Badertscher, and Imboden 1989; Traniello and Leuthold 2000). Provided with constant warmth and humidity as well as competitor-free conditions inside their nest, the fungal cultivar thrives and, together with the termites and their gut symbionts, decomposes plant biomass almost completely (reviewed in Rouland-Lefèvre 2000; Poulsen 2015; da Costa et al. 2019a). By feeding on the fungal cultivar, and especially the protein-rich nodules that contain asexual spores, the termites may profit from fungal decomposing enzymes, enhanced plant degradation, and increased protein/nitrogen availability (Martin and Martin 1978; Rouland-Lefèvre 2000; Hyodo et al. 2003). Low nitrogen content of dead plant material has been discussed as a limiting factor of termites' growth rates (Higashi, Abe, and Burns 1992; Shellman-Reeve 1994). Accordingly, the ecological success of fungus-growing termites can be attributed to improved protein/nitrogen provisioning by the fungal symbiont. The fast conversion of carbohydrate (C)-rich but protein (N)-poor dead plant material into N-enriched fungal comb and termite biomass can be considered a major factor driving plant decomposition and nutrient cycling in African savannahs where limited rainfall and humidity constrains decomposition by soil microbes (e.g., Deshmukh 1989). Similar to the way that the termite gut is regarded as a "bioreactor" (Brune 1998), the nests of the Macrotermitinae can be considered as "glass-houses" that facilitate fast decomposition, which is advantageous especially under dry conditions. Thus the symbiosis can explain the dominance, high biomass, and ecological importance of Macrotermitinae that function as drivers of savannah ecosystem processes.

In this chapter, I discuss some ideas about (1) how living in a symbiosis may lead to niche expansion but probably also constrains it, by using a specific termite-fungus association as an example, (2) how the fungal symbiosis may mediate coexistence of closely related species through niche differentiation, (3) how evolutionary theory about symbiosis fits with termite-fungus ecology, and (4) what we can learn from the Macrotermitinae-*Termitomyces* symbiosis more generally. I end with a comparison to human agriculture.

Niche Expansion and Contraction of *Macrotermes bellicosus*—Ecological Consequences of a Close Association?

In natural West African savannahs, *Macrotermes bellicosus* is the most dominant termite species. It constructs mounds of a height of up to 6 meters (figure 6.1a), has colony sizes of several million individuals, and can reach densities of around 50 mounds per hectare (living mounds only: up to 17 mounds per hectare) (Collins 1981; Lepage 1984; Tano and Lepage 1989; Korb and Linsenmair 2001a). It is an important ecosystem engineer that enhances key processes in ecosystems (Traniello and Leuthold 2000; Erpenbach et al. 2012, 2017). It improves soil properties such as water-holding potential and fertility. It increases animal as well as plant diversity by, for instance, creating habitat heterogeneity and providing new habitats.

Applying evolutionary reasoning, we can attribute the dominance of *Macrotermes bellicosus* to its obligate and specific symbiosis. In the Comoé National Park (CNP) in Ivory Coast, *M. bellicosus* lives in a 1:1 association with a single *Termitomyces* species/lineage (Aanen et al. 2002; Nobré et al. 2011a) that is inherited vertically via uniparental transmission through the dispersing male (Grasse and Noirot 1955; Johnson 1981; Johnson et al. 1981; Korb and Aanen 2003). Such strict 1:1 associations are rare, and in Macrotermitinae most *Termitomyces* symbionts are horizontally transmitted (Johnson et al. 1981; Korb and Aanen 2003) (discussed further below). All else being equal, uniparental, vertical transmissions with strict 1:1 associations are predicted to result in associations with less conflict and more cooperation between partners (Frank 1994, 1996, 1997, 1998; chapters 3, 4, and 5, this volume). This may lead, for instance, to a "finer-tuning" of division of labor in the decomposition process (Rouland, Lenoir, and Lepage 1991; Rouland-Lefèvre 2000; Korb and Aanen 2003). Thus, it is tempting to argue that *M. bellicosus* owes its ecological dominance to its specific symbiotic association, which allowed, for example, the evolution of a more elaborate (i.e., more complementary and more efficient) plant decomposition system, a hypothesis that requires testing.

Although dominant in savannahs, *M. bellicosus* is largely absent from forests or dense thickets interspersed within savannahs (Korb and Linsenmair 1998a). The distribution pattern of *M. bellicosus* and its potential causes have been studied in the CNP (8° 45'N, 3° 47'W; elevation around 250 m) in northeastern Ivory Coast for more than 25 years. The CNP (size: 11.500 km²) is characterized by a steep climatic gradient with dry Sudan savannah in the northeast and moist Guinea savannah in the south (for more details see Poilecot 1991). The vegetation consists of a mosaic of patches of open to dense tree and shrub savannah, where large areas are annually burned, and of forest islands. Gallery forests occur along the main rivers. The area has marked dry and rainy seasons of rather unpredictable lengths. Typically, the rainy season lasts from April to October, peaking in August and September, with a relatively dry period in May (total annual rainfall 700–1170 mm; Korb 1997).

Distribution Pattern in the Savannah

M. bellicosus mounds have a patchy distribution in the savannah of the CNP, with dead mounds (i.e., mounds where the constructing *M. bellicosus* colony had died) occurring in

Figure 6.1
Macrotermes bellicosus mounds from the Comoé National Park in Ivory Coast. (A) A more than 6-meter-tall mound. (B) A "cathedral" mound from the savannah with many ridges, thin walls, and high surface complexity. (C) A dome-shaped mound from an open forest stand with thick walls and less complex architecture. (D) A mound from an open forest stand after experimental removal of shading trees with increased mound complexity compared to typical forest mounds. Photos: (A): Volker Salewski (used with permission), (B–D): Judith Korb.

varying densities everywhere except in seasonally flooded areas (Korb 1997; Korb and Linsenmair 1998a, 2001a). Living mounds (i.e., mounds containing a living *M. bellicosus* colony) are less common but have a corresponding distribution pattern. The higher density of dead than live mounds is due to the stability of dead mounds. Large mounds with a height of more than 3 meters can last for decades, maybe even centuries, after the constructing *M. bellicosus* colony has died (pers. obs.). These dead mounds can be reoccupied by a new *M. bellicosus* colony when a royal pair founds its new colony inside after the nuptial flight (pers. obs.). The preexisting mound structures and especially the more favorable homeostatic conditions seem to make founding success higher in old mounds compared with the surrounding soil (unpublished data).

Within the savannah, the patchy distribution pattern of living mounds can be explained by stochastic events (Korb and Linsenmair 1999b, 2001a). Whether colony establishment is successful within an area or not depends on rainfall patterns at the beginning of the rainy season. Rainfall can be very patchy on a small spatial scale of a few hundred meters. Long dry spells after the nuptial flight hinder colony establishment as founding reproductives die due to desiccation. In contrast, mortality rates of established colonies are largely due to stochastic predation events. When army ants enter a patch, they kill all *M. bellicosus* colonies regardless of colony size. Thus, the overall system can be described by metapopulation dynamics driven mainly by (limited) dispersal/colonization and local extinction.

Within a patch, the small-scale dynamics and distribution of living *M. bellicosus* mounds appears to be influenced by intraspecific competition as inter-mound distances increase and mound distribution becomes more over-dispersed with increasing colony sizes (Korb and Linsenmair 2001a). Additionally, pattern analyses indicate that more closely spaced colonies have a higher likelihood of dying (Korb and Linsenmair 2001a).

Considering its obligate dependence on *M. bellicosus*, the distribution pattern of the fungal *Termitomyces* symbiont is expected to be limited by the same abiotic and biotic factors that determine and regulate the distribution of its host.

Occurrence between Habitats

Despite its common occurrence in the savannah, *M. bellicosus* is largely missing from forest habitats, such as gallery forests, of the CNP, where it is restricted to forest edges or stands with an open canopy (Korb 1997; Korb and Linsenmair 1998a). The lack of *M. bellicosus* in denser forest areas is not caused by limited dispersal, as swarming alates and founding couples are commonly found in the forest (pers. obs.). Rather, cooler ambient temperatures and "physiological" constraints associated with fungus cultivation seem to limit *M. bellicosus*'s occurrence in forests (Korb 1997; Korb and Linsenmair 1998a). Like other *Macrotermes* species (Zimmermann and Darlington 1987; Darlington 1989; Darlington et al. 1997), *M. bellicosus* colonies require high exchange rates of respiratory gases due to their large colony sizes of up to a few million individuals, and especially due to the high metabolism of the cultivated fungus (Lüscher 1955, 1956). A mature *Macrotermes* nest can contain over 4 kg dry weight of termites and roughly twice as much fungus comb (Darlington 1989). Thus, CO_2 needs to be exchanged for O_2. The mounds of *M. bellicosus* are closed (i.e., they do not have open exits, such as chimneys) (Korb 2011) (figure 6.1). Hence, gas exchange occurs through the mound walls. In the savannah, mounds have thin walls with a large surface complexity (i.e., the ratio between the real and the minimal possible surface—that of a cone—is high) due to the occurrence of many ridges ("cathedral" mounds) (Korb and Linsenmair 1999a) (figure 6.1B). The ventilation of air through these ridges is driven by temperature gradients between the inner parts of the mound and the ambient temperatures (Korb and Linsenmair 1999a, 2000). They result in efficient gas exchange, reflected in CO_2 concentrations inside the mound/nest that are as low as in the ambient air (Korb and Linsenmair 1999a, 2000).

By contrast, in the colder open forest stands, mounds are dome-shaped with low surface complexity and thick walls with a higher thermal capacity (Korb and Linsenmair 1998b) (figure 6.1c) as well as suboptimal low nest temperatures and less developed temperature gradients (Korb and Linsenmair 1999a, 2000). Accordingly, ventilation is less efficient, and CO_2 concentrations inside the mound are significantly increased compared with the ambient air (Korb and Linsenmair 1999a, 2000). These suboptimal conditions are associated with reduced reproductive success of *M. bellicosus* colonies from open forest stands compared with those in the savannah (Korb and Linsenmair 1999b). A question may thus arise: Why does *M. bellicosus* build such apparently suboptimal mounds in the forest? Manipulation experiments in which shading trees were cut in open forest stands suggest that mound architecture is an adaptation to ambient temperature conditions (Korb and Linsenmair 1998a). After removal of trees, ambient temperatures increased, and the termites rebuilt the mounds toward more cathedral-like structures with higher surface complexities (figure 6.1D). Associated with this, mound temperatures increased, and CO_2 concentrations dropped (Korb and Linsenmair 1998a, 1999a). These results imply that *M. bellicosus* is excluded from (dense) forests because of the relatively low ambient temperatures in forests. Under such conditions, *M. bellicosus* colonies must insulate their nests by constructing mounds with thick walls and a minimal surface that does not allow sufficient ventilation for gas exchange (Korb and Linsenmair 1998a, 1999a, 1999b). Laboratory data suggest that the *Termitomyces* fungus of *M. bellicosus* is more temperature sensitive than the termites (pers. obs.). Additionally, the high metabolism of the fungal symbiont demands high rates of gas exchange. Therefore, one can argue that the distribution of *M. bellicosus* is limited to the savannah because of its symbionts' requirements.

These results indicate how the intricate interdependency of both partners—*M. bellicosus* and its *Termitomyces* symbionts—determine their distribution (figure 6.2). Accordingly, they are restricted mainly to savannah habitats because of the fungal symbionts' physiological requirements. Within the savannah, the requirements of not completely dry conditions during nest foundation (before fungus cultivation starts) and high predation pressure by army ants on the *M. bellicosus* colonies also constrain the distribution of the fungal symbiont. Thus, the symbiosis not only extends each species' niche, making *M. bellicosus* an abundant and important ecosystem engineer, but may also constrain it.

Niche Differentiation through Fungal Symbionts?

Food Niches of Termites

Among Macrotermitinae, different species are supposed to occupy similar fundamental niches. All fungus-growing termites require very similar ambient conditions and rely on the same dead plant material for food. Accordingly, all Macrotermitinae are grouped into a single feeding group, fungus-growing wood- or leaf-litter feeders (i.e., group IIF, according to Donovan, Eggleton, and Bignell 2001; Eggleton and Tayasu 2001). The substrate of a specific species can depend on food availability and thus season and location (e.g., Boutton, Arshad, and Tieszen 1983; Lepage, Abbadie, and Mariotti 1993). Comparative studies are still hampered by taxonomic problems of incongruent species identification, because many Macrotermitinae cannot be identified reliably using existing morphological keys (Hausberger et al. 2011; Korb et al. 2019). A recent review that summarizes substrate use by Macrotermitinae concludes that they are generalists with high dietary flexibility, "which may well contribute to termite abundance and their prominent role in nutrient recycling in African savannah ecosystems" (da Costa et al. 2019a, 4). Thus, keeping the taxonomic caveat in mind, we currently expect strong niche overlap among Macrotermitinae.

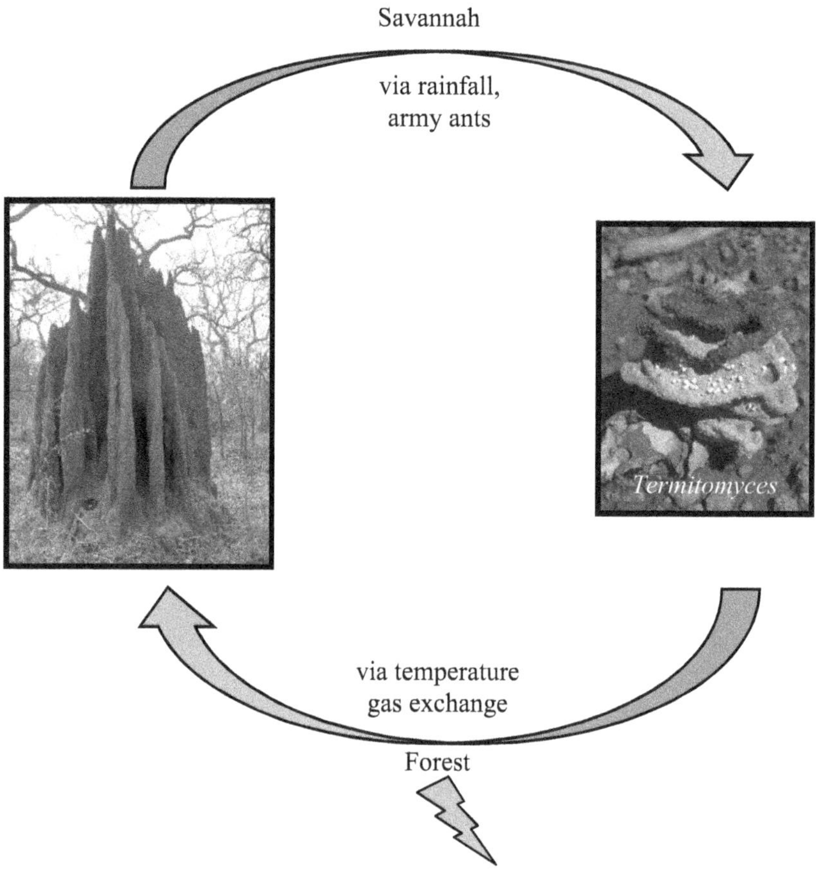

Figure 6.2
Interdependency of *Macrotermes bellicosus* and its fungal symbiont. The *Termitomyces* symbiont seems to exclude *M. bellicosus* from dense forest stands because of its requirement for high gas exchange and its temperature sensitivity. *M. bellicosus* limits the distribution of its symbiont within the savannah as it can establish colonies only under certain rainfall conditions and is preyed upon intensively by army ants. Photos: Judith Korb.

Community Ecology of Fungus-Growing Termites

According to classical community ecology, strong niche overlap is predicted to result in strong interspecific competition that could lead to competitive exclusion under equilibrium conditions. Hence not many fungus-growing termite species are expected to coexist, especially as food seems to be limiting. Although dead plant material appears as a superabundant resource, it is restricted, at least seasonally, due to annual fires in West African savannahs, as has been shown for *M. bellicosus* colonies in the CNP (Korb and Linsenmair 2001a). Therefore, we expect interspecific competition to be common and *M. bellicosus*, as the dominant fungus grower, to have a negative impact on other Macrotermitinae (Korb and Linsenmair 2001b). Yet, for the protected area of the Pendjari National Park (PNP) in Benin community analyses that had reliably identified species using morphological as well as genetic means imply the contrary. In areas with living *M. bellicosus* mounds, coexisting termite species were significantly more closely related (e.g., more Macrotermitinae species co-occurred) than in areas without living mounds (Hausberger and Korb

2015). How can this unexpected result be explained? A termite community study in the Oti Keran National Park in Togo that reliably identified species as in the Benin study (Schyra and Korb 2019; Schyra, Gbenyedji, and Korb 2019) may provide some hints. Analyses of stable isotopes are powerful tools to investigate arthropod food webs. Such analyses that covered all co-occurring termite species imply that there might be subtle niche differentiation between fungus-growing termites (Schyra, Scheu, and Korb 2019). The signature for the stable isotopes $\delta^{13}C$ and $\delta^{15}N$ was complementary between co-occurring species, with *M. bellicosus* having the widest carbon and the lowest nitrogen spectrum (figure 6.3a). This suggests that *M. bellicosus* has a nitrogen spectrum complementary to that of other fungus growers and that it is very versatile in the use of its carbon source. Hence, it is tempting to speculate that coexistence is achieved through subtle niche differentiation, which may even be mediated through different fungal cultivars. Interestingly, besides the special *M. bellicosus* signature, the four co-occurring *Microtermes* species had especially different $\delta^{13}C$ values (figure 6.3b). According to Aanen et al. (2002), different *Microtermes* species have species-specific fungal symbionts. Unfortunately, a 1:1 comparison of the species between both studies is impossible due to taxonomic problems in assigning species names in this genus (Korb et al. 2019).

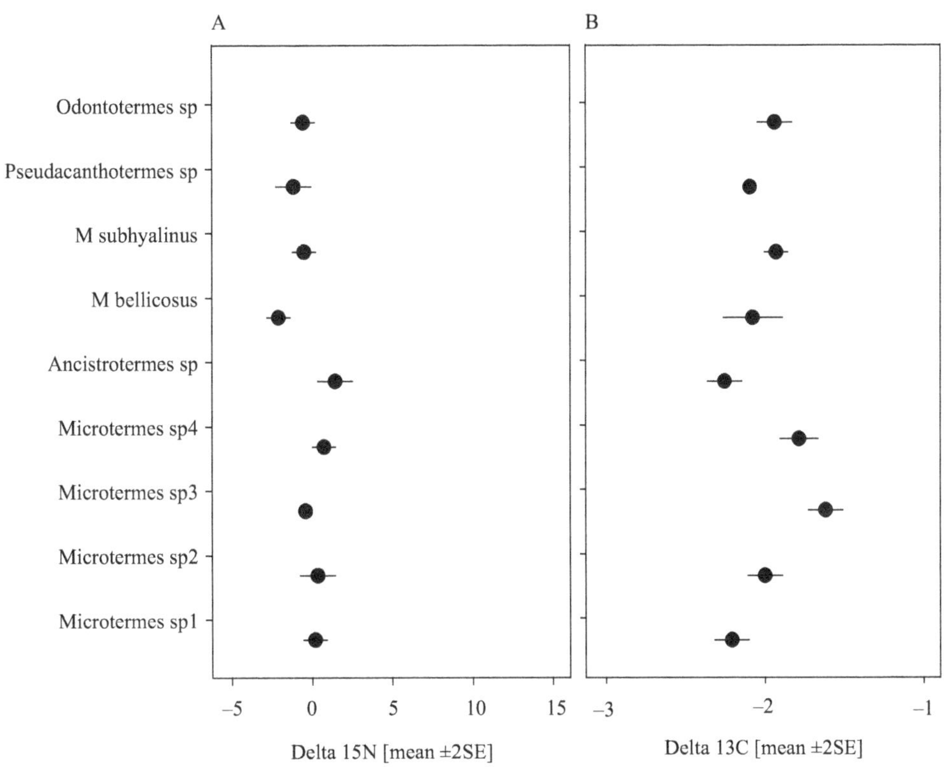

Figure 6.3
Isotope signatures of co-occurring Macrotermitinae from a savannah in Togo. a = $\delta^{15}N$ values, b = $\delta^{13}C$ values. Microtermes 1: *Microtermes subhyalinus*, Microtermes 2: *Microtermes lepidus*, Ancistrotermes: *Ancistrotermes* spec., *M. bellicosus*: *Macrotermes bellicosus*, *M. subhyalinus*: *Macrotermes subhyalinus*, Pseudacanthotermes sp: *Pseudacanthotermes militaris*, Odontotermes: Odontotermes spec. From Schyra, Scheu, and Korb 2019b; reproduced with permission from Wiley.

Comparative studies on fungal cultivars to test this hypothesis are rare. A few studies showed that within a species $\delta^{13}C$ values increase from food to fungus comb to termites reflecting nutritional processing (Boutton, Arshad, and Tieszen 1983; Tayasu, Hyodo, and Abe 2002; Hyodo et al. 2003). Studies analyzing fungal combs of different Macrotermitinae may provide some support that isotope signatures between species differ. For instance, fungus combs seem to differ in $\delta^{13}C$ values between *Pseudacanthotermes militaris*, *Microtermes toumodiensis*, *Ancistrotermes cavithorax*, and an undescribed *Odontotermes*, though habitat signatures and seasonal differences also cause variation (Lepage et al. 1993). Also, *Macrotermes muelleri* seems to have a $\delta^{15}N$ signature different from *Macrotermes gilvus*, *Macrotermes annandalei*, and *Macrotermes carbonarius*, though intercolonial variability in *M. gilvus* is very high (Tayasu, Hyodo, and Abe 2002). A similar but less pronounced pattern occurred for these *Macrotermes* species for $\delta^{13}C$. In Kenya, unidentified *Odontotermes* differed in $\delta^{15}N$ from unidentified *Macrotermes*, though their $\delta^{13}C$ values were very similar (De Visser, Freymann, and Schnyder 2008). None of these studies was designed to test for interspecific differences. Thus, it is unclear whether samples were derived from coexisting single versus multiple species. Clearly more studies are needed to address the potential role of the fungal symbiont in niche differentiation.

Termite-Termitomyces Associations: Evolutionary Considerations

A Bit of Theory

The evolution of stable, mutualistic symbioses is predicted to be affected by two, often linked, factors: (1) transmission mode (e.g., Frank 1994, 1996, 1997, 1998; Foster and Wenseleers 2006; Leeks, dos Santos, and West 2019) and (2) partner choice (e.g., Bull and Rice 1991; Herre et al. 1999; Sachs et al. 2004).

Transmission Mode

In associations with strict vertical transmission, the offspring of the symbiont can be propagated only with the offspring of the host. Thus the reproductive success of both partners is linked. The degree of potential conflict between host and symbiont is reduced in that—put simply—"harming" the host also reduces the symbiont's fitness. Additionally, the degree of potential conflict is further reduced especially if there is uniparental (instead of biparental), vertical transmission (i.e., by one sex only) because this reduces potential conflict among different symbiotic lineages within a host. Thus both factors, (1) alignment of fitness interests of host and symbiont and (2) increased genetic homogeneity (i.e., relatedness) of symbionts within hosts, are expected to lead to symbionts being more beneficial to their hosts (chapters 3, 4, and 5, this volume). A recent modeling and simulation study implies that both relatedness and fitness alignment influence the level of cooperation evolving between hosts and symbionts but that relatedness is more important than fitness alignment to explain symbiont cooperation (Leeks, dos Santos, and West 2019). Uniparental, vertical transmission contrasts with horizontal transmission when both partners can reproduce independently in the sense that the symbiont can produce offspring that spread independently from the (parental) host's offspring to new hosts. Under these conditions "harming" the host does not necessarily reduce symbiont fitness, and "parasitic" associations are more likely to evolve. Additionally, horizontal transmission is often associated with the acquisition of multiple symbiont strains, reducing genetic homogeneity.

However, such considerations often neglect potential differences in power between host and symbionts. They may arise due to incomplete transmission linkage, as when one partner is less dependent on its current partner than the other. For instance, even with uniparental, vertical transmission, a host might have options to take up new symbionts or symbionts may also be able to spread horizontally. Additionally, the optimal number of symbionts transmitted by hosts during vertical transmission may differ between host and symbiont. Under such conditions, the relationship between host and symbiont becomes asymmetric with regard to higher power.

Partner Choice

Partner choice is linked with transmission mode but it is not completely congruent. It may include choice of a specific species in more generalist associations and/or choosing specific individuals within a species. Even in obligate species-specific symbioses selecting a specific partner among several potential partners is possible, especially when there is horizontal transmission. With vertical transmission specific partners can also be chosen—for instance, when there is biparental transmission of symbionts or when dependency during transmission is asymmetric and the host thus has more power. Through partner choice, a symbiont can stabilize cooperation/mutualistic associations by switching partners if one is cheating (i.e., exploiting the partner by taking more than its fair share), so that only associations of cooperators occur. Yet, it can also destabilize cooperation, when "cheaters" select cooperators or, more generally, when exploiting a partner has minimal consequences for a symbiont's own fitness—for instance, because the reproduction of both partners is not linked (as is the case with horizontal transmission). We would predict cheating to evolve when potential partners are common and a switch to a new partner is not costly. Thus it is expected to evolve in the species that is less common as its few members can select among many potential partners. Cheated individuals can "retaliate" and sanction cheaters by, for instance, switching partners. Thus, partner choice can be regarded as a dynamic mechanism that may evolve when potential partners can change. Hence, it can play a role whenever there is no strict uniparental, vertical transmission in which host and symbiont offspring reciprocally depend on each other.

Theoretical Considerations Applied to Fungus-Growing Termites

Surprisingly, and in contrast to fungus-growing ants (chapters 8 and 14, this volume), most fungus-growing termites have horizontal transmission. *Termitomyces* produces its sexual fruiting bodies (mushrooms) outside the termite nest. The fruiting bodies thus release spores to the environment, which are picked up and brought back to the nest by foraging termite workers (Johnson et al. 1981). Studies from Ivory Coast and Nigeria suggest that fructification of *Termitomyces* coincides species-specifically with the occurrence of the first foraging workers after colony foundation (Johnson et al. 1981; Koné et al. 2011). Only *M. bellicosus* and *Microtermes* species are known to have vertical transmission (Grassé and Noirot 1955; Johnson 1981; Johnson et al. 1981; Korb and Aanen 2003). Strikingly, it is always uniparental, although biparental transmission would be the default option as termite colonies are founded by a male and a female (in contrast to social Hymenoptera, in which females generally found colonies alone, and accordingly uniparental inheritance is the default option in fungus-farming ants [Nobré, Rouland-Lefèvre, and Aanen 2011]). In *M. bellicosus* (predominantly) the winged male sexuals (future king) carry fungal spores while it is the female (future queen) in *Microtermes* species (Grassé and Noirot 1955; Johnson 1981; Johnson et al. 1981; Korb and Aanen 2003; Nobré et al. 2011a).

Vertical Transmission

As outlined above, all else being equal, theoretical considerations predict that associations with uniparental, vertical transmission will be more cooperative and have a closer coevolution than those with horizontal transmission. This may result, for instance, in more finely tuned plant decomposition systems with complementary degrading enzymes (Rouland-Lefèvre 2000; da Costa et al. 2019b) and an associated higher ecological success. That reduction in conflict can result in higher ecological success is suggested by an experimental termite study in which lower intracolonial conflict resulted in interspecific competitive dominance (Korb and Foster 2010).

For *M. bellicosus* this may be the case. As described above, in regions where it occurs, it is an ecologically dominant species with high biomass. Whether this is indeed due to more complementary plant degradation systems warrants more studies (da Costa et al. 2019a, 2019b). Phylogenetic analyses based on molecular markers imply that across its geographic range *M. bellicosus* has several *Termitomyces* strains that are not shared with other fungus-growing termites (Aanen et al. 2002). In the CNP in Ivory Coast, *M. bellicosus* is associated with a single *Termitomyces* strain that has low genetic polymorphism (Nobré et al. 2011a). However, in Senegal two *Termitomyces* strains have been found associated with *M. bellicosus* (Nobré et al. 2011a). This study also suggests that it is only in the PNP in Benin that *M. bellicosus* can share its fungal symbiont with *Macrotermes subhyalinus*. Strikingly, in the PNP (but not, for instance, in the CNP or other West African regions) *M. subhyalinus* often inhabits dead *M. bellicosus* mounds (pers. obs.). Thus, it is tempting to speculate that *M. subhyalinus* acquires the typical *M. bellicosus* fungal symbiont via spores left behind in the mound by *M. bellicosus*.

For *Microtermes* species, however, the prediction of an ecologically "more efficient" and evolutionarily "closer" association under conditions of uniparental, vertical transmission seems less clear. We currently lack systematic analyses at the ecological level to evaluate their ecological success. At the evolutionary scale, a species-specific coevolutionary signal is missing in phylogenies. Aanen et al. (2002) showed that *Microtermes sp. 3*, for instance, has two fungal symbiont strains, while *Microtermes sp. 1* and *Microtermes sp. 2* share a symbiont strain with several *Ancistrotermes* species. In general, there seems to be a coevolutionary pattern that implies ancient co-diversification and recent symbiont dispersal across hosts, often associated with specificity at the genus level (Aanen et al. 2002; Aanen and Eggleton 2005; Aanen et al. 2007; Nobré et al. 2011b; Nobré, Rouland-Lefèvre, and Aanen 2011).

Horizontal Transmission

As mentioned above, the most common transmission mode in fungus-growing termites is horizontal transmission, and such associations can be highly ecologically successful (e.g., *Macrotermes natalensis* in southern Africa, several *Ancistrotermes* and *Odontotermes* species). Also, no association between transmission mode and interaction specificity was found for eight Macrotermitinae and their symbionts from South Africa (Aanen et al. 2007). *M. natalensis* was associated with a single unique fungal haplotype despite having horizontal transmission, whereas *Microtermes* species had low specificity and shared a pool of symbionts, although they most likely have uniparental, vertical transmission (Aanen et al. 2007). An explanation for the commonness of horizontal transmission in Macrotermitinae-*Termitomyces* associations—despite theoretically predicted high potential for conflict—is a mechanism that reduces fungal heterogeneity within termite colonies. Genetic bottlenecks and positive frequency-dependent selection/choice of the fungal cultivar by the termites during colony ontogeny could result in fungal clonality (Aanen 2006). Within a colony, *Termitomyces* clones that are more productive are more likely to be used to inoculate new fungus combs. This, together with genetic bottlenecks during fungal inoculation and fungal propagation within the colony, leads to selection of the most productive fungal clone, which becomes the single cultivar within a colony. Thus high relatedness (clonality) and an absence of conflict within the fungal cultivar follow, although the colony might have started with several genetically different cultivar strains. This elegant and plausible explanation is in line with current data. *M. natalensis* as well as two *Odontotermes* species, which all have horizontal transmission, almost always cultivate single clones within established colonies in South Africa (Aanen et al. 2009). Additionally, cultivation experiments that simulated within-nest propagation of *Termitomyces* in the lab are in line with the proposed mechanism that positive frequency-dependent selection together with genetic bottlenecks can lead to high fungal relatedness and productivity (Aanen et al. 2009). This nicely pinpoints a mechanism of how, starting with genetic heterogeneity, high relatedness can be obtained by partner choice. According to Leeks, dos Santos, and West (2019), high relatedness among symbionts associated with a host is more important for explaining symbiont cooperation than is fitness alignment between host and symbiont caused by strict vertical transmission. Thus high fungal relatedness can be sufficient to explain the evolutionary stability and ecological success of the termite-fungus symbiosis regardless of transmission mode.

Some questions remain. First, is the mechanism shown in the lab by Aanen et al. (2009) also working in nature? Unfortunately, time-series data on the genetic diversity of the fungal cultivar during colony establishment in the field are missing. The mechanism rests on the assumption that during colony foundation several clones are used to inoculate the fungus garden of a colony. This is not known. There might be other mechanisms in place that prevent inoculation with several cultivars in the first place. Second, the proposed mechanism prevents potential conflict between fungal strains within single colonies, but it does not prevent potential conflict between host (termites) and symbionts (*Termitomyces*), as their fitness interests are less well aligned in associations with horizontal compared with those with strict uniparental, vertical transmission. It also does not explain the observed phylogenetic pattern of some degree of co-cladogenesis (Nobré et al. 2011b), which suggests that there is some specificity in selecting specific spores to inoculate colonies. Evidence for this exists (N'golo Koné, pers. comm., J. K. unpubl. data). Therefore, it may be interesting to test, for instance, whether fructification time or cohabitation are factors explaining the occurrence of shared fungal strains at the termite species level.

What Can We Learn from the Macrotermitinae-*Termitomyces* Symbiosis?

The examples above illustrate eco-evolutionary feedbacks in the termite-fungus symbiosis. Lessons that might be learned from fungus-growing termites and their symbionts point to some general, often overlooked issues during discussion of symbiotic associations:

• Uniparental, vertical transmission does not necessarily mean strict codependency and complete alignment of fitness interests of species. In *M. bellicosus*, despite its strict uniparental, vertical co-transmission of host and symbiont, conflict is not completely absent. Fungal symbionts still "try" to produce fruiting bodies that would "allow" them to spread horizontally. This is not in the fitness interest of the termites, as, from their point of view, fruiting bodies would result in a loss of resources. Thus sanctioning mechanisms, such as consumption of fungal nodules in *M. bellicosus*, may be important to "enforce" alignment of fitness interests.

• Horizontal transmission does not necessarily mean that conflict will be prevalent in an association. If mechanisms evolve that guarantee within-host genetic homogeneity (high relatedness) of the symbiont, within-symbiont conflict (hereafter: intra-partner conflict) is reduced. Additionally, lifetime commitment (i.e., a lack of options for partner change), together with reciprocal dependency of partners, is expected to result in the prevalence of cooperation (chapter 5, this volume). Here one has to note that in associations with horizontal transmission, lifetime commitment does not apply for the symbionts as *Termitomyces* spread horizontally during the lifetime of a colony.

• Reduction of intra-partner conflict (due to genetic homogeneity) can be caused by (1) uniparental, vertical transmission, (2) repeated genetic bottlenecks, (3) mechanisms such as positive frequency-dependent selection, and/or (4) partner choice. Yet, a lack of intra-partner conflict does not imply a lack of conflict between host and symbiont (inter-partner conflict) and the alignment of fitness interests. Reduction of intra-partner conflict also can hardly explain patterns of co-cladogenesis.

Comparison with Human Agriculture

We saw that the symbiosis between termites and their fungus seems to extend their biological niche, making them ecologically dominant ecosystem engineers. However, in 1:1 associations such as that of *M. bellicosus* and its symbiont, it may also restrict the niche (in this case, to the savannah). In human agriculture, crops are also ecologically very successful in the sense that they are very abundant, often dominating whole landscapes. Yet they are doomed to extinction when humans do not care for or abandon them. Vice versa, in historical times the well-being of humans also depended on single crops (e.g., the Irish potato famine; Salaman and Burton 1985). Yet this seems to have changed with global interconnectedness, which currently offers humans a choice between crops worldwide. Such partner choice leads to humans having more power than their crops because of asymmetric interdependency.

Concerning transmission mode, crops are horizontally transmitted by humans as in most termite-fungus associations. Yet, there is a major difference. Horizontal transmission in Macrotermitinae is associated with the ability of the symbiont to reproduce and be "picked up" by another host. While this might have been the case formerly when certain crops were able to spread, this is no longer the case for our "modern" crops. Thus, fitness interests are not reciprocally aligned. Humans are largely independent of a certain crop, while the crop's fitness is fully dependent on humans.

In accordance with the fungus-growing termites, humans nowadays cultivate crops largely as monocultures of low genetic heterogeneity. While this reduces conflict within the cultivar, and thus is expected to result in more stable cooperation (chapters 3, 4, and 5, this volume), humans face problems of reduced pathogen resistance, among other things. We do not know how exactly fungus-growing termites keep fungal pathogens at bay, but most certainly strict barriers that prevent pathogens as well as competitors from entering, together with immune defenses, play important roles. Additionally, the fungal selection process that takes place during each colony foundation may result in the selection of the cultivar that is most appropriate at that time.

Overall, these considerations illustrate that compared with human agricultural systems, Macrotermitinae-*Termitomyces* associations show major differences.

Acknowledgments

I wish to thank Peter Peregrine, Ted Schultz, and Rick Gawne for organizing and inviting me to the interdisciplinary workshop on the evolution of agriculture at the Konrad Lorenz Institute (Klosterneuburg, Austria) and for realizing this book project. Many thanks—I learned a lot! I further thank Ted Schultz and Rick Gawne for comments on an earlier draft, and K. Eduard Linsenmair for founding and establishing the Comoé Research Station and for his constant commitment to keep it going. I am grateful for support in the field over many years by David Kouassi, Aboulaye Kouakou Kouadio, and Koffi Kouadio.

References

Aanen, D. K. 2006. "As You Reap, So Shall You Sow: Coupling of Harvesting and Inoculating Stabilizes the Mutualism between Termites and Fungi." *Biology Letters* 2:209–212.

Aanen, D. K., and P. Eggleton. 2005. "Fungus-Growing Termites Originated in African Rain Forest." *Current Biology* 15:851–855.

Aanen, D. K., P. Eggleton, C. Rouland-Lefevre, T. Guldberg-Froslev, S. Rosendahl, and J. J. Boomsma. 2002. "The Evolution of Fungus-Growing Termites and Their Mutualistic Fungal Symbionts." *Proceedings of the National Academy of Sciences* 99:14887–14892.

Aanen, D. K., H. H. de Fine Licht, A. J. M. Debets, N. A. G. Kerstes, R. F. Hoekstra, and J. J. Boomsma. 2009. "High Symbiont Relatedness Stabilizes Mutualistic Cooperation in Fungus-Growing Termites." *Science* 326:1103–1106.

Aanen, D. K., V. I. D. Ros, H. H. D. Licht, J. Mitchell, Z. W. de Beer, et al. 2007. "Patterns of Interaction Specificity of Fungus-Growing Termites and Termitomyces Symbionts in South Africa." *BMC Evolutionary Biology* 7:e115.

Badertscher, S., C. Gerber, and R. H. Leuthold. 1983. "Polyethism in Food Supply and Processing in Termite Colonies of *Macrotermes subhyalinus* (Isoptera)." *Behavioural Ecology and Sociobiology* 12:115–119.

Bignell, D. E., and P. Eggleton. 2000. "Termites in Ecosystems." In *Termites: Evolution, Sociality, Symbiosis and Ecology*, edited by T. Abe, D. E. Bignell, and M. Higashi, 363–387. Dordrecht, The Netherlands: Kluwer Academic Publishers.

Boutton, T. W., M. A. Arshad, and L. L. Tieszen. 1983. "Stable Isotope Analysis of Termite Food Habits in East African Grasslands." *Oecologia* 59:1–6.

Brune, A. 1998. "The Smallest Bioreactor in the World." *Biofutur* 138:40–43.

Bull, J. J., and W. R. Rice. 1991. "Distinguishing Mechanisms for the Evolution of Cooperation." *Journal of Theoretical Biology* 149:63–74.

Collins, N. M. 1981. "Populations, Age Structure and Survivorship of Colonies of *Macrotermes bellicosus* (Isoptera: Macrotermitinae)." *Journal of Animal Ecology* 50:293–311.

da Costa, R. R., H. Hu, H. Li, and M. Poulsen. 2019a. "Symbiotic Plant Biomass Decomposition in Fungus-Growing Termites." *Insects* 10:e87.

da Costa, R. R., H. Hu, B. Pilgaard, S. M. E Vreeburg, J. Schückel, et al. 2019b. "Enzyme Activities at Different Stages of Plant Biomass Decomposition in Three Species of Fungus-Growing Termites." *Applied and Environmental Microbiology* 84:e01815–e01817.

Darlington, J. P. E. C. 1989. "Ventilation Systems in Termite Nests in Kenya." *Sociobiology* 15:263–264.

Darlington, J. P. E. C., P. R. Zimmerman, J. Greenberg, C. Westberg, and P. Bakwin. 1997. "Production of Metabolic Gases by Nests of the Termite *Macrotermes jeanneli* in Kenya." *Journal of Tropical Ecology* 13:491–510.

Deshmukh, I. 1989. "How Important Are Termites in the Production Ecology of African Savannas?" *Sociobiology* 15:155–168.

De Visser, S. N., B. P. Freymann, and H. Schnyder. 2008. "Trophic Interactions among Invertebrates in Termitaria in the African Savanna: A Stable Isotope Approach." *Ecological Entomology* 33:758–764.

Donovan, S. E., P. Eggleton, and D. Bignell. 2001. "Gut Content Analysis and a New Feeding Group Classification of Termites." *Ecological Entomology* 26:356–366.

Eggleton, P., and I. Tayasu. 2001. "Feeding Groups, Lifetypes and the Global Ecology of Termites." *Ecological Research* 16:941–960.

Erpenbach, A., M. Bernhardt-Römermann, R. Wittig, and K. Hahn. 2017. "The Contribution of Termite Mounds to Landscape-Scale Variation in Vegetation in a West African National Park." *Journal of Vegetation Science* 28:105–116.

Erpenbach, A., M. Bernhardt-Römermann, R. Wittig, A. Thiombiano, and K. Hahn. 2012. "The Influence of Termite-Induced Heterogeneity on Savanna Vegetation along a Climatic Gradient in West Africa." *Journal of Tropical Ecology* 29:11–23. doi:10.1017/S0266467412000703.

Foster, K. R., and T. Wenseleers. (2006). "A General Model for the Evolution of Mutualisms." *Journal of Evolutionary Biology* 19:1283–1293.

Frank, S. A. 1994. "Kin Selection and Virulence in the Evolution of Protocells and Parasites." *Proceedings of the Royal Society of London B—Biological Sciences* 258:153–161.

Frank, S. A. 1996. "Models of Parasite Virulence." *Quarterly Review of Biology* 71:37–78.

Frank, S. A. 1997. "Models of Symbiosis." *American Naturalist* 150:S80–S99.

Frank, S. A. 1998. *Foundations of Social Evolution*. Princeton: Princeton University Press.

Grassé, P. P., and C. Noirot. 1955. "La Fondation de nouvelles societes par *Bellicositermes natalensis* Hav." *Insectes Sociaux* 2:213–219.

Hausberger, B., and J. Korb. 2015. "A Phylogenetic Community Approach for Studying Termite Communities in a West African Savanna." *Biology Letters* 11:20150625.

Hausberger, B., A. van Neer, D. Kimpel, and J. Korb. 2011. "Uncovering Cryptic Species Diversity of a Termite Community in a West African Savanna." *Molecular Phylogenetics and Evolution* 61:964–969.

Herre, E. A., N. Knowlton, U. G. Mueller, and S. A. Rehner. 1999. "The Evolution of Mutualisms: Exploring the Paths between Conflict and Cooperation." *Trends in Ecology and Evolution* 14:49–53.

Higashi, M., T. Abe, and T. P. Burns. 1992. "Carbo-Nitrogen Balance and Termite Ecology." *Proceedings of the Royal Society of London B—Biological Sciences* 249:303–308.

Hyodo, F., I. Tayasu, T. Inoue, J.-I. Azuma, T. Kudo, et al. 2003. "Differential Role of Symbiotic Fungi in Lignin Degradation and Food Provision for Fungus-Growing Termites (Macrotermitinae: Isoptera)." *Functional Ecology* 17:186–193.

Johnson, R. A. 1981. "Colony Development and Establishment of the Fungus Comb in *Microtermes sp. nr. umbaricus* (Sjöstedt) (Isoptera: Macrotermitinae) from Nigeria." *Insectes Sociaux* 28:3–12.

Johnson, R. A., R. J. Thomas, T. G. Wood, and M. J. Swift. 1981. "The Inoculation of the Fungus Comb in Newly Founded Colonies of Some Species of the Macrotermitinae (Isoptera) from Nigeria." *Journal of Natural History* 15:751–756.

Koné, N. A., K. Dosso, S. Konaté, J. Y. Kouadio, and K. E. Linsenmair. 2011. "Environmental and Biological Determinants of *Termitomyces* Species Seasonal Fructification in Central and Southern Côte d'Ivoire." *Insectes Sociaux* 58:371–382.

Korb, J. 1997. *Lokale und regionale Verbreitung von Macrotermes bellicosus (Isoptera; Macrotermitinae): Stochastik oder Deterministik.* Berlin: Wissenschaft und Technik Verlag.

Korb, J. 2011. "Termite Mound Architecture, from Function to Construction." In *Biology of Termites: A Modern Synthesis*, edited by D. E. Bignell, Y. Roisin, and N. Lo., 349–376. Dordrecht: Springer.

Korb, J., and D. K. Aanen. 2003. "The Evolution of Uniparental Transmission of Fungal Symbionts in Fungus-Growing Termites (Macrotermitinae)." *Behavioral Ecology and Sociobiology* 53:65–71.

Korb, J., and K. R. Foster. 2010. "Ecological Competition Favours Cooperation in Termite Societies." *Ecology Letters* 13:754–760.

Korb, J., B. D. Kasseney, Y. T. Cakpo, R. H. Casalla, J. N. B. K. Gbenyedji, et al. 2019. "Termite Taxonomy, Challenges and Prospects: West Africa, a Case Example." *Insects* 10:e32.

Korb, J., and K. E. Linsenmair. 1998a. "The Effects of Temperature on the Architecture and Distribution of *Macrotermes bellicosus* (Isoptera: Macrotermitinae) Mounds in Different Habitats of a West African Guinea Savanna." *Insectes Sociaux* 45:51–65.

Korb, J., and K. E. Linsenmair. 1998b. "Experimental Heating of *Macrotermes bellicosus* (Isoptera, Macrotermitinae) Mounds: What Role Does Microclimate Play in Influencing Mound Architecture?" *Insectes Sociaux* 45:335–342.

Korb, J., and K. E. Linsenmair. 1999a. "The Architecture of Termite Mounds: A Result of a Trade-Off between Thermoregulation and Gas Exchange?" *Behavioral Ecology* 10:312–316.

Korb, J., and K. E. Linsenmair. 1999b. "Reproductive Success of *Macrotermes bellicosus* (Isoptera: Macrotermitinae) in Two Neighbouring Habitats." *Oecologia* 118:183–191.

Korb, J., and K. E. Linsenmair. 2000. "Ventilation of Termite Mounds: New Results Require a New Model." *Behavioral Ecology* 11:486–494.

Korb, J., and K. E. Linsenmair. 2001a. "The Causes of Spatial Patterning of Mounds of a Fungus-Cultivating Termite: Results from Nearest-Neighbour Analysis and Ecological Studies." *Oecologia* 127:324–333.

Korb, J., and K. E. Linsenmair. 2001b. "Resource Availability and Distribution Patterns, Indicators of Competition between *Macrotermes bellicosus* and Other Macro-Detritivores in the Comoé National Park, Côte d'Ivoire." *African Journal of Ecology* 39:257–265.

Leeks, A., M. dos Santos, and S. A. West. 2019. "Transmission, Relatedness, and the Evolution of Cooperative Symbionts." *Journal of Evolutionary Biology* 32:1036–1045.

Lepage, M. 1984. "Distribution, Density and Evolution of *Macrotermes bellicosus* Nests (Isoptera: Macrotermitinae) in the North-East of Ivory Coast." *Journal of Animal Ecology* 53:107–117.

Lepage, M., L. Abbadie, and A. Mariotti. 1993. "Food Habits of Sympatric Termite Species (Isoptera, Macrotermitinae) as Determined by Stable Carbon Isotope Analysis in a Guinean Savanna (Lamto, Côte d'Ivoire)." *Journal of Tropical Ecology* 9:303–311.

Leuthold, R. H., S. Badertscher, and H. Imboden. 1989. "The Inoculation of Newly Formed Fungus Combs with Termitomyces in Macrotermes Colonies (Isoptera, Macrotermitinae)." *Insectes Sociaux* 36:328–338.

Lüscher, M. 1955. "Der Sauerstoffverbrauch bei Termiten und die Ventilation des Nestes bei *Macrotermes natalensis* (Haviland)." *Acta Tropica* 12:289–307.

Lüscher, M. 1956. "Die Lufterneuerung im Nest der Termite *Macrotermes natalensis* (Haviland)." *Insectes Sociaux* 3:273–276.

Martin, M. M., and J. S. Martin. 1978. "Cellulose Digestion in the Midgut of the Fungus-Growing Termite *Macrotermes natalensis*: The Role of Acquired Digestive Enzymes." *Science* 199:1453–1455.

Nobré, T., C. Fernandes, J. J. Boomsma, J. Korb, and D. K. Aanen. 2011a. "Farming Termites Determine the Genetic Population Structure of Termitomyces Fungal Symbionts." *Molecular Ecology* 20:2023–2033.

Nobré, T., N. A. Kone, S. Konate, K. E. Linsenmair, and D. K. Aanen. 2011b. "Dating the Fungus-Growing Termites' Mutualism Shows a Mixture between Ancient Co-Diversification and Recent Symbiont Dispersal across Divergent Hosts." *Molecular Ecology* 20:2619–2627.

Nobré, T., C. Rouland-Lefèvre, and D. K. Aanen. 2011. "Comparative Biology of Fungus Cultivation in Termites and Ants." In *Biology of Termites: A Modern Synthesis*, edited by D. E. Bignell, Y. Roisin, and N. Lo, 193–210. Dordrecht, The Netherlands: Springer.

Poilecot, P. (1991). *Un Écosysteme de Savane Soudanienne: Le Parc National de la Comoé (Côte d'Ivoire).* Paris: UNESCO.

Poulsen, M. 2015. "Towards an Integrated Understanding of the Consequences of Fungus Domestication on the Fungus Growing Gut Microbiota." *Environmental Microbiology* 17:2562–2572.

Rouland, C., F. Lenoir, and M. Lepage. 1991. "The Role of the Symbiotic Fungus in the Digestive Metabolism of Several Species of Fungus-Growing Termites." *Comparative Biochemistry and Physiology* 99:657–663.

Rouland-Lefèvre, C. 2000. "Symbiosis with Fungi." In *Termites: Evolution, Sociality, Symbioses, Ecology*, edited by T. Abe, D. E. Bignell, and M. Higashi, 289–306. Dordrecht, The Netherlands: Kluwer Academic Publishers.

Sachs, J. L., U. G. Mueller, T. P. Wilcox, and J. J. Bull. 2004. "The Evolution of Cooperation." *Quarterly Review of Biology* 79:135–160.

Salaman, R. N., and W. G. Burton. 1985. *The History and Social Influence of the Potato*. Cambridge: Cambridge University Press.

Schyra, J., and J. Korb. 2019. "Termite Communities along a Disturbance Gradient in a West African Savanna." *Insects* 10:e17.

Schyra, J., J. N. B. K. Gbenyedji, and J. Korb. 2019. "A Comparison of Termite Assemblages from West African Savannah and Forest Ecosystems Using Morphological and Molecular Markers." *PLoS ONE* 14:e0216986.

Schyra, J., S. Scheu, and J. Korb. 2019. "Cryptic Niche Differentiation in West African Savannah Termites as Indicated by Stable Isotopes." *Ecological Entomology* 44:190–196.

Shellman-Reeve, J. S. 1994. "Limited Nutrients in a Dampwood Termite: Nest Preference, Competition and Cooperative Nest Defence." *Journal of Animal Ecology* 63:921–932.

Tano, Y., and M. Lepage. 1989. "Abundance and Distribution of *Macrotermes bellicosus* Mounds (Isoptera: Macrotermitinae) in Sudan Savanna, Ivory Coast." *Sociobiology* 15:275.

Tayasu, I., F. Hyodo, and T. Abe. 2002. "Caste-Specific N and C Isotope Ratios in Fungus-Growing Termites with Special Reference to Uric Acid Preservation and Their Nutritional Interpretation." *Ecological Entomology* 27:355–361.

Traniello, J. F., and R. H. Leuthold. 2000. "Behavior and Ecology of Foraging in Termites." In *Termites: Evolution, Sociality, Symbiosis and Ecology*, edited by T. Abe, D. E. Bignell, and M. Higashi, 141–168. Dordrecht, The Netherlands: Kluwer Academic Publishers.

Zimmermann, P. R., and J. P. E. C. Darlington. 1987. "Methane and Other Metabolic Gases Produced by Termites." In *Chemistry and Biology of Social Insects*, edited by J. Eder and H. Rembold, 637. Munich: Verlag J. Peperny.

7 Mycangia Define the Diverse Ambrosia Beetle–Fungus Symbioses

Chase G. Mayers, Thomas C. Harrington, and Peter H. W. Biedermann

Ambrosia beetles (Coleoptera: Curculionidae: Platypodinae and multiple lineages in the Scolytinae) bore tunnels (galleries) into the sapwood of trees (figure 7.1A, B). There they inoculate and cultivate lush gardens of fungi (ambrosia fungi) on which the adults and larvae feed (figure 7.1C). This obligate mutualism (ambrosia symbiosis) allows both beetle and fungus to carve a unique niche in a nutrient-poor substrate (Hubbard 1897a, 1897b; Baker 1963; Francke-Grosmann 1963, 1967; Beaver 1989; Six 2003, 2012; Harrington 2005; Hulcr and Stelinski 2017). Ambrosia fungi have not been found free-living and are poorly equipped for independent dispersal, yet when adult ambrosia beetles tunnel into new trees, they introduce the same fungal cultivars they fed upon as larvae. In the 1950s and 1960s, a diversity of pocket-like spore-carrying organs (mycangia or mycetangia) were found to be the facilitators of this cultivar persistence (Francke-Grosmann 1956a, 1963, 1967; Batra 1963), and the multiple origins of these organs now appear to be major events in the evolution of the ~3,500 recognized species of ambrosia beetles.

Ambrosia beetles, defined here as fungus-farming beetles in xylem (sapwood, heartwood, or pith), have convergently evolved many times in the true weevil family (Curculionidae) including the entire subfamily Platypodinae (Jordal 2015) and in more than a dozen separate clades comprising almost half the total species in the subfamily Scolytinae (Johnson et al. 2018, 2020). With the exception of ambrosia beetles and a small number of seed and pith feeders, the remaining species of Scolytinae are bark beetles, which feed primarily in the inner bark or phloem tissue (Harrington 2005; Six 2012). Ambrosia beetles farm truly domesticated fungal cultivars that are adapted to cultivation and can no longer live independently, which are called *primary ambrosia fungi* or simply *ambrosia fungi* (Batra 1985; Mayers et al. 2020a). Ambrosia fungi are carried in mycangia, benefitting from free transport to new substrate, and like the beetles, they are polyphyletic (table 7.1). Each lineage of ambrosia beetles developed a specific type of mycangium (Hulcr and Stelinski 2017) whose origin predates or closely coincides with the domestication or acquisition of its current fungal cultivars (Vanderpool, Bracewell, and McCutcheon 2017; Mayers et al. 2020a). Fungal cultivars are closely tied to the mycangium type of their host beetles, with various levels of specificity and phylogenetic congruence (table 7.1; Mayers et al. 2015, 2018, 2020a, 2020b; Hulcr and Stelinski 2017; Skelton et al. 2018).

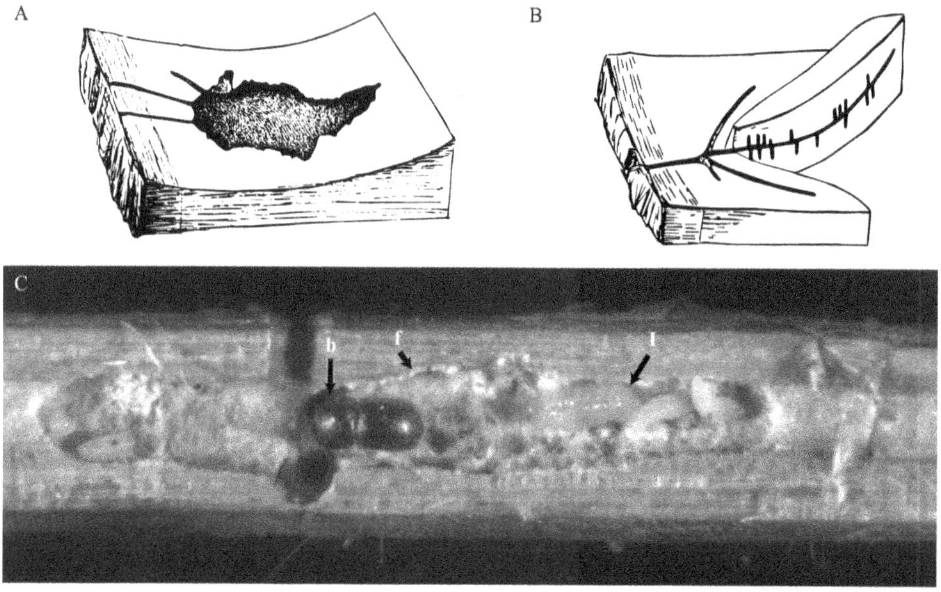

Figure 7.1
Ambrosia beetle galleries. (A) Communal gallery with shared chambers. (B) Gallery with individual side rooms ("cradles") for larvae. (C) Exposed gallery in twig with adult mother *Xylosandrus compactus* (b) and larvae (l) feeding on ambrosia fungus *Ambrosiella xylebori* (f). A and B modified from Hubbard (1897).

Research interest in the ambrosia symbiosis has increased substantially over the past decades, and the field currently stands at a turning point from describing the symbiosis to exploring theoretical concepts and specific mechanisms (Hulcr et al. 2020). Recent reviews have treated ambrosia beetles from an entomological perspective (e.g., Hulcr et al. 2015; Kirkendall, Biedermann, and Jordal 2015) or broadly reviewed the ambrosia symbiosis (Hulcr and Stelinski 2017; Biedermann and Vega 2020), but as interest in mycangia increases, a modern and comprehensive review of these organs is needed. This chapter intends to (1) build on previous reviews of bark and ambrosia beetle mycangia (Baker 1963; Francke-Grosmann 1963, 1967; Beaver 1989; Six 2003) to summarize the history and current knowledge of these organs in ambrosia beetles and (2) discuss why mycangia are so important to the ambrosia symbiosis. Ambrosia beetle mycangia facilitate the cultivation, domestication, and preservation of fungal crops, convergently solving challenges also faced by other fungus-farming insects and humans. Bark beetle mycangia are not comprehensively covered here, but they have been treated elsewhere (Six 2003, 2012; Harrington 2005; Bracewell and Six 2015).

Mycangia Set the Ambrosia Symbiosis Apart

As discussed in other chapters of this book, fungus farming and the domestication of fungal cultivars by insects is not unique to ambrosia beetles. Besides ambrosia beetles, advanced farming (sensu Biedermann and Vega 2020) has also convergently evolved in fungus-growing ants (Hymenoptera: Myrmicinae: tribe Attini) (Mueller et al. 2018; chapters 8 and 14, this volume) and fungus-growing termites (Blattodea: Macrotermitinae) (Aanen

Table 7.1
Mycangium types and mycangial fungi of ambrosia beetles

Beetle tribe	Ambrosia beetle clade	Mycangium type[a]	Primary mycangial fungi[b]
Platypodinae: Platypodini	*Platypus*	Pronotal pits (A), generally numerous, and procoxal (K)/mesocoxal	*Raffaelea* sensu lato (Asco: Ophiostomatales)
	Austroplatypus, Euplatypus, Megaplatypus, Myoplatypus, Oxoplatypus, Treptoplatypus	Pronotal pits (A), various sizes and arrangements	*Raffaelea* sensu lato and (in *Oxoplatypus*) *R. sulphurea* complex (Asco: Ophiostomatales)
	Carchesiopygus, Cylindropalpus[c], *Dinoplatypus, Epiplatypus, Neotrachyostus, Peroplatypus*	Pronotal pits (A), various sizes and arrangements	Unknown
	Crossotarsus niponicus	Cranial sphere (C)	Unknown thick white fungus
	Baiocis, Costaroplatus, Doliopygus, Mesoplatypus[c], *Pereioplatypus, Platyphysus, Teloplatypus, Trachyostus, Triozastus*[c]	Unknown, but pronotal pits absent	
Platypodinae: Tesserocerini	*Cenocephalus*[c], *Chaetastus, Genyocerus, Mitosomac, Spathidicerus*	Pronotal pits (A), various sizes and arrangements	Unknown
	Diapus	Pronotal pits (A) in transverse band, and pronotal transverse crevices	Unknown
	Notoplatypus	Unknown, but pronotal pits absent	Unknown, except *N. elongatus* and *Raffaelea* sensu lato (Asco: Ophiostomatales)
Scolytinae: Bothrosternini	*Cnesinus*[d]	Setose patch (E) on both sides of pronotum	*Geosmithia cnesini* (Asco: Hypocreales)[e]
	Eupagiocerus[d]	Setose patch (E) on both sides of pronotum	*Geosmithia eupagioceri* (Asco: Hypocreales)[e]
	Bothrosternus[d]	Setose patch (E) on both sides of pronotum	Unknown
Scolytinae: Corthylini	*Gnathotrichus*	Procoxal (K)	*Raffaelea* sensu lato (Asco: Ophiostomatales)
	Monarthrum (most studied species)	Procoxal (K)	*Raffaelea* sensu lato (Asco: Ophiostomatales)
	Monarthrum bicallosum	Single oral pocket (J)	Unknown
	Corthylus	Prothoracic tubes (L)	*Meredithiella* (Asco: Microascales)
	Microcorthylus	Prothoracic tubes (L)	*Geosmithia microcorthyli* (Asco: Hypocreales)[e]
Scolytinae: Cryphalini	*Hypothenemus curtipennis*	Setose patch (E) on both sides of pronotum	Unknown
Scolytinae: Hexacolini	*Scolytodes unipunctatus*	Unknown	*Raffaelea scolytidis* (Asco: Ophiostomatales)[e]
	Scolytodes culcitatus	Setose patch (E) on both sides of pronotum	Unknown
Scolytinae: Hylesinini	*Phloeoborus*[d]	Setose patch (E) on both sides of pronotum	Unknown
	Dactylipalpus[d]	Transverse slit on top of pronotum	Unknown

(*continued*)

Table 7.1
(continued)

Beetle tribe	Ambrosia beetle clade	Mycangium type[a]	Primary mycangial fungi[b]
Scolytinae: Hyorrhynchini	*Sueus*	Unknown	*Diatrypella japonica* (Asco: Xylariales)[e]
Scolytinae: Ipini	*Premnobius*	Dual preoral pockets (F)	*Afroraffaelea ambrosiae* (Asco: Ophiotomatales)
Scolytinae: Scolytini	*Camptocerus*	Genal brush (crescent, circular, or triangular)	Unknown
Scolytinae: Scolytoplatypodini	African *Scolytoplatypus*	Pronotal disk (D), with simple hairs	*Wolfgangiella* (Asco: Microascales)
	Asian *Scolytoplatypus*	Pronotal disk (D), with cone-base hairs	*Toshionella* (Asco: Microascales)
	Remansus mutabilis	Pronotal disk (D), with hybrid hairs	*Ambrosiella remansi* (Asco: Microascales)
Scolytinae: Xyleborini	*Amasa, Cyclorhipidion, Euwallacea, Xyleborus*, and others	Dual preoral pockets (F)	*Raffaelea* sensu lato (Asco: Ophiostomatales)
	Some *Euwallacea*	Dual preoral pockets (F)	Ambrosia *Fusarium* Complex ("AFC") (Asco: Hypocreales)
	Ambrosiodmus, Ambrosiophilus	Dual preoral pockets (F)	*Flavodon subulatus* (Basidio: Polyporales)
	Xyleborinus, and presumably *Coptodryas, Cryptoxyleborus, Heteroborips, Microperus, Taurodemus*	Elytral notches (G)	*Raffaelea* sensu lato (Asco: Ophiostomatales) (in *Xyleborinus*)
		Gut (in *Xyleborinus*)	*Raffaelea sulphurea* complex (Asco: Ophiostomatales)
	Cnestus, Eccoptopterus, Hadrodemius, Xylosandrus	Mesonotal pouch, lobed (H)	*Ambrosiella* (Asco: Microascales)
	Anisandrus	Mesonotal pouch, unlobed (I)	*Ambrosiella* (Asco: Microascales)
Scolytinae: Xyloterini	*Xyloterinus politus*	Dual preoral pockets (F)	*Raffaelea* cf. *canadensis* RNC5 (Asco: Ophiostomatales)
		Prothoracic basins (M)	*Kaarikia abrahamsonii* (Asco: genus incertae sedis)
	Indocryphalus pubipennis	Pleural cavities (fin) (N)	*Toshionella* (Asco: Microascales)
	Other *Indocryphalus*	Vertical or horizontal entrances, unknown anatomy	Unknown
	Trypodendron	Pleural cavities (horseshoe) (O)	*Phialophoropsis* (Asco: Microascales)

Notes:

[a]Letter in parentheses refers to mycangium silhouette in figure 7.4.

[b]Asco = Ascomycota; Basidio = Basidiomycota.

[c]Based on observations by R. Beaver.

[d]Obligate fungus farming not confirmed.

[e]Inferred from gallery dominance; mycangial transport not confirmed.

et al. 2002; Roberts et al. 2016; chapter 6, this volume). The fourth group of animals to develop true agriculture are humans (Mueller et al. 2005; chapters 1, 2, 10, and 12, this volume). Primitive farming may exist in ship-timber beetles (Lymexylidae) and burying beetles (Silphidae), among other insect groups (Biedermann and Vega 2020).

Ambrosia beetles are also not the only insects that use mycangia to carry fungal symbionts. Nutritional fungal mutualists are carried in mycangia of some bark beetles (Batra 1966; Francke-Grosmann 1967; Six 2003, 2012; Harrington 2005; Hofstetter et al. 2015; Raffa, Grégoire, and Lindgren 2015; Hulcr and Stelinski 2017; Six and Elser 2020), as well as in mycangia of ship-timber beetles (Lymexylidae) (Batra and Francke-Grosmann 1961, 1964; Francke-Grosmann 1967) and the lizard beetle *Doubledaya bucculenta* (Erotylidae: Languriinae) (Toki et al. 2012). Mycangia of wood wasps and horntails (Siricidae and Xiphydriidae) carry wood-decay fungi (Baker 1963; Francke-Grosmann 1963, 1967; Talbot 1977) that act as an external gut; the larvae feed only on the liquid fraction of fungus-digested xylem (Thompson et al. 2014). Non-nutritional fungal symbionts are carried in the mycangia of several groups of Coleoptera, including Lucanidae stag beetles (Tanahashi et al. 2010; Tanahashi and Fremlin 2013; Tanahashi and Hawes 2016; Roets and Oberlander 2020), leaf-rolling weevils (*Euops*, Attelabidae) (Sawada and Morimoto 1986; Kobayashi et al. 2008; Li et al. 2016), and potentially others (Grebennikov and Leschen 2010).

What sets ambrosia beetles apart from all of these other organisms is that, with the exception of a small minority of phloem-feeding bark beetles (as discussed later in this chapter) and the possible exception of the understudied ship-timber beetles, only ambrosia beetles rely both on farming reciprocally-obligate fungal cultivars and on using mycangia to maintain those cultivars.

Brief History of Mycangia

Hartig (1844, 1872a, 1872b) discovered that “ambrosia,” as named by Schmidberger (1836), was fungal. Hartig assumed that the ambrosia appeared spontaneously, but it was later realized that ambrosia beetles must introduce the fungus to their tunnels (Goethe 1895; Hubbard 1897a, 1897b). The method of transport remained elusive for some time. Early hypotheses were that the fungi were carried superficially (Strohmeyer 1911) or in the gut (Schneider-Orelli 1911, 1913), and gut transmission remained the leading theory until Hadorn (1933) demonstrated that some overwintering female adults had empty guts but still successfully transmitted ambrosia fungi. Some ambrosia beetles (e.g., *Xyleborinus saxesenii* and some Platypodinae) can indeed transport fungal cultivars in their guts, but these species also have mycangia for fungal transport (Baker 1963; Francke-Grosmann 1967, 1975; Biedermann et al. 2013; Gharabigloozare 2015).

Nunberg (1951) correctly hypothesized that the large organs he discovered in *Trypodendron* and other genera, as well as the pits observed by Beeson (1917) in *Genyocerus* and the large organ illustrated by Berger and Cholodkovsky (1916) in *Scolytoplatypus*, might serve a common purpose: storing and transporting fungal propagules. Francke-Grosmann (1956a) independently discovered and characterized such organs for the first time and showed that they did indeed store masses of fungal propagules (Francke-Grosmann 1956a, 1956b, 1958). More mycangia were discovered in the following decade, especially by Schedl (1962), as summarized and categorized by Francke-Grosmann (1967). Later reviews and classifications of mycangia were made by Beaver (1989), Six (2003), and Hulcr and Stelinski (2017).

Definition and General Features of Mycangia

Following Francke-Grosmann's (1956) discovery of the organs, Batra (1963) coined the term "mycangium," defined as sac- or cup-shaped cavities on the exterior of ambrosia beetles that hold growing fungal propagules. Francke-Grosmann had first called them Hautdrüsen or Hohldrüsen (loosely translated as cuticular glands or folliculose glands) but later suggested the term "mycetangium" (Francke-Grosmann 1967). The term "mycetangium" has not been broadly accepted (especially in English) but may be the more correct term (Vega and Biedermann 2019). The term was eventually broadened over time to include any physical structure that holds spores, even simple tufts of hairs (Farris and Funk 1965; Livingston and Berryman 1972; Nakashima 1975); Furniss et al. (1987, 692) defined mycangium as "any repository of the insect cuticle that is adapted for the transport of fungus." Six (2003, 101) defined mycangium as "any structure that consistently functions to transport specific fungi" and proposed a hierarchical classification of mycangia that first separated them into "pit," "sac," or "setal-brush," mycangia, next into "glandular" or "nonglandular," and then into their locations on the body. In more recent times, "mycangium" has sometimes been restricted to refer only to glandular mycangia (Hulcr et al. 2015; Hulcr and Stelinski 2017).

Mycangia vary dramatically in size, shape, and location on the beetles' bodies (figure 7.2; table 7.1), but they all share the common function of storing fungal propagules and facilitating their transfer to new trees (Francke-Grosmann 1956a, 1963, 1967; Schneider 1976, 1991; Beaver 1989; Hulcr and Stelinski 2017). Most mycangia are ectodermal invaginations of the exoskeleton with varying levels of depth and complexity, ranging from simple pits (figure 7.2A) or shallow bowls lined with hairs (figure 7.2B, L) in the surface of the exoskeleton to winding cylindrical tubes and folded cavities (figure 7.2D, E, K) or membranous pouches (figure 7.2F–H) within the body. Typically, each ambrosia beetle species has only one mycangium type, only one sex (usually the female) has mycangia, and the sex with mycangia initiates gallery construction (Francke-Grosmann 1967), but there are exceptions to each of these generalities.

Mycangial Glands

As a rule, ambrosia beetle mycangia are associated with an abundance of gland cells. The glands are embedded in the mycangium wall, nearby tissue, or near the mycangium opening (Francke-Grosmann 1956a, 1963; Schedl 1962; Batra 1963; Farris and Funk 1965; Schneider 1976; Beaver 1989). Mycangial gland cell anatomy has been described in detail for multiple ambrosia beetles (Happ, Happ, and French 1976; Cassier et al. 1996; Schneider and Rudinsky 1969b; Stone et al. 2007). The glands produce secretions that fill the mycangium lumen and support active fungal growth. These secretions have been referred to as "clear," "oily," "waxy," and "slightly acidic" (Francke-Grosmann 1956a; Batra 1963;

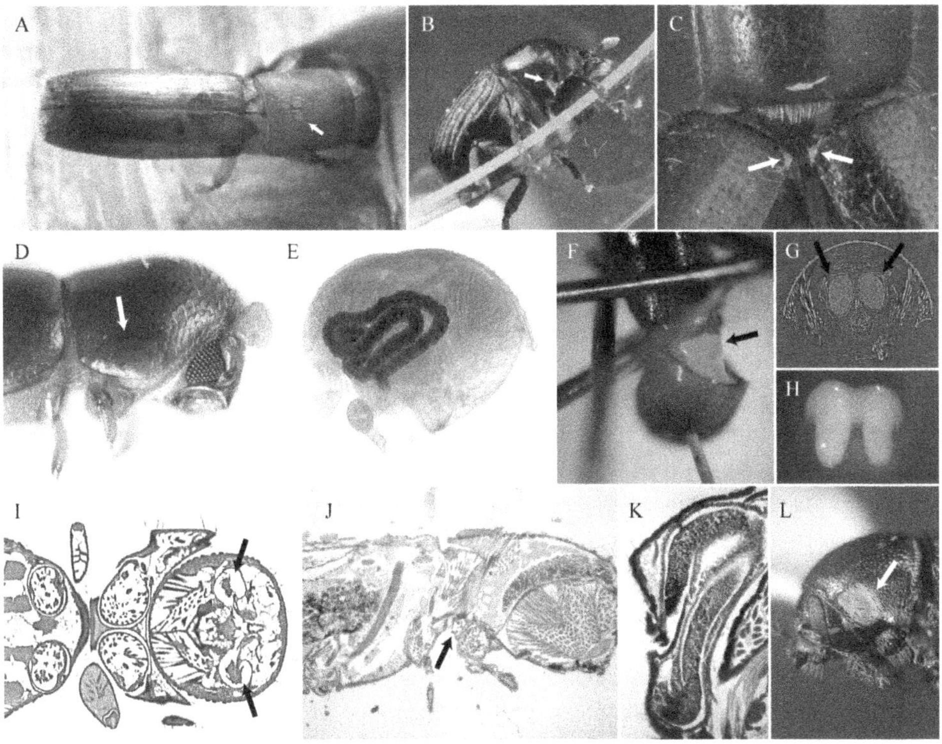

Figure 7.2
Different types of mycangia and mycangium imaging techniques. (A) Pronotal pit mycangia of *Oxoplatypus quadridentatus*. (B) Setose patch mycangia of *Phloeoborus* cf. *niger*. (C) Elytral notch mycangia of *Xyleborinus saxesenii*, with overflowing hyphae of ambrosia fungus *Raffaelea*. (D, E) Prothoracic tube mycangia of *Corthylus punctatissimus*. (D) Mycangium visible through exoskeleton. (E) Mycangium 3D model constructed from Micro-CT scanning. (F–H) Mesonotal pouch mycangium of *Xylosandrus amputatus* (F, G) and *X. germanus* (H). (F) "Draculian method" to expose mycangium (arrow). (G) Dual lobes visible in cross section using Micro-CT. (H) Excised mycangium/spore mass. (I) Dual preoral pocket mycangia of *Premnobius cavipennis*. (J) Prothoracic basin mycangia of *Xyloterinus politus*. (K) Prothoracic horseshoe cavity mycangia of *Trypodendron betulae*. (L) Setose patch mycangium of unidentified Bothrosternini. When present, arrows indicate mycangia. E, G, I by Micro-CT. J, K by microtome sectioning. A, B, D-G, I provided by Craig Bateman, You Li, and Jiri Hulcr. J, K provided by Lawrence Abrahamson. L provided by Luana de Souza Covre.

Francke-Grosmann 1963; Schneider 1976; Beaver 1989), but the secretions have not been fully characterized. Whether they directly feed the fungi or merely produce a favorable environment for selective fungal growth is unclear. Lipids produced by the gland cells may protect against desiccation rather than serve as fungal nutrients (Kok 1979; Beaver 1989; Six 2003). The mycangium lumen has been shown to include fatty acids, phospholipids, free sterols, sterol esters, and triglycerides, as well as an abundance of amino acids such as alanine, valine, and especially proline, but these compounds may or may not be directly produced and introduced by the gland cells (Abrahamson 1969; Norris 1979). Some nutrients such as nitrogen may be derived from the breakdown of flight muscles during and/or after flight (Chapman 1956) or other body reserves and enter the mycangium from free haemolymph (Abrahamson 1969; Abrahamson and Norris 1970; French and Roeper 1973). Mycangia are often surrounded by a conspicuously high number of tracheae, at least in the large mycangia of *Xylosandrus* (Stone et al. 2007), *Corthylus* (Mayers,

Bateman, and Harrington 2018), and *Trypodendron* (Lehenberger et al. 2019). The tracheae may serve the function of aerating the mycangial glands or mycangium lumen, but this has not been tested.

As originally proposed by Francke-Grosmann (1966, 1967), mycangial glands may have evolved from glands that originally served other purposes, such as lubrication or waterproofing of the exoskeleton. Lubrication glands have been observed near the joints of different insects (Billen 2009; Billen and Vander Plancken 2014; Nijs and Billen 2015), and many mycangia are located near joints, particularly mouthparts and coxae. Many insects (including beetles) have ubiquitous epidermal glands (Kendall 1972; Noirot and Quennedey 1974; Billen 2009) that are thought to secrete the lipid and protein components of the waxy, hydrophobic layer on the outside of the exoskeleton (Hughes et al. 2011; Mitov, Soldan, and Balor 2018). The cytology of epidermal glands (Kölsch 2000) is remarkably similar to that of mycangial glands (Cassier et al. 1996), and epidermal glands in other beetles have evolved to serve other specialized roles such as adhesion or defense (Pasteels et al. 1989; Kölsch 2000). Ambrosia fungi in the Ophiostomatales and Ceratocystidaceae (Microascales) are derived from insect-dispersed wood saprophytes with adaptations for cuticular transmission via sticky spores (Harrington 2005; Six 2012; Vanderpool, Bracewell, and McCutcheon 2017; Mayers et al. 2020a). It is intriguing to speculate that these ancestral fungi were adapted to surviving on the waxy cuticle or consuming components secreted by epidermal glands of bark beetles, predisposing them for mycangial selection. If true, this could help explain the mystery of why ambrosia fungi were separately domesticated multiple times within both the Ophiostomatales and Ceratocystidaceae (Vanderpool, Bracewell, and McCutcheon 2017; Mayers et al. 2020a).

Mycangium Propagules

Ambrosia fungi are sensitive to desiccation (Batra 1963, 1985; Zimmerman 1973; Zimmerman and Butin 1973), thus necessitating transport in mycangia. Fungi may enter the mycangium through active or passive methods. In *Trypodendron*, fungal propagules are transferred by friction from the gallery walls to the mycangium entrances when young adults first emerge from pupae (Schneider and Rudinsky 1969a). Schedl (1962), Batra (1966), and Batra and Batra (1967) reported that callow adults of many species perform distinct rocking movements to force spores into their mycangia, and Stone et al. (2007) reported similar movements in *Cnestus mutilatus*. Kaneko (1967) observed that when *Xylosandrus germanus* performed such movements, its large mycangium turned inside-out, collected spores from the gallery walls, and was then pulled back in with spores attached. This would be impossible for mycangia of the related *Anisandrus*, whose mycangia are smaller and have membranes that are too closely attached to the inside of the beetle's exoskeleton (Francke-Grosmann 1956a, 1956b). However, these mycangia are accompanied by unique pores on the top of the scutellum adorned with directional hairs that may help fungi reach the mycangium opening as the beetle articulates (Mayers, Harrington, and Ranger 2017). In *Austroplatypus incompertus*, similar hairs on the rim of mycangial pits guide single spores into each pit as the beetle scrapes the gallery walls (Kent 2008).

All mycangial ambrosia fungi are dimorphic in that they take on different growth forms in wood and in mycangia (Batra 1963, 1966; Batra and Downing Michie 1963;

Francke-Grosmann 1967; Beaver 1989; Blackwell 2017). Ambrosia fungi switch to their mycangial forms shortly after entering the mycangium (Batra 1966; Schneider and Rudinsky 1969a; Schneider 1976). In contrast to the filamentous (hyphal) gallery form, mycangial propagules are either (1) true budding yeasts, as in *Raffaelea* (Francke-Grosmann 1956a, 1958; Lhoste and Roche 1959; Batra 1967; Gebhardt et al. 2004; Harrington, Fraedrich, and Aghayeva 2008; Harrington, Aghayeva, and Fraedrich 2010) and potentially in ambrosial *Fusarium* (Kasson et al. 2013), or (2) fragmenting, branching, thallic-arthric hyphae, as in Ceratocystidaceae genera (Harrington et al. 2014; Mayers et al. 2015, 2017, 2018, 2020a, 2020b) *Afroraffaelea* (Bateman et al. 2017) and *Flavodon subulatus* (Kasson et al. 2013; Jusino et al. 2020). The mycangial propagules are adapted to multiply and fill mycangia when mycangial gland cells are activated (Schneider and Rudinsky 1969a). The yeast-phase/hyphal-phase dimorphism is evident in axenic cultures of *Raffaelea*, and similar dimorphism has been studied in its relatives in the Ophiostomatales, such as the human pathogen *Sporothrix schenkii* and the related Dutch elm disease pathogens *Ophiostoma ulmi* and *O. novo-ulmi* (Berrocal et al. 2012; Gauthier 2015; Sil and Andrianopoulos 2015; Nigg and Bernier 2016). In these fungi, switches are due to changes in temperature or cell density (quorum sensing), and dense concentrations of spores of *Raffaelea* favor yeast growth over hyphal growth in vitro (Harrington, unpublished observations). Ambrosia fungi are also thought to respond to external beetle secretions when growing in the gallery or in culture, triggering lush sporulation and mounds of edible cells that are difficult to reproduce in culture (Batra and Downing Michie 1963; French and Roeper 1972). This abundance of edible spores, visible as conspicuous moist, wet carpets in galleries, is sometimes referred to as the “ambrosial state,” “sprout cells,” or described as “yeast-like” but should not be confused with the mycangial state.

Fungal propagules regularly ooze from mycangium openings by growing and overflowing from the mycangium lumen (Mayers, Harrington, and Ranger 2017; Mayers et al. 2020b; figure 7.2B, C). There are conflicting reports on whether propagules are expelled actively by internal muscular contractions of the mycangium or by pressing internally with other body parts (Schedl 1962; Francke-Grosmann 1967; Schneider 1976; Li et al. 2018b). Fungal propagules can be expelled from larger mycangia without harming the beetle by physical manipulation in the laboratory (Skelton et al. 2019; Mayers et al. 2020b; figure 7.2F). In some types of mycangia, bristles or ridges on moving parts near the mycangium opening are thought to scrape and dispense fungal propagules (Farris 1965; Schneider and Rudinsky 1969b), and in others conspicuous tufts of hairs near the mycangium opening may contribute to spore entry and/or exit (Mayers, Harrington, and Ranger 2017). The propagules of Ceratocystidaceae start to germinate and produce hyphae as soon as they leave the mycangium, without necessarily contacting wood (Mayers, Harrington, and Ranger 2017; Mayers et al. 2020b).

The Mycangium Cycle

The combined results of decades of research paints a picture of an evolved “mycangium cycle” (figure 7.3) in ambrosia beetles that favors the growth and retention of specialized ambrosia fungi over other fungi, as first posited by Batra (1966), observed in *Trypodendron* by Schneider and Rudinsky (1969a), explored in detail in multiple species by Schneider (1975), and experimentally demonstrated by Skelton et al. (2019). Using evidence from a variety of ambrosia beetles and their fungal symbionts, the generalized stages of the cycle are as follows:

1. Immediately after emerging from pupae or leaving pupal cradles, the newly matured beetles acquire ambrosia fungus propagules, often with other fungi and debris, into their empty mycangia from the gallery walls (Schneider and Rudinsky 1969a, 1969b; Nakashima 1979; Kinuura, Hijii, and Kanamitsu 1991; Kajimura and Hijii 1992; Kinuura 2002).
2. Gland cells lining the mycangium activate at or just before the time beetles depart their birth gallery (Schneider 1976). When activated, the gland cells grow larger and fill the mycangium with their secretions (Schneider and Rudinsky 1969a, 1969b; Schneider 1976), which promote the growth of the primary ambrosia fungi over other fungi (Schneider and Rudinsky 1969a; Nakashima 1975; Kajimura and Hijii 1992; Skelton et al. 2019).
3. The propagules of the faster-growing ambrosia fungi outcompete other fungi, eventually filling the lumen with a pure or nearly pure fungal mass that overflows during dispersal (flight) and while boring through wood (Schneider and Rudinsky 1969a, 1969b; Schneider 1976; Roeper 1988; Kinuura, Hijii, and Kanamitsu 1991; Kinuura 2002).
4. During early gallery construction and egg laying, ambrosia fungi are expelled from the mycangium either actively or passively, inoculating the walls of the freshly bored galleries (Francke-Grosmann 1967; Schneider 1976; Kinuura, Hijii, and Kanamitsu 1991; Six 2003). Whether the gland cells deactivate only a few days after gallery foundation or remain active for a prolonged period appears to vary among species (Schneider 1976).
5. Gland cells deactivate and fungal growth diminishes, eventually resulting in an empty and deteriorated mycangium (Schneider and Rudinsky 1969a). If the beetle overwinters, the mycangium becomes fallow and other fungi or debris may reside in it (Schneider and Rudinsky 1969a; Kinuura, Hijii, and Kanamitsu 1991), but a small reserve of the ambrosia fungus may remain and resume growth to fill the mycangium again after hibernation (Schneider and Rudinsky 1969a).

The exclusion of "weed" fungi from the mycangium in stage 3 can be inferred by the pure masses of primary ambrosia fungi that are consistently observed during dispersal and gallery initiation compared to the mixed mycangium contents before the glands activate (Kajimura and Hijii 1992). Mycangia support preferential growth of primary ambrosia fungi over non-ambrosial fungi (Batra 1966) and—at least in *Xylosandrus* and their *Ambrosiella* cultivars—enforce preference for specific ambrosia fungus species over other related ambrosia fungi (Skelton et al. 2019). The process is not likely due to antifungal products in the mycangium, though Schneider (1975) did not specifically discount this possibility. Rather, it is likely due to a process termed "biological screening" (Archetti et al. 2011), in which the host creates a nutritionally ideal environment that promotes the growth of the specific mutualist over other symbionts. The mycangium cycle appears to be generalizeable for well-studied mycangia but remains to be tested for other types—particularly those smaller and more difficult to study.

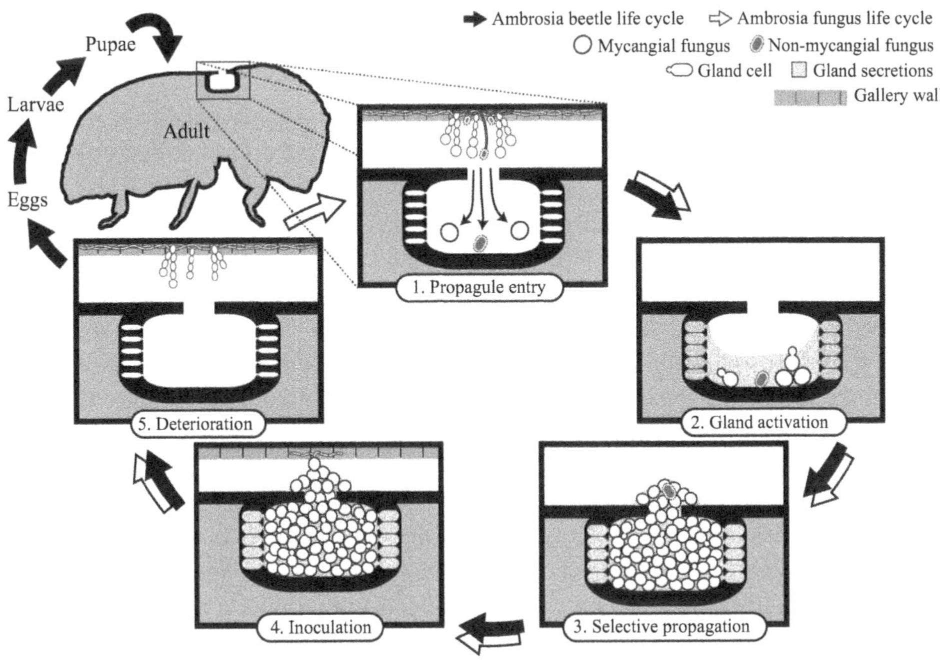

Figure 7.3
The mycangium cycle. Steps numbered as detailed in the text.

Mycangia Preserve Fungal Cultivars

Insects that depend on beneficial fungal symbionts evolve measures to maintain those valuable symbionts over time (Biedermann and Vega 2020). Compared with other insects with fungal mutualists, ambrosia beetles face a number of unique challenges in maintaining their special cultivars. First, ambrosia beetles typically must leave their galleries and establish fungus farms in new trees every generation, unlike other fungus-farming insects that maintain long-lived colonies (chapter 5, this volume). Second, ambrosia beetles coexist with a large diversity of wood microorganisms, including those that are well adapted for external transmission by bark and ambrosia beetles (Batra 1963; Francke-Grosmann 1967; Paine, Raffa, and Harrington 1997; Harrington 2005; Six 2012; Bateman et al. 2016; Skelton et al. 2018). As the beetles tunnel through the phloem to get to the sapwood, they may pick up additional superficial commensals. Third, the purity of the microflora of the gallery and mycangium changes over the course of the beetles' life stages and seasons (Batra 1963, 1966; Francke-Grosmann 1967; Schneider and Rudinsky 1969a, 1969b; Roeper 1988; Schneider 1991; Kinuura, Hijii, and Kanamitsu 1991; Kajimura and Hijii 1992; Kinuura 1995, 2002; Yang, Ye, and Zhang 2008; Ibarra-Juarez et al. 2020), and mycangia can be contaminated by mites, nematodes, and assorted debris (Batra 1966; Schneider and Rudinsky 1969b). Fourth, the mycangium may include a myriad of non-ambrosial, auxiliary passengers that travel alongside the ambrosia fungus as mycangium parasites or have secondarily beneficial roles. Yeasts (Batra 1963; Baker and Kreger-van Rij 1964; Roeper 1988; further reviewed by Mayers, Bateman, and Harrington 2018), filamentous fungi (Batra 1963; Francke-Grosmann 1967; Kinuura 2002; Biedermann et al. 2013; Bateman et al. 2016; Freeman et al. 2016; Lynch et al. 2016), and bacteria (Hulcr et al. 2012) have all been noted as auxiliary passengers.

In light of these challenges, it is remarkable that during flight and gallery establishment ambrosia beetle mycangia are almost always full of essentially pure cultures of ambrosia fungi (Francke-Grosmann 1956a, 1958; Abrahamson et al. 1967; Schneider and Rudinsky 1969a; Nakashima 1979; Gebhardt et al. 2004; Harrington, Aghayeva, and Fraedrich 2010; Harrington et al. 2014; Mayers et al. 2015, 2020a, 2020b; Mayers, Harrington, and Ranger 2017; Mayers, Bateman, and Harrington 2018) and that many ambrosia beetle lineages appear to have maintained the same fungal cultivars for many millions of years or cleanly replaced ancestral cultivars with newly domesticated or horizontally transferred ones (Farrell et al. 2001; O'Donnell et al. 2015; Vanderpool, Bracewell, and McCutcheon 2017; Mayers et al. 2020a, 2020b). Mycangia may carry small numbers of other fungi (Nakashima, Goto, and Iizuka 1987; Kinuura, Hijii, and Kanamitsu 1991; Bateman et al. 2016), but as long as the primary ambrosia fungi dominate the mycangium and the gallery inoculation, social or individual behaviors (e.g., weeding) by the beetles can minimize the presence of detrimental interlopers (Francke-Grosmann 1967; Biedermann and Taborsky 2011; Nuotclà, Biedermann, and Taborsky 2019). The paired mycangia typically observed in ambrosia beetles may be another advantage. Schneider and Rudinsky (1969b) observed ambrosia beetles captured in flight that had a nematode infesting one of the mycangia. Glands in the infested mycangium were deactivated, but the other mycangium had active glands and a pure mass of ambrosia fungus. Thus, one mycangium can be irreversibly compromised by a parasite while the other functions normally. This risk mitigation through mycangium compartmentalization is somewhat analogous to the risk mitigation provided by sequestered fungus-growing chambers in leaf-cutter ants and separated fungus combs in fungus-farming termites (Hart and Ratnieks 2001; Aanen et al. 2009; Mueller et al. 2010; Kirkendall, Biedermann, and Jordal 2015) and may also exist in the branching tunnel systems of many ambrosia beetles.

Undoubtedly, other members of the greater microbiome contribute to the biology of ambrosia beetles, whether positively, neutrally, or negatively (Baker and Norris 1968; Haanstad and Norris 1985; Hofstetter et al. 2015; Bateman et al. 2016; Vissa and Hofstetter 2017; Skelton et al. 2018; Ibarra-Juarez et al. 2020), and the ambrosia symbiosis exists within a complicated network of other interactions with gradients of benefits, specificity, and relevance. Some of the "auxiliary fungi" (Batra 1985; Mayers et al. 2020a) may supplement ambrosia beetle nutrition, as do similar fungi that supplement the nutrition of phloem-feeding bark beetles (Harrington 2005). However, ambrosia beetles have lower fecundity, survival rates, and smaller adult sizes when they feed on fungi other than their primary ambrosia fungus (Freeman et al. 2012; Saucedo et al. 2018; Carrillo et al. 2020). Ambrosia beetles have maintained fungal cultivars for tens of millions of years of coevolution over a background diversity of other microbes and frequent changes in environment. This would likely not be possible without mycangia and the mycangium cycle.

Feedbacks among Fungal Cultivars, Mycangia, and Cooperative Farming

There appear to be patterns between mycangia and levels of beetle sociality in that mycangium complexity may balance social complexity. Social evolution in insects is typically driven by foraging and collaborative nest defense (Korb and Heinze 2016), yet ambrosia beetles do neither and are among the most highly social beetles (Kirkendall, Biedermann, and Jordal 2015). Collaborative cultivar management may drive sociality in ambrosia beetles (Biedermann and Rohlfs 2017; Nuotclá, Biedermann, and Taborsky 2019). However, high sociality is found in relatively few ambrosia beetle taxa. Only one species of ambrosia beetle (*Austroplatypus incompertus*; Platypodinae) is known to exhibit true obligate eusociality with a single lifetime mother, a worker caste of unmated females, and dispersing males (Kent and Simpson 1992; Smith et al. 2018; chapter 5, this volume). Other examples of lower, but still highly social, behavior are in the haplo-diploid Xyleborini. Some species in *Xyleborus* and *Xyleborinus* practice alloparental care and cooperative breeding, with division of labor, nest-cleaning offspring, and multigenerational interbreeding (Biedermann and Taborsky, 2011; Biedermann, Peer, and Taborsky 2012; Kirkendall, Biedermann, and Jordal 2015).

In the highly social systems that have been studied, beetles have relatively small and less selective mycangia and *Raffaelea* cultivars that are generally associated with a broad range of beetle lineages. These taxa may rely on cooperative farming behavior (e.g., weeding) to establish and maintain the dominance of specific fungi over other, auxiliary fungi within their nests. In contrast, species in the *Xylosandrus* complex have relatively large, highly selective mycangia and specific *Ambrosiella* symbionts. These *Xylosandrus*-type beetles are subsocial, with no cooperative farming by adults and offspring, though adult daughters often delay dispersal from the natal nest (Bischoff 2004; Biedermann 2012).

Many other subsocial species have relatively small mycangia and non-Microascales cultivars, and this may be related to their degree of gregariousness. Some subsocial species are gregarious (all life stages interact and move freely within large communal galleries; figure 7.1 A), whereas others are solitary (each offspring develops in its own cradle with limited contact with the mother beetle; figure 7.1B) (Kirkendall, Biedermann, and Jordal 2015). Typically, inbreeding (sibling-mating) species such as the Xyleborini are gregarious, and outbreeding species are solitary (Biedermann 2012). In the outbreeding and nongregarious taxa studied in the Corthylini (*Gnathotrichus* and *Monarthrum*) and Xyloterini (*Trypodendron*) mycangial glands remain active for prolonged periods of time within the nest, thus ensuring continuous inoculation of the gallery with the primary fungus, whereas mycangial gland cells of Xyleborini deteriorate rapidly within the first days after gallery foundation (Schneider 1976). Gregarious species can control their fungal communities through weeding and have less need for active gland cells during brood development (Biedermann and Rolfs 2017).

Mycangia Influence Cultivar Choice and Flexibility and Define the Ambrosia Symbioses

Hubbard (1897a, 1897b) was the first to realize that different ambrosia beetle genera farmed different fungi, but he noted that more closely related beetles had more similar fungi. Roeper (1996, 2011) recognized distinct patterns in mycangium types and the ambrosia fungi they carry and that larger mycangia tended to carry certain fungi, later recognized as *Ambrosiella* sensu lato in the Microascales (Harrington et al. 2014; Mayers et al. 2015). Mycangia appear to be the most important driver for which fungi are associated with which beetle lineages (Mayers et al. 2020a, 2020b).

The best-studied primary ambrosia fungi are found in two orders of Ascomycota. In the Ophiostomatales, the polyphyletic *Raffaelea* (Ophiostomatales) includes the ambrosia fungi of all studied Platypodinae and most of the Scolytinae with relatively small mycangia (Harrington, Aghayeva, and Fraedrich 2010; Simmons et al. 2016a; Vanderpool, Bracewell, and McCutcheon 2017). Another Ophiostomatales, *Afroraffaelea ambrosiae*, is the ambrosia fungus of *Premnobius cavipennis* (Bateman et al. 2017). Other genera of the Ophiostomatales are common associates of bark beetles, and the two or more adaptations to ambrosia beetle cultivation that have arisen in the Ophiostomatales over the last 100 million years are likely due to convergent domestications that enabled niche shifts to the nutrient-poor sapwood from the nutrient-rich inner bark (Harrington 2005; Vanderpool, Bracewell, and McCutcheon 2017). Multiple domestications of ambrosia fungi are also apparent in the Ceratocystidaceae (Microascales) (Mayers et al. 2020a, 2020b), a family that is otherwise largely composed of bark beetle associates (De Beer et al. 2014). These ambrosial lineages probably followed evolutionary histories similar to that of the Ophiostomatales ambrosia fungi but appear to be closely tied to the evolution of relatively large mycangia in at least four tribes of Scolytinae; this, in turn, suggests that the maintenance of ambrosia fungi in this family requires (for unknown reasons) larger mycangia (Mayers et al. 2020a, 2020b). The other discovered ambrosia fungi include confirmed mycangial ambrosia fungi in the ambrosia *Fusarium* clade (AFC) (Kasson et al. 2013; O'Donnell et al. 2015; Aoki et al. 2019), *Kaarikia abrahamsonii* (Mayers et al. 2020b), and the wood decay basidiomycete *Flavodon subulatus* (Li et al. 2016, 2017; Kasson et al. 2016; Simmons et al. 2016b; Jusino et al. 2020) as well as putative ambrosia fungi in *Geosmithia* (Kolařík and Kirkendall 2010; Kolařík, Hulcr, and Kirkendall 2015) and *Diatrypella japonica* (Li et al. 2020), all of which are farmed by significantly narrower clades of ambrosia beetles (table 7.1).

Some of the broad ambrosia beetle symbioses appear to have remained remarkably stable over evolutionary time, such as *Raffaelea* cultivars and the Platypodinae (Vanderpool, Bracewell, and McCutcheon 2017), as well as most ambrosial genera of Ceratocystidaceae with their Scolytinae genera (Mayers et al. 2020a). Other symbioses appear to be examples of recent replacements of ancestral cultivars with newly domesticated cultivars (O'Donnell et al. 2015). There is evidence that ancient horizontal cultivar exchange allowed the creation of new ambrosia beetle lineages, which developed new mycangia to accommodate those cultivars (Vanderpool, Bracewell, and McCutcheon 2017; Mayers et al. 2020a, 2020b). The dozen or more separately evolved lineages of ambrosia beetles (Jordal 2015; Johnson et al. 2018, 2020) farm at least five different orders of fungi (table 7.1), with multiple domestication events within some of the fungal orders (Vanderpool, Bracewell, and McCutcheon 2017; Mayers et al. 2020a). This starkly contrasts with the single evolutionary origin of fungus farming in ants with their three fungal lineages in a single order (Schultz and Brady 2008; Branstetter et al. 2017; Mueller et al. 2018; chapter 14, this volume) and termites with their single fungal genus (Aanen et al. 2002; Aanen et al. 2009; Nobre et al. 2011; chapter 6, this volume).

Ambrosia fungi are not always species-specific in their beetle relations, as can be seen in numerous examples of *Raffaelea* (e.g., Gebhardt et al. 2004; Harrington and Fraedrich 2010; Harrington, Aghayeva, and Fraedrich 2010; Bateman et al. 2015; Li et al. 2018a; Saucedo-Carabez et al. 2018), ambrosia fungi in the Ceratocystidaceae (Lin et al. 2017;

Mayers, Bateman, and Harrington 2018; Mayers et al. 2020a, 2020b; Skelton et al. 2019), and the ambrosia *Fusarium* clade (AFC) (Carrillo et al. 2020). Rather, there are persistent, broad, mycangium-cultivar associations, and various genera of ambrosia fungi appear to have adapted to specific types of mycangia. These different mycangia types impose different levels of selection and symbiont fidelity. For example, the relatively small mycangia that carry *Raffaelea* (Ophiostomatales) can generally carry multiple *Raffaelea* species, even in a single mycangium (Harrington, Aghayeva, and Fraedrich 2010; Vanderpool, Bracewell, and McCutcheon 2017). In contrast, the relatively large mycangium types found in some tribes of Scolytinae appear to be strictly associated with a single genus of Ceratocystidaceae (Microascales) and carry pure cultures of a single species (Mayers et al. 2015, 2018, 2020a, 2020b), which may be facilitated by mycangium-mediated selection (Skelton et al. 2019).

Ambrosia symbioses can be sorted into ecological categories based on the fungi they farm, such as Ophiostomatales farmers, Ceratocystidaceae farmers, *Fusarium* farmers, and *Flavodon subulatus* farmers (table 7.1). There are likely important differences between these symbioses due to differences in the biology and ecology of the fungal cultivars. However, ambrosia fungi are difficult to isolate and identify accurately (Francke-Grossman 1956a, 1967; Baker 1963; Batra 1963, 1967, 1985; Harrington, Aghayeva, and Fraedrich 2010), about half of the known ambrosia beetle lineages lack clearly identified primary ambrosia fungi, and some have not had their mycangia characterized in detail (Hulcr and Stelinski 2017). It seems certain that there are additional undiscovered ambrosia symbioses with other diverse characteristics waiting to be studied. Even in the studied symbioses, ecologies may differ substantially among beetles that farm related fungi. Conclusions based on a single beetle-fungus system should be applied carefully to other ambrosia beetles.

Confusing the matter further is the blurred line between ambrosia beetles and their bark beetle relatives. "*Ambrosia beetle*" as defined by Hubbard (1897a, 1897b) refers to beetles that, specifically in contrast to bark beetles and other wood insects, (1) make galleries that are wholly within wood (i.e., sapwood and/or heartwood), uniform in size, and free of frass, and (2) do not eat wood but instead feed on fungi that they propagate themselves on the gallery walls. The term has been used inconsistently ever since, and modern uses of "ambrosia beetle" increasingly consider (2) while disregarding (1). Therefore, the distinction between ambrosia beetles and bark beetles is made complicated by widespread fungus feeding in bark beetles (Harrington 2005; Hulcr and Stelinski 2017; Six 2012, 2020; Six and Elser 2020). Despite practicing phloeophagy (phloem-feeding) in bark, most bark beetles also (with or without mycangia) obligately or facultatively feed on mainly ascomycetous and some basidiomycetous fungi as significant or supplemental components of their diets (Harrington 2005). These bark beetles should not be universally discounted as having less developed, less specific, or less evolved symbioses than ambrosia beetles. Their relationships with fungi exist on a gradient. At one extreme, some species of *Dendroctonus* have prothoracic mycangia that are as large and complex as the largest mycangia of ambrosia beetles (Yuceer et al. 2011; Six 2012; Bracewell and Six 2015) and possess equally developed gland cells (Barras and Perry 1971; Happ, Happ, and Barras 1971; Six 2003; Yuceer et al. 2011). These mycangia select for both basidiomycete and ascomycete fungi (which may occur simultaneously in the mycangia), which are fed upon by the developing beetles, and most are unable to disperse independently (Barras and Perry 1971; Paine and Birch 1983; Harrington 2005; Six and Bracewell 2015; Six and Elser

2019; Six 2020). If it took place in the sapwood, the *Dendroctonus*-fungus symbiosis would certainly be considered an ambrosial symbiosis. Additional types of bark beetle mycangia are found in other *Dendroctonus* species (glandular maxillary pouches or pit mycangia), in species of *Dryocoetes* and *Ips* (mandibular pockets), in *Tomicus* (long crevice between the elytra), in *Pityoborus* (shallow depressions and brushes of hairs), and in *Scolytus ventralis* (glandular pits). These mycangia all harbor, with various specificities, a variety of wood-decaying basidiomycetes and/or Ophiostomatales (Beaver 1989; Harrington 2005; Six 2012, 2020). Other bark beetles have no discernable mycangia and transmit fungi only superficially or in the gut, but they still display a range of symbioses with free-living fungi (Harrington 2005; Six 2020; Six and Elser 2020). In reality, species of bark and ambrosia beetles likely exist on a gradient between reciprocally obligate sapwood/heartwood fungus farming (ambrosia beetles) and pure phloem feeding in bark (bark beetles, though few to none have phloem diets without fungal supplementation) (Six 2012). Their strict separation into distinct groups is probably an artificial one, and a more nuanced form of nomenclature would be useful to properly classify their ecologies (Hulcr et al. 2020). Regardless, bark beetles are associated with different fungi and have sufficiently different ecologies to consider ambrosia beetles and the ambrosia symbioses as separate and distinct from them.

Types of Ambrosia Beetle Mycangia

Mycangia were originally categorized by location on the beetle—for example, mesonotal mycangium or pronotal mycangium (Francke-Grosmann 1963, 1967). However, these anatomic terms are too imprecise to differentiate different mycangia that develop in the same body part, such as the multiple prothoracic and pronotal mycangia (table 7.1). Recently, we (C.G.M. and T.C.H.) have referred to some of the mycangium types by their location and a simple noun, as in "mesonotal pouch mycangium" (Mayers et al. 2015, 2017). Similar names are employed here for all known types of mycangia.

Pronotal Pit Mycangia

These mycangia (figures 7.2A, 7.4A, B) are small, scattered pits, pores, or semi-spherical cavities of the exoskeleton on the top of the pronotum found in many genera of Platypodinae (Beeson 1917; Roche and Lhoste 1960; Lhoste and Roche 1961; Farris and Funk 1965; Francke-Grosmann 1967; Browne 1971; Nakashima 1971, 1972, 1975, 1979, 1982; Wood 1993; Cassier et al. 1996; Beaver 1998; Beaver and Liu 2007, 2018; Kent 2008, 2010; Moon et al. 2008, 2012; Belhoucine et al. 2013; Beaver and Sanguansub 2015; Beaver 2016; Li et al. 2018b; Mueller 2019). As described in these references, but reviewed in detail by Nakashima (1975) and further described in later studies (Beaver 1998; Wood 1993; Beaver and Liu 2007; Beaver and Sanguansub 2015), the number, size, position, and arrangement of pits vary greatly within genera and even within species (figure 7.4B). For example, *Oxoplatypus quadridentalus* has only two large pits that are each large enough to hold many spores (Li et al. 2018a), whereas *Austroplatypus incompertus* has as many as 80 tiny pits that each hold a single spore (Kent 2008), and species of *Platypus* can have as many as 650 tiny pits (Nakashima 1975). The pit arrangement can be asymmetrical (Beaver and Sanguansub 2015) or a mix of dramatically differently sized pits (Nakashima 1975). Unlike mycangia of other ambrosia beetles, the pits may be present in both sexes but larger in the female and reduced in the male, or present only in the female, though the male typically initiates gallery construction (Farris and Funk 1965; Francke-Grosmann 1967; Kent 2008; Moon et al. 2008; Belhoucine et al. 2013). Pronotal pit mycangia can be well developed, with gland cells that feed secretions through small tubules to each pit embedded in the exoskeleton (Lhoste and Roche 1961; Farris and Funk 1965; Cassier et al. 1996; Kent 2008). Sometimes the pits are accompanied by a single hair that holds the fungal propagules in place (Kent 2008), and some of these hairs are quite elaborate, such as the fan-like pit hairs of *Diapus quinquespinatus* (Nakashima 1975). These pits are clearly distinct from the so-called nonglandular pit mycangia that are found on many bark and ambrosia beetles and that may or may not have important roles in fungal transport (Six 2003). Many Platypodinae genera, and individual species within genera, have no pronotal pits at all (table 7.1), which suggests that pronotal pits evolved or were lost multiple separate times. Such species probably have other mycangia or transmission methods. *Raffaelea* species have been consistently associated with all studied Platypodinae, including many species of Platypodini (Guerrero 1966; Batra 1967, Cassier et al. 1996, Harrington, Aghayeva, and Fraedrich 2010; Vanderpool, Bracewell, and McCutcheon 2017; Li et al. 2018a; Mueller 2019) and only one species of *Notoplatypus* in Tesserocerini (Mueller 2019), but many genera remain unstudied (table 7.1). Several genera that lack pronotal pit mycangia previously said to be members of the Platypodinae by Wood (1993), e.g., *Coptonotus, Mecopelmus, Protohylastes, Protoplatypus, Schedlarius*, and *Scolytotarsus*, are now known not to be platypodines or ambrosia beetles at all (R. Beaver, pers. comm.).

Cranial Sphere Mycangia

This unique mycangium (figure 7.4C), a space between a spherical body of tissue in the center of the head and a thin membrane wrapped around it, has been described in only one species, *Crossotarsus niponicus* (Platypodinae) (Nakashima 1971, 1975, 1979, 1982). The space, referred to as an "epistomal cavity" by Hulcr and Stelinski (2017), is connected broadly to the mouth cavity. Before it flies, the beetle secretes and fills its mouth with a mucous-like substance to protect the spores inside (Nakashima et al. 1979). Glandular tissue can be found directly behind the mycangium (Nakashima et al. 1979). Cranial sphere mycangia might also exist in other *Crossotarsus* and in the related *Doliopygus*, which do not have pronotal pit mycangia (R. Beaver, pers. comm.), but require examination. *Raffaelea* ambrosia fungi have been isolated from another *Crossotarsus* species (Gebhardt and Oberwinkler 2005) as well as by Musvuugwa et al. (2015) from a *Doliopygus* species, as later identified by B. Jordal (pers. comm.). This suggests that the thick white ambrosia observed in galleries of *C. niponicus* (Nakashima, Goto, and Iizuka 1987) might also be *Raffaelea*.

Dual Preoral Pocket Mycangia

This mycangium type (figures 7.2I, 7.4F), previously called "oral mycangia" (Francke-Grosmann 1967) and sometimes inaccurately called "mandibular mycangia" as discussed

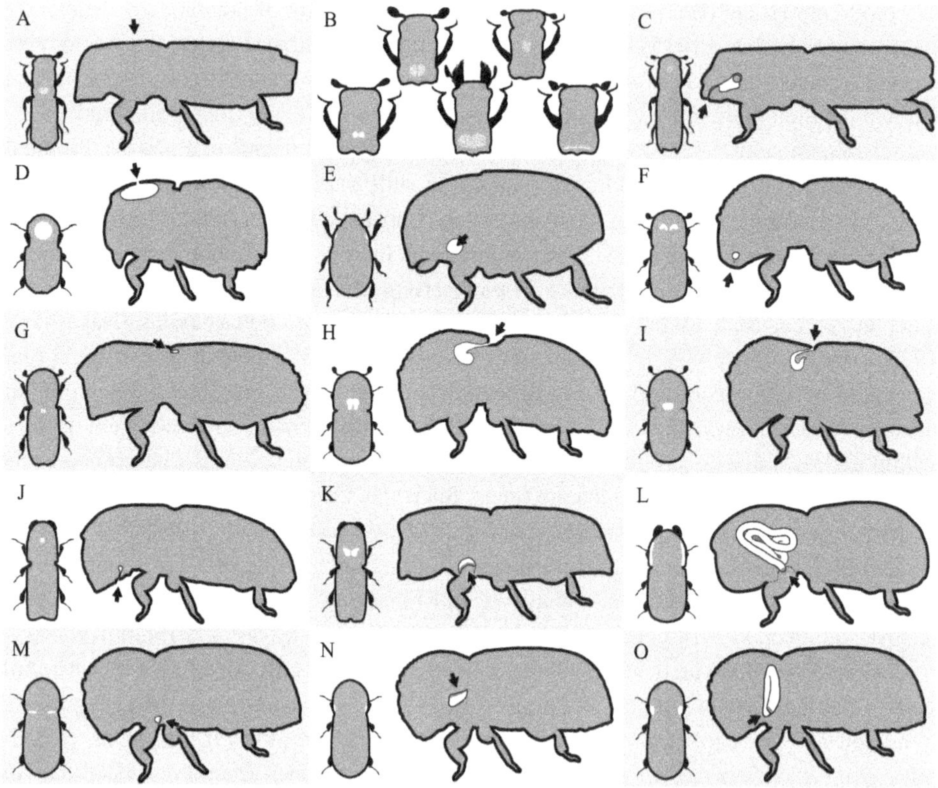

Figure 7.4
Beetle silhouettes showing representative mycangium types in white. (A) Pronotal pits (*Platypus*). (B) Pronotal pit arrangements. From left to right: *Oxoplatypus quadridentatus, Platypus cylindrus, Ganyocerus papuanus, Austroplatypus incompertus, Diapus quinquespinatus* (accompanied by pronotal crevices). (C) Cranial sphere (*Crossotarsus niponicus*). (D) Pronotal disk (*Scolytoplatypus*). (E). Setose patch (*Phloeoborus*). (F). Dual preoral pockets (*Euwallacea*). (G) Elytral notches (*Xyleborinus*). (H) Mesonotal pouch, lobed (*Xylosandrus*). (I) Mesonotal pouch, unlobed (*Anisandrus*). (J) Single oral pocket (*Monarthrum bicallosum*). (K) Procoxal; dotted black line represents procoxal cavity (*Monarthrum*). (L) Prothoracic tubes; dotted black line represents procoxal cavity (*Corthylus*). (M) Prothoracic basins (*Xyloterinus politus*). (N) Pleural cavities (fin type) (*Indocryphalus*). (O) Pleural cavities (horseshoe type) (*Trypodendron*). Beetle body silhouettes are traced from photographs of whole beetles; legs are generic illustrations. Black arrows point to the locations of mycangia and generally follow the direction of propagule entry. Mycangium shapes and locations are approximate.

by Hulcr and Stelinski (2017), are small pockets on both sides of the esophagus near the mouth (Beaver 1989; Hulcr et al. 2015; Hulcr and Stelinski 2017). These pockets are found in many genera of Xyleborini, including *Amasa*, *Ambrosiodmus*, *Ambrosiophilus*, *Cyclorhipidion*, *Euwallacea*, and *Xyleborus* (Fernando 1959; Francke-Grosmann 1963, 1967; Schedl 1962, 1963; Takagi 1967; Nakashima 1982; Schneider 1987; Hulcr et al. 2007; Fraedrich et al 2008; Kasson et al. 2013, 2016; Li et al. 2015; Freeman et al. 2016; Hulcr and Stelinski 2017; Jiang et al. 2019). Similar, but slightly differently positioned, oral mycangia convergently evolved in *Premnobius cavipennis* (Ipini) (Schedl 1962; Bateman et al. 2017) and in both sexes of *Xyloterinus politus* (Xyloterini) (Abrahamson and Norris 1966). Some *Euwallacea* have been shown to have two pairs (four total) of preoral pocket mycangia (Goto 1998; Spahr, Kasson, and Kijimoto 2020). *Raffaelea* species are the most common inhabitants of preoral pocket mycangia (Harrington, Aghayeva, and Fraedrich

2010; Dreaden et al. 2014; Bateman et al. 2015; Simmons et al. 2016a; Mayers et al. 2020b), but exceptions include some *Euwallacea* species with *Fusarium* ambrosia fungi (AFCs) (Freeman et al. 2013; O'Donnell et al. 2015; Kasson et al. 2016; Aoki et al. 2018, 2019), *P. cavipennis* with *Afroraffaelea ambrosiae* (Bateman et al. 2017), and *Ambrosiodmus/Ambrosiophilus* with *Flavodon subulatus* (Kasson et al. 2016; Jusino et al. 2020).

Elytral Notch Mycangia

Previously called "elytral mycangia" (Francke-Grosmann 1963), elytral notch mycangia (figures 7.2C, 7.4G) are small notches in the leading edges of the elytra and are fenced by protective hairs and lined with gland cells. Elytral notch mycangia have been identified in the *Xyleborinus* (Xyleborini) species *X. saxesenii*, *X. gracilis*, *X. schreineri*, and *X. sentosus* (Francke-Grosmann 1956a, 1956b, 1967; Schedl 1962; Hulcr et al. 2007; Biedermann, Klepzig, and Taborsky 2011) and inferred in *X. attenuatus* (Gharabigloozare 2015). They have also been observed in *Taurodemus* (C. Bateman, unpublished observation). They are presumably present in other *Xyleborinus* species, and are apparently present in other xyleborine genera (*Coptodryas*, *Cryptoxyleborus*, *Heteroborips*, and *Microperus*) but not yet studied in detail (Hulcr et al. 2007; Hulcr and Stelinski 2017; Mandelshtam et al. 2019). The mycangia appear to vary in size, as those of *X. gracilis* (Schedl 1962) appear to be much larger than those of *X. saxesenii* (Francke-Grosmann 1956a, 1956b, 1967). All studied elytral notch mycangia have carried *Raffaelea* ambrosia fungi (Biedermann et al. 2013; Bateman et al. 2015; Gharabigloozare 2015; Simmons et al. 2016a). The larger mycangia of *X. gracilis* can carry a variety of *Raffaelea* ambrosia fungi (Harrington, unpublished data). In contrast, the smaller mycangia of *X. saxesenii* appear to always carry a specific species, *R. canadensis*, but *X. saxesenii* also has a second ambrosia fungus, *R. sulphurea*, which does not form a yeast state in culture and is uniquely gut-transmitted (Francke-Grosmann 1975; Biedermann et al. 2013; Gharabigloozare 2015).

Mesonotal Pouch Mycangia

These mycangia (figures 7.2F–H, 7.4H, I), previously called "pro-mesonotal mycangia" (Francke-Grosmann 1963) or "mesonotal mycangia," are large, elastic cavities in the prothorax formed by invaginations of the flexible membrane that connects the top side of the prothorax to the scutellum (elytra hinge) (Francke-Grosmann 1956a, 1956b, 1967; Lhoste and Roche 1959; Schedl 1962; Stone et al. 2007; Li et al. 2019). This mycangium type is found in all genera of the monophyletic *Xylosandrus* complex except *Diuncus* (Johnson et al. 2018; Skelton et al. 2019), and mycangium anatomy differs among them. In *Xylosandrus* the mycangium branches into two large lobes (on the left and right) connected by a "mycangium bridge" (figure 7.4H) and the majority of the organ is in front of the scutellum (Francke-Grosmann 1956a, 1956b, 1958, 1963; Lhoste and Roche 1959; Schedl 1962; Kaneko 1967; Harrington et al. 2014; Mayers et al. 2015; Li et al. 2019). The thin, elastic mycangium membrane expands as the beetle matures and the mycangium fills with growing fungal propagules (Li et al. 2019). The mycangium of *Eccoptopterus* similarly has dual lobes (Mayers et al. 2015), but the mycangium bridge and scutellum are tightly wound into a spiral (Francke-Grosmann 1958, 1963, 1967). The mycangium of *Cnestus* also has dual lobes, but the mycangium bridge is placed farther back under the scutellum and spirals slightly, and the mycangium wall appears stiff and reticulated (Stone et al. 2007). In *Anisandrus* the mycangium is unlobed (figure 7.4I), much smaller, and attached directly to the bottom of the scutellum (Francke-Grosmann 1956a, 1956b, 1958, 1967; Mayers, Harrington, and Ranger 2017). Gland cells are found on the portions of the scutellum abutting the mycangium lumen (Francke-Grosmann 1963, 1967), but in *Cnestus mutilatus* the gland cells are also interspersed across the entire reticulated wall of the organ (Stone et al. 2007). The mycangium appears to have been lost or reduced in *Diuncus*, likely as a result of specialization on fungus-stealing from nearby galleries of other ambrosia beetles (Hulcr and Cognato 2010; Skelton et al. 2019). *Ambrosiella* (Ceratocystidaceae) species are consistently the ambrosia fungi carried in mesonotal pouch mycangia (Mayers et al. 2015, 2017; Lin et al. 2017; Skelton et al. 2019; Nel et al. 2020).

Single Oral Pocket Mycangia

In contrast to the dual oral pocket mycangia found in many ambrosia beetles, a single oral pocket (a "pregular" pocket; Hulcr and Stelinski 2017) positioned in the center of the head (figure 7.4J) was described by Schedl (1962) in *Pterocyclon bicallosum* (now *Monarthrum bicallosum*, Corthylina; Wood and Bright 1992). The tissue below the mycangium is lined with gland cells (Schedl 1962). Schedl specifically did *not* find procoxal mycangia in this species, as he did in other *Monarthrum*. This unusual report suggests either that there is unexplored diversity of mycangia in *Monarthrum* or that *M. bicallosum* is misplaced in the genus, which should be investigated further. The ambrosia fungus carried in this mycangium type is unknown.

Coxal Mycangia

These mycangia (figure 7.4K) are broad enlargements of the coxal cavities (leg sockets) in which fungus spores persist as a "loosely compressed hemispherical cake" (Farris 1965). Coxal mycangia are found in the front coxae (procoxae) in the Corthylini genera *Gnathotrichus* (Farris 1963; Francke-Grosmann 1963, 1967; Schneider and Rudinsky 1969a, 1969b) and *Monarthrum* (Schedl 1962; Farris 1965; Francke-Grosmann 1963; Lowe, Giese, and McManus 1967), where they were previously called "prosternal-subcoxal mycangia" (Francke-Grosmann 1963), and in both the front and middle (mesocoxae) in some Platypodinae (Nakashima 1972, 1975, 1982). In both Corthylini genera, the cavities are lined with gland cells (Schedl 1962; Farris 1965; Schneider and Rudinsky 1969a, 1969b), and in *Gnathotrichus* the glands feed secretions into the cavities via small channels (Schneider and Rudinsky 1969b). The mycangia are in males of *Gnathotrichus* but females of *Monarthrum*, and are more toward the front in *Monarthrum* (Schneider and Rudinsky 1969b); that these genera may not be sister clades (Gohli et al. 2017; Johnson et al. 2018; Pistone, Gohli, and Jordal 2018) implies that the mycangia may have evolved separately. Both Corthylini genera have *Raffaelea* ambrosia fungi (Farris 1965; Funk 1970; Batra 1967; Harrington, Aghayeva, and Fraedrich 2010).

Prothoracic Tube Mycangia

These mycangia (figures 7.2D, E, 7.4L) are long, winding or spiraling, hose-like tubes in both sides of the prothorax that exit into the front leg sockets of male *Corthylus* (Schedl 1962; Finnegan 1963; Giese 1967; Nord 1972; Orañegui and Atkinson 1984; Mayers, Bateman, and Harrington 2018) and *Microcorthylus* (Schedl 1962). The mycangium wall is intricately reticulated (Finnegan 1963; Mayers, Bateman, and Harrington 2018) and presumably holds gland cells, but they have not been characterized. Large tubular tracheae in the vicinity of the mycangia can easily be mistaken for the mycangium itself, but tracheae are ribbed, branching, and flexible rather than reticulated and rigid. The anatomy and degree of winding of the coiling tube differs among *Corthylus* species (Mayers, Bateman, and Harrington 2018), but they probably evolved from less elaborate invaginations of enlarged coxal cavity mycangia similar to those of *Gnathotrichus*, whose mycangia are placed in a similar position and are also found exclusively in males. The ambrosia fungi of *Corthylus*, however, are species of *Meredithiella* (Ceratocystidaceae) (Mayers, Bateman, and Harrington 2018) rather than the *Raffaelea* found in the simpler mycangia of *Gnathotrichus*.

Prothoracic Basin Mycangia

Xyloterinus politus (Xyloterini) is unique among Scolytinae in having two distinctly different types of mycangia. Both sexes have dual oral pocket mycangia, but females also have prothoracic basin mycangia (figures 7.2J, 7.4M), which are shallow elongated bowl-like depressions on the rear, bottom side of the prothorax and are rimmed with protective fences of hairs (Francke-Grosmann 1963, 1967; Abrahamson and Norris 1966; MacLean and Giese 1968; Mayers et al. 2020b). These prothoracic basins may be a homologous and reduced form of the pleural cavity mycangia of other Xyloterini (Mayers et al. 2020b), but the basins are located on different parts of the prothorax and are shallow basins rather than true internal cavities. The glandular nature of prothoracic basin mycangia has not been explored, and active growth was not seen in mycangial propagules of the ambrosia fungus they carry, *Kaarikia abrahamsonii* (Abrahamson and Norris 1969; Mayers et al. 2020b), but a distinct matrix forms in the basins that holds the fungal propagules in place (R. Roeper, pers. comm.).

Pleural Cavity Mycangia

Previously called "prothoracic pleural mycangia" (Francke-Grosmann 1963), pleural cavity mycangia are cavities in the pleura (left and right sides of the prothorax) of *Trypodendron* and *Indocryphalus* in the Xyloterini. The entrances to these cavities are small slits fenced with hairs on both sides of the prothorax, and although these slits can be either vertically or horizontally oriented, referring to the mycangia themselves as "vertical" or "horizontal" (e.g., Cognato et al. 2015; Johnson et al. 2020) is incorrect since—unlike the "basin" mycangia of *X. politus*—the internal orientation of the mycangium does not necessarily correlate with the orientation of its entrance. Two types of pleural cavity mycangia are known.

"Horseshoe"-type pleural cavity mycangia of *Trypodendron* (figures 7.2K, 7.4O) are inverted-U-shaped cavities that begin at horizontal slit entrances, rise upward, turn, and travel downward, generally ending in a tight "crook" (Nunberg 1951; Francke-Grosmann 1956a, 1956b, 1958, 1959; Abrahamson et al. 1967; Schneider and Rudinsky 1969a; Nakashima 1982; Lehenberger et al. 2019). The cavities in *Trypodendron* have been called "tubes," "tubules," or "tubular," but they are not uniformly cylindrical like the mycangia of *Corthylus* (Corthylini). The walls of the pleural mycangia are hard and reticulated (Francke-Grosmann 1956a; Abrahamson et al. 1967) and lined with gland cells (Schneider and Rudinsky 1969a). Horseshoe pleural cavity mycangia are found in every studied species of *Trypodendron* and vary somewhat in the size and shape of the ascending portion of the cavity, how extensively the terminal crook curls, and the distance from entrance to crook (Francke-Grosmann 1956a; Lehenberger et al. 2019). The entrance to the mycangium opens and closes via movement of the leg muscles (Francke-Grosmann 1963). *Trypodendron* mycangia exclusively carry *Phialophoropsis* ambrosia fungi (Lehenberger, Biedermann, and Benz 2019; Mayers et al. 2020b).

"Fin"-type pleural cavity mycangia (figure 7.4N) are like front pants pockets, hanging down in a fin-shaped, tapered cavity from wide, thin horizontal entrances. Despite having entrances in somewhat similar locations (Wood 1957; Beaver 2000; Cognato et al. 2015), fin-type pleural mycangia are anatomically distinct from and much smaller than the horseshoe type pleural mycangia of *Trypodendron*. The fin pleural cavity mycangium has been reported only in *Indocryphalus pubipennis* (Xyloterini), where it carries a *Toshionella* ambrosia fungus (Mayers et al. 2020b). Other *Indocryphalus* with horizontal mycangium openings (Cognato et al. 2015) likely have similar mycangium anatomy and ambrosia fungi. However, *Indocryphalus* is probably polyphyletic (Gohli et al. 2017; Pistone, Gohli, and Jordal 2018; Johnson et al. 2020), and species with vertical mycangium openings may have prothoracic basin mycangia like *X. politus* or entirely different mycangia and different mutualists.

Pronotal Disk Mycangia

Found in both genera of the tribe Scolytoplatypodini are large disk- or saucer-shaped cavities in the top of the prothorax (figure 7.4D), which have a single blowhole-like pore where the spores enter and exit (Berger and Cholodkovsky 1916; Schedl 1962; Nakashima 1982; Beaver and Gebhardt 2006; Mayers et al. 2020a). The organ is surrounded by a multitude of gland cells, and the inside of the organ may be filled with cones and/or special hairs that grow from the inside surface of the cavity and support the organ's shape, help guide spores to the exit, and/or help bathe mycangial spores in glandular secretions (Schedl 1962; Mayers et al. 2020a). Tribe Scolytoplatypodini may represent the earliest ambrosia beetle symbiosis in the Scolytinae, and since both genera share the same mycangium type, pronotal disk mycangia may have been the first scolytine mycangia to evolve (Jordal 2013; Pistone, Gohli, and Jordal 2018; Mayers et al. 2020a). The cones and hairs differ significantly among the three major lineages of the tribe (Schedl 1962; Mayers et al. 2020a), which correspond to three different groups of ambrosia fungi in the Ceratocystidaceae (table 7.1).

Setose Patch Mycangia

A number of unrelated beetles have "setose patch" mycangia (Hulcr and Stelinski 2017) that are conspicuous yellowish patches of dense setae on both sides of the pronotum (figures 7.2B, 7.2L, 7.4E). The setae are generally found in shallow, wide depressions and may serve the purpose of keeping spores inside the depressions rather than collecting spores that stick to the setae themselves. The setose patch mycangia of *Phloeoborus* (figures 7.2B, 7.4E; Hulcr and Stelinski 2017) were first described by Nunberg (1951) as having glands that fill the depressions with an unknown liquid, and the mycangia can be observed to overflow with fungal growth (figure 7.2B) presumably supported by these secretions. Though they live in xylem, *Phloeoborus* are not confirmed to be ambrosia beetles, as they are not thought to feed exclusively on fungi but instead to use fungal symbionts to make boring easier and to supplement their wood diet (Wood 2007). Setose patch mycangia in other genera are generally larger (in proportion to the beetle's size) than those of *Phloeoborus*. The setose patch mycangia of female *Hypothenemus curtipennis* (Beaver 1976, 1986; Vega, Infante, and Johnson 2015) have been observed to hold enough fungal propagules to appear white (Beaver 1986). No other species of *Hypothenemus* (or the entire tribe Cryphalini) is known to be ambrosial or to have mycangia (Vega, Infante, and Johnson 2015). Certain species in the genera *Bothrosternus*, *Eupagiocerus*, and *Cnesinus* (tribe Bothrosternini) have similar setose patch mycangia (figure 7.2L) (Hulcr and Stelinski 2017), but none of these genera is composed solely of ambrosia beetles, and the ambrosial status of certain species remains ambiguous (Wood 2007). *Scolytodes culcitatus* has similar impressions covered in dense hairs on both sides of the pronotum, but besides *S. unipunctatus* the genus is otherwise not known to be ambrosial (Jordal and Kirkendall 2019). All of these setose patch mycangia are very similar to those found in *Pityoborus* bark beetles (Furniss et al. 1987). It is unclear whether all setose patch mycangia are associated with glands as in *Phloeoborus*, and none of the fungi carried by these mycangia have been characterized.

Other Understudied Mycangia

Unique mycangia are found on the gena ("cheeks" on both sides of the head) of *Camptocerus* (Scolytini), which are fungus-farmers (Wood 2007). The mycangia come in three types. "Crescent" types are shallow setae-lined furrows, whereas "circular" and "triangle" types are patches of setae in the respective shapes (Smith and Cognato 2010; Smith 2013; Petrov 2017). These mycangia can perhaps be collectively referred to as "genal brush" mycangia. The three types may correspond with multiple evolutionary origins of fungus-farming in the genus (Cognato, Smith, and Jordal 2021).

All known females of *Dactylipalpus* species (Hylesinini) have long transverse crevices in the front side of the pronotum (Blandford 1896; Nunberg 1951; Wood 1961; Hulcr and Stelinski 2017), which were hypothesized to be mycangia by Nunberg (1951) but remain uninvestigated. The ecology of *Dactylipalpus* is unknown but is likely the same as *Phloeoborus* (R. Beaver, pers. comm.). Several Platypodinae, including species of *Diapus*, also have (usually in addition to pronotal pit mycangia) long, deep transverse crevices on the top of the prothorax (figure 7.4B) (Nunberg 1951; Nakashima 1975, 1982; Beaver 1989, 2002).

Some simple cuticular pits are hypothesized to transport fungi but are known or assumed to lack glands. Neither the presence of an exoskeleton cavity in a mycophagous beetle nor the circumstantial presence of fungal spores in such a cavity is adequate evidence for its utility as a mycangium (Grebennikov and Leschen 2010), and such "nonglandular mycangia" (though few have been proven to lack glands) have sometimes been called "pseudomycangia" (Stone et al. 2007) or excluded as mycangia (Hulcr et al. 2015; Hulcr and Stelinski 2017). Simple nonglandular pits may transmit fungal propagules in a similar way to sporothecae in mites or other animals (Moser 1985; Ebermann et al. 2013; Vissa and Hofstetter 2017), but they are not as likely to allow for selective transmission of specialized fungal symbionts without glandular secretions. Some Platypodinae have nonglandular pits on the pronotum, especially when the other sex has glandular pronotal pit mycangia (Lhoste and Roche 1961; Nakashima 1975). *Scolytodes unipunctatus*, which has a *Raffaelea* ambrosia fungus, has shallow, nonglandular pits and no other known mycangia (Hulcr, Kolařík, and Kirkendall, 2007; Kolařík and Hulcr 2009). As one of two putative ambrosia beetles in its genus (Jordal and Kirkendall 2019), its biology needs further study (Hulcr and Stelinski 2017). *Cnestus mutilatus* (Stone et al. 2007) and *Anisandrus* spp. (Mayers, Harrington, and Ranger 2017) have nonglandular pits in the scutellum that have been observed to hold ambrosia propagules, but these species also have mesonotal mycangia that carry *Ambrosiella* symbionts. *Corthylus zulmae* has pits on the front side of the coxae near the mycangium entrance that were observed to hold spores (Penagos 2015), but this species likely also has the prothoracic tube mycangia typical for the genus.

Mycangia have not yet been described for some known or assumed ambrosia beetles (Hulcr and Stelinski 2017). As discussed above, many Platypodinae species have no known mycangia. Another example is *Sueus* (Hyorrhynchini), which, despite having no known mycangia (Hulcr and Stelinski 2017), display obvious ambrosial habits (Beaver 1984) and have galleries consistently dominated by the putative ambrosia fungus *Diatrypella japonica* (Ascomyocta: Xylariales) (Li et al., 2020). Classically, mycangia were discovered by careful microtome sectioning (figure 7.2J, K) or hand dissections (figure 7.2F, H), but new techniques such as noninvasive micro-CT scans and X-ray computed tomography (figure 7.2E, G, I; Bateman et al. 2017; Li et al. 2018b, 2019; Mayers, Bateman, and Harrington 2018; Jiang et al. 2019; Spahr, Kasson, and Kijimoto 2020) will likely aid in identifying new mycangia.

Mycangia Facilitate the Convergent Evolution of Agriculture

As discussed in the other chapters of this volume, agriculture presents challenges that have been faced and convergently overcome by other fungus-farming insects and by humans. Chief among these challenges is crop domestication (chapters 2 and 10, this volume), which requires changes in the cultivar that benefit the farmer but would reduce the fitness of the cultivar in its natural environment. Farmers must discourage selfish evolution in their cultivars, because the development of intracultivar genetic diversity could lead to a reduction in cultivar benefits and a breakdown of the symbiosis (chapter 4, this volume). Farmers can avoid selfish evolution by reducing the genetic diversity inherent in the cultivar and/or reducing reproductive independence (i.e., sexual reproduction) in the cultivar (chapter 5, this volume).

Mycangia allow individual ambrosia beetles to preserve their valuable cultivars and transport them to new trees. Whether the ancestral associations between ambrosia beetles and ambrosia fungi began with the fungi hitchhiking on beetle surfaces (transmission-first model) or the beetles feeding on commensal wood fungi (consumption-first model) (Six 2012), the original pressure for mycangia to evolve was probably simply to improve fungus transmission. What appears to be an inevitable side effect of this improved transmission is the facilitation of coevolution and, eventually, domestication. In ambrosia beetles, cultivar selection may be facilitated by mycangia, which enforce vertical transmission and produce genetic bottlenecks in every generation via the mycangium cycle (figure 7.3). Young adult beetles take a small number of asexual fungal propagules into their mycangia from the ambrosia lining their natal galleries (Schneider 1976; Beaver 1989), and their mycangia are mitotically filled with the same fungal genotype originally introduced to the galleries in the mycangia of their parents. This links the ancestry of ambrosia beetle farmers and their fungal cultivars through vertical inheritance, which allows for progressive domestication of the fungi over many generations. Direct kin selection may occur by favoring dominant cultivar genotypes before galleries are established, even if rare horizontal transfer of cultivars between colonies has occurred. Direct kin selection in the mycangium has been demonstrated in the *Xylosandrus* complex, where mycangia can select for specific cultivar species when multiple species are present (Skelton et al. 2019). Indirect evidence for kin selection is seen in the ambrosia fungus lawns that appear to be clonal in both native and introduced populations of ambrosia beetles (Ito and Kajimura 2017; Van de Peppel, Aanen, and Biedermann 2018), and ambrosia beetles maintain coarse mycangium-cultivar association patterns on a global scale (Mayers et al. 2015, 2018, 2020a, 2020b; Skelton et al. 2019). Predominantly vertical inheritance with rare horizontal exchange is also seen in fungus-farming ants (Howe, Schiøtt, and Boomsma 2019) but stands in contrast to the frequent horizontal transfer of fungus-farming termite cultivars (Van de Peppel and Aanen 2020).

Ambrosia fungi were long assumed to be strictly asexual lineages, which would eliminate the risk of sexual recombination and the possible development of selfish fungal cultivars that threaten the symbiosis. However, sexual states (Mayers, Harrington, and Ranger 2017, 2020a; Jusino et al. 2020) and genetic evidence for sex (Wuest et al. 2017) have since been discovered in mycangial ambrosia fungi, and many ambrosia fungi may practice rare sexual reproduction. Regardless, the sexual states of some ambrosia fungi may not be able to disperse freely because their sticky-spored sexual states are trapped in tunnels (Mayers, Harrington, and Ranger 2017; Mayers et al. 2020a) and are not as readily dispersed as the wind-dispersed sexual spores in the other fungus-farming systems (Aanen 2006). In the case of some *Ambrosiella* species, the ascospores are enclosed in sexual fruiting bodies fed upon by the beetles, and the ascospores are likely passed through the gut for dispersal, rather than dispersed via the mycangia (Mayers, Harrington, and Ranger 2017). An exception is *Flavodon subulatus*, an ambrosia fungus that produces wind-dispersed sexual spores on the outside of wood (Jusino et al. 2020), but these spores disperse separately from asexual mycangium inoculum. Even if sexual reproduction did rarely lead to mixed genotypes in the galleries of the next generation of beetles, the brood would again acquire only a few asexual spores of the dominant genotype in the gallery, and their mycangia would select for the dominant genotypes growing in the mycangium. Therefore, the risk of sexual reproduction in ambrosia fungi is not as great of a threat as that posed by sexual reproduction and wind-dispersed sexual spores in the cultivars of fungus-growing ants and termites.

Humans frequently exchange cultivars horizontally (chapter 10, this volume), but they convergently learned how to domesticate crops through successive bottlenecks. From the origin of their agriculture ~10,000 years ago to the modern age, humans have rapidly domesticated a variety of crops by selecting only a few individuals with desirable traits to produce subsequent generations (Doebley, Gaut, and Smith 2006). The closest analogue in human agriculture to fungal domestication by ambrosia beetles is probably the domestication of clonally propagated plants (McKey et al. 2010; chapter 10, this volume). This selection of a few individuals to clonally propagate the next crop planting is analogous to passage through mycangia. Besides selecting for rapid growth and high yield, humans intelligently select cultivars based on evaluated characteristics such as taste, appearance, or nutritional value (i.e., artificial selection) over short time scales a fraction of a single individual's lifespan, and this allows rapid responses to local environmental conditions (chapter 5, this volume). In contrast, each ambrosia beetle generation is committed to a single strain of ambrosia fungus, which started from clonal or practically clonal inoculum asexually reproducing in the mycangium. Even if meaningful variation in cultivar quality existed within a single gallery, the beetles likely cannot make meaningful selections for superior cultivars. Ambrosia beetles are not known to segregate cultivar genotypes within their galleries, and probability of propagule entry into mycangia is likely based on the type of sporulation and abundance of spores in the gallery.

The mycangium cycle selects for mycangium growth and the overflow of propagules from the mycangium, not necessarily superior cultivars. Therefore, although mycangia enforce coarse phylogenetic congruence between beetle and fungus through adaptations for mycangial growth (Skelton et al. 2019), there is also natural selection for ambrosia beetle lineages that successfully carry superior cultivars. Rather than selection within individual galleries, cultivar variation would exist across populations, with cultivars differing slightly through genetic drift and imparting slightly different levels of fitness on their farmers. Ambrosia beetles may have a much less direct influence over their cultivars than do humans, but mycangia force the clonal propagation that avoids the selfish evolution of their cultivars. The effectiveness and necessity of mycangia are evidenced by the numerous times mycangia and fungus farming have evolved, and further by the persistence of these coevolving partnerships for tens of millions of years.

The multiple beetle lineages that adapted to fungus farming and the separate domestications of their fungal cultivars make it difficult to derive broad conclusions about the ambrosia beetle-fungus symbiosis. Rather, the "ambrosia symbiosis" should instead be considered many distinct ambrosia *symbioses* (Hulcr and Stelinski 2017), where each distinct symbiosis (beetle, mycangium, and fungus) is appreciated as a particular system with unique ecological characteristics. In this respect ambrosia symbioses are more similar to the many convergently evolved agricultural systems of humans, which involve a wide variety of unrelated plant lineages, than to the other insect fungiculturalists that are specialized on only one or a few lineages of fungi. Regardless, one conclusion that can generally be drawn for all ambrosia beetles is that mycangia are essential drivers of their ecology and evolution, and mycangia should be appreciated for their role in not only facilitating, but maintaining, these incredible symbioses.

Acknowledgments

Roger Beaver, Richard Roeper, and James Skelton graciously read preliminary drafts of the manuscript and gave detailed and valuable critical feedback and suggestions for revisions. Diana Six provided valuable conceptional input and advice on bark beetles. Maximilian Lehenberger, James Bickerstaff, and Cassandra Wattenburger kindly read the text and provided additional helpful feedback. Sarah Smith and You Li gave useful notes on *Camptocerus* and Platypodini, respectively. Craig Bateman, You Li, Robert Mueller,

Lawrence Abrahamson, Luana de Souza Covre, and Bjarte Jordal provided unpublished images and observations included in the chapter. Beetle silhouettes were traced with permission from the high-resolution photographs of Jiri Hulcr and Andrew Johnson (http://ambrosiasymbiosis.org/parade/) and Tom Atkinson (http://www.barkbeetles.info) as well as photos from Bjarte Jordal and Udo Schmidt. Johan Billen provided helpful insight on lubrication glands in other insects. P.H.W.B. acknowledges the invitation by Ted Schultz to join the 38th Altenberg Workshop in Theoretical Biology and funding by the German Science Foundation (DFG; Emmy Noether Grant BI 1956/1–1).

References

Aanen, D. K. 2006. "As You Reap, So Shall You Sow: Coupling of Harvesting and Inoculating Stabilizes the Mutualism between Termites and Fungi." *Biology Letters* 22:209–212.

Aanen, D. K., H. Henrik, A. J. Debets, N. A. Kerstes, R. F. Hoekstra, et al. 2009. "High Symbiont Relatedness Stabilizes Mutualistic Cooperation in Fungus-Growing Termites." *Science* 326:1103–1106.

Aanen, D. K., P. Eggleton, C. Rouland-Lefevre, T. Guldberg-Frøslev, S. Rosendahl, et al. 2002. "The Evolution of Fungus-Growing Termites and Their Mutualistic Fungal Symbionts." *Proceedings of the National Academy of Sciences* 99:14887–14892.

Abrahamson, L. P. 1969. "Physiological Interrelationships between Ambrosia Beetles and Their Symbiotic Fungi." PhD dissertation, University of Wisconsin, Madison.

Abrahamson, L. P., H. M. Chu, and D. M. Norris. 1967. "Symbiontic Interrelationships between Microbes and Ambrosia Beetles. II. The Organs of Microbial Transport and Perpetuation in *Trypodendron betulae* and *T. retusum* (Coleoptera: Scolytidae)." *Annals of the Entomological Society of America* 60:1107–1110.

Abrahamson, L. P., and D. M. Norris. 1966. "Symbiontic Interrelationships between Microbes and Ambrosia Beetles. I. The Organs of Microbial Transport and Perpetuation of *Xyloterinus politus*." *Annals of the Entomological Society of America* 59:877–880.

Abrahamson, L. P., and D. M. Norris. 1969. "Symbiontic Interrelationships between Microbes and Ambrosia Beetles IV. Ambrosial Fungi Associated with *Xyloterinus politus*." *Journal of Invertebrate Pathology* 14:381–385.

Abrahamson, L. P., and D. M. Norris. 1970. "Symbiontic Interrelationships between Microbes and Ambrosia Beetles (Coleoptera: Scolytidae). V. Amino Acids as a Source of Nitrogen to the Fungi in the Beetle." *Annals of the Entomological Society of America* 63:177–180.

Aoki, T., M. T. Kasson, M. C. Berger, S. Freeman, D. M. Geiser, et al. 2018. "*Fusarium oligoseptatum* sp. nov., a Mycosymbiont of the Ambrosia Beetle *Euwallacea validus* in the Eastern US and Typification of *F. ambrosium*." *Fungal Systematics and Evolution* 1:23–39.

Aoki, T., J. A. Smith, M. T. Kasson, S. Freeman, D. M. Geiser, et al. 2019. "Three Novel Ambrosia *Fusarium* Clade Species Producing Clavate Macroconidia Known (*F. floridanum* and *F. obliquiseptatum*) or Predicted (*F. tuaranense*) to Be Farmed by *Euwallacea* spp. (Coleoptera: Scolytinae) on Woody Hosts." *Mycologia* 111:919–935.

Archetti, M., I. Scheuring, M. Hoffman, M. E. Frederickson, N. E. Pierce, et al. 2011. "Economic Game Theory for Mutualism and Cooperation." *Ecology Letters* 14:1300–1312.

Baker, J. M. 1963. "Ambrosia Beetles and Their Fungi, with Particular Reference to *Platypus cylindrus* Fab." In *Symbiotic Associations*, edited by R. Dubos and A. Kiessler, 232–265. Cambridge: Cambridge University Press.

Baker, J. M., and N. J. W. Kreger-van Rij. 1964. "*Endomycopsis platypodis* sp. n. (Ascomycetes): An Auxiliary Ambrosia Fungus of *Platypus cylindrus* Fab. (Col. Platypodidae)." *Antonie van Leeuwenhoek* 30:433–441.

Baker, J. M., and D. M. Norris. 1968. "A Complex of Fungi Mutualistically involved in the Nutrition of the Ambrosia Beetle *Xyleborus ferrugineus*." *Journal of Invertebrate Pathology* 11:246–250.

Barras, S. J., and T. Perry. 1971. "Gland Cells and Fungi Associated with Prothoracic Mycangium of *Dendroctonus adjunctus* (Coleoptera: Scolytidae)." *Annals of the Entomological Society of America* 64:123–126.

Bateman, C., Y. T. Huang, D. R. Simmons, M. T. Kasson, E. L. Stanley, et al. 2017. "Ambrosia Beetle *Premnobius cavipennis* (Scolytinae: Ipini) Carries Highly Divergent Ascomycotan Ambrosia Fungus, *Afroraffaelea ambrosiae* gen. nov. et sp. nov. (Ophiostomatales)." *Fungal Ecology* 25:41–49.

Bateman, C., P. E. Kendra, R. Rabaglia, and J. Hulcr. 2015. "Fungal Symbionts in Three Exotic Ambrosia Beetles, *Xylosandrus amputatus*, *Xyleborinus andrewesi*, and *Dryoxylon onoharaense* (Coleoptera: Curculionidae: Scolytinae: Xyleborini) in Florida." *Symbiosis* 66:141–148.

Bateman, C., M. Šigut, J. Skelton, K. E. Smith, and J. Hulcr. 2016. "Fungal Associates of the *Xylosandrus compactus* (Coleoptera: Curculionidae, Scolytinae) Are Spatially Segregated on the Insect Body." *Environmental Entomology* 45:883–890.

Batra, L. R. 1963. "Ecology of Ambrosia Fungi and Their Dissemination by Beetles." *Transactions of the Kansas Academy of Science* 66:213–236.

Batra, L. R. 1966. "Ambrosia Fungi: Extent of Specificity to Ambrosia Beetles." *Science* 153:193–195.

Batra, L. R. 1967. "Ambrosia Fungi—A Taxonomic Revision, and Nutritional Studies of Some Species." *Mycologia* 59:976–1017.

Batra, L. R. 1985. "Ambrosia Beetles and Their Associated Fungi: Research Trends and Techniques." *Proceedings of the Indian Academy of Science (Plant Sciences)* 94:137–148.

Batra, L. R, and M. Downing Michie. 1963. "Pleomorphism in Some Ambrosia and Related Fungi." *Transactions of the Kansas Academy of Science* 66:470–481.

Batra, L. R., and H. Francke-Grosmann. 1961. "Contributions to Our Knowledge of Ambrosia Fungi. I. *Asocidea hylecoeti* sp. nov. (Ascomycetes)." *American Journal of Botany* 42:453–456.

Batra, L. R., and H. Francke-Grosmann H. 1964. "Two New Ambrosia Fungi: *Ascoidea asiatica* and *A. africana*." *Mycologia* 56:632–636.

Batra, S. W. T., and L. R. Batra. 1967. "The Fungus Gardens of Insects. *Scientific American* 217:112–120.

Beaver, R. A. 1976. "The Biology of Samoan Scolytidae and Platypodidae (Coleoptera)." *Bulletin of Entomological Research* 65:531–548.

Beaver, R. A. 1984. "Biology of the Ambrosia Beetle, *Sueus niisimai* (Eggers) (Coleoptera: Scolytidae), in Fiji." *Entomologists' Monthly Magazine* 120:99–102.

Beaver, R. A. 1986. "The Taxonomy, Mycangia and Biology of *Hypothenemus curtipennis* (Schedl), the First Known Cryphaline Ambrosia Beetle (Coleoptera: Scolytidae)." *Insect Systematics and Evolution* 17:131–135.

Beaver, R. A. 1989. "Insect-Fungus Relationships in the Bark and Ambrosia Beetles." In *Insect-Fungus Interactions*, edited by N. Wilding, N. M. Collins, P. M. Hammond, and J. F. Webber, 121–143. London: Academic Press.

Beaver, R. A. 1998. "Three New *Spathidicerus* Chapuis (Col., Platypodidae) with a Key to Species." *Entomologists Monthly Magazine* 134:285–291.

Beaver, R. A. 2000. "The Ambrosia Beetle Genus *Indocryphalus* Eggers (Coleoptera: Scolytidae): A New Species from Peninsular Malaysia, New Synonymy and a Key to Species." *Serangga* 5:165–179.

Beaver, R. A. 2002. "Studies on the Genus *Diapus* Chapuis (Coleoptera: Platypodidae): Keys to Males and Females, a New Species and a New Synonym." *Serangga* 7:245–260.

Beaver, R. A. 2016. "The Platypodine Ambrosia beetles of Laos (Coleoptera: Curculionidae: Platypodinae)." *Entomologica Basiliensia et Collectionis Frey* 35:487–504.

Beaver, R. A., and H. Gebhardt. 2006. "A Review of the Oriental Species of *Scolytoplatypus* Schaufuss (Coleoptera, Curculionidae, Scolytinae)." *Deutsche Entomologische Zeitschrift* 53:155–178.

Beaver, R. A., and L. Y. Liu. 2007. "A Review of the Genus *Genyocerus* Motschulsky (Coleoptera: Curculionidae: Platypodinae), with New Synonymy and Keys to Species." *Zootaxa* 1576:25–56.

Beaver, R. A., and L. Y. Liu. 2018. "A Review of the Genus *Baiocis* Browne, 1962 (Coleoptera: Curculionidae: Platypodinae), with New Species, New Synonymy and a Key to Males." *Zootaxa* 4434:481–501.

Beaver, R. A., and S. Sanguansub. 2015. "A Review of the Genus *Carchesiopygus* Schedl (Coleoptera: Curculionidae: Platypodinae), with Keys to Species." *Zootaxa* 3931:401–412.

Beeson, C. F. C. 1917. "The Life-History of *Diapus furtivus*, Sampson (Platypodidae)." *Indian Forest Records* 6:1–38.

Belhoucine, L., R. T. Bouhraoua, J. M. Harrak, A. Viñolas, A. Equihua-Martinez, et al. 2013. "New Contribution to Knowledge of Mycangia in *Platypus cylindrus* (Fabricius, 1792), and Comments about the Variation of Some Morphological Structures in Mediterranean Isolated Populations (Col.: Curculionidae, Platypodinae)." *Boletín de la Sociedad Entomológica Aragonesa* 53:125–134.

Berger, B., and N. Cholodkovsky. 1916. "On Some Points of Biology and Anatomy of the Genus *Scolytoplatypus* Blandford (Coleoptera, Ipidae)." *Revue Russe d'Entomologie* 16: 1–7.

Berrocal, A., J. Navarrete, C. Oviedo, and K. W. Nickerson. 2012. "Quorum Sensing Activity in *Ophiostoma ulmi*: Effects of Fusel Oils and Branched Chain Amino Acids on Yeast-Mycelial Dimorphism." *Journal of Applied Microbiology* 113:126–314.

Bickerstaff, J. R. 2017. "Morphological and Molecular Characterisation of Australian Pinhole Borers (Coleoptera: Curculionidae, Platypodinae)." MS thesis, Western Sydney University.

Biedermann, P. H., K. D. Klepzig, and M. Taborsky M. 2011. "Costs of Delayed Dispersal and Alloparental Care in the Fungus-Cultivating Ambrosia Beetle *Xyleborus affinis* Eichhoff (Scolytinae: Curculionidae)." *Behavioral Ecology and Sociobiology* 65:1753–1761.

Biedermann, P. H., K. D. Klepzig, M. Taborsky, and D. L. Six. 2013. "Abundance and Dynamics of Filamentous Fungi in the Complex Ambrosia Gardens of the Primitively Eusocial Beetle *Xyleborinus saxesenii* Ratzeburg (Coleoptera: Curculionidae, Scolytinae)." *FEMS Microbiology Ecology* 83:711–723.

Biedermann, P. H., K. Peer, and M. Taborsky. 2012. "Female Dispersal and Reproduction in the Ambrosia Beetle *Xyleborinus saxesenii* Ratzeburg (Coleoptera; Scolytinae)." *Mitteilungen der Deutschen Gesellschaft für Allgemeine und Angewandte Entomologie* 18:231–235.

Biedermann, P. H., and M. Rohlfs. 2017. "Evolutionary Feedbacks between Insect Sociality and Microbial Management." *Current Opinion in Insect Science* 22:92–100.

Biedermann, P. H., and M. Taborsky. 2011. "Larval Helpers and Age Polyethism in Ambrosia Beetles." *Proceedings of the National Academy of Sciences* 108:17064–17069.

Biedermann, P. H., and F. E. Vega. 2020. "Ecology and Evolution of Insect–Fungus Mutualisms." *Annual Review of Entomology* 65:431–455.

Biedermann, P. H. W. 2012. "The Evolution of Cooperation in Ambrosia Beetles." PhD dissertation, University of Bern.

Billen, J. 2009. "Occurrence and Structural Organization of the Exocrine Glands in the Legs of Ants." *Arthropod Structure & Development* 38:2–15.

Billen, J., and L. Vander Plancken. 2014. "Exocrine Glands in the Legs of the Stingless Bee *Frieseomelitta varia* (Lepeletier) (Apidae: Meliponini)." *Sociobiology* 61:386–392.

Bischoff, L. L. 2004. "The Social Structure of the Haplodiploid Bark Beetle, *Xylosandrus germanus*." Diploma thesis, Zoological Institute, University of Bern.

Blackwell, M. 2017. "Yeasts in Insects and Other Invertebrates." In *Yeasts in Natural Ecosystems: Diversity*, edited by P. Buzzini, M. A. Lachance, and A. Yurkov, 397–433. Cham: Springer.

Blandford, W. F. 1896. "XLIV.—On the Genus *Dactylipalpus*, Chapuis, and Two New Genera of Scolytidæ from Africa." *Journal of Natural History* 17:319–325.

Bracewell, R. R., and D. L. Six, . 2015. "Experimental Evidence of Bark Beetle Adaptation to a Fungal Symbiont." *Ecology and Evolution* 5:5109–5119.

Branstetter, M. G., A. Ješovnik, J. Sosa-Calvo, M. W. Lloyd, B. C. Faircloth, et al. 2017. "Dry Habitats Were Crucibles of Domestication in the Evolution of Agriculture in Ants." *Proceedings of the Royal Society B—Biological Sciences* 284:20170095.

Browne, F. G. 1971. "The Genus *Chaetastus* Nunberg (Coleoptera: Platypodidae)." *Journal of Entomology Series B, Taxonomy* 40:7–20.

Carrillo J. D., C. Dodge, R. Stouthamer, and A. Eskalen. 2020. "Fungal Symbionts of the Polyphagous and Kuroshio Shot Hole Borers (Coleoptera: Scolytinae, *Euwallacea* spp.) in California Can Support Both Ambrosia Beetle Systems on Artificial Media." *Symbiosis* 7:1–4.

Cassier, P., J. Lévieux, M. Morelet, and D. Rougon. 1996. "The Mycangia of *Platypus cylindrus* Fab. and P. oxyurus Dufour (Coleoptera: Platypodidae). Structure and Associated Fungi." *Journal of Insect Physiology* 42:171–179.

Chapman, J. A. 1956. "Flight-Muscle Changes during Adult Life in a Scolytid Beetle." *Nature* 177:1183.

Cognato, A. I., S. M. Smith, and T. H. Pham. 2015. "Cladistic Analysis of *Indocryphalus* Eggers (Coleoptera: Curculionidae: Scolytinae: Xyloterini) and Description of a New Species from Vietnam." *Insect Systematics & Evolution* 46:493–506.

Cognato, A. I., S. M. Smith, and B. H. Jordal. 2021. "Patterns of Host Tree Use within a Lineage of Saproxlic Snout-less Weevils (Coleoptera: Curculionidae: Scolytinae: Scolytini)." *Molecular Phylogenetics and Evolution* 159:107107.

De Beer, Z. W., T. A. Duong, I. Barnes, W. D. Wingfield, and M. J. Wingfield. 2014. "Redefining *Ceratocystis* and Allied Genera." *Studies in Mycology* 79:187–219.

Doebley, J. F., B. S. Gaut, and B. D. Smith. 2006. "The Molecular Genetics of Crop Domestication." *Cell* 127:1309–1321.

Dreaden, T. J., J. M. Davis, Z. W. De Beer, R. C. Ploetz, P. S. Soltis, et al. 2014. "Phylogeny of Ambrosia Beetle Symbionts in the Genus *Raffaelea*." *Fungal Biology* 118:970–978.

Ebermann, E., M. Hall, U. Hausl-Hofstätter, J. M. Jagersbacher-Baumann, R. Kirschner, T. Pfingstl, et al. 2013. "A New Phoretic Mite Species with Remarks to the Phenomenon "Sporothecae" (Acari, Scutacaridae; Hymenoptera, Aculeata)." *Zoologischer Anzeiger—A Journal of Comparative Zoology*. 252:234–242.

Farrell, B. D., A. S. Sequeira, B. C. O'Meara, B. B. Normark, J. H. Chung, et al.. 2001. "The Evolution of Agriculture in Beetles (Curculionidae: Scolytinae and Platypodinae)." *Evolution* 55:2011–2027.

Farris, S. H. 1963. "Ambrosia Fungus Storage in Two Species of *Gnathotrichus* Eichhoff (Coleoptera: Scolytidae)." *Canadian Entomologist* 95:257–259.

Farris, S. H. 1965. "Repositories of Symbiotic Fungus in the Ambrosia Beetle *Monarthrum scutellare* Lec. (Coleoptera: Scolytidae)." *Journal of the Entomological Society of British Columbia* 62:30–33.

Farris, S. H., and A. Funk. 1965. "Repositories of Symbiotic Fungus in the Ambrosia Beetle *Platypus wilsoni* Swaine (Coleoptera: Platypodidae)." *Canadian Entomologist* 97:527–532.

Fernando, E. F. W. 1959. "A Note on the Storage of the Ambrosia Fungus by the Shot-Hole Beetle (*Xyleborus fornicatus* Eichh.)." *Tea Quarterly* 30:50.

Finnegan R, J. 1963. "The Storage of Ambrosia Fungus Spores by the Pitted Ambrosia Beetle, *Corthylus punctatissimus* Zimm. (Coleoptera: Scolytidae)." *Canadian Entomologist* 95:137–139.

Fraedrich, S.W., T. C. Harrington, R. J. Rabaglia, M. D. Ulyshen, A. E. Mayfield, et al. 2008. "A Fungal Symbiont of the Redbay Ambrosia Beetle Causes a Lethal Wilt in Redbay and Other Lauraceae in the Southeastern United States." *Plant Disease* 92:215–224.

Francke-Grosmann, H. 1956a. "Hautdrüsen als Träger der Pilzsymbiose bei Ambrosiakäfern." *Zeitschrift für Morphologie und Ökologie der Tiere* 45:275–308.

Francke-Grosmann, H. 1956b. "Grundlagen der Symbiose bei pilzzüchtenden Holzinsekten." *Verhandlungen der Deutschen Zoologischen Gesellschaft Hamburg (Liepzig)*, 112–118.

Francke-Grosmann, H. 1958. "Über die Ambrosiazucht holzbrütender Ipiden im Hinblick auf das System." *Verhandlungen Der Deutschen Gesellschaft für Angewandte Entomologie* 14:139–144.

Francke-Grosmann, H. 1959. "Beiträge zur Kenntnis der Ubertragungsweise von Pflanzenkrankheiten durch Käfer." *Proceedings of the 4th International Congress of Plant Protection, Hamburg, 1957,* 1:805–809.

Francke-Grosmann, H. 1963. "Some New Aspects in Forest Entomology." *Annual Review of Entomology* 8:415–438.

Francke-Grosmann, H. 1966. "Über Symbiosen von xylomycetophagen und phloeophagen Scolytoidea mit holzbewohnenden Pilzen." *Material und Organismen* 1:503–522.

Francke-Grosmann, H. 1967. "Ectosymbiosis in Wood-Inhabiting Insects." In *Symbiosis*, edited by S. M. Henry, 142–206. New York: Academic Press.

Francke-Grosmann, H. 1975. "Zur epizoischen und endozoischen Übertragung der symbiotischen Pilze des Ambrosiakäfers *Xyleborus saxeseni* (Coleoptera: Scolytidae)." *Entomologica Germanica* 1:279–292.

Freeman, S., A. Protasov, M. Sharon, K. Mohotti, M. Eliyahu, et al. 2012. "Obligate Feed Requirement of *Fusarium* sp. nov., an Avocado Wilting Agent, by the Ambrosia Beetle *Euwallacea* aff. *fornicata*." *Symbiosis* 58:245–251.

Freeman, S., M. Sharon, M. Dori-Bachash, M. Maymon, E. Belausov, et al. 2016. "Symbiotic Association of Three Fungal Species throughout the Life Cycle of the Ambrosia Beetle *Euwallacea* nr. *fornicatus*." *Symbiosis* 68:115–128.

Freeman, S., M. Sharon, M. Maymon, Z. Mendel, A. Protasov, et al. 2013. "*Fusarium euwallaceae* sp. nov.—A Symbiotic Fungus of *Euwallacea* sp., an Invasive Ambrosia Beetle in Israel and California." *Mycologia* 105:1595–1606.

French, J. R., and R. A. Roeper. 1972. "Interactions of the ambrosia Beetle, *Xyleborus dispar* (Coleoptera: Scolytidae), with its symbiotic Fungus *Ambrosiella hartigii* (Fungi Imperfecti)." *Canadian Entomologist* 104:1635–1641.

French, J. R., and R. A. Roeper. 1973. "Patterns of Nitrogen Utilization between the Ambrosia Beetle *Xyleborus dispar* and Its Symbiotic Fungus." *Journal of Insect Physiology* 19:593–605.

Furniss, M. M., J. Y. Woo, M. A. Deyrup, and T. H. Atkinson. 1987. "Prothoracic Mycangium on Pine-Infesting *Pityoborus* spp. (Coleoptera: Scolytidae)." *Annals of the Entomological Society of America* 80:692–696.

Gauthier, G. M. 2015. "Dimorphism in Fungal Pathogens of Mammals, Plants, and Insects." *PLoS Pathogens* 11: e1004608.

Gebhardt, H., D. Begerow, and F. Oberwinkler. 2004. "Identification of the Ambrosia Fungus of *Xyleborus monographus* and *X. dryographus* (Coleoptera: Curculionidae, Scolytinae)." *Mycological Progress* 3:95–102.

Gebhardt, H., and F. Oberwinkler. 2005. "Conidial Development in Selected Ambrosial Species of the Genus *Raffaelea*." *Antonie van Leeuwenhoek* 88:61–65.

Gharabigloozare, Y. 2015. "*Raffaelea* spp. from Five Ambrosia Beetles in the Genera *Xyleborinus* and *Cyclorhipidion* (Coleoptera: Curcurlionidae: Scolytinae: Xyleborini)." MS thesis, Iowa State University, Ames.

Giese, R. L. 1967. "The Columbian Timber Beetle, *Corthylus columbianus* (Coleoptera: Scolytidae): V. A Description of the Mycetangia." *Canadian Entomologist* 99:54–58.

Goethe R. 1895. Der ungleiche Borkenkäfer, *Xyleborus (Bostrichus) dispar Faber. In: Berichte der Königliche Lehranstalt für Obst-, Wein-, und Gartenbau an Geisenheim am Rhein für das Etatsjahr 1894/1895*, edited by R. Goethe, 25–26. Wiesbaden: Rudolf Bechthold & Komp.

Gohli, J., L. R. Kirkendall, S. M. Smith, A. I. Cognato, J Hulcr, et al. 2017. "Biological Factors Contributing to Bark and Ambrosia Beetle Species Diversification." *Evolution* 71:1258–1272.

Goto, H. 1998. "A New Type of Mycangia Found in the Genus *Euwallacea* (Coleoptera, Scolytidae)." In *Book of Abstracts 1, Proceedings of the Sixth European Congress of Entomology*, Ceske Budejovice, Czech Republic, August 23–29, 1998, 340.

Grebennikov, V. V., and R. A. Leschen. 2010. "External Exoskeletal Cavities in Coleoptera and Their Possible Mycangial Functions." *Entomological Science* 13:81–98.

Guerrero, R. T. 1966. "Una nueva especie de hongo inperfecto asociado con el coleoptero *Platypus sulcatus* Chapius." *Revista de Investigaciones Agropecuarias* 53:97–103.

Haanstad, J. O., and D. M. Norris. 1985. "Microbial Symbiotes of the Ambrosia Beetle *Xyloterinus politus*." *Microbial Ecology* 11:267–276.

Hadorn, C. 1933. "Recherches sur la morphologie, les stades évolutifs et l'hivernage du Bostryche liseré (*Xyloterus lineatus* Oliv.)." *Supplément aux Organes de la Société Forestière Suisse, Bern* 11:1–120.

Happ, G. M., C. M. Happ, and S. J. Barras. 1971. "Fine Structure of the Prothoracic Mycangium, a Chamber for the Culture of Symbiotic Fungi, in the Southern Pine Beetle, *Dendroctonus frontalis*." *Tissue and Cell* 3:295–308.

Happ, G. M., C. M. Happ, and J. R. French. 1976. "Ultrastructure of the Mesonotal Mycangium of an Ambrosia Beetle, *Xyleborus dispar* (F.) (Coleoptera: Scolytidae)." *International Journal of Insect Morphology and Embryology* 5:381–391.

Harrington, T. C. 2005. "Ecology and Evolution of Mycophagous Bark Beetles and Their Fungal Partners. In *Insect-Fungal Associations: Ecology and Evolution*, edited by F. E. Vega and M. Blackwell, 257–291. New York: Oxford University Press.

Harrington, T. C, D. N. Aghayeva, and S. W. Fraedrich. 2010. "New Combinations in *Raffaelea*, *Ambrosiella*, and *Hyalorhinocladiella*, and four New Species from the Redbay Ambrosia Beetle, *Xyleborus glabratus*." *Mycotaxon* 111:337–361.

Harrington, T. C., and S. W. Fraedrich. 2010. "Quantification of Propagules of the Laurel Wilt Fungus and Other Mycangial Fungi from the Redbay Ambrosia Beetle, *Xyleborus glabratus*." *Phytopathology* 100:1118–1123.

Harrington, T. C., S. Fraedrich, and D. N. Aghayeva. 2008. "*Raffaelea lauricola*, a New Ambrosia Beetle Symbiont and Pathogen on the Lauraceae." *Mycotaxon* 104:399–404.

Harrington, T. C., D. McNew, C. Mayers, S. W. Fraedrich, and S. E. Reed. 2014. "*Ambrosiella roeperi* sp. nov. Is the Mycangial Symbiont of the Granulate Ambrosia Beetle, *Xylosandrus crassiusculus*." *Mycologia* 106:835–845.

Hart, A. G., and F. L. Ratnieks. 2001. "Task Partitioning, Division of Labour and Nest Compartmentalisation Collectively Isolate Hazardous Waste in the Leafcutting Ant *Atta cephalotes*." *Behavioral Ecology and Sociobiology* 49:387–392.

Hartig, T. 1844. "Ambrosia des Bostrichus dispar." *Allgemeine Forst- und Jagdzeitung* 13:73–74.

Hartig, T. 1872a. "Der Fichten-Splintkäfer Bostrichus (Xyloterus) lineatus." *Allgemeine Forst- und Jagdzeitung* 48:181–183.

Hartig, T. 1872b. "Der Buchen-Splintkäfer Bostrichus (Xyloterus) domesticus." *Allgemeine Forst- und Jagdzeitung* 48:183–184.

Hofstetter, R. W., J. Dinkins-Bookwalter, T. S. Davis, and K. D. Klepzig. 2015. "Symbiotic Associations of Bark Beetles." In *Bark Beetles*, edited by F. E. Vega and R. W. Hofstetter, 209–245. London: Academic Press.

Howe, J., M. Schiøtt, and J. J. Boomsma. 2019. "Horizontal Partner Exchange Does Not Preclude Stable Mutualism in Fungus-Growing Ants." *Behavioral Ecology* 30:372–382.

Hubbard, H. G. 1897a. "Ambrosia Beetles." *Yearbook of the United States Department of Agriculture* 1896:421–430.

Hubbard, H. G. 1897b. "The Ambrosia Beetles of the United States." *Bulletin of the U.S. Department of Agriculture Division of Entomology*, new series, 7:9–30.

Hughes, G. P., A. E. Spikes, J. D. Holland, and M. D. Ginzel. 2011. "Evidence for the Stratification of Hydrocarbons in the Epicuticular Wax Layer of Female *Megacyllene robiniae* (Coleoptera: Cerambycidae)." *Chemoecology* 21:99–105.

Hulcr, J., T. H. Atkinson, A. I. Cognato, B. H. Jordal, and D. D. McKenna. 2015. "Morphology, Taxonomy, and Phylogenetics of Bark Beetles." In *Bark Beetles*, edited by F. E. Vega and R. W. Hofstetter, 41–84. London: Academic Press.

Hulcr, J., I. Barnes, Z. W. De Beer, T. A. Duong, R, Gazis, et al. 2020. "Bark Beetle Mycobiome: Collaboratively Defined Research Priorities on a Widespread Insect-Fungus Symbiosis." *Symbiosis* 81:101–113.

Hulcr, J., and A. I. Cognato. 2010. "Repeated Evolution of Crop Theft in Fungus-Farming Ambrosia Beetles." *Evolution* 64:3205–3212.

Hulcr, J., S. A. Dole, R. A. Beaver, and A. I. Cognato. 2007. "Cladistic Review of Generic Taxonomic Characters in Xyleborina (Coleoptera: Curculionidae: Scolytinae)." *Systematic Entomology* 32:568–584.

Hulcr, J., M. Kolařík, and L. P. Kirkendall. 2007. "A New Record of Fungus-Beetle Symbiosis in *Scolytodes* Bark Beetles (Scolytinae, Curculionidae, Coleoptera)." *Symbiosis* 43:151–159.

Hulcr, J., N. R. Rountree, S. E. Diamond, L. L. Stelinski, N. Fierer, et al. 2012. "Mycangia of Ambrosia Beetles Host Communities of Bacteria. *Microbial Ecology* 64:784–793.

Hulcr, J., and L. L. Stelinski. 2017. "The Ambrosia Symbiosis: From Evolutionary Ecology to Practical Management." *Annual Review of Entomology* 62:285–303.

Ibarra-Juarez, L. A., M. A. Burton, P. H. Biedermann, L. Cruz L, D. Desgarennes, et al. 2020. "Evidence for Succession and Putative Metabolic Roles of Fungi and Bacteria in the Farming Mutualism of the Ambrosia Beetle *Xyleborus affinis*." *mSystems* 5:e00541–20.

Ito, M., and H. Kajimura. 2017. "Landscape-Scale Genetic Differentiation of a Mycangial Fungus Associated with the Ambrosia Beetle, *Xylosandrus germanus* (Blandford) (Curculionidae: Scolytinae) in Japan." *Ecology and Evolution* 7:9203–9221.

Jiang, Z. R., S. I. Kinoshita, O. Sasaki, A. I. Cognato, and H. Kajimura. 2019. "Non-Destructive Observation of the Mycangia of *Euwallacea interjectus* (Blandford) (Coleoptera: Curculionidae: Scolytinae) Using X-Ray Computed Tomography." *Entomological Science* 22:173–181.

Johnson, A. J., J. Hulcr, M. Knížek, T. H. Atkinson, M. Y. Mandelshtam, et al. 2020. "Revision of the Bark Beetle Genera within the Former Cryphalini (Curculionidae: Scolytinae)." *Insect Systematics and Diversity* 4:1–81.

Johnson, A. J., D. D. McKenna, B. H. Jordal, A. I. Cognato, S. M. Smith, et al. 2018. "Phylogenomics clarifies Repeated Evolutionary Origins of Inbreeding and Fungus Farming in Bark Beetles (Curculionidae, Scolytinae)." *Molecular Phylogenetics and Evolution* 127:229–238.

Jordal, B. H. 2013. "Deep Phylogenetic Divergence between *Scolytoplatypus* and *Remansus*, a New Genus of Scolytoplatypodini from Madagascar (Coleoptera, Curculionidae, Scolytinae)." *ZooKeys* 352:9–33.

Jordal, B. H. 2015. "Molecular Phylogeny and Biogeography of the Weevil Subfamily Platypodinae Reveals Evolutionarily Conserved Range Patterns." *Molecular Phylogenetics and Evolution* 92:294–307.

Jordal, B. H., and L. R. Kirkendall. 2019. "Rainforest and Cloud Forest *Scolytodes* (Curculionidae, Scolytinae, Hexacolini) from the ALAS Inventory in Costa Rica: New Species, New Synonymy, New Records." *ZooKeys* 863:1–34.

Jusino, M. A., J. Skelton, C. Chen, J. Hulcr, and M. E. Smith. 2020. "Sexual Reproduction and Saprotrophic Dominance by the Ambrosial Fungus *Flavodon subulatus* (= *Flavodon ambrosius*)." *Fungal Ecology* 47:100979.

Kajimura, H., and N. Hijii. 1992. "Dynamics of the Fungal Symbionts in the Gallery System and the Mycangia of the Ambrosia Beetle, *Xylosandrus mutilatus* (Blandford) (Coleoptera: Scolytidae) in Relation to Its Life History." *Ecological Research* 7:107–117.

Kaneko, T. 1967. "Shot-Hole Borer of Tea Plant in Japan." *Japan Agricultural Research Quarterly* 22:19–21.

Kasson, M. T., K. O'Donnell, A. P. Rooney, S. Sink, R. C. Ploetz, et al. 2013. "An Inordinate Fondness for *Fusarium*: Phylogenetic Diversity of Fusaria Cultivated by Ambrosia Beetles in the Genus *Euwallacea* on Avocado and Other Plant Hosts." *Fungal Genetics and Biology* 56:147–157.

Kasson, M. T., K. L. Wickert, C. M. Stauder, A. M. Macias, M. C. Berger, et al. 2016. "Mutualism with Aggressive Wood-Degrading *Flavodon ambrosius* (Polyporales) Facilitates Niche Expansion and Communal Social Structure in *Ambrosiophilus* Ambrosia Beetles." *Fungal Ecology* 23:86–96.

Kendall, D. A. 1972. "The Dermal Glands of Some Adult Beetles." *Journal of Entomology Series A, General Entomology* 46:153–159.

Kent, D. S. 2008. "Mycangia of the Ambrosia Beetle, *Austroplatypus incompertus* (Schedl) (Coleoptera: Curculionidae: Platypodinae)." *Austral Entomology* 47:9–12.

Kent, D. S. 2010. "The External Morphology of *Austroplatypus incompertus* (Schedl) (Coleoptera, Curculionidae, Platypodinae)." *ZooKeys* 56:121–140.

Kent, D. S., and J. A. Simpson. 1992. "Eusociality in the Beetle *Austroplatypus incompertus* (Coleoptera: Curculionidae)." *Naturwissenschaften* 79:86–87.

Kinuura, H. 1995. "Symbiotic Fungi Associated with Ambrosia Beetles." *Japan Agricultural Research Quarterly* 29:57–63.

Kinuura, H. 2002. "Relative Dominance of the Mold Fungus, *Raffaelea* sp., in the Mycangium and Proventriculus in Relation to Adult Stages of the Oak Platypodid Beetle, *Platypus quercivorus* (Coleoptera; Platypodidae)." *Journal of Forest Research* 7:7–12.

Kinuura, H., N. Hijii, and K. Kanamitsu. 1991. "Symbiotic Fungi Associated with the Ambrosia Beetle, *Scolytoplatypus mikado* Blandford (Coleoptera: Scolytidae)—Succession of the Flora and Fungal Phases in the Gallery System and the Mycangium in Relation to the Developmental Stages of the Beetle." *Journal of the Japanese Forestry Society* 73:197–205.

Kirkendall, L. R., P. H. Biedermann, and B. H. Jordal. 2015. "Evolution and Diversity of Bark and Ambrosia Beetles." In *Bark Beetles*, edited by F. E. Vega and R. W. Hofstetter, 85–156. London: Academic Press.

Kobayashi, C., Y. Fukasawa, D. Hirose, and M. Kato. 2008. "Contribution of Symbiotic Mycangial Fungi to Larval Nutrition of a Leaf-Rolling Weevil." *Evolutionary Ecology* 22:711–722.

Kok, L. T. 1979. "Lipids of Ambrosia Fungi and the Life of Mutualistic Beetles." In *Insect-Fungus Symbiosis: Nutrition, Mutualism and Commensalism*, edited by L. R. Batra, 33–52. New York: Wiley.

Kolařík, M., and J. Hulcr. 2009. "Mycobiota Associated with the Ambrosia Beetle *Scolytodes unipunctatus* (Coleoptera: Curculionidae, Scolytinae)." *Mycological Research* 113:44–60.

Kolařík, M., and L. R. Kirkendall. 2010. "Evidence for a New Lineage of Primary Ambrosia Fungi in *Geosmithia* Pitt (Ascomycota: Hypocreales)." *Fungal Biology* 114:676–689.

Kolařík, M. I., J. I. Hulcr, and L. R. Kirkendall. 2015. "New Species of *Geosmithia* and *Graphium* Associated with Ambrosia Beetles in Costa Rica." *Czech Mycology* 67:29–35.

Kölsch, G. 2000. "The Ultrastructure of Glands and the Production and Function of the Secretion in the Adhesive Capture Apparatus of *Stenus* Species (Coleoptera: Staphylinidae)." *Canadian Journal of Zoology* 78:465–475.

Korb, J., and J. Heinze. 2016. "Major Hurdles for the Evolution of Sociality." *Annual Review of Entomology* 61:297–316.

Lehenberger, M., P. H. Biedermann, and J. P. Benz. 2019. "Molecular Identification and Enzymatic Profiling of *Trypodendron* (Curculionidae: Xyloterini) Ambrosia Beetle–Associated Fungi of the Genus *Phialophoropsis* (Microascales: Ceratocystidaceae)." *Fungal Ecology* 38:89–97.

Lehenberger, M., M. Thalhammer, J. P. Benz, and P. H. W. Biedermann. 2019. "Take Good Care of My Fungi—Fungus-Spore Carrying Organs in *Tryopodendron* Ambrosia Beetles." *Mitteilungen der Deutschen Gesellschaft für allgemeine und angewandte Entomologie* 22:195–198

Lhoste, J., and A. Roche. 1959. "Contributionàla connaissance de l'anatomie interne de *Xyleborus morsatti* Haged." *Café, Cacao, Thé* 3:278–286.

Lhoste, J., and A. Roche. 1961." Anatomie comparée des organes transporteurs de champignons chez quelques Scolytoidea." *Verhandlungen: XI Internationaler Kongress für Entomologie, Wien* 1:385–387.

Li, X., W. Guo, Y. Wen, M. K. Solanki, and J. Ding. 2016. "Transmission of Symbiotic Fungus with a Nonsocial Leaf-Rolling Weevil." *Journal of Asia-Pacific Entomology* 19:619–624.

Li, Y., C. C. Bateman, J. Skelton, M. A. Jusino, Z. J. Nolen, et al. 2017. "Wood Decay Fungus *Flavodon ambrosius* (Basidiomycota: Polyporales) Is Widely Farmed by Two Genera of Ambrosia Beetles." *Fungal Biology* 121:984–989.

Li, Y., Y. T. Huang, M. T. Kasson, A. M. Macias, J. Skelton, et al. 2018a. "Specific and Promiscuous Ophiostomatalean Fungi Associated with Platypodinae Ambrosia Beetles in the Southeastern United States." *Fungal Ecology* 35:42–50.

Li, Y., Y. Ruan, M. T. Kasson, E. L. Stanley, C. P. Gillett, A. J. Johnson, et al. 2018b. "Structure of the Ambrosia Beetle (Coleoptera: Curculionidae) Mycangia Revealed through Micro-Computed Tomography." *Journal of Insect Science* 18:13.

Li ,Y., Y. Y. Ruan, E. L. Stanley, J. Skelton, and J. Hulcr. 2019. "Plasticity of Mycangia in *Xylosandrus* Ambrosia Beetles." *Insect Science* 26:732–742.

Li, Y., D. R. Simmons, C. C. Bateman, D. P. Short, M. T. Kasson, R. J. Rabaglia, et al. 2015. "New Fungus-Insect Symbiosis: Culturing, Molecular, and Histological Methods Determine Saprophytic Polyporales Mutualists of *Ambrosiodmus* Ambrosia Beetles." *PLoS ONE* 10:e0137689.

Li, Y., J. Skelton, S. Adams, Y. Hattori, M. Smith, and J. Hulcr. 2020. "The Ambrosia Beetle *Sueus niisimai* (Scolytinae: Hyorrhynchini) Is Associated with the Canker Disease Fungus *Diatrypella japonica* (Xylariales)." *Plant Disease* 104:3143–3150.

Lin, Y. T., H. H. Shih, J. Hulcr, C. S. Lin, S. S.Lu, et al. 2017. "*Ambrosiella* in Taiwan including One New Species." *Mycoscience* 58:242–252.

Livingston, R. L., and A. A. Berryman. 1972. "Fungus Transport Structures in the Fir Engraver, *Scolytus ventralis* (Coleoptera: Scolytidae)." *Canadian Entomologist* 104:1793–1800.

Lowe, R. E., R. L. Giese, and M. L. McManus. 1967. "Mycetangia of the Ambrosia Beetle *Monarthrum fasciatum*." *Journal of Invertebrate Pathology* 9:451–458.

Lynch, S. C., M. Twizeyimana, J. S. Mayorquin, D. H. Wang, F. Na, et al. 2016. "Identification, Pathogenicity and Abundance of *Paracremonium pembeum* sp. nov. and *Graphium euwallaceae* sp. nov.—Two Newly Discovered Mycangial Associates of the Polyphagous Shot Hole Borer (*Euwallacea* sp.) in California." *Mycologia* 108:313–329.

MacLean, D. B., and R. L. Giese. 1968. "Fungi Associated with *Xyloterinus politus* (Say) (Coleoptera: Scolytidae)." *Journal of Invertebrate Pathology* 10:185–189.

Mandelshtam, M. Y., A. V. Petrov, S. M. Smith, and A. I. Cognato. 2019. "Resurrection of *Heteroborips* Reitter, 1913 (Coleoptera: Curculionidae: Scolytinae) from Synonymy with *Xyleborus* Eichhoff, 1864." *Coleopterists Bulletin* 73:387–394.

Mayers, C. G., C. C. Bateman, and T. C. Harrington. 2018. "New *Meredithiella* Species from Mycangia of *Corthylus* Ambrosia Beetles Suggest Genus-Level Coadaptation but Not Species-Level Coevolution." *Mycologia* 110:63–78.

Mayers, C. G., T. C. Harrington, H. Masuya, B. H. Jordal, D. L. McNew, et al. 2020a. "Patterns of Coevolution between Ambrosia Beetle Mycangia and the Ceratocystidaceae, with Five New Fungal Genera and Seven New Species." *Persoonia* 44:41–66.

Mayers, C. G., T. C. Harrington, D. L. McNew, R. A. Roeper, P. H. W. Biedermann, et al. 2020b. "Four Mycangium Types and Four Genera of Ambrosia Fungi Suggest a Complex History of Fungus Farming in the Ambrosia Beetle Tribe Xyloterini." *Mycologia* 112:1104–1137.

Mayers, C. G., T. C. Harrington, and C. M. Ranger. 2017. "First Report of a Sexual State in an Ambrosia Fungus: *Ambrosiella cleistominuta* sp. nov. Associated with the Ambrosia Beetle *Anisandrus maiche*." *Botany* 95:503–512.

Mayers, C. G., D. L. McNew, T. C. Harrington, R. A. Roeper, S. W. Fraedrich, et al. 2015. "Three Genera in the Ceratocystidaceae Are the Respective Symbionts of Three Independent Lineages of Ambrosia Beetles with Large, Complex Mycangia." *Fungal Biology* 119:1075–1092.

McKey, D., M. Elias, B. Pujol, and A. Duputié. 2010. "The Evolutionary Ecology of Clonally Propagated Domesticated Plants." *New Phytologist* 186:318–332.

Mitov, M., V. Soldan, and S. Balor. 2018. "Observation of an Anisotropic Texture inside the Wax Layer of Insect Cuticle." *Arthropod Structure & Development* 47:622–626.

Moon, M. J., J. G. Park, E. Oh, and K. H. Kim. 2008. "External Microstructure of the Ambrosia Beetle *Platypus koryoensis* (Coleoptera: Curculionidae: Platypodinae)." *Entomological Research* 38:202–210.

Moon, M. J., J. G. Park, S. T. Seo, and K. H. Kim. 2012. "Microstructure of the Prothoracic Mycangia in the Ambrosia Beetle *Platypus koryoensis* (Coleoptera: Curculionidae: Platypodinae)." *Journal of Asia-Pacific Entomology* 15:51–57.

Moser, J. C. 1985. "Use of Sporothecae by Phoretic *Tarsonemus* Mites to Transport Ascospores of Coniferous Bluestain Fungi." *Transactions of the British Mycological Society* 84:750–753.

Mueller, R. 2019. "Characterisation of Fungal Symbionts and Microbial Communities of *Austroplatypus incompertus* (Platypodinae) and Other Australian Ambrosia Beetle Species." PhD dissertation, Western Sydney University.

Mueller, U. G., N. M. Gerardo, D. K. Aanen, D. L. Six, and T. R. Schultz. 2005. "The Evolution of Agriculture in Insects." *Annual Review of Ecology, Evolution, and Systematics* 36:563–595.

Mueller, U. G., M. R. Kardish, H. D. Ishak, A. M. Wright, S. E. Solomon, et al. 2018. "Phylogenetic Patterns of Ant-Fungus Associations Indicate that Farming Strategies, Not Only A Superior Fungal Cultivar, Explain the Ecological Success of Leafcutter Ants." *Molecular Ecology* 27:2414–2434.

Mueller, U.G., J. J. Scott, H. D. Ishak, M. Cooper, and A. Rodrigues. 2010. "Monoculture of Leafcutter Ant Gardens." *PLoS ONE* 5:e12668.

Musvuugwa, T., Z. W. De Beer, T. A. Duong, L.L. Dreyer, K.C. Oberlander, and F. Roets. 2015. "New Species of Ophiostomatales from Scolytinae and Platypodinae Beetles in the Cape Floristic Region, Including the Discovery of the Sexual State of *Raffaelea*." *Antonie van Leeuwenhoek* 108:9339–9350.Nakashima, T. 1971. "Notes on the Associated Fungi and the Mycetangia of the Ambrosia Beetle, *Crossotarsus niponicus* Blandford (Coleoptera: Platypodidae)." *Applied Entomology and Zoology* 6:131–137.

Nakashima, T. 1972. "Notes on the Mycetangia of the Ambrosia Beetles *Platypus severini* Blandford and *P. calmus* Blandford (Coleoptera: Platypodidae)." *Applied Entomology and Zoology* 7:217–225.

Nakashima, T. 1975. "Several Types of the Mycetangia Found in Platypodid Ambrosia Beetles (Coleoptera: Platypodidae)." *Insecta Matsumurana* 7:1–69.

Nakashima, T. 1979. "Notes on the Time When the New Female Adults of the Ambrosia Beetle *Crossotarsus niponicus* Blandford (Coleoptera: Platypodidae) Harvest Their Symbiotic Fungi into Their Mycetangia." *Insecta Matsumurana* 17:1–9.

Nakashima, T. 1982. "Function and Location of Mycetangia in Ambrosia Beetles." In *The Ultrastructure and Functioning of Insect Cells*, edited by H. Akai, R. C. King, and S. Morohoshi, 87–90. Sapporo: Society for Insect Cells.

Nakashima, T, C. Goto, and T. Iizuka. 1987. "The Primary and Auxiliary Ambrosia Fungi Isolated from the Ambrosia Beetles, *Scolytoplatypus shogun* Blandford (Coleoptera: Scolytidae) and *Crossotarsus niponicus* Blandford (Coleoptera: Platypodidae)." *Journal of the Faculty of Agriculture, Hokkaido University* 63: 185–208.

Nel, W. J., Z. W. De Beer, M. J. Wingfield, and T. A. Duong. 2020. "The Granulate Ambrosia Beetle, *Xylosandrus crassiusculus* (Coleoptera: Curculionidae, Scolytinae), and Its Fungal Symbiont Found in South Africa." *Zootaxa* 4838:427–435.

Nigg, M., and L. Bernier. 2016. "From Yeast to Hypha: Defining Transcriptomic Signatures of the Morphological Switch in the Dimorphic Fungal Pathogen *Ophiostoma novo-ulmi*." *BMC Genomics* 17:920.

Nijs, C., and J. Billen. 2015. "Exocrine Glands in the Legs of the Social Wasp *Vespula vulgaris*." *Arthropod Structure & Development* 44:433–443.

Nobre, T., N. A. Koné, S. Konaté, K. E. Linsenmair, and D. K. Aanen. 2011. "Dating the Fungus-Growing Termites' Mutualism Shows a mixture between Ancient Codiversification and Recent Symbiont Dispersal across Divergent Hosts." *Molecular Ecology* 20:2619–2627.

Noirot, C., and A. Quennedey. 1974. "Fine Structure of Insect Epidermal Glands." *Annual Review of Entomology* 19:61–80.

Nord, J. C. 1972. "Biology of the Columbian Timber Beetle, *Corthylus columbianus* (Coleoptera: Scolytidae), in Georgia." *Annals of the Entomological Society of America* 65:350–358.

Norris, D. M. 1979. "The Mutualistic Fungi of Xyleborini Beetles." In *Insect-Fungus Symbiosis: Nutrition, Mutualism and Commensalism*, edited by L. R. Batra, 53–63. New York: Wiley.

Nunberg, M. 1951. "Contribution to the Knowledge of Prothoracic Glands of Scolytidae and Platypodidae (Coleoptera)." *Annales Musei Zoologici Polonici* 14:261–267.

Nuotclà, J. A., P. H. Biedermann, and M. Taborsky. 2019. "Pathogen Defence Is a Potential Driver of Social Evolution in Ambrosia Beetles." *Proceedings of the Royal Society B—Biological Sciences* 286:20192332.

O'Donnell, K., S. Sink, R. Libeskind-Hadas, J. Hulcr, M. T. Kasson, et al. 2015. "Discordant Phylogenies Suggest Repeated Host Shifts in the *Fusarium–Euwallacea* Ambrosia Beetle Mutualism." *Fungal Genetics and Biology* 30:277–290.

Orañegui, F. B., and T. H. Atkinson. 1984. "Biología del barrenador de las ramas del peral *Corthylus fuscus* Blandford (Coleoptera: Scolytidae), en el norte del Estado de Morelos." *Folia Entomolágim Mexicana* 60: 83–101.

Paine, T. D., and M. C. Birch. 1983. "Acquisition and Maintenance of Mycangial Fungi by *Dendroctonus brevicomis* LeConte (Coleoptera: Scolytidae)." *Environmental Entomology* 12:1384–1386.

Paine, T. D., K. F. Raffa, and T. C. Harrington. 1997. "Interactions among Scolytid Bark Beetles, Their Associated Fungi, and Live Host Conifers." *Annual Review of Entomology* 42:179–206.

Pasteels, J. M., M. Rowell-Rahier, J. C. Braekman, D. Daloze, and S. Duffey. 1989. "Evolution of Exocrine Chemical Defense in Leaf Beetles (Coleoptera: Chrysomelidae)." *Experientia* 45:295–300.

Penagos, C. M. O. 2015. "Interacción entre *Corthylus zulmae* Wood y hongos ambrosiales y su relación en la muerte descendente de árboles de *Alnus acuminata* HBK." MS thesis, Universidad Nacional de Colombia, Bogotá.

Petrov, A. B. 2017. "Morfologičeskie i Povedenčeskie Osobennosti Koroedov Triby Scolytini (Coleoptera, Curculionidae, Scolytinae), Associirovannyh s Mikobiotoj." *Izvestija Sankt-Peterburgskoj Lesotehničeskoj Akademii* 220:20–32.

Pistone, D., J. Gohli, and B. H. Jordal. 2018. "Molecular Phylogeny of Bark and Ambrosia Beetles (Curculionidae: Scolytinae) Based on 18 Molecular Markers." *Systematic Entomology* 43:387–406.

Raffa, K. F., J. Grégoire, and B. S. Lindgren. 2015. "Natural History and Ecology of Bark Beetles." In *Bark Beetles*, edited by F. E. Vega and R. W. Hofstetter, 1–40. London: Academic Press.

Roberts, E. M., C. N. Todd, D. K. Aanen, T. Nobre, H. L. Hilbert-Wolf, et al. 2016. "Oligocene Termite Nests with in Situ Fungus Gardens from the Rukwa Rift Basin, Tanzania, Support a Paleogene African Origin for Insect Agriculture." *PloS ONE* 11:e0156847.

Roche, A., and J. Lhoste. 1960. "Descriptions d'organes adaptés à la dissémination des champignons chez les Scolytoidea (Coléoptères)." *Comptes Rendus Hebdomadaires des Séances de l'Académie des Sciences* 250:2056–2058.

Roeper, R. A. 1988. "Interaction between the Ambrosia Beetle *Corthylus punctatissimus* and Its Mutualistic Fungi." *Proceedings of the XVIII International Congress of Entomology*, Vancouver, BC, Canada, July 3–9, 1988, 410.

Roeper, R. A. 1996. "Ambrosia Beetles of the Continental United States and Canada and Status of Knowledge of Their Associated Primary Symbiotic Fungi." *Newsletter of the Michigan Entomological Society* 41:12–14.

Roeper, R. A. 2011. "Relationship between Mycangia and Symbiotic Fungi of the Ambrosia Beetles in the USA." *Newsletter of the Michigan Entomological Society* 56:32–33.

Roets, F., and K. C. Oberlander. 2020. "Symbiotic Yeasts from the Mycangium, Larval Gut and Woody Substrate of an African Stag Beetle *Xiphodontus antilope* (Coleoptera: Lucanidae)." *Antonie van Leeuwenhoek* 113:1123–1134.

Saucedo, J. R., R. C. Ploetz, J. L. Konkol, M. Ángel, J. Mantilla, et al. . 2018. "Nutritional Symbionts of a Putative Vector, *Xyleborus bispinatus*, of the Laurel Wilt Pathogen of Avocado, *Raffaelea lauricola*." *Symbiosis* 75:29–38.

Saucedo-Carabez, J. R., R. C. Ploetz, J. L. Konkol, D. Carrillo, and R. Gazis. 2018. "Partnerships between Ambrosia Beetles and Fungi: Lineage-Specific Promiscuity among Vectors of the Laurel Wilt Pathogen, *Raffaelea lauricola*." *Microbial Ecology* 76:925–940.

Sawada, Y., and K. Morimoto. 1986. "The Mycetangia and the Mode of the Fungus Transmission in the Weevil Genus *Euops* (Coleoptera: Attelabidae)." *Science Bulletin of the Faculty of Agriculture, Kyushu University* 49:197–205.

Schedl, W. 1962. "Ein Beitrag zur Kenntnis der Pilzübertragungsweise bei xylomycetophagen Scolytiden (Coleoptera)." *Sitzungsberichten der Österreichischen Akademie der Wissenschaften* 171:363–387.

Schedl, W. 1963. "Biologie des gehöckerten Eichenholzbohrers, *Xyleborus monographus* Fab. (Scolytidae, Coleoptera)." *Journal of Applied Entomology* 53:411–428.

Schmidberger, J. 1836. "Naturgeschichte des Apfelborkenkäfers *Apate dispar*." *Beiträge zur Obstbaumzucht und zur Naturgeschichte der den Obstbäumen schädlichen Insekten* 4: 213–230.

Schneider, I. 1976. "Untersuchungen über die biologische bedeutung der Mycetangien bei einigen Ambrosiakäfern." *Material und Organismen* 3:489–497.

Schneider, I. 1987. "Verbreitung, Pilzübertragung und Brutsystem des Ambrosiakäfers *Xyleborus affinis* im Vergleich mit *X. mascarensis* (Coleoptera: Scolytidae)." *Entomologia Generalis* 12:267–275.

Schneider, I. 1991. "Einige ökologische Aspekte der Ambrosiasymbiose." *Anzeiger für Schädlingskunde, Pflanzenschutz, Umweltschutz* 64:41–45.

Schneider, I., and J. A. Rudinsky. 1969a. "Anatomical and Histological Changes in Internal Organs of Adult *Trypodendron lineatum*, *Gnathotrichus retusus*, and *G. sulcatus* (Coleoptera: Scolytidae)." *Annals of the Entomological Society of America* 62:995–1003.

Schneider, I. A., and J. A. Rudinsky. 1969b. "Mycetangial Glands and Their Seasonal Changes in *Gnathotrichus retusus* and *G. sulcatus*." *Annals of the Entomological Society of America* 62:39–43.

Schneider-Orelli, A. 1911. "Übertragung und Keimung des Ambrosiapilzes von *Xyleborus* (*Anisandrus*) *dispar* F." *Naturwissenschaftliche Zeitschrift für Forst- und Landwirtschaft* 9:186–192.

Schneider-Orelli, A. 1913. "Untersuchungen über den pilzzüchtenden Obstbaumborkenkäfer, *Xyleborus* (*Anisandrus*) *dispar* und seinen Nährpilz." *Central-Blatt fur Bakteriologie, Abt. II* 38:25–110.

Schultz, T. R., and S. G. Brady. 2008. Major Evolutionary Transitions in Ant Agriculture." *Proceedings of the National Academy of Sciences* 105:5435–5440.

Sil, A., and A. Andrianopoulos. 2015." Thermally Dimorphic Human Fungal Pathogens—Polyphyletic Pathogens with a Convergent Pathogenicity Trait." *Cold Spring Harbor Perspectives in Medicine* 5:a019794.

Simmons, D. R., Z. W. de Beer, Y. T. Huang, C. Bateman, A. S. Campbell, et al. 2016a. "New *Raffaelea* Species (Ophiostomatales) from the USA and Taiwan Associated with Ambrosia Beetles and Plant Hosts." *IMA Fungus* 7:265–273.

Simmons, D. R., Y. Li, C. C. Bateman, and J. Hulcr. 2016b. "*Flavodon ambrosius* sp. nov., a Basidiomycetous Mycosymbiont of *Ambrosiodmus* Ambrosia Beetles." *Mycotaxon* 131:277–285.

Six, D. L. 2003. "Bark Beetle–Fungus Symbioses." In: *Insect Symbiosis*, edited by K. Bourtzis and T. A. Miller, 99–116. New York: CRC Press.

Six, D. L. 2012. "Ecological and Evolutionary Determinants of Bark Beetle–Fungus Symbioses." *Insects* 3:339–366.

Six, D. L. 2020. "Niche Construction Theory Can Link Bark Beetle-Fungus Symbiosis Type and Colonization Behavior to Large Scale Causal Chain-Effects." *Current Opinion in Insect Science* 39:27–34.

Six, D. L, and R. Bracewell. 2015. "*Dendroctonus*." In *Bark Beetles*, edited by F. E. Vega and R. W. Hofstetter, 305–350. London: Academic Press.

Six, D. L., and J. J. Elser. 2020. "Mutualism Is Not Restricted To Tree-Killing Bark Beetles and Fungi: The Ecological Stoichiometry of Secondary Bark Beetles, Fungi, and a Scavenger." *Ecological Entomology*, June 20. doi:10.1111/een.12897.

Skelton, J., A. J. Johnson, M. A. Jusino, C. C. Bateman, Y. Li, et al. 2019. "A Selective Fungal Transport Organ (Mycangium) Maintains Coarse Phylogenetic Congruence between Fungus-Farming Ambrosia Beetles and Their Symbionts." *Proceedings of the Royal Society B* 286:20182127.

Skelton, J., M. A. Jusino, Y. Li, C. Bateman, P. H. Thai, et al. 2018. "Detecting Symbioses in Complex Communities: The Fungal Symbionts of Bark and Ambrosia Beetles within Asian Pines." *Microbial Ecology* 24:1–2.

Smith, S. M. 2013. "Phylogenetics of the Scolytini (Coleoptera: Curculionidae: Scolytinae) and Host-Use Evolution." PhD dissertation, Michigan State University, Lansing.

Smith, S. M., and A. I. Cognato. 2010. "A Taxonomic Revision of *Camptocerus* Dejean (Coleoptera: Curculionidae: Scolytinae)." *Insecta Mundi* 0148:1–88.

Smith, S. M., D. S. Kent, J. J. Boomsma, and A. J. Stow. 2018. "Monogamous Sperm Storage and Permanent Worker Sterility in a Long-Lived Ambrosia Beetle." *Nature Ecology & Evolution* 2:1009–1018.

Spahr, E., M. T. Kasson, and T. Kijimoto. 2020. "Micro-Computed Tomography Permits Enhanced Visualization of Mycangia across Development and between Sexes in Euwallacea Ambrosia Beetles." *PLoS ONE* 15:e0236653.

Stone, W. D., T. E. Nebeker, W. A. Monroe, and J. A. MacGown. 2007. "Ultrastructure of the Mesonotal Mycangium of *Xylosandrus mutilatus* (Coleoptera: Curculionidae)." *Canadian Journal of Zoology* 85:232–238.

Strohmeyer, H. 1911. "Die biologische Bedeutung sekundärer Geschlechtscharaktere am Kopfe weiblicher Platypodiden." *Entomologische Blätter* 7:103–107.

Takagi, K. 1967. "The Storage Organ of Symbiotic Fungus in the Ambrosia Beetle *Xyleborus rubricollis* Eichhoff: Coleoptera: Scolytidae." *Applied Entomology and Zoology* 2:168–170.

Talbot, P. H. 1977. "The *Sirex-Amylostereum-Pinus* Association." *Annual Review of Phytopathology* 15:41–54.

Tanahashi, M, and M. Fremlin. 2013. "The Mystery of the Lesser Stag Beetle *Dorcus parallelipipedus* (L.) (Coleoptera: Lucanidae) Mycangium Yeasts." *Bulletin of the Amateur Entomological Society* 72:146–152.

Tanahashi, M., and C. J. Hawes. 2016. "The Presence of a Mycangium in European *Sinodendron cylindricum* (Coleoptera: Lucanidae) and the Associated Yeast Symbionts." *Journal of Insect Science* 16:76.

Tanahashi, M., K. Kubota, N. Matsushita, and K. Togashi. 2010. "Discovery of Mycangia and the Associated Xylose-Fermenting Yeasts in Stag Beetles (Coleoptera: Lucanidae)." *Naturwissenschaften* 97:311–317.

Thompson, B. M., J. Bodart, C. McEwen, and D. S. Gruner. 2014. "Adaptations for Symbiont-Mediated External Digestion in *Sirex noctilio* (Hymenoptera: Siricidae)." *Annals of the Entomological Society of America* 107:453–460.

Toki, W., M. Tanahashi, K. Togashi, and T. Fukatsu. 2012. "Fungal Farming in a Non-Social Beetle." *PLoS ONE* 7:e41893.

Van de Peppel, L. J., and D. K. Aanen. 2020. "High Diversity and Low Host-Specificity of *Termitomyces* Symbionts Cultivated by *Microtermes* spp. Indicate Frequent Symbiont Exchange." *Fungal Ecology* 45:100917.

Van de Peppel, L. J., D. K. Aanen, and P. H. Biedermann. 2018. "Low Intraspecific Genetic Diversity Indicates Asexuality and Vertical Transmission in the Fungal Cultivars of Ambrosia Beetles." *Fungal Ecology* 32:57–64.

Vanderpool, D., R. R. Bracewell, and J. P. McCutcheon. 2017. "Know Your Farmer: Ancient Origins and Multiple Independent Domestications of Ambrosia Beetle Fungal Cultivars. *Molecular Ecology,* October 31. doi.org/10.1111/mec.14394.

Vega, F. E., and P. H. Biedermann. 2019. "On Interactions, Associations, Mycetangia, Mutualists and Symbiotes in Insect-Fungus Symbioses." *Fungal Ecology* 24:100909.

Vega, F. E., F. Infante, and A. J. Johnson. 2015. "The Genus *Hypothenemus*, with Emphasis on *H. hampei*, the Coffee Berry Borer." In *Bark Beetles*, edited by F. E. Vega and R. W. Hofstetter, 427–494. London: Academic Press.

Vissa, S., and Hofstetter, R. W. 2017. "The Role of Mites in Bark and Ambrosia Beetle–Fungal Interactions." In *Insect Physiology and Ecology*, edited by V. D. C. Shields, 135–156. Zagreb: InTech.

Wood, S. L. 1957. "Ambrosia Beetles of the Tribe Xyloterini (Coleoptera: Scolytidae) in North America." *Canadian Entomologist* 89:337–354.

Wood, S. L. 1961. "A New *Dactylipalpus* (Coleoptera: Scolytidae) from the Philippine Islands." *Great Basin Naturalist* 21:8–9.

Wood, S. L. 1993. "Revision of the Genera of Platypodidae (Coleoptera)." *Great Basin Naturalist* 53:259–281.

Wood, S. L. 2007. *Bark and Ambrosia Beetles of South America (Coleoptera, Scolytidae)*. Provo, UT: Monte L. Bean Life Science Museum, Brigham Young University.

Wood, S.L. and D.E. Bright. 1992. "A Catalog of Scolytidae and Platypodidae (Coleoptera), Part 2. Taxonomic Index." *Great Basin Naturalist Memoirs* 13:1–1553.

Wuest, C. E., T. C. Harrington, S. W. Fraedrich, H. Y. Yun, and S. S. Lu. 2017. "Genetic Variation in Native Populations of the Laurel Wilt Pathogen, *Raffaelea lauricola*, in Taiwan and Japan and the Introduced Population in the United States." *Plant Dsease* 101:619–628.

Yang, W. F., H. Z. Ye, and M. Zhang. 2008. "Composition and Variety of the Ambrosia Fungi Associated with Ambrosia Beetle, *Xylosandrus germanus* (Blandford) (Coleoptera: Scolytidae)." *Acta Entomologica Sinica* 51:595–600.

Yuceer, C., C. Y. Hsu, N. Erbilgin, and K. D. Klepzig. 2011. "Ultrastructure of the Mycangium of the Southern Pine Beetle, *Dendroctonus frontalis* (Coleoptera: Curculionidae, Scolytinae): Complex Morphology for Complex Interactions." *Acta Zoologica* 92:2162–2124.

Zimmermann, V. G. 1973. "Vergleichende ökologisch-physiologische Untersuchungen an Ambrosiapilzen, assoziierten Bläuepilzen und Luftbläuepilzen." PhD dissertation, University of Göttingen, Germany.

Zimmermann, V. G., and H. Butin. 1973. "Untersuchungen über die Hitze- und Trockenresistenz holzbewohnender Pilze." *Flora* 162:393–419.

8 Agricultural and Proto-Agricultural Symbioses in Ants

Ana Ješovnik and Ted R. Schultz

Across most terrestrial habitats, ants engage in complicated relationships with other organisms. Ants are the only entirely eusocial family of insects: all 13,628 described species of Formicidae live in organized societies divided into reproductive and worker castes (Hölldobler and Wilson 1990; Bolton 2019). They have profound impacts on their habitats not only through their sheer numbers, but also due to the roles they play in food webs and nutrient cycling, as well as in the diverse symbioses they form with other organisms. The majority of these relationships—with other animals and with plants, fungi, and microorganisms—are poorly understood. Some of those that are better researched are shown to share striking similarities with human interspecies interactions, including agriculture (Schultz et al. 2005). This, in turn, suggests that agriculture is not a unique human behavior but that it has instead evolved convergently in very distantly related organisms (chapters 1 and 2, this volume).

A discussion of the interactions of ants with other organisms could be structured according to any number of organizational frameworks. For example, we could categorize such interactions based on whether they represent parasitism, commensalism, or mutualism. Here we have chosen to broadly survey multiple, mostly nutritional symbioses of ants, organized by symbiont taxon, and to call attention to the cases most similar to agricultural symbioses in humans. For readers interested in more detailed accounts of particular nutritional symbioses in which ants engage, we recommend the many available focused reviews (Way 1963; Weber 1972a; Pierce et al. 2002; Rico-Gray and Oliveira 2007; Stadler and Dixon 2008; Mehdiabadi and Schultz 2010; Orivel and Leroy 2011; Voglmayr et al. 2011; Mayer et al. 2014; Ivens 2015; Chomicki, Janda, and Renner 2017).

To describe the diversity of agricultural or agriculture-like behavior in ants we will use the definitions outlined in the introduction to this book: *Cultivation* is a set of practices that one species (the farmer) may perform to promote the growth of another species (the cultivated symbiont)—for example, preparing the substrate for growing crops, fertilizing, tilling, planting, pest control, protecting, and sheltering; these may be carried out on small scales and in the absence of domestication. *Domestication* is a genetic change in one species (the domesticate) in response to the influence of another species (the farmer) that benefits the farmer but that would reduce the fitness of the domesticate in its original niche. *Agriculture* is cultivation on a landscape-altering scale, characterized by economic (i.e., obligate) reliance. Although our definition of agriculture does not require domestication, in human agriculture domestication preceded agriculture in most cases (Fuller et al. 2014; chapter 10, this volume), and, in nonhuman agriculture, most cases involve a one-way or a two-way obligate mutualism between the farming species and a domesticated symbiont. Various ecological interactions, such as central-place foraging, niche construction, and incidental cultivation, have been suggested as consistent precursors to agriculture (Smith 2011a, 2011b; Allaby et al. 2015; chapter 14, this volume). Here we will use the umbrella term *proto-agriculture* to refer to such preagricultural ecologies and behaviors.

Ants and Other Animals

Many distantly related groups of ant species have independently evolved relationships with insects that feed on plant sap (figure 8.1A–C), mostly homopterans such as aphids and scale insects (Way 1963; Hölldobler and Wilson 1990). These sap-sucking insects provide ants with an excreted liquid called *honeydew*, which is rich in sugars and amino acids. Ants, in turn, provide sap-sucking insects with protection from predators and parasites, sanitation by removing honeydew from the insects' bodies, and, in some systems, transport and/or shelter. The majority of ant species that tend sap-feeding insects belong to the ant subfamilies Formicinae, Dolichoderinae, and Myrmicinae. The insects that engage in such symbioses with ants occur in the homopteran families Aphidae, Cercopidae, Cicadellidae, Coccidae, Fulgoridae, Membracidae, Pseudococcidae, and Psyllidae, but also in other insect groups such as the lepidopteran families Lycaenidae and Riodinidae (Schultz and McGlynn 2000; Fiedler 2001). About 1,000 of the 4,000 known aphid species, family Aphididae (figure 8.1A), are tended by ants (Bristow 1991; Ivens 2015), and at least 483 species (out of approximately 6,000 total) in the lepidopteran family Lycaenidae (figure 8.1B) are associated with ants (Pierce et al. 2002).

Ant-homopteran mutualisms range from facultative—that is, occurring optionally in response to circumstances—to obligate, in which one or both partners cannot survive outside of the symbiosis. The majority of homopteran-tending ant species are facultative mutualists, with an omnivorous diet that also includes, in addition to honeydew, animal prey, plant nectar, and plant seeds. The proportion of honeydew in ant diets varies greatly between species, from minor in otherwise mainly predatory species such as *Daceton armigerum*, to over 80% in, for example, the red wood ants of the *Formica rufa* species group (Hölldobler and Wilson 1990; Offenberg 2001; Cannon and Fell 2002; Paris and Espadaler 2009; Dejean et al. 2012; Seifert 2018). It also varies within species and within colonies, depending on seasonal or other changes in the availability of other food sources. Further, in addition to deriving nutrition from honeydew, when prey is scarce, the ants consume some of their sap-feeding insect symbionts as an additional protein source (Pontin 1958; Ivens 2015). When tended by ants, some species of ant-associated aphids express different behavioral and physiological traits, such as increased feeding rates and changes in the composition of the honeydew, relative to their phenotypes when unattended by ants (Banks and Nixon 1958; Yao and Akimoto 2001; Stadler and Dixon 2005; Ivens 2015). The dependence in ant-homopteran associations is asymmetric; ant-tended insects are more likely to be obligate symbionts than are their ant hosts, which are more likely to be facultative symbionts (Stadler and Dixon 2008). The distributions of obligate ant-tended mutualists are restricted not by the presence of their host plants but by the presence of the tending ant species (Hopkins and Thacker 1999). One of the largest species of aphids, *Stomaphis quercus*, lives only on oak trees within the foraging territory of the highly territorial European ant species *Lasius fuliginosus*, its obligate ant host (Goidanich 1957; Hopkins et al. 1999; Depa, Mroz, and Szawaryn 2012). The protection provided by actively tending ants accords well with the definition of cultivation discussed earlier. The morphological, behavioral, and physiological traits present in some ant-tended insects accord with the definition of domestication, as does the facultative/obligate ant/homopteran asymmetry of many of the mutualisms. As per our definition, fully developed agriculture requires a critical level of reliance on the part of the farmer. Because in most of the ant/homopteran systems the homopterans are not the only source of the ants' nutrition, they probably qualify as cultivation rather than as fully developed agriculture.

There are some notable exceptions. At least 12 ant species in the *Dolichoderus cuspidatus* species group, a clade of closely related species widely distributed in tropical Southeast Asia, engage in obligate symbioses with at least 35 different mealybug species from the tribe Allomyrmococcini (Homoptera: Pseudococcidae) (Dill and Maschwitz 1994). All of the ant species in this group depend on the honeydew produced by mealybugs and display a collection of behavioral traits that have earned them the name "migrating herdsmen" (Maschwitz and Hänel 1985). Mealybugs feed on the sap of very young plant parts, so once they have exhausted local resources the ants carry them to a new part of the plant or to a new tree, and they also keep them in their nests during the night. To be able to quickly migrate to new locations and to avoid wasting energy on nest construction, *D. cuspidatus*–group ants create temporary nests (so-called bivouacs) formed by their bodies, keeping the ant brood and the mealybug domesticates in the center. Some other ant species such as army ants form similar bivouacs, but ants of the *D. cuspidatus* species group are the only ant species that have evolved a mobile-nesting behavior as an adaptation to the requirements of their nutritional mutualists. In addition to highly efficient colony relocation, *D. cuspidatus*–group ants have evolved other behaviors and morphological features that facilitate mealybug transport, and, unlike most other ants, they reproduce by budding, in which a young daughter queen and a subset of the workers in her mother's colony depart to begin a new colony, instead of through mating flights (Dill and Maschwitz 1994). In the *D. cuspidatus* group, budding necessarily includes carrying along part of the colony's mealybug herd. Experimental evidence strongly indicates that the mealybugs in this system have evolved behavioral and morphological traits adapted for ant tending and that they are unable to survive without their ant hosts, per our definition of domestication. The adaptations of the host ants further indicate that they obligately rely on the relationship with their mealybugs, per our definition of agriculture.

Additional examples of obligate nutritional mutualism exist in other ant-homopteran symbioses. Ants from the genus *Acropyga* and some species of the ant genus *Lasius* (e.g., *Lasius flavus*) live in highly specialized and mutually obligate nutritional symbioses with root mealybugs and root aphids, respectively (Flanders 1957; Way 1963; Pontin 1978; LaPolla, Cover, and Mueller 2002; Schneider and LaPolla 2011; Seifert 2018). An *Acropyga* daughter queen departs on her mating flight carrying a mealybug symbiont in her mandibles (figure 8.1C) to serve as the "seed" individual for a clonally produced herd that will provide the root-honeydew-derived nutrition for the queen's new ant colony (Silvestri

1924; Wheeler 1935). As in the case of *Dolichoderus cuspidatus* and their mealybugs, these symmetrically obligate ant-homopteran associations involve traits associated with cultivation, domestication, and agriculture.

Ant-Lepidoptera associations, which include mostly lepidopteran larvae or pupae in the families Lycaenidae and Riodinidae, share similarities with ant-homopteran associations. They vary from facultative to obligate for the Lepidoptera but are facultative for the ants, and they depend on nutritious liquids produced by specialized organs of the lepidopteran larvae. In exchange for this nutrition, ants protect the larvae from predators and parasitoids (Hölldobler and Wilson 1990; Fiedler 2001, 2012; Pierce et al. 2002). Ant-tended lycaenid and riodinid larvae have thick cuticles and specialized acoustical structures that produce sounds, which enable them to exploit the ant social communication system and to attract the ants (DeVries 1991; Fiedler 2012; Tartally et al. 2019). A small percentage of ant-Lepidoptera associations have shifted from mutualism to parasitism in a direction opposite to that found in ant-homopteran associations. In these instances parasitic lepidopteran larvae are carried back to the nest, where they feed on ant brood or are fed by worker ants through trophallaxis (Fiedler, Hölldobler, and Seufert 1996; Pierce et al. 2002; Fiedler 2012; Tartally et al. 2019). The nutritious secretions provided by the lepidopteran larvae never comprise the main food resource of the ants, which also depend on plant nectar and homopteran honeydew (Fiedler 2001). Thus, although lepidopteran larvae are clearly genetically modified for life with ants and are thus domesticated, ant-lepidopteran associations better fit the definition of cultivation than of true agriculture.

Thousands of other animals, so-called myrmecophiles, are adapted to the predictable ecological niches created by ant colonies, inhabiting their nests in a spectrum of facultative to obligate symbioses (Hölldobler and Wilson 1990; Navarrete-Heredia 2001; Parmentier, Dekoninck, and Wenseleers 2014). In the majority of cases, rather than benefiting their ant hosts, myrmecophiles are parasites or commensals, so that, although many have evolved complex adaptations for life with ants, these mutualisms in general are best described as proto-agricultural.

Ants and Plants

Ants participate in a diverse array of symbiotic relationships with plants, ranging from diffuse to highly specific (Rico-Gray and Oliveira 2007; New 2017). Foremost among ant-plant associations is the dispersal of seeds by ants (myrmecochory), which occurs in some 11,000 species of plants (Lengyel et al. 2010). In the second-most diverse category of ant-plant associations, plants provide shelter and food and ants provide protection against herbivores and competing plants. Nearly 700 species in 50 different families of vascular plants have evolved "domatia," preformed cavities within plant stems, thorns, leaf petioles, and leaf pouches in which ants build their nests (Nelsen, Ree, and Moreau 2018). Classic ant-plant systems include, in the Neotropics, *Pseudomyrmex* ants and *Acacia* plants (figure 8.1D) (Janzen 1966, 1967) and *Azteca* ants and *Cecropia* plants (Janzen 1969) and in the Paleotropics, *Crematogaster* ants and *Acacia* plants in Africa (Stanton and Palmer 2011) and *Crematogaster* ants and *Macaranga* plants in Asia (Heil et al. 1998, 2001; Feldhaar et al. 2003). In all these examples, the plant provides the nesting site in the form of hollow structures and produces food in the form of either or both extrafloral nectaries and specialized,

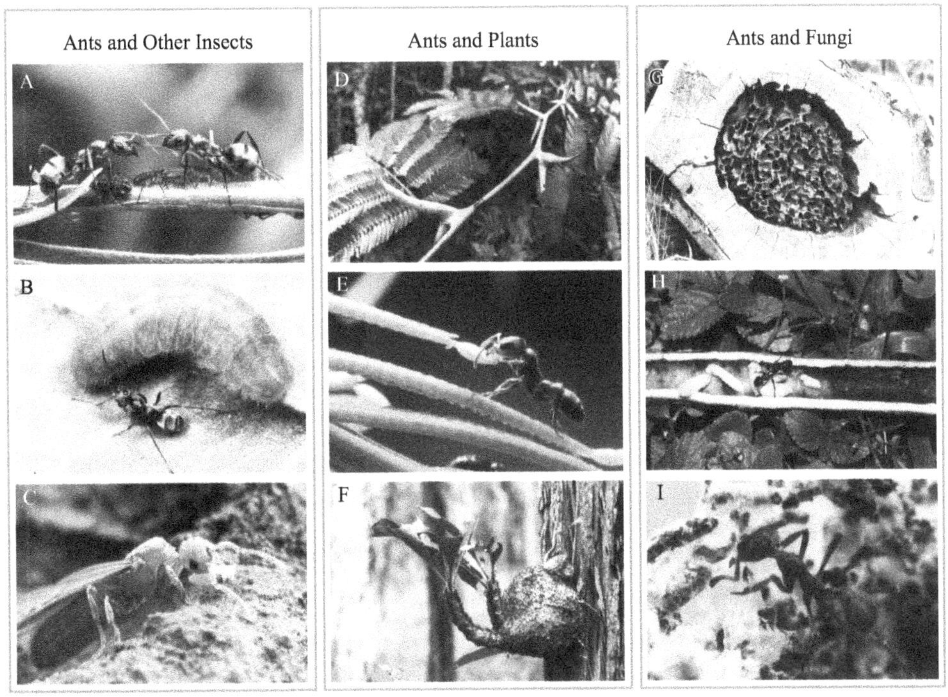

Figure 8.1
Ants and other insects: (A) two workers of the ant species *Formica aerata* tending aphids on a willow branch, California, USA; (B) the ant *Formica francoeuri* tending a larva of the Great Copper Butterfly (*Lycaena xanthoides*), California, USA; (C) the queen of the ant species *Acropyga goeldii* carrying a mealybug in her mandibles, Minas Geiras, Brazil. Ants and plants: (D) worker of the ant species *Pseudomyrmex peperi* on an *Acacia* plant with swollen, hollow thorns that serve as ant nesting sites, Cayóo District, Belize; (E) a worker of the ant species *Pseudomyrmex peperi* harvesting a lipid-rich Beltian body from the leaf tips of an *Acacia* plant, Armenia, Belize; (F) a *Myrmecodia* ant plant that is commonly inhabited by *Philidris* ants, Queensland, Australia. Ants and fungi: (G) a carton nest of the ant species *Lasius fuliginosus* inside of a tree trunk, eastern Austria; (H) nest of the ant species *Tetraponera aethiops* in a twig of *Barteria fistulosa*, with a patch of black fungus of the order Chaetothyriales, Southwest Region, Cameroon; (I) the fungus-farming ant *Apterostigma collare* on its fungus garden, Costa Rica. Photos reproduced with permission from: Alex Wild (A–F, I), Birgit C. Schlick-Steiner & Florian M. Steiner (G), Rumsais Blatrix (H).

protein- and/or lipid-rich food bodies: Beltian bodies in *Acacia* (figure 8.1E), Mullerian bodies in *Cecropia*, and Beccarian bodies in *Macaranga* (Bischof et al. 2013; New 2017). The plant-inhabiting ants are fiercely aggressive and protect their host plants from arthropod and large vertebrate herbivores, including elephants in the African savannas (Goheen and Palmer 2010). They also monitor the area surrounding their host plant and kill competing plants (Janzen 1969; Schupp 1986) and fungal pathogens (Letourneau 1998). Some evidence suggests that *Azteca* ants also provide *Cecropia* with a source of nitrogen (Sagers, Ginger, and Evans 2000). The domatia-inhabiting ants therefore exhibit behaviors associated with cultivation, as they protect, weed, and to some extent manure their plants. However, in the majority of cases they do not play a role in the propagation or dispersal of domatia-bearing host plants, and in many cases they do not obligately depend on the plant for nutrition because they also actively prey on insects. The majority of these ant-plant symbioses may be characterized as shelter and defense mutualisms, although nutrition plays a significant role in a subset of such associations.

Another category of ant-plant relationships is that of arboreally nesting ants and epiphytic plants, known as ant gardens (AGs) (Ule 1901; Janzen 1974; Orivel and Leroy 2011). We will refer to both the plants and the ants in these association as AG plants and AG ants, consistent with the existing literature (Orivel and Leroy 2011). In both Neotropical and Paleotropical forests, diverse epiphyte species grow in arboreal ant nests constructed from soil or carton (a fibrous, paper-like material constructed by ants from plant material, soil, wood, and other sources). Approximately 137 species of epiphytes associate to some degree with ants, but here we will briefly discuss only the smaller number of plants involved in obligate associations with AG ants (i.e., those that are never found free-living outside of the symbiosis) because those associations share the most traits with human agriculture (Orivel and Leroy 2011). A more detailed discussion of plant-farming by ants is provided by Chomicki (chapter 9, this volume). The greatest number of species known to be obligate AG ants occur in the genera *Philidris*, *Camponotus*, and *Crematogaster*, but many also occur in *Pheidole*, *Diacamma*, *Azteca*, *Odontomachus*, and *Pachycondyla* (Orivel and Leroy 2011; Chomicki, Janda, and Renner 2017). True ant garden epiphytes (figure 8.1F), with seeds that are consistently collected by ants and planted in their carton nests, belong to at least 14 plant families, but are most concentrated in the families Rubiaceae, Apocynaceae, Asclepiadaceae, and Gesneriaceae (Orivel and Leroy 2011; Chomicki, Janda, and Renner 2017). AG ants recognize the seeds of their obligate plant symbionts by chemical cues (Youngsteadt et al. 2008, 2010; Chomicki and Renner 2016b), collect them, incorporate them into their nests shortly after nest foundation, and continue to collect and plant new seeds throughout the life of the colony, constantly maintaining and expanding the garden. The plants provide shelter: in Neotropical ant gardens, the roots serve as a scaffold for the carton or soil nest, whereas in Australasia and Southeast Asia the ants nest inside the plant in enclosed nesting spaces, or domatia, formed from the hypocotyl (Chomicki et al. 2019; chapter 9, this volume). AG plants provide protection from abiotic factors such as rain and wind and food in the form of fruits, nectaries, and elaiosome-bearing seeds. In most cases these do not constitute the major portion of the ant diet, because the ants also forage for other food (Orivel and Leroy 2011). In addition to dispersing them, ants actively protect their AG plants from herbivores and provide them with nutrients. Canopy-inhabiting plants are particularly nutrient-deprived relative to ground-living plants, especially in the seedling phase. The ant nest, containing soil, vertebrate feces, ant feces, and various other organic materials, provides a continuous source of nutrients for the plants (so-called myrmecotrophy).

Obligate AG symbioses possess many of the traits of cultivation, but for most cases data are insufficient for drawing conclusions about whether domestication has occurred. One of the best-studied examples, however, convincingly demonstrates the domestication of a clade of plants by an ant species (Chomicki and Renner 2016a, 2016b; Chomicki et al. 2019, 2020; chapter 9, this volume). In this system, from Fiji, colonies of a single species of ant, *Philidris nagasau*, cultivate a clade of six closely related plant species in the genus *Squamellaria*, sometimes incorporating up to three species at a time into their ant gardens. Research has established the obligate dispersal and active planting of seeds by ants, fertilization of plants by ant fecal droplets, protection from herbivores, crop yield optimization by growing plants in full sun, and dependency of ants on the plants' fruits for food (Chomicki et al. 2020). Both symbionts (ants and plants) have evolved phenotypic traits unique to the symbiosis (e.g., loss of carton-nest building in the ants, specialized root structures that enable rapid development of the first domatia in the plants). This has led researchers to consider this a case of co-evolution. By the definition used in this book, such genetic change constitutes domestication. In Fiji, AGs can become very large, containing over 50 individual *Squamellaria* plants, a productivity comparable to that of early human agriculture (Chomicki et al. 2016b, 2020).

A different type of ant "garden" results from the association of the Amazon ant species *Myrmelachista schumanni* and several ground-living myrmecophytic plants, most notably *Duroia hirsuta* (family Rubiaceae) (Morawetz, Henzl, and Wallnofer 1992; Renner and Ricklefs 1998; Frederickson, Greene, and Gordon 2005; Salas-Lopez, Talaga, and Lalagüe 2016). *Myrmelachista* ants nest in domatia within *D. hirsuta* and systematically kill all other plants in the vicinity by injecting them with formic acid (Frederickson, Greene, and Gordon 2005). By killing competing plants, *Myrmelachista* ants promote the growth and spread of their host plant, which in turn expands the number of nesting sites for their colonies. Such colonies can become very large and contain up to 3 million workers and 15,000 queens distributed across monocultural stands of up to 600 *Duroia* trees (Frederickson, Greene, and Gordon 2005). Such stands are known colloquially as "devil's gardens" and are believed by local people to be inhabited by evil forest spirits. Until recently, devil's gardens were attributed solely to allelopathic compounds produced by the *Duroia* plants (Campbell, Richardson, and Rosas 1989; Pfannes and Baier 2002). It has alternatively been suggested that, rather than produced by *D. hirsuta*, the "domatia" are in fact the products of galling by the *M. schumanni* ants (Edwards et al. 2009).

A similar system is that of the ant *Pseudomyrmex triplarinus* and the host plant *Triplaris americana* (Polygonaceae), which provides the ants with domatia but no nutrition such as extafloral nectaries. The *P. triplarinus* ants, which feed on homopteran honeydew, remove seeds and prune seedlings, producing cleared areas around their host plants (Larrea-Alcázar and Simonetti 2007). Given that there is a cost of increased herbivory in both systems (Pfannes and Baier 2002; Frederickson and Gordon 2007), that there are no plant-generated food rewards, and that the *D. hirsuta* "domatia" may instead be ant-induced galls, it remains unclear whether or not the plants in these systems receive any net benefits or are domesticated. Given this uncertainty, "devil's gardens" may be interpreted as cultivation rather than agriculture.

Ants and Fungi

Symbiotic associations between ants and fungi are far less frequent than the associations of ants with animals and plants. As far as is known, ants rarely feed on fungi (Orr and Charles 1994; Tobin 1994; Epps and Penick 2018) or disperse them (Schultz and McGlynn 2000), although fascinating and poorly investigated exceptions exist (Witte and Maschwitz 2008; Von Beeren, Mair, and Witte 2014). Indeed, ants, which live mostly in the soil, limit their exposure to fungi via antibiotic secretions from the metapleural, mandibular, and other glands and via behaviors such as self- and allogrooming. In spite of this general avoidance of fungi, ant-fungus symbioses include some of the most remarkable examples of agriculture in ants.

Carton-Nest Builders

Similar to shelter-providing symbioses seen in most ant gardens, some carton-building ant species use fungal hyphae for structural reinforcement of carton nests (figure 8.1G) (Hölldobler and Wilson 1990; Schlick-Steiner et al. 2008; Mayer et al. 2014) and carton galleries that are used as prey traps (Dejean et al. 2005; Mayer and Voglmayr 2009). One such example is the European ant species *Lasius fuliginosus*, which lives in symbiosis with a fungus that is found only in ant nests and is most commonly referred to as *Cladosporium myrmecophilum* (Maschwitz and Hölldobler 1970; Hölldobler and Wilson 1990; Seifert 2018). Interestingly, genetic data from Eastern Austria suggests that *Lasius fuliginosus* and three other *Lasius* species can host in their colonies at least two fungal species at the same time and that a total of four fungal species are associated with *L. fuliginosus* nests (Schlick-Steiner et al. 2008). The carton nests are built in soil cavities and tree trunks from chewed-up particles of wood, soil, and plant matter mixed with ant saliva, excretions, and aphid honeydew (figure 8.1G). The fungal hyphae grow inside of the walls of the carton nest, providing scaffolding and reinforcement for the sponge-like carton galleries. Ants manure the walls with their fecal droplets and with collected honeydew, providing nutrition for the fungus. They also continuously crop the fungus to prevent it from overgrowing the nest. At least one of the fungal species they grow is transmitted vertically: when virgin daughter ant queens fly out to mate and start new colonies, they carry a small fungal inoculum from the parental nest inside their infrabuccal pockets (Hölldobler and Wilson 1990; Schlick-Steiner et al. 2008). Because ants disperse, inoculate, manure, provide nutrients for, and crop the fungus, we consider the *Lasius*-wall fungus mutualism to represent an obvious example of cultivation, and, because the fungus is obligately dependent on the ants (and, apparently, the ants are also dependent on the fungus), the association fulfills our definition of domestication. It does not constitute a nutritional agricultural symbiosis, but instead is akin to human non-nutritional, "material-goods" agricultures such as the cultivation of bottle gourds, lumber, and cotton. The symbioses between carton-building ants and fungi can also be viewed as an extension of the ant-aphid farming mutualism. The honeydew collected from tended aphids is used as a food source both for the ants and for the ants' structural symbiont, the fungus.

Black-Yeast Growers

Few ant-fungus nutritional agricultural systems are known. One broad category, the product of multiple independent evolutionary origins, involves domatia-bearing plants, domatia-inhabiting ants, and ascomycete "black yeast" fungi from the order Chaetothyriales (figure 8.1H). The most well-known example was first observed more than a century ago (Miehe 1911; Bailey 1920), but properly described and recognized as a nutritional mutualism only recently (Defossez et al. 2009, 2011; Blatrix et al. 2012). In this system, the African ant species *Petalomyrmex phylax* participates in a mutualism with the domatia-forming plant *Leonardoxa africana*. A small patch of black-yeast fungus grows on the inner surface of most domatia, where it is apparently nourished by the ants with feces and arthropod parts. The ants, in turn, use the black yeast as a food source, at least for their larvae (Blatrix et al. 2012). Similar domatia-inhabiting black-yeast fungal species are now known to associate with at least 19 plant and 10 ant genera, and these associations have apparently arisen independently in at least three geographically distinct plant-ant linages, including a subset of Neotropical *Cecropia-Azteca* symbioses, in which the fungi are transmitted vertically by *Azteca* foundress queens (Voglmayr et al., 2011; Blatrix et al., 2012, 2013; Mayer et al., 2014; Nepel et al., 2016; Mayer et al. 2018). Each domatia-inhabiting black-yeast species seems to be specific to a particular ant-plant symbiosis (Defossez et al. 2011; Voglmayr et al. 2011). This potentially widespread and complex system clearly needs to be further studied, especially the nutrient flow between ants, fungi, and plants. The current limited data indicate that the *Petalomyrmex-Leonardoxa* and *Azteca-Cecropia* systems involve most of the traits of cultivation, domestication, and agriculture (Voglmayr et al. 2011; Mayer et al. 2018).

The Fungus Farmers

The best known fungus-farming ants are the 245 extant species belonging to a single Neotropical clade, subtribe Attina (so-called attine ants) (Schultz and Brady 2008; Branstetter et al. 2017; chapter 14, this volume). All attine-ant species are obligate fungus-farmers, unable to live without their fungus gardens (figure 8.11). The cultivated fungi (phylum Basidiomycota, order Agaricales, families Pterulaceae and Agaricaceae) are grown in monocultures inside ant nests, in most species in underground nest chambers, on substrates of organic materials collected by the ants (Möller 1893; Weber 1972b; Mehdiabadi and Schultz 2010). Attine ants disperse their fungal crops; when leaving from the maternal nest to mate and start a new colony, daughter queens carry in their infrabuccal pocket a fungal pellet from the maternal garden. Attine ants also nourish their fungal cultivars, manure them with their fecal droplets, weed out microbial pathogens, and practice biochemical pest control by culturing antibiotic-producing bacteria in specialized crypts on their bodies (Weber 1972b; Bass and Cherrett 1996; Currie and Stuart 2001; Li et al. 2018). Attine-ant agriculture can be broadly divided into "lower" and "higher" ant agricultural systems. In lower ant agriculture, the fungal cultivars are facultative symbionts that can apparently live freely without the ants. Although it remains possible that lower-attine fungi possess adaptations for life with ants, this is a poorly explored area of inquiry, and so, in the case of lower ant agriculture, there is currently no evidence for domestication of fungi by ants (Mueller et al. 1998; Schultz and Brady 2008). In higher ant agriculture, practiced by 110 species of attine ants, the fungal cultivars are obligate symbionts, never found outside of the symbiosis (Mehdiabadi and Schultz 2010). Higher-attine fungi possess clear adaptations for life with ants: they are all polyploid, they consistently produce nutritious food bodies called "gongylidia" that are preferentially eaten by the ants, and they are significantly different from lower-attine fungi in expressed enzyme and amino acid profiles (Möller 1893; De Fine Licht and Boomsma 2014; Kooij et al. 2014, 2015). Higher-attine agriculture possesses all of the traits of domestication and agriculture (Schultz et al. 2005). The scale of agriculture practiced by the leaf-cutting ants, a subset of the higher-attine ants, is comparable to human industrial-scale agriculture (chapter 14, this volume), and, in terms of number of agricultural adaptations, more complex (chapter 1, this volume).

Leaf-cutting ants, unlike all other fungus-farming ants, have evolved the ability to cut and process fresh vegetation (living leaves, flowers, and grasses) as a fungal substrate. This ability has enabled leaf-cutter ant colonies to evolve into "superorganisms," consisting of 5 million or more individuals and dominating New World grassland and forest ecosystems (Schultz and Brady 2008). A mature nest of a leaf-cutting *Atta* species is the ecological equivalent of a large vertebrate herbivore in terms of biomass, life span, and consumption: a medium-sized colony collects almost one ton of leaves per year (Wirth et al. 1997, 2003; Hölldobler and Wilson 2010). *Atta* nests can be up to 7 meters deep and include more than 7,000 garden chambers, separate waste-disposal chambers, and a ventilation system (Moreira et al. 2004; Forti et al. 2017). Leaf-cutter nests alter the chemical and physical properties of the soil, contribute substantially to carbon turnover in Neotropical ecosystems, and represent an important resource for many other symbiotic species (Navarrete-Heredia 2001; Ballari, Farji-Brener, and Tadey 2007; Farji-Brener and Werenkraut 2017). As in human industrial-scale agriculture, leaf-cutter agriculture is achieved through finely scaled division of labor, accomplished through a highly polymorphic worker-caste task force in which the largest individuals (the "soldier" caste) weigh 200 times more than the smallest individuals (the gardener-nurses) (Wilson 1980; Hart and Ratnieks 2001; Garrett et al. 2016).

Remarkably, in at least some higher-attine ant species, it appears that higher agriculture has produced an agricultural surplus that ants can "trade" for other goods, similar to exchanges in human societies and to the earlier example in which the ant species *Lasius fuliginosus* uses surplus honeydew collected from aphids to nourish its structurally beneficial fungal symbiont. Relative to when it is not supporting a commensal ant symbiont, when it is hosting a colony of the socially parasitic, non-fungus-growing ant *Megalomyrmex symmetochus*, a colony of the higher-attine ant species *Sericomyrmex amabilis* increases the productivity of its fungus gardens by foraging for higher-protein substrates (Shik et al. 2018). Most of the time, *Megalomyrmex symmetochus* parasitizes the *S. amabilis* nest, feeding on its fungus garden and brood (Adams et al. 2000). *Sericomyrmex amabilis* is not very effective at colony defense, whereas *M. symmetochus*, which possesses volatile alkaloid venom, is very efficient at defense. When the *S. amabilis* colony is attacked by other ants, particularly by the raiding "agro-predatory" species *Gnamptogenys hartmani*, *M. symmetochus* defends the nest and usually rescues the colony from complete devastation (Adams et al. 2013). The ability of *Sericomyrmex amabilis* to upregulate the nutritional content of its fungus garden provides it with excess agricultural goods that are used to compensate *M. symmetochus*, which can behave as a simple social parasite or an efficient mercenary defender, depending on context.

Conclusion

Ants are one of the nine phylogenetic lineages of animals that have independently evolved similar sets of agricultural behavioral traits, although with varying levels of complexity (chapter 1, this volume). The specific ant behaviors that have converged on human practices of cultivation, domestication, and agriculture have been discussed throughout this chapter. In general, ants most commonly display behaviors comparable to human cultivation practices, such as fertilizing, planting, protecting, harvesting, and sheltering, while examples of convergence on domestication and agriculture are less common. In terms of kingdoms of life, ants frequently associate with animals and plants, and those associations range from facultative cultivation, such as tending homopteran insects or weeding and manuring epiphytic plants, to obligate dependence on domesticates, such as the ant *Dolichoderus cuspidatus* converging on the behavior of human nomadic herdsman, which migrate because of their cattle, and *Philidris nagasau* ants planting their obligate crop plants closer to the sun, convergent with various human practices to maximize crop yield. The known agricultural symbioses of ants and fungi, less common than those with plants, are apparently obligate for the ant species engaged in them. They also share similarities with human agriculture: *Lasius fuliginosus* obligately cultivates its fungus for a structural rather than a nutritional purpose, similar to human cultivation of wood as a building material, while the fungus-farming attine ants independently evolved planting, substrate preparation, manuring, weeding, harvesting, and elaborate division of labor, as well as economic (i.e., obligate) reliance on their crop.

To conclude this review of the diversity of ant symbioses and the degree of convergence with human agricultural practices, we wish to consider if there are any lessons to be learned from ant-farming systems in the light of the urgent agricultural crisis currently experienced by humans. Unsustainable human agricultural practices on a global scale have altered planetary-level ecological processes through land-use change, carbon release, chemical pollution, biodiversity loss, and soil erosion (Skinner et al. 1997; Stoate et al. 2001; German, Thompson, and Benton 2017; IPBES 2019). Because of our reliance on industrial-scale monocultures, our crops are genetically less diverse and subsequently less resilient to unpredictable random catastrophic events (Dunn 2017; Nabhan et al. 2012). In our comparisons of traits of ant and human agriculture, an obvious lack of convergence exists in some of the traits relevant to agricultural sustainability. For example, colonies of *Atta* leaf-cutter ants, which harvest large amounts of plant material (e.g., 132 kg of biomass/ha/year), rarely exhaust local resources of native vegetation (Cherrett 1968; Wirth et al. 2003; Herz, Thompson, and Benton 2007), have positive impacts on the productivity of their local environments, and occupy keystone positions in local ecologies (Farji-Brener and Illes 2000; Farji-Brener and Werenkraut 2015). Also, despite growing a single fungal clone in a monoculture, vulnerable to the specialized, coevolved parasitic fungus *Escovopsis*, attine ants practice efficient pest management (Currie 2006; Mehdiabadi et al. 2010; chapter 11, this volume) by growing antibiotic-producing bacteria in specialized structures on their cuticles (Li et al. 2018). As a consequence of 60 million years of coevolution with their crops, crop pathogens, and antibiotic-producing bacteria, ants have developed efficient methods of biocontrol that do not adversely affect the sustainability of the system (Pathak, Kett, and Marvasi 2019).

Fungus farming in ants originated shortly after the end-of-Cretaceous mass-extinction-event known as the KT boundary (Branstetter et al. 2017). This was a period in earth history when organisms that depended on photosynthesis and those that depended on those organisms were threatened with extinction (Janzen 1995). Because fungi are saprobes, which can grow on dead plant and animal matter, this was also a period when fungivory and associating with fungi provided a stable food source for the ancestors of attine ants. From environmentally catastrophic conditions and in the midst of mass extinction, fungus-farming attine ants emerged as an ecologically successful radiation.

For human agriculture to survive analogous climate-related changes, we likewise need to find ways to diversify our resources and use them sustainably, rather than continuing to rely on those that are likely to be highly sensitive to climate change and associated natural disasters and that are dependent on large quantities of chemical pesticides and fertilizers (Dunn 2017; Poux and Aubert 2018). Our current approach to agriculture is much less resilient than those of the various ant farmers considered in this review. A shift to sustainable agricultural practices such as organic, conservation, or regenerative agriculture will require radical changes to present agricultural policies, a strong emphasis on conservation of wild relatives of cultivated plants and on the genetic diversity of existing cultivars, a shift in human diet, and mainstreaming of biodiversity-oriented policies into nonagricultural sectors (IPBES 2019). As the authors of this review find ants to be a continuous source of inspiration, we hope that the diversity and complexity of the most successful organisms on earth will inspire the reader to reconsider the ways of the ants (King Solomon, Proverbs 6:6), to discover in their study the motivation for much-needed change, and to continue to look for new solutions in nature.

Acknowledgments

A. J. is grateful to KLI and the workshop organizers for the opportunity to attend this workshop, and to all of the workshop participants for engaging discussions and good company. T. R. S. is likewise grateful to KLI for their support of our workshop and to the workshop participants for continuing intellectually stimulating interactions.

References

Adams, R. M. M., J. Liberti, A. A. Illum, T. H. Jones, D. R. Nash, and J. J. Boomsma. 2013. "Chemically Armed Mercenary Ants Protect Fungus-Farming Societies." *Proceedings of the National Academy of Sciences* 110(39): 15752–15757.

Adams, R. M. M., U. G. Mueller, T. R. Schultz, and B. Norden. 2000. "Agro-Predation: Usurpation of Attine Fungus Gardens by *Megalomyrmex* Ants." *Die Naturwissenschaften* 87(12): 549–554.

Allaby, R. G., L. Kistler, R. M. Gutaker, R. Ware, J. L. Kitchen, et al. 2015. "Archaeogenomic Insights into the Adaptation of Plants to the Human Environment: Pushing Plant–Hominin Co-Evolution Back to the Pliocene." *Journal of Human Evolution* 79:150–157.

Bailey, I. W. 1920. "Some Relations between Ants and Fungi." *Ecology* 1(3): 174–189.

Ballari, S., A. G. Farji-Brener, and M. Tadey. 2007. "Waste Management in the Leaf-Cutting Ant *Acromyrmex lobicornis*: Division of Labour, Aggressive Behaviour, and Location of External Refuse Dumps." *Journal of Insect Behavior* 20(1): 87–98.

Banks, C. J., and H. L. Nixon. 1958. "Effects of the Ant, *Lasius niger* L., on the Feeding and Excretion of the Bean Aphid, *Aphis fabae* Scop." *Journal of Experimental Biology* 35(4): 703–711.

Bass, M., and J. M. Cherrett. 1996. "Leaf-Cutting Ants (Formicidae, Attini) Prune Their Fungus to Increase and Direct Its Productivity." *Functional Ecology* 10(1): 55–61.

Bischof, S., M. Umhang, S. Eicke, S. Streb, W. Qi, et al. 2013. "*Cecropia peltata* Accumulates Starch or Soluble Glycogen by Differentially Regulating Starch Biosynthetic Genes." *Plant Cell* 25(4): 1400–1415.

Blatrix, R., S. Debaud, A. Salas-Lopez, C. Born, L. Benoit, et al. 2013. "Repeated Evolution of Fungal Cultivar Specificity in Independently Evolved Ant-Plant-Fungus Symbioses." *PLoS ONE* 8(7): e68101.

Blatrix, R., C. Djiéto-Lordon, L. Mondolot, P. La Fisca, H. Voglmayr, et al. 2012. "Plant-Ants Use Symbiotic Fungi as a Food Source: New Insight into the Nutritional Ecology of Ant-Plant Interactions." *Proceedings of the Royal Society B— Biological Sciences* 279(1744): 3940–3947.

Bolton, B. 2019. "An Online Catalog of the Ants of the World." http://antcat.org.

Branstetter, M. G., A. Ješovnik, J. Sosa-Calvo, M. W. Lloyd, B. C. Faircloth, et al. 2017. "Dry Habitats Were Crucibles of Domestication in the Evolution of Agriculture in Ants." *Proceedings of the Royal Society B—Biological Sciences* 284:20170095.

Bristow, C. M. 1991. "Why Are So Few Aphids Ant-Tended?" In *Ant-Plant Interactions*, edited by C. R. Huxley and D. F. Cutler, 104–119. Oxford: Oxford University Press.

Campbell, D. G., P. M. Richardson, and A. Rosas. 1989. "Field Screening for Allelopathy in Tropical Forest Trees, Particularly *Duroia hirsuta*, in the Brazilian Amazon." *Biochemical Systematics and Ecology* 17(5): 403–407.

Cannon, C. A., and R. D. Fell. 2002. "Patterns of Macronutrient Collection in the Black Carpenter Ant, *Camponotus pennsylvanicus* (De Geer) (Hymenoptera: Formicidae)." *Environmental Entomology* 31(6): 977–981.

Cherrett, J. M. 1968. "The Foraging Behaviour of *Atta cephalotes* L. (Hymenoptera, Formicidae)." *Journal of Animal Ecology* 37(2): 387.

Chomicki, G., M. Janda, and S. S. Renner. 2017. "The Assembly of Ant-Farmed Gardens: Mutualism Specialization Following Host Broadening." *Proceedings of the Royal Society B—Biological Sciences* 284:20161759.

Chomicki, G., G. Kadereit, S. S. Renner, and E. T. Kiers. 2020. "Tradeoffs in the Evolution of Plant Farming by Ants." *Proceedings of the National Academy of Sciences* 117(5): 2535–2543.

Chomicki, G., and S. S. Renner. 2016a. "Evolutionary Relationships and Biogeography of the Ant-Epiphytic Genus *Squamellaria* (Rubiaceae: Psychotrieae) and Their Taxonomic Implications." *PLoS ONE* 11(3): e0151317.

Chomicki, G., and S. S. Renner. 2016b. "Obligate Plant Farming by a Specialized Ant." *Nature Plants* 2(12): 16181.

Chomicki, G., C. J. Thorogood, A. Naikatini, and S. S. Renner. 2019. "*Squamellaria*: Plants Domesticated by Ants." *Plants, People, Planet* 1:302–305.

Currie, C. R. 2006. "Coevolved Crypts and Exocrine Glands Support Mutualistic Bacteria in Fungus-Growing Ants." *Science* 311(5757): 81–83.

Currie, C. R., and A. E. Stuart. 2001. "Weeding and Grooming of Pathogens in Agriculture by Ants." *Proceedings of the Royal Society B—Biological Sciences* 268:1033–1039.

De Fine Licht, H. H., and J. J. Boomsma. 2014. "Variable Interaction Specificity and Symbiont Performance in Panamanian *Trachymyrmex* and *Sericomyrmex* Fungus-Growing Ants." *BMC Evolutionary Biology* 14(1): 244.

Defossez, E., C. Djiéto-Lordon, D. McKey, M.-A. Selosse, and R. Blatrix. 2011. "Plant-Ants Feed Their Host Plant, but Above All a Fungal Symbiont to Recycle Nitrogen." *Proceedings of the Royal Society B—Biological Sciences* 278:1419–1426.

Defossez, E., M.-A. Selosse, M.-P. Dubois, L. Mondolot, A. Faccio, et al. 2009. "Ant-Plants and Fungi: A New Three-Way Symbiosis." *New Phytologist* 182(4): 942–949.

Dejean, A., J. H. C. Delabie, B. Corbara, F. Azémar, S. Groc, et al. 2012. "The Ecology and Feeding Habits of the Arboreal Trap-Jawed Ant *Daceton armigerum*." *PLoS ONE* 7(5): e37683.

Dejean, A., P. J. Solano, J. Ayroles, B. Corbara, and J. Orivel. 2005. "Arboreal Ants Build Traps to Capture Prey." *Nature* 434:973.

Depa, L., E. Mroz, and K. Szawaryn. 2012. "Molecular Identity of *Stomaphis quercus* (Hemiptera: Aphidoidea: Lachnidae) and Description of a New Species." *European Journal of Entomology* 109(3): 435–444.

DeVries, P. 1991. "Call Production by Myrmecophilous Riodinid and Lycaenid Butterfly Caterpillars (Lepidoptera): Morphological, Acoustical, Functional, and Evolutionary Patterns." *American Museum Novitates* 3025:1–23.

Dill, M., and U. Maschwitz. 1994. "The Migrating Herdsmen Ants of Southeast Asia: A Species-Rich Complex of Trophobiotic Symbioses of Dolichoderus Ants and Their Mealybugs (Homoptera: Pseudococcidae: Allomyrmococcini)." In *Les Insectes sociaux. 12ème Congrès de l'Union Internationale pour l'Étude des Insectes Sociaux*, edited by A. Lenoir, G. Arnold, and M. Lepage, 199. Paris: Université Paris Nord.

Dunn, R. 2017. *Never out of Season: How Having the Food We Want When We Want It Threatens Our Food Supply and Our Future*. New York: Little, Brown.

Edwards, D. P., M. E. Frederickson, G. H. Shepard, and D. W. Yu. 2009. "A Plant Needs Ants like a Dog Needs Fleas: *Myrmelachista schumanni* Ants Gall Many Tree Species to Create Housing." *American Naturalist* 174(5): 734–740.

Epps, M. J., and C. A. Penick. 2018. "Facultative Mushroom Feeding by Common Woodland Ants (Formicidae, *Aphaenogaster* spp.)." *Food Webs* 14:9–13.

Farji-Brener, A. G., and A. E. Illes. 2000. "Do Leaf-Cutting Ant Nests Make 'Bottom-Up' Gaps in Neotropical Rain Forests? A Critical Review of the Evidence." *Ecology Letters* 3(3): 219–227.

Farji-Brener, A. G., and V. Werenkraut. 2015. "A Meta-Analysis of Leaf-Cutting Ant Nest Effects on Soil Fertility and Plant Performance." *Ecological Entomology* 40(2): 150–158.

Farji-Brener, A. G., and V. Werenkraut. 2017. "The Effects of Ant Nests on Soil Fertility and Plant Performance: A Meta-Analysis." *Journal of Animal Ecology* 86(4): 866–877.

Feldhaar, H., B. Fiala, R. B. Hashim, and U. Maschwitz. 2003. "Patterns of the *Crematogaster-Macaranga* Association: The Ant Partner Makes the Difference." *Insectes Sociaux* 50(1): 9–19.

Fiedler, K. 2001. "Ants That Associate with Lycaeninae Butterfly Larvae: Diversity, Ecology and Biogeography." *Diversity and Distributions* 7(1–2): 45–60.

Fiedler, K. 2012. "The Host Genera of Ant-Parasitic Lycaenidae Butterflies: A Review." *Psyche: A Journal of Entomology* 2012:1–10.

Fiedler, K., B. Hölldobler, and P. Seufert. 1996. "Butterflies and Ants: The Communicative Domain." *Experientia* 52(1): 14–24.

Flanders, S. E. 1957. "The Complete Interdependence of an Ant and a Coccid." *Ecology* 38(3): 535.

Forti, L. C., A. P. P. de Andrade, R. da S. Camargo, N. Caldato, and A. A. Moreira. 2017. "Discovering the Giant Nest Architecture of Grass-Cutting Ants, *Atta capiguara* (Hymenoptera, Formicidae)." *Insects* 8(2): 1–13.

Frederickson, M. E., and D. M. Gordon. 2007. "The Devil to Pay: A Cost of Mutualism with *Myrmelachista schumanni* Ants in 'Devil's Gardens' Is Increased Herbivory on *Duroia hirsuta* Trees." *Proceedings of the Royal Society B—Biological Sciences* 274:1117–1123.

Frederickson, M. E., M. J. Greene, and D. M. Gordon. 2005. "'Devil's Gardens' Bedeviled by Ants." *Nature* 437(7058): 495–496.

Fuller, D. Q., T. Denham, M. Arroyo-Kalin, L. Lucas, C. J. Stevens, et al. 2014. "Convergent Evolution and Parallelism in Plant Domestication Revealed by an Expanding Archaeological Record." *Proceedings of the National Academy of Sciences* 111(17): 6147.

Garrett, R. W., K. A. Carlson, M. S. Goggans, M. H. Nesson, C. A. Shepard, et al. 2016. "Leaf Processing Behaviour in *Atta* Leafcutter Ants: 90% of Leaf Cutting Takes Place inside the Nest, and Ants Select Pieces That Require Less Cutting." *Royal Society Open Science* 3(1): 150111.

German, R. N., C. E. Thompson, and T. G. Benton. 2017. "Relationships among Multiple Aspects of Agriculture's Environmental Impact and Productivity: A Meta-Analysis to Guide Sustainable Agriculture." *Biological Reviews* 92(2): 716–738.

Goheen, J. R., and T. M. Palmer. 2010. "Defensive Plant-Ants Stabilize Megaherbivore-Driven Landscape Change in an African Savanna." *Current Biology* 20(19): 1768–1772.

Goidanich, A. 1957. "Le migrazioni coatte mirmecogene dello *Stomaphis quercus* Linnaeus, Afide olociclico monoico omotopo (Hemiptera Aphidoidea Lachnidae)." *Bollettino dell'Istituto di Entomologia della Università degli Studi di Bologna* 23:93–131.

Hart, A. G., and F. L. W. Ratnieks. 2001. "Task Partitioning, Division of Labour and Nest Compartmentalisation Collectively Isolate Hazardous Waste in the Leafcutting Ant *Atta cephalotes*." *Behavioral Ecology and Sociobiology* 49(5): 387–392.

Heil, M., B. Fiala, W. Kaiser, and K. E. Linsenmair. 1998. "Chemical Contents of *Macaranga* Food Bodies: Adaptations to Their Role in Ant Attraction and Nutrition." *Functional Ecology* 12(1): 117–122.

Heil, M., B. Fiala, U. Maschwitz, and K. E. Linsenmair. 2001. "On Benefits of Indirect Defence: Short- and Long-Term Studies of Antiherbivore Protection via Mutualistic Ants." *Oecologia* 126(3): 395–403.

Herz, H., W. Beyschlag, and B. Hölldobler. 2007. "Herbivory Rate of Leaf-Cutting Ants in a Tropical Moist Forest in Panama at the Population and Ecosystem Scales." *Biotropica* 39(4): 482–488.

Hölldobler, B., and E. O. Wilson. 1990. *The Ants*. Berlin: Springer-Verlag.

Hölldobler, B., and E. O. Wilson. 2010. *The Leafcutter Ants: Civilization by Instinct*. New York: W. W. Norton.

Hopkins, G. W., and J. I. Thacker. 1999. "Ants and Habitat Specificity in Aphids." *Journal of Insect Conservation* 3(1): 25–31.

IPBES (Intergovernmental Science-Policy Platform for Biodiversity and Ecosystem Services). 2019. *Global Assessment Report on Biodiversity and Ecosystem Services of the Intergovernmental Science-Policy Platform on Biodiversity and Ecosystem Services*, edited by E. S. Brondizio, J. Settele, S. Diaz, and H. T. Ngo. Bonn: IPBES secretariat.

Ivens, A. B. F. 2015. "Cooperation and Conflict in Ant (Hymenoptera: Formicidae) Farming Mutualisms—A Review." *Myrmecological News* 21:19–36.

Janzen, D. H. 1966. "Coevolution of Mutualism between Ants and Acacias in Central America." *Evolution* 20(3): 249.

Janzen, D. H. 1967. "Interaction of the Bull's-Horn Acacia (*Acacia cornigera* L.) with an Ant Inhabitant (*Pseudomyrmex ferrugineus* F. Smith) in Eastern Mexico." *Kansas University Science Bulletin* 47:315–558.

Janzen, D. H. 1969. "Allelopathy by Myrmecophytes: The Ant *Azteca* as an Allelopathic Agent of *Cecropia*." *Ecology* 50(1): 147–153.

Janzen, D. H. 1974. "Swollen-Thorn Acacias of Central America." *Smithsonian Contributions to Botany* 13:1–131.

Janzen, D. H. 1995. "Who Survived the Cretaceous?" *Science* 268(5212): 785.

Kooij, P. W., D. K. Aanen, M. Schiøtt, and J. J. Boomsma. 2015. "Evolutionarily Advanced Ant Farmers Rear Polyploid Fungal Crops." *Journal of Evolutionary Biology* 28(11): 1911–1924.

Kooij, P. W., A. Rogowska-Wrzesinska, D. Hoffmann, P. Roepstorff, J. J. Boomsma, et al. 2014. "*Leucoagaricus gongylophorus* Uses Leaf-Cutting Ants to Vector Proteolytic Enzymes towards New Plant Substrate." *ISME Journal* 8(5): 1032–1340.

LaPolla, J. S., S. P. Cover, and U. G. Mueller. 2002. "Natural History of the Mealybug-Tending Ant, *Acropyga epedana*, with Descriptions of the Male and Queen Castes." *Transactions of the American Entomological Society* 128:367–76.

Larrea-Alcázar, D. M., and J. A. Simonetti. 2007. "Why Are There Few Seedlings beneath the Myrmecophyte *Triplaris americana*?" *Acta Oecologica* 32(1): 112–118.

Lengyel, S., A. D. Gove, A. M. Latimer, J. D. Majer, and R. R. Dunn. 2010. "Convergent Evolution of Seed Dispersal By Ants, and Phylogeny and Biogeography in Flowering Plants: A Global Survey." *Perspectives in Plant Ecology, Evolution and Systematics* 12(1): 43–55.

Letourneau, D. K. 1998. "Ants, Stem-Borers, and Fungal Pathogens: Experimental Tests of a Fitness Advantage in Piper Ant-Plants." *Ecology* 79(2): 593.

Li, H., J. Sosa-Calvo, H. A. Horn, M. T. Pupo, J. Clardy, et al. 2018. "Convergent Evolution of Complex Structures for Ant-Bacterial Defensive Symbiosis in Fungus-Farming Ants." *Proceedings of the National Academy of Sciences* 115(42): 10720–10725.

Maschwitz, U., and H. Hänel. 1985. "The Migrating Herdsman *Dolichoderus* (*Diabolus*) *cuspidatus*: An Ant with a Novel Mode of Life." *Behavioral Ecology and Sociobiology* 17:171–184.

Maschwitz, U., and B. Hölldobler. 1970. "Der Kartonnestbau bei *Lasius fuliginosus* Latr. (Hym. Formicidae)." *Zeitschrift Fur Vergleichende Physiologie* 66(2): 176–189.

Mayer, V. E., M. E. Frederickson, D. McKey, and R. Blatrix. 2014. "Current Issues in the Evolutionary Ecology of Ant-Plant Symbioses." *New Phytologist* 202(3): 749–764.

Mayer, V. E., M. Nepel, R. Blatrix, F. B. Oberhauser, K. Fiedler, et al. 2018. "Transmission of Fungal Partners to Incipient *Cecropia*-Tree Ant Colonies." *PLoS ONE* 13(2): e0192207.

Mayer, V. E., and H. Voglmayr. 2009. "Mycelial Carton Galleries of *Azteca brevis* (Formicidae) as a Multi-Species Network." *Proceedings of the Royal Society B—Biological Sciences* 276(1671): 3265–3273.

Mehdiabadi, N. J., and T. R. Schultz. 2010. "Natural History and Phylogeny of the Fungus-Farming Ants (Hymenoptera: Formicidae: Myrmicinae: Attini)." *Myrmecological News* 13:37–55.

Miehe, H. 1911. "Untersuchungen über die javanische *Myrmecodia*." *Abhandlungen Der Mathematisch-Physischen Klasse Der Königlich-Sächsischen Gesellschaft Der Wissenschaften* 32:312–361.

Möller, A. 1893. "Die Pilzgärten Einiger Südamerikanischer Ameisen." *Botanische Mittheilungen Aus Den Tropen* 6:1–127.

Morawetz, W., M. Henzl, and B. Wallnofer. 1992. "Tree Killing by Herbicide Producing Ants for the Establishment of Pure *Tococa occidentalis* Populations in the Peruvian Amazon." *Biodiversity and Conservation* 1(1): 19–33.

Moreira, A. A., L. C. Forti, M. A. C. Boaretto, A. P. P. Andrade, J. F. S. Lopes, et al. 2004. "External and Internal Structure of *Atta bisphaerica* Forel (Hymenoptera: Formicidae) Nests." *Journal of Applied Entomology* 128(3): 204–211.

Mueller, U. G., S. A. Rehner, and T. R. Schultz. 1998. "The Evolution of Agriculture in Ants." *Science* 281(5385): 2034–2038.

Nabhan, G. P., K. Wilson, O. Aknazarov, K.-A. Kassam, L. Monti, et al. 2012. "Agrobiodiversity Shifts on Three Continents since Vavilov and Harlan: Assessing Causes, Processes, and Implications for Food Security." In *Biodiversity in Agriculture*, edited by P. Gepts, T. R. Famula, R. L. Bettinger, S. B. Brush, A. B. Damania, et al., 407–425. Cambridge: Cambridge University Press.

Navarrete-Heredia, J. L. 2001. "Beetles Associated with *Atta* and *Acromyrmex* Ants (Hymenoptera: Formicidae: Attini)." *Transactions of the American Entomological Society* 127:381–429.

Nelsen, M. P., R. H. Ree, and C. S. Moreau. 2018. "Ant-Plant Interactions Evolved through Increasing Interdependence." *Proceedings of the National Academy of Sciences* 115(48): 12253–12258.

Nepel, M., H. Voglmayr, R. Blatrix, J. T. Longino, K. Fiedler, et al. 2016. "Ant-Cultivated *Chaetothyriales* in Hollow Stems of Myrmecophytic *Cecropia* sp. Trees—Diversity and Patterns." *Fungal Ecology* 23:131–140.

New, T. R. 2017. "Classic Themes: Ants, Plants and Fungi." In *Mutualisms and Insect Conservation*, by T. R. New, 63–103. Cham: Springer.

Offenberg, J. 2001. "Balancing between Mutualism and Exploitation: The Symbiotic Interaction between *Lasius* Ants and Aphids." *Behavioral Ecology and Sociobiology* 49(4): 304–310.

Orivel, J., and C. Leroy. 2011. "The Diversity and Ecology of Ant Gardens (Hymenoptera: Formicidae; Spermatophyta: Angiospermae)." *Myrmecological News* 14:73–85.

Orr, A. G., and J. K. Charles. 1994. "Foraging in the Giant Forest Ant, *Camponotus gigas* (Smith) (Hymenoptera: Formicidae): Evidence for Temporal and Spatial Specialization in Foraging Activity." *Journal of Natural History* 28(4): 861–872.

Paris, C. I., and X. Espadaler. 2009. "Honeydew Collection by the Invasive Garden Ant *Lasius neglectus* versus the Native Ant *L. grandis*." *Arthropod-Plant Interactions* 3(2): 75–85.

Parmentier, T., W. Dekoninck, and T. Wenseleers. 2014. "A Highly Diverse Microcosm in a Hostile World: A Review on the Associates of Red Wood Ants (*Formica rufa* Group)." *Insectes Sociaux* 61(3): 229–237.

Pathak, A., S. Kett, and M. Marvasi. 2019. "Resisting Antimicrobial Resistance: Lessons from Fungus Farming Ants." *Trends in Ecology & Evolution* 34(11): 974–976.

Pfannes, K. R., and A. C. Baier. 2002. "'Devil's Gardens' in the Ecuadorian Amazon—Association of the Allelopathic Tree *Duroia hirsuta* (Rubiaceae) and Its 'Gentle' Ants." *Revista de Biologia Tropical* 50(1): 293–301.

Pierce, N. E., M. F. Braby, A. Heath, D. J. Lohman, J. Mathew, et al. 2002. "The Ecology and Evolution of Ant Association in the Lycaenidae (Lepidoptera)." *Annual Review of Entomology* 47(1): 733–771.

Pontin, A. J. 1958. "A Preliminary Note on the Eating of Aphids by Ants of the Genus *Lasius* (Hym., Formicidae)." *Entomologist Monthly Magazine* 94:9–11.

Pontin, A. J. 1978. "The Numbers and Distribution of Subterranean Aphids and Their Exploitation by the Ant *Lasius flavus* (Fabr.)." *Ecological Entomology* 3(3): 203–207.

Poux, X., and P. M. Aubert. 2018. *An Agroecological Europe in 2050: Multifunctional Agriculture for Healthy Eating—Findings from the Ten Years for Agroecology (TYFA) Modelling Exercise.* Paris: IDDRI, 1–74.

Renner, S. S., and R. E. Ricklefs. 1998. "Herbicidal Activity of Domatia-Inhabiting Ants in Patches of *Tococa guianensis* and *Clidemia heterophylla.*" *Biotropica* 30(2): 324–327.

Rico-Gray, V., and P. S. Oliveira. 2007. *The Ecology and Evolution of Ant-Plant Interactions.* Chicago: University of Chicago Press.

Sagers, C. L., S. M. Ginger, and R. D. Evans. 2000. "Carbon and Nitrogen Isotopes Trace Nutrient Exchange in an Ant-Plant Mutualism." *Oecologia* 123(4): 582–586.

Salas-Lopez, A., S. Talaga, and H. Lalagüe. 2016. "The Discovery of Devil's Gardens: An Ant–Plant Mutualism in the Cloud Forests of the Eastern Amazon." *Journal of Tropical Ecology* 32(3): 264–268.

Schlick-Steiner, B. C., F. M. Steiner, H. Konrad, B. Seifert, E. Christian, et al. 2008. "Specificity and Transmission Mosaic of Ant Nest-Wall Fungi." *Proceedings of the National Academy of Sciences* 105(3): 940–943.

Schneider, S. A., and J. S. LaPolla. 2011. "Systematics of the Mealybug Tribe Xenococcini (Hemiptera: Coccoidea: Pseudococcidae), with a Discussion of Trophobiotic Associations with *Acropyga* Roger Ants." *Systematic Entomology* 36(1): 57–82.

Schultz, T. R., and S. G. Brady. 2008. "Major Evolutionary Transitions in Ant Agriculture." *Proceedings of the National Academy of Sciences* 105(14): 5435–5440.

Schultz, T. R., and T. McGlynn. 2000. "The Interaction of Ants with Other Organisms." In *Ants: Standard Methods for Measuring and Monitoring Biodiversity*, edited by D. Agosti, J. D. Majer, L. E. Alonso, and T. R. Schultz, 35–44. New York: Smithsonian Institute Press.

Schultz, T. R., U. G. Mueller, C. R. Currie, and S. A. Rehner. 2005. "Reciprocal Illumination: A Comparison of Agriculture in Humans and Ants." In *Ecological and Evolutionary Advances in Insect-Fungal Associations*, edited by F. Vega, and M. Blackwell, 149–190. New York: Oxford University Press.

Schupp, E. W. 1986. "*Azteca* Protection of *Cecropia*: Ant Occupation Benefits Juvenile Trees." *Oecologia* 70(3): 379–385.

Seifert, B. 2018. *The Ants of Central and Northern Europe.* Boxberg, Germany: Lutra Verlags und Vertriebsgesellschaft.

Shik, J. Z., A. Concilio, T. Kaae, and R. M. M. Adams. 2018. "The Farming Ant *Sericomyrmex amabilis* Nutritionally Manages Its Fungal Symbiont and Its Social Parasite." *Ecological Entomology* 43(4): 440–446.

Silvestri, F. 1924. "A New Myrmecophilous Genus of Coccidae from India." *Record of the Indian Museum* 26:311–315.

Skinner, J. A., K. A. Lewis, K. S. Bardon, P. Tucker, J. A. Catt, et al. 1997. "An Overview of the Environmental Impact of Agriculture in the U.K." *Journal of Environmental Management* 50(2): 111–128.

Smith, B. D. 2011a. "A Cultural Niche Construction Theory of Initial Domestication." *Biological Theory* 6(3): 260–271.

Smith, B. D. 2011b. "General Patterns of Niche Construction and the Management of "Wild" Plant and Animal Resources by Small-Scale Pre-Industrial Societies." *Philosophical Transactions of the Royal Society B—Biological Sciences* 366(1566): 836–848.

Stadler, B., and A. F. G. Dixon. 2005. "Ecology and Evolution of Aphid-Ant Interactions." *Annual Review of Ecology, Evolution, and Systematics* 36(1): 345–372.

Stadler, B., and A. F. G. Dixon. 2008. *Mutualism: Ants and Their Insect Partners.* Cambridge: Cambridge University Press.

Stanton, M. L., and T. M. Palmer. 2011. "The High Cost of Mutualism: Effects of Four Species of East African Ant Symbionts on Their Myrmecophyte Host Tree." *Ecology* 92(5): 1073–1082.

Stoate, C., N. Boatman, R. Borralho, C. R. Carvalho, G. R. De Snoo, et al. 2001. "Ecological Impacts of Arable Intensification in Europe." *Journal of Environmental Management* 63(4): 337–365.

Tartally, A., J. A. Thomas, C. Anton, E. Balletto, F. Barbero, et al. 2019. "Patterns of Host Use by Brood Parasitic *Maculinea* Butterflies across Europe." *Philosophical Transactions of the Royal Society B—Biological Sciences* 374(1769): 20180202.

Tobin, J. E. 1994. "Ants as Primary Consumers: Diet and Abundance in the Formicidae." In *Nourishment and Evolution in Insect Societies*, edited by J. H. Hunt and C. A. Nalepa, 279–307. Boulder, CO: Westview Press.

Ule, E. 1901. "Ameisengarten im Amazonasgebiet." *Botanische Jahrbücher für Systematik, Pflanzengeschichte und Pflanzengeographie* 68:45–52.

Voglmayr, H., V. Mayer, U. Maschwitz, J. Moog, C. Djieto-Lordon, et al. 2011. "The Diversity of Ant-Associated Black Yeasts: Insights into a Newly Discovered World of Symbiotic Interactions." *Fungal Biology* 115(10): 1077–1091.

Von Beeren, C., M. M. Mair, and V. Witte. 2014. "Discovery of a Second Mushroom Harvesting Ant (Hymenoptera: Formicidae) in Malayan Tropical Rainforests." *Myrmecological News* 20:37–42.

Way, M. J. 1963. "Mutualism between Ants and Honeydew-Producing Homoptera." *Annual Review of Entomology* 8(1): 307–344.

Weber, N. A. 1972a. *Gardening Ants, the Attines*. Philadelphia: American Philosophical Society.

Weber, N. A. 1972b. "The Attines: The Fungus-Culturing Ants: The World's Finest Gardeners Have Attained Skill in Maintaining Flourishing Fungus Cultures That Humans Might Envy." *American Scientist* 60(4): 448–456.

Wheeler, W. M. 1935. "Ants of the Genus *Acropyga* Roger, with Description of a New Species." *Journal of the New York Entomological Society* 43:321–329.

Wilson, E. O. 1980. "Caste and Division of Labor in Leaf-Cutter Ants (Hymenoptera: Formicidae: *Atta*): I. The Overall Pattern in *A. sexdens*." *Behavioral Ecology and Sociobiologyal* 7:143–156.

Wirth, R., W. Beyschlag, R. J. Ryel, and B. Hölldobler. 1997. "Annual Foraging of the Leaf-Cutting Ant *Atta colombica* in a Semideciduous Rain Forest in Panama." *Journal of Tropical Ecology* 13(5): 741–757.

Wirth, R., H. Herz, R. J. Ryel, W. Beyschlag, and B. Hölldobler. 2003. "Herbivory of Leaf-Cutting Ants: A Case Study on *Atta colombica* in the Tropical Rainforest of Panama." *Ecological Studies* 164:230.

Witte, V., and U. Maschwitz. 2008. "Mushroom Harvesting Ants in the Tropical Rain Forest." *Naturwissenschaften* 95(11): 1049–1054.

Yao, I., and S. Akimoto. 2001. "Ant Attendance Changes the Sugar Composition of the Honeydew of the Drepanosiphid Aphid *Tuberculatus quercicola*." *Oecologia* 128(1): 36–43.

Youngsteadt, E., P. Guerra Bustios, and C. Schal. 2010. "Divergent Chemical Cues Elicit Seed Collecting by Ants in an Obligate Multi-Species Mutualism in Lowland Amazonia." *PLoS ONE* 5(12): e15822.

Youngsteadt, E., S. Nojima, C. Haberlein, S. Schulz, and C. Schal. 2008. "Seed Odor Mediates an Obligate Ant-Plant Mutualism in Amazonian Rainforests." *Proceedings of the National Academy of Sciences* 105(12): 4571–4575.

9 Plant Farming by Ants: Convergence and Divergence in the Evolution of Agriculture

Guillaume Chomicki

Diverse forms of cultivation have evolved across the tree of life in systems as varied as the cultivation of bacteria by crabs (Thurber et al. 2011), amoebae (Brock et al. 2011), and fungi (Pion et al. 2013); the cultivation of algae by three-toed sloths (Pauli et al. 2014) and damselfish (Hata and Kato 2006); and the cultivation of fungi by snails (Silliman and Newell 2003). In contrast to these cultivation mutualisms, which can involve some form of planting or cultivation such as weeding and harvesting of "crops," true agriculture involves a unique set of behaviors and traits that occur sequentially in which a "farmer" farms a "crop" species. Four key steps define true agriculture—namely habitual planting, cultivation, harvest, and dependence of the farmer on the crop (Mueller et al. 2005). Outside of humans, agriculture is restricted to social insects (ants, termites, beetles) cultivating fungi (Farrell et al. 2001; Aanen et al. 2002; Mueller et al. 2005; Schultz and Brady 2008) and ants cultivating plants (Chomicki and Renner 2016a). The iconic fungus-farming agricultural symbioses have been studied for decades, with a focus on their evolution (Mueller et al. 1998; Farrell et al. 2001; Aanen et al. 2002; Mueller and Gerardo 2002; Schultz and Brady 2008), stability (Poulsen and Boomsma 2005; Aanen et al. 2009), cultivation behavior (Van Bael et al. 2009; Fernández-Marín et al. 2009; Biedermann et al. 2009; Katariya et al. 2017), genomics (Poulsen et al. 2014; Nygaard et al. 2016; Vanderpool, Bracewell, and McCutcheon 2018), and control of fungal parasites (Currie et al. 2003, 2006; Li et al. 2018). Plant farming by ants was discovered only in 2016, and hence relatively little is known about this agricultural system. In this chapter, I describe what is known about the evolution and ecology of plant farming by ants, highlighting convergences and divergences with fungus farming.

The Different Forms of Plant Cultivation by Ants

Ant/Plant Symbioses: No Cultivation

About 700 plant species in 159 genera and 50 families and some 113 ant species in 29 genera and five subfamilies engage in sometimes very specialized ant-plant symbioses (Chomicki and Renner 2015). Those symbioses are primarily defensive, in which the plant provides a nesting structure—termed a "domatium"—in which the ants nest in return for protection against herbivores. The plant host sometimes provides food rewards, as in the case of the iconic Central American *Acacia* (*Vachellia*)-*Pseudomyrmex* symbioses or the Neotropical *Cecropia-Azteca* mutualism (Chomicki, Ward, and Renner 2015; Gutiérrez-Valencia et al. 2017). In some symbioses, such as the one between central American *Piper* and *Pheidole bicornis* ants, the ants provide both defense and nutrients to their plant host (Fischer et al. 2003). In all these cases, even the most specialized systems that involve ants obligately nesting in plants, both partners disperse independently, and the symbiosis reassembles at each generation (horizontal transmission). Hence, these are not farming mutualisms.

Ant Gardens: Cultivation Mutualisms but Not True Agricultures

In 1901, Ule described a new type of mutualism from the New World in which arboreal ants plant seeds in their carton nests—nests built in trees mainly with plant parts and ant secretions—in rainforest canopies (figure 9.1A, B, arrows). These mutualisms, termed "ant gardens" (AGs), involve a form of cultivation in which arboreal ants use epiphytic plants to stabilize their nests. These gardens typically involve diffuse associations between several ant species and several plant species (Davidson 1988; figure 9.1A). Cases in which two ant species participate in a mutualism, called "parabiosis," and cultivate the same ant garden often involve species of the ant genera *Camponotus* and *Crematogaster* (figure 9.1A; Orivel, Errard, and Dejean 1997). The main function of the AG plants is to maintain a sturdy nest structure (Kaufmann 2002). Some AG plant species also provide extrafloral nectaries (EFNs), small nectar glands outside flowers. EFNs serve in so-called indirect plant defense, in which ants are attracted to EFNs and deter herbivores in return for this reward. EFNs occur in over 3,000 plant species (Weber and Keeler 2013), typically alone but sometimes in the context of AG plants or ant-plant symbioses. Neotropical AGs are cultivation mutualisms but not true agricultures. Habitual planting and cultivation are fulfilled, although neither partner is likely to be obligately dependent on the other. Unless the AG plants have extrafloral nectaries, there is no harvest, and the farmer has no obligate dependence on the crop. Thus, AG mutualisms are not true agricultures. AGs of another type were later described in Southeast Asia and encompass a wide range of both ant and plant species (figure 9.1B; Kaufmann 2002; Kaufmann and Maschwitz 2006). Like Neotropical AGs, Southeast Asian AGs involve habitual planting and cultivation and sometimes harvest, but all these are likely facultative, as is dependence. This means that such AG-forming ants have flexible diets and behaviors. For example, in several ant species, including *Philidris cordata*, forming AGs (i.e., habitual planting) is facultative. The conditions driving AG-forming behavior, and whether this is a trait that varies at the colony level, are unknown. In this case the four defining steps of agriculture are not fulfilled (table 9.1). The main difference between Neotropical and Southeast Asian AGs is that the latter frequently involve plants with derived structures for ant nesting (domatia), such as *Myrmecodia* and *Hydnophytum* (Rubiaceae), *Dischidia* (Apocynaceae), or *Lecanopteris* (Polypodiaceae) (figure 9.1B). In both Neotropical and Southeast Asian AGs, ants build carton nests that serve as substrates for AG plants. Because Southeast Asian AGs involve plant assemblages, often including species with domatia, the colony lives both in the carton nest and in the domatia, although the major part of the colony (including the queen) is typically found in the carton nest (Davidson and Epstein 1989).

Figure 9.1
The different forms of plant cultivation by ants. (A) Neotropical AG at Petit Saut, Sinnamary (Lat. 5.053, Long. –52.976), French Guiana, between several unrelated plant species (*Aechmea mertensii* [Bromeliaceae], *Codonanthe calcarata* [Gesneriaceae], and *Philodendron* sp. [Araceae]) cultivated by the two parabiotic ant species (*Camponotus femoratus* and *Crematogaster levior*); arrow points to a carton nest. (B) South-East Asian AG, here in Papua New Guinea, involving a mix of plants with domatia (here *Myrmecodia tuberosa*) and without (here *Dischidia nummularia*); arrows point to carton nest. (C) The ant-plant agricultural symbiosis between *Philidris nagasau* and *Squamellaria*; this farming ant colony contains dozens of *Squamellaria* crops, here *S. major* and *S. thekii*, overlooking a bay in Taveuni, Fiji. Photos: A: Jérôme Orivel, B: Milan Janda, C: Guillaume Chomicki.

The Fijian Farming Symbiosis between *Squamellaria* and *Philidris nagasau* Is the Only True Ant/Plant Agriculture Demonstrated to Date

The Fijian symbiosis between the plant *Squamellaria* (Hydnophytinae, Rubiaceae) and the ant *Philidris nagasau* drastically differs from Neotropical and Southeast Asian AGs. The four defining characteristics of farming are met, and the association is strictly obligate for both the farming ant and plant crop partners (Chomicki and Renner 2016a, 2019). The farming ant *Philidris nagasau* obligately nests in *Squamellaria* domatia—a dependence driven by the evolutionary loss of the carton nest-building behavior (Chomicki and Renner 2016a). It also differs from AGs in scale: a single “farm” frequently involves dozens of *Squamellaria* plant crops (figure 9.1C). Is the Fijian mutualism between *Squamellaria* and

Table 9.1
The diversity of plant cultivation mutualisms by ants

	Neotropical ant gardens	Southeast Asian ant gardens	Fijian *Squamellaria*/ *Philidris nagasau*
Habitual planting	Yes	Yes	Yes
Cultivation	Yes, but no monitoring	Yes, but monitoring probably rare	Yes, continuous monitoring of seeds and seedlings
Harvest	In general no	In general no	Yes, continuous
Dependence of farmer	Facultative	Facultative	Obligate
Nest structure	Carton nest stabilized by plant roots	Carton nest stabilized by plant roots + domatia of some AG plants	*Squamellaria* domatia exclusively
Queen nesting	Carton nest	Carton nest	Domatium

Philidris nagasau the only true plant-farming agricultural system in ants? So far, true agriculture has been demonstrated only in this system. But this plant clade—the Hydnophytinae—contains over 100 species, and three lineages other than *Squamellaria* have been independently colonized by *Philidris* or related dolichoderine ants, creating highly specialized cultivation symbioses (Chomicki and Renner 2017; Chomicki et al. 2020). These symbioses are concentrated in New Guinea (both Indonesian Papua and Papua New Guinea), in many cases in highly inaccessible locations, and have thus been very little studied. It is thus likely that other true agricultures have evolved in the Hydnophytinae.

Biology of the Fijian Farming Symbiosis

It has recently been shown that nonhuman agriculture is not restricted to fungus farming, but also involves plant farming by the ant *Philidris nagasau* (subfamily Dolichoderinae) of *Squamellaria* species in Fiji (Chomicki and Renner 2016a; figures 9.2, 9.3). This ant species forms large "farms" with dozens of *Squamellaria* plants cultivated high up in the canopy (figure 9.1C). The ants collect seeds in the fruits before they are fully ripe (which in turn prevents their removal by birds) and plant them in pockets under tree bark (figures 9.2A, 9.3). They plant up to 10 seeds and a small group of workers stays under the bark with the seeds, presumably to prevent seed predation. To emerge from their germination sites under the tree bark, the seedlings have evolved an elongated "foot" of the hypocotyl (the part of the stem below the cotyledons), which delays domatium development until the seedling has escaped from the bark (Chomicki and Renner 2016a). The ant planting behavior and plant escape traits likely result from adaptive coevolution. Once the seedling has an ~1.5 cm large domatium with its first entrance hole, ant workers enter the seedling and actively fertilize it by defecation (figures 9.2B, C, 9.3; Chomicki and Renner 2016a). Fertilization by ants in *Squamellaria* continues throughout the life of the plant and occurs exclusively on hyper-absorptive structures inside the domatia called the "warts" (figures 9.2D, 9.3), similar to the fertilization by *Philidris* ants of other specialized Hydnophytinae such as *Myrmecodia* species (Huxley 1978). Sexually mature farmed *Squamellaria* produce food rewards consisting of nectaries in old flowers from which the corolla has dropped

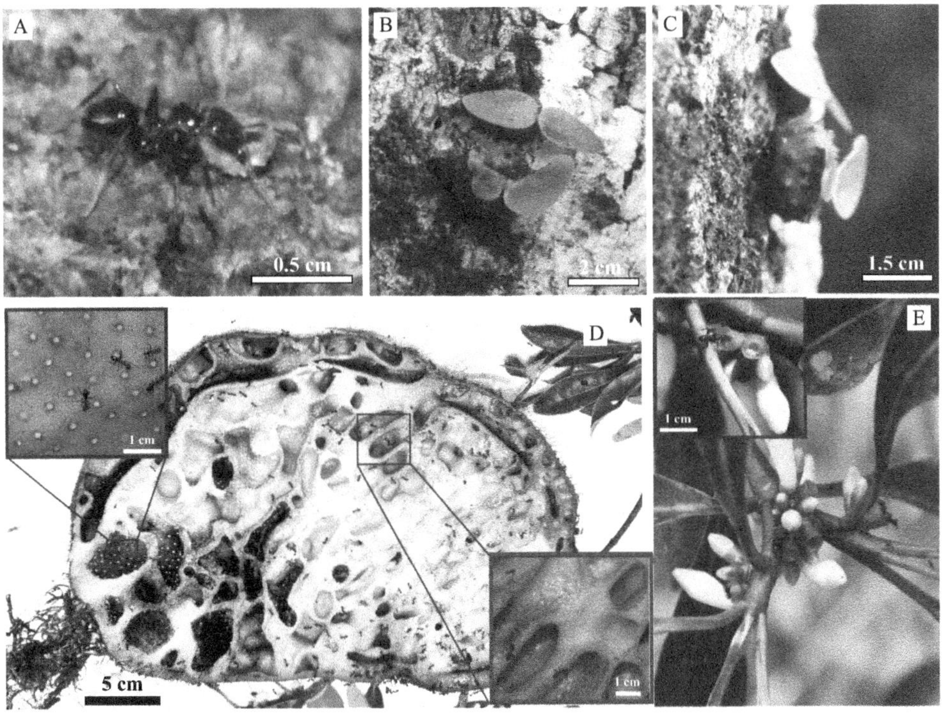

Figure 9.2
Illustration of the agricultural symbiosis between *Philidris nagasau* ants and *Squamellaria*. (A) *P. nagasau* worker dispersing a *Squamellaria* seed. (B) *Squamellaria* "nursery" where a site has been planted with seeds and two seedlings are emerging. Note the hole in the bark where a worker emerges; bark removal revealed that many seeds were present, guarded by *P. nagasau* workers. (C) *P. nagasau* worker visiting the small (1.5 cm in diameter) domatia of a *Squamellaria* seedling to fertilize it. (D) Longitudinal section through a large *Squamellaria wilsonii* plant crop, revealing two types of inner wall surfaces: the warted walls (left inset) where ants defecate and the smooth walls (right inset) where ants live and rear their brood. (E) *Squamellaria* crops (here *S. wilsonii*) produce food rewards that consist of old flowers from which the white corollas have fallen; in these post-anthetic nectaries, sugar is concealed, and the farming ants have to bite through the thick epidermis to obtain it, which prevents exploitation by opportunistic ants (Chomicki et al. 2016). Photos: Guillaume Chomicki.

(so-called post-anthetic nectaries), which conceal a sugar- and amino acid–rich sap (figures 9.2E, 9.3; Chomicki et al. 2016a). The planting of several generations of plants makes it possible for *Philidris nagasau* to avoid the reestablishment of the symbiosis, a critical step in other ant-plant symbioses involving bird-dispersed trees or shrubs that need to be located in every new ant generation by foundress queens.

Comparative Ecology of Plant and Fungus Farming in Social Insects and Humans

A tabulation of 15 agricultural behavioral traits in insects and humans reveals that five trait states are shared by all fungal and plant agricultures, two are shared by plant farming by ants and a subset of fungus-farming agricultures, one is shared by plant farming by ants and humans, two are unique, and seven are yet unknown in plant farming by ants (table 9.2).

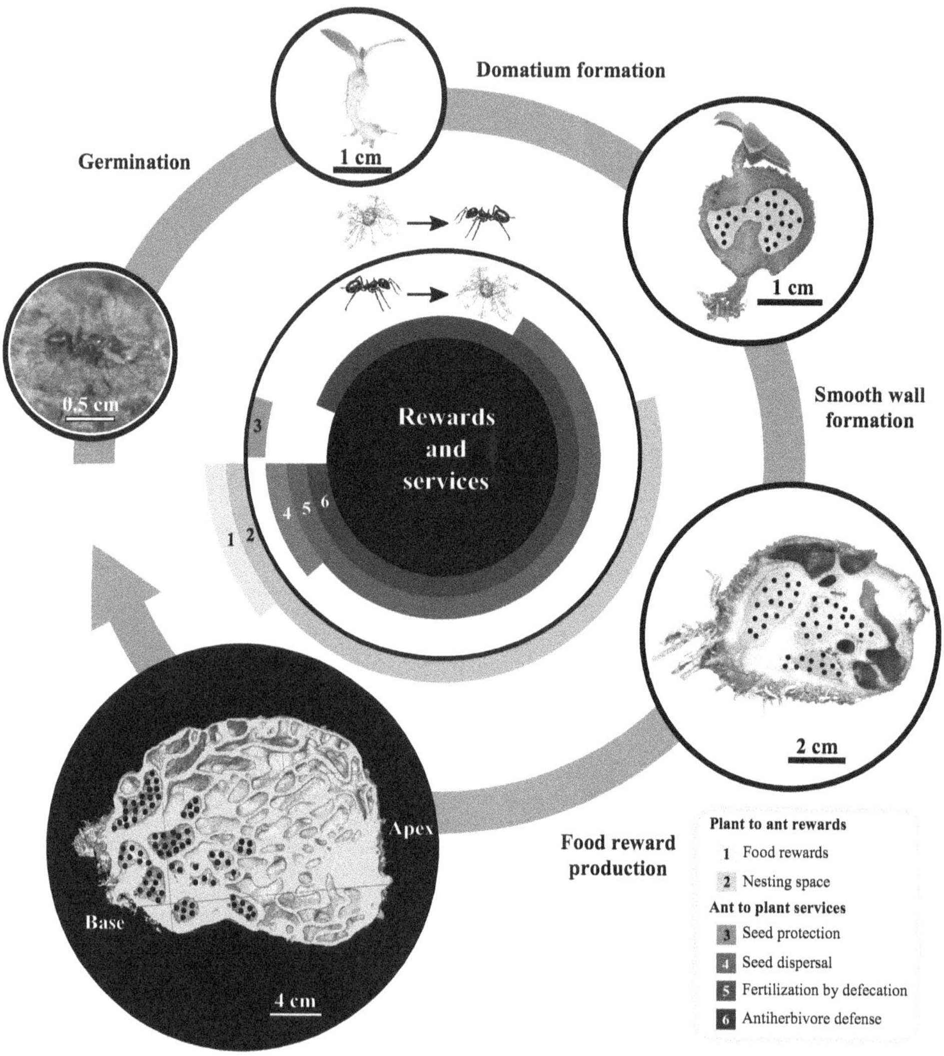

Figure 9.3
Rewards and services exchanged between *Squamellaria* and *Philidris nagasau* mapped onto plant ontogeny. In the domatium longitudinal section, white-dotted cavities point to warty surfaces where fertilization by defecation occurs and blackish cavities show smooth surfaces where farming ants rear their brood and live. Adapted from Chomicki and Renner (2019).

These diverse agricultures share a number of fundamental characteristics. All farmers engineer crop growth conditions to increase their productivity. In plant farming by ants, this involves the preferential planting of *Squamellaria* crops in full sun, which maximizes crop yield at the expense of lower nitrogen in their plants' diets (Chomicki et al. 2020). It also includes a tight nutrient recycling system wherein the farming ant *Philidris nagasau* continuously and exclusively defecates on hyper-absorptive warts (Chomicki and Renner 2019). Similarly, fecal manure is regularly applied in attine-ant fungiculture (De Fine Licht et al. 2013), and predigested material is continuously added to the combs in termite fungiculture (Johnson 1981). Sustainable harvesting of the crop for food is also common to all

Table 9.2
Comparative ecology of plant and fungus farming

	Fungus farming			Plant farming	
Agricultural behavior	Attine ants	Macrotermitine termites	Xyleborine ambrosia beetles	*Philidris nagasau*	Humans
Dependency on crop for food	Obligate	Obligate	Obligate	Facultative	Facultative
Dependency on crop for nesting	Absent	Absent	Absent	Obligate	Absent
Engineering of optimal crop growth	Present	Present	Present	Present	Present
Planting of crop on improved substrate	Present	Present	Present	Absent	Present
Continuous monitoring of growth and disease status of all crops	Present	Present	Present	Unknown	Absent
Sustainable harvesting of crop for food	Present	Present	Present	Present	Present
Protection of crop from disease and consumers	Present	Present	Present	Present	Present
Weeding of alien organisms invading the garden	Present	Present	Unknown	Unknown	Present
Use of chemical herbicides to combat pests	Present	Unknown	Unknown	Unknown	Present
Use of microbial symbionts for nutrient procurement for the crop	Present	Unknown	Unknown	Unknown	Present
Use of disease-suppressant microbes for biological pest control	Present	Unknown	Unknown	Unknown	Absent
Sociality	Eusocial	Eusocial	Subsocial or communal[a]	Eusocial	Social
Task partitioning in agricultural processes	Present	Unknown	Present	Unknown	Present
Application of artificial selection for crop improvement	Unknown, but ants exert symbiont choice	Unknown	Unknown	Unknown	Present
Learning and cultural transmission of agricultural innovations	Absent	Absent	Absent	Absent	Present

Note: [a]One ambrosia beetle is eusocial; all other species seem to be subsocial or communal.

Source: Comparison of fungus farming and human farming are adapted from Mueller et al. (2005).

agricultures. However, only in fungus farming is harvesting correlated with a specialized diet, such that farmers obligately feed on the fungal crops they farm. In contrast, both ants that farm plants and humans who farm plants and animals can feed on other resources and are thus not obligately reliant on farming for food (Chomicki et al. 2020). Crop protection from disease and competing consumers in fungus farming and herbivores in plant farming takes place in all agriculture systems. In fungus farming, crop protection typically involves a combination of monitoring, mechanical and chemical application, and, in the case of attine ants, reliance on specific symbiotic bacteria (reviewed by Mueller et al. 2005). In plant farming by ants, it involves defense against herbivores driven by food rewards (Chomicki et al. 2020). *Philidris nagasau* farming ants likely also play a role in disease protection of the plant, but this is not yet understood. Ant-exclusion experiments of farmed versus nonfarmed closely related *Squamellaria* species revealed that farmed species without ants suffered high pathogen attacks and rapidly died from these pathogens (Chomicki and Renner 2019). This suggests that ants play a role in reducing the pathogen load, either through chemicals or via microbial associates. That the experimental exclusion of the farming ants from the crops led to a high mortality apparently driven by pathogens suggest that the ant farmers may provide an obligate anti-pathogen defense role to the crop. Such an obligate dependence might have evolved via the evolutionary loss of defense genes in plants, compensated by the farmer anti-pathogen defense role.

Two linked aspects are unique to Fijian ant-plant agriculture. While all farmers—both fungus farmers and human plant farmers—improve the substrate in which they plant their crops, *Philidris nagasau* plants *Squamellaria* seeds in soilless pockets under tree bark. In contrast, other AG ants form a nutrient-rich carton nest in which they grow their plants (see above). In response to being sown in these harsh, sunken sites, *Squamellaria* crops have evolved a specific coevolved trait—an elongated hypocotyl—which allows them to escape these deep pockets before forming their domatia (Chomicki and Renner 2016a). The second aspect unique to this Fijian agriculture is that the crop provides a nesting structure to the farmer on which it is obligately dependent. This is because *Philidris nagasau* ants have lost nest-building behavior over evolutionary time (Chomicki and Renner 2016a).

Because plant farming by ants was discovered only in 2016, much of the biology remains unknown. In particular, as discussed above, there is indirect evidence that farming ants provide a pathogen-suppressing function (Chomicki and Renner 2019), but how this occurs is unknown. Similarly, it is unknown whether *Philidris nagasau* removes alien vegetation. *Squamellaria* "farms" tend to be found on trees with low epiphyte and vine covers (G. Chomicki, pers. obs.), but active physical weeding has not been observed. Task partitioning is also unknown. How *Philidris* domesticated *Squamellaria* and whether it performs artificial selection are unknown, although *Squamellaria* species differ in abundance, even sometimes within the same ant colony, in which *P. nagasau* may farm up to three separate *Squamellaria* species at once (Chomicki and Renner 2016a), possibly reflecting preferences for certain plant species.

Comparative Analysis of the Evolution of Agricultures

Similar to agriculture in attine ants and macrotermitine termites, current data pinpoint a single origin of agriculture in *Philidris* ants (Chomicki and Renner 2016a; Chomicki,

Janda, and Renner 2017). On the plant crop side, the Fijian farming ant *Philidris nagasau* has domesticated a single plant lineage (*Squamellaria*). However, three additional plant lineages (genus *Myrmecodia*, subclades of *Anthorrhiza* and of *Myrmephytum*) form highly specialized symbioses in which they are cultivated by other dolichoderine ants (mostly other *Philidris* species). Their occurrence in remote areas of New Guinea has precluded in-depth studies so far. *Philidris nagasau—Squamellaria* agriculture is restricted to two islands of the Fijian archipelago: Vanua Levu and Taveuni. Dated phylogenies of both the ant and plant clades converge to an age of 3–4 million years for the origin of this agriculture (Chomicki and Renner 2016a, 2016b; Chomicki, Janda, and Renner 2017). Taveuni island is only circa. 800,000 years old while Vanua Levu is thought to date to about 4 million years old, suggesting an origin of plant agriculture in the rainforests of Vanua Levu (Chomicki, Janda, and Renner 2017).

The mode of crop transmission is a critical step in any agriculture. Because crops typically contain traits that result from domestication, they are typically vertically transmitted from one generation to the next (table 9.3; Mueller et al. 2005). Macrotermitine termites are the outlier, in that most termite species acquire their crops horizontally (from the environment) via wind-mediated spore dispersal. However, termites ensure a high relatedness of fungal crops by propagating single variants (Aanen et al. 2009). The situation in the plant-farming ants is quite unique. On the one hand, overlapping *Squamelleria* crop generations—where the newer generations are offspring of older plants from the same colony—are found within the same ant colony, and crops are dispersed exclusively by farming ants. On the other hand, as far as we know, reproductive ants have no role in the crop's reproduction; rather, workers only disperse and propagate the crops by sowing seeds. This implies that new colonies likely emerge by "budding" or "splitting." This could happen when *Philidris nagasau* ants from a single colony spanning several trees become separated by a tree fall. Colonies are monogynous, but there is a high number of female alates that can potentially become queens if the plant with the queen becomes isolated. Fine-scale mapping of a mature colony has revealed that all plant crops within one colony are connected by ant trails, but that those trails are structured hierarchically and centered on the oldest plant—that is, the plant with the domatium that contains the queen (Chomicki and Renner 2016a). We can speculate that both the large number of female alates and the presence of a single type of worker performing all farming tasks can readily promote autonomous new colonies when physical isolation occurs. As in many human plant crops, *Squamellaria* crop reproduction occurs sexually, via the ant farmer sowing seeds. Sexual reproduction is also dominant in the cultivated fungi of termites, but in that case spore dispersal is mediated by wind rather than by the farmer. In contrast, attine ants and ambrosia beetles disperse their crops mostly asexually (table 9.3; Mueller et al. 2005).

At the genus level, clade-clade correspondence occurs between interacting farmers and crops in the attine ants, xyleborine beetles, and farming ants but not in fungus-farming termites (table 9.3; Mueller et al. 2005; Chomicki, Janda, and Renner 2017). This reflects the predominance of vertical versus lateral crop transfer. Another interesting trait of plant-farming agriculture that differs from fungus-farming but is shared with human farming is variable cultivation by monoculture, with polyculture also frequent. This contrasts with fungus farming where crop monoculture is constant. As in human plant farming, monoculture occurs in plant farming by ants. However, mixed cultivation of up

Table 9.3
Crop ecology and evolution in ant, termite, beetle, and human agriculture

	Fungus farming			Plant farming	
Crop ecology & evolution	Attine ants	Macrotermitine termites	Xyleborine ambrosia beetles	*Philidris nagasau*	Humans
Number of inferred evolutionary origins of agricultural behavior	Single origin	Single origin	Single origin in Xyleborini; six independent origins in ambrosia beetles	Single origin; but four crop lineages colonized (demonstration of true agriculture only in one)	Multiple independent origins
Estimated date and region of origin of agricultural behavior	45–65 mya in Amazonian rainforests	24–34 mya in African rainforests	30–40 mya in xyleborine beetles; region of origin unknown 21–60 mya in the other six ambrosia beetle lineages	3–4 million years old in Fijian rainforests (Vanua Levu island)	~12,000 years old, various places
Number of crop clades cultivated	Multiple cultivar clades: one in Pterulaceae; at least three additional ones in Lepiotaceae	Single cultivar clade (*Termitomyces*)	Multiple cultivar clades *Ambrosiella* and *Raffaellea*; several origins within *Ophiostoma* and *Ceratocystis*	Single clade in Fiji (*Squamellaria*); three related additional clades likely in New Guinea and SE Asia	Multiple lineages and geographic origins
Vertical crop transmission (from generation to generation)	Present	Variable (most lineages acquire cultivars horizontally via spore wind dispersal)	Present	Present (see text)	Present
Specialization on crops at higher phylogenetic level	Clade-clade correspondence	Clade-clade correspondence	No strict clade-clade correspondence	Clade-clade correspondence	Not applicable
Specialization at level of farmer species	Present	Present	Present	Unknown	Absent
Lateral exchange of cultivars within the same farmer species (within the same farmer generation)	Present	Unknown	Unknown	Unknown	Present
Cultivar exchange between farmer species	Present	Present	Rare	Not applicable	Not applicable
Genetic exchange between domesticated and wild crop populations	Variable	Absent	Absent	Absent	Absent
Crop monoculture	Present	Present	Present	Variable	Variable
Crop sexuality	Variable, but sexual recombination is rare	Variable, but most cultivars are propagated sexually from spores	Variable, but sexual reproduction is probably rare	Sexual reproduction exclusively	Variable, many are propagated sexually (e.g., most vegetables) but others clonally (e.g., fruit trees by grafting)

Source: Comparisons of fungus farming and human farming are adapted from Mueller et al. (2005).

to three *Squamellaria* species also occurs (Chomicki and Renner 2016a). Statistically, monoculture is dominant, but it is not clear whether it reflects local crop species abundance or farmer preferences (which are likely linked since farming ants are the only dispersing vectors of farmed *Squamellaria* species). The two most common species (*Squamelleria imberbis* on Vanua Levu and its sister species *S. wilsonii* on Taveuni island) have higher altitudinal ranges, linked to their C3 photosynthesis metabolism, while crassulacean-acid metabolism (CAM; a photosynthetic metabolism by which plants open their stomata for gas exchange only at night to save water) species have thick succulent leaves and are restricted to lower elevations where climate is drier (Chomicki and Renner 2016b; Chomicki et al. 2020).

Can Plant Farming by Ants Be Useful for Understanding Human Agriculture?

A timely question is whether we can learn from these insect agricultural systems to improve the sustainability of our own farming practices. One major insight from the study of insect farming is the role of bacterial associates in disease resistance. Studies of insect farming, and in particular of attine ants, have been a key motivation for studying the microbiomes of crop plants. Studies have shown how human farmers can select microbiomes that improve their crop's health and nutrition—hence productivity in just a few steps (Scheuring and Yu 2012; Mueller and Sachs 2015).

A unique aspect of ant-plant agriculture is that it takes place in a typically nutrient-limited environment—the forest canopy. *Philidris nagasau* ants preferentially plant *Squamellaria* in full sun, which maximizes crop yield, food rewards, and defense (since defense is mediated by food rewards). This suggests that ants will eat more food rewards (carbon-based diet), spend less time hunting insects, and thus have a diet with less nitrogen. As a result, full-sun cultivation leads to less nitrogen-rich fertilization of the plants (Chomicki et al. 2020). This tradeoff between nutrition and defense (i.e., with the best defense coinciding with the poorest nutrition, when the crop's productivity in terms of food rewards is maximal) is balanced because of the tight nitrogen (and phosphorus) recycling system. Farming ants defecate exclusively on hyper-absorptive warts inside the domatia and so there is no nutrient loss in the system (Chomicki and Renner 2019).

Genetic engineering approaches to increasing nitrogen use efficiency have met with limited success (Schroeder et al. 2013; chapter 3, this volume). One reason is that plants can modify their nitrogen uptake efficiency mechanisms in response to the available nitrogen level (Noguero and Lacombe 2016). Here, farming ants have practiced a form of "precision agriculture" for millions of years, wherein they actively and exclusively provide nitrogen-rich feces to warts, which have evolved an extreme nitrogen uptake rate (Chomicki and Renner 2019). This unique combination of extreme nitrogen uptake efficiency with high nitrogen concentration in fecal fertilizer raises new possibilities in the race to genetically engineer plants with increased nitrogen uptake efficiency. We are currently investigating the genes up-regulated in the hyper-absorptive warts, and promising candidates will be studied further.

Conclusion

The agriculture practiced by Fijian farming ants shares a number of traits with both fungus-farming in other insects and human agriculture. Habitual planting, optimization of the best growth conditions to maximize crop yield, continuous fertilization of the crops, constant defense against pests (here insect herbivores), sustainable harvesting, and dependence are the key features that are convergent across these systems. Yet there are also a number of divergences—features that are unique to this agriculture. *Philidris nagasau* ant farmers do not improve the substrate in which they plant their crops due to an evolutionary adaptation to the sowing mode. The crops provide a highly complex nesting site for the farmer on which the farmer obligately depends for survival, but the dependence of the farmer on the crop as food is only facultative. Because of its recent discovery, much about ant-plant agriculture remains unknown. In particular, the potential role of ants in disease suppression in the plants, and whether this involves symbiotic microbes, remains to be investigated. The extreme recycling of nitrogen in the ant-plant agricultural system, accomplished by nitrogen hyper-absorption by specialized tissues, might be transferable to human crops and improve their yields.

References

Aanen, D. K., H. De Fine Licht, A. J. M. Debets, N. A. G. Kerstes, R. F. Hoekstra, J. J. Boomsma. 2009. "High Symbiont Relatedness Stabilizes Mutualistic Cooperation in Fungus-Growing Termites." *Science* 326(5956): 1103–1106.

Aanen, D. K., P. Eggleton, C. Rouland-Lefevre, T. Guldberg-Frøslev, S. Rosendahl, et al. 2002. "The Evolution of fungus-Growing Termites and Their Mutualistic Fungal Symbionts." *Proceedings of the National Academy of Sciences* 99(23): 14887–14892.

Biedermann, P. H. W., K. D. Klepzig, and M. Taborsky. 2009. "Fungus Cultivation by Ambrosia Beetles: Behavior and Laboratory Breeding Success in Three Xyleborine Species." *Environmental Entomology* 38(4): 1096–1105.

Brock, D. A., T. E. Douglas, D. C. Queller, and J. E. Strassmann. 2011. "Primitive Agriculture in a Social Amoeba." *Nature* 469(7330): 393–396.

Chomicki, G., M. Janda, and S. S. Renner. 2017. "The Assembly of Ant-Farmed Gardens: Mutualism Specialization Following Host Broadening." *Proceedings of the Royal Society B—Biological Sciences* 284(1850): 20161759.

Chomicki, G., G. Kadereit, S. S. Renner, and E. T. Kiers. 2020. "Tradeoffs in the Evolution of Plant Farming by Ants." *Proceedings of the National Academy of Sciences* 117(5): 2535–2543.

Chomicki, G., and S. S. Renner. 2015. "Phylogenetics and Molecular Clocks Reveal the Repeated Evolution of Ant-Plants after the Late Miocene in Africa and the Early Miocene in Australasia and the Neotropics." *New Phytologist* 207(2): 411–424.

Chomicki, G., and S. S. Renner. 2016a. "Obligate Plant Farming by a Specialized Ant." *Nature Plants* 2(12): 1–4.

Chomicki, G., and S. S. Renner. 2016b. "Evolutionary Relationships and Biogeography of the Ant-Epiphytic Genus *Squamellaria* (Rubiaceae: Psychotrieae) and Their Taxonomic Implications." *PloS ONE* 11(3): e0151317.

Chomicki, G., and S. S. Renner. 2017. "Partner Abundance Controls Mutualism Stability and the Pace of Morphological Change over Geologic Time." *Proceedings of the National Academy of Sciences* 114(15): 3951–3956.

Chomicki, G., and S. S. Renner. 2019. "Farming by Ants Remodels Nutrient Uptake in Epiphytes." *New Phytologist* 223(4): 2011–2023.

Chomicki, G., Y. M. Staedler, J. Schönenberger, and S. S. Renner. 2016. "Partner Choice through Concealed Floral Sugar Rewards Evolved with the Specialization of Ant-Plant Mutualisms." *New Phytologist* 211(4): 1358–1370.

Chomicki, G., P. S. Ward, and S. S. Renner. 2015. "Macroevolutionary Assembly of Ant/Plant Symbioses: *Pseudomyrmex* Ants and Their Ant-Housing Plants in the Neotropics." *Proceedings of the Royal Society B—Biological Sciences* 282(1819): 20152200.

Currie, C. R., M. Poulsen, J. Mendenhall, J. J. Boomsma, and J. Billen. 2006. "Coevolved Crypts and Exocrine Glands Support Mutualistic Bacteria in Fungus-Growing Ants." *Science* 311(5757): 81–83.

Currie, C. R., B. Wong, A. E. Stuart, T. R. Schultz, S. A. Rehner, et al. 2003. "Ancient Tripartite Coevolution in the Attine Ant-Microbe Symbiosis." *Science* 299(5605): 386–388.

Davidson, D. W. 1988. "Ecological Studies of Neotropical Ant Gardens." *Ecology* 69(4): 1138–1152.

Davidson, D. W., and W. W. Epstein. 1989. "Epiphytic Associations with Ants." In *Vascular Plants as Epiphytes*, edited by U. Lüttge, 200–233. Berlin: Springer.

De Fine Licht, H., M. Schiøtt, A. Rogowska-Wrzesinska, S. Nygaard, P. Roepstorff, et al. 2013. "Laccase Detoxification Mediates the Nutritional Alliance between Leaf-Cutting Ants and Fungus-Garden Symbionts." *Proceedings of the National Academy of Sciences* 110(2): 583–587.

Farrell, B. D., A. S. Sequeira, B. C. O'Meara, B. B. Normark, J. H. Chung, et al. 2001. "The Evolution of Agriculture in Beetles (Curculionidae: Scolytinae and Platypodinae)." *Evolution* 55(10): 2011–2027.

Fernández-Marín, H., J. K. Zimmerman, D. R. Nash, J. J. Boomsma, and W. T. Wcislo. 2009. "Reduced Biological Control and Enhanced Chemical Pest Management in the Evolution of Fungus Farming in Ants." *Proceedings of the Royal Society B—Biological Sciences* 276(1665): 2263–2269.

Fischer, R. C., W. Wanek, A. Richter, and V. Mayer. 2003 "Do Ants Feed Plants? A 15N Labelling Study of Nitrogen Fluxes from Ants to Plants in the Mutualism of Pheidole and Piper." *Journal of Ecology* 91(1): 126–134.

Gutiérrez-Valencia, J., G. Chomicki, and S. S. Renner. 2017. "Recurrent Breakdowns of Mutualisms with Ants in the Neotropical Ant-Plant Genus Cecropia (Urticaceae)." *Molecular Phylogenetics and Evolution* 111:196–205.

Hata, H., and M. Kato. 2006. "A Novel Obligate Cultivation Mutualism between Damselfish and *Polysiphonia* Algae." *Biology Letters* 2(4): 593–596.

Huxley, C. R. 1978. "The Ant-Plants *Myrmecodia* and *Hydnophytum* (Rubiaceae), and the Relationships between Their Morphology, Ant Occupants, Physiology and Ecology." *New Phytologist* 80(1): 231–268.

Johnson, R. A. 1981. "Colony Development and Establishment of the Fungus Comb in *Microtermes* sp. nr. Usambaricus (Sjöstedt) (Isoptera: Macrotermitinae) from Nigeria." *Insectes Sociaux* 28(1): 3–12.

Katariya, L., P. B. Ramesh, T. Gopalappa, S. Desireddy, J.-M.Bessière, et al. 2017. "Fungus-Farming Termites Selectively Bury Weedy Fungi That Smell Different from Crop Fungi." *Journal of Chemical Ecology* 43(10): 986–995.

Kaufmann, E. 2002. "Southeast Asian Ant-Gardens: Diversity, Ecology, Ecosystematic Significance, and Evolution of Mutualistic Anti-epiphyte Associations." PhD dissertation, Goethe University, Frankfurt.

Kaufmann, E., and U. Maschwitz. 2006. "Ant-Gardens of Tropical Asian Rainforests." *Naturwissenschaften* 93(5): 216.

Li, H., J. Sosa-Calvo, H. A. Horn, M. T. Pupo, J. Clardy, et al. 2018. "Convergent Evolution of Complex Structures for Ant-Bacterial Defensive Symbiosis in Fungus-Farming Ants." *Proceedings of the National Academy of Sciences* 115(42): 10720–10725.

Mueller, U. G., and N. Gerardo. 2002. "Fungus-Farming Insects: Multiple Origins and Diverse Evolutionary Histories." *Proceedings of the National Academy of Sciences* 99(24): 15247–15249.

Mueller, U. G., N. M. Gerardo, D. K. Aanen, D. L. Six, and T. R. Schultz. 2005. "The Evolution of Agriculture in Insects." *Annual Review of Ecology, Evolution and Systematics* 36: 563–595.

Mueller, U. G., S. A. Rehner, and T. R. Schultz. 1998. "The Evolution of Agriculture in Ants." *Science* 281(5385): 2034–2038.

Mueller, U. G., and J. L. Sachs. 2015. "Engineering Microbiomes to Improve Plant and Animal Health." *Trends in Microbiology* 23(10): 606–617.

Noguero, M., and B. Lacombe. 2016. "Transporters Involved in Root Nitrate Uptake and Sensing by Arabidopsis." *Frontiers in Plant Science* 7:1391.

Nygaard, S., H. Hu, C. Li, M. Schiøtt, Z. Chen, et al. 2016. "Reciprocal Genomic Evolution in the Ant-Fungus Agricultural Symbiosis." *Nature Communications* 7(1): 1–9.

Orivel, J., C. Errard, and A. Dejean. 1997. "Ant Gardens: Interspecific Recognition in Parabiotic Ant Species." *Behavioral Ecology and Sociobiology* 40(2): 87–93.

Pauli, J. N., J. E. Mendoza, S. A. Steffan, C. C. Carey, P. J. Weimer, et al. 2014. "A Syndrome of Mutualism Reinforces the Lifestyle of a Sloth." *Proceedings of the Royal Society B—Biological Sciences* 281(1778): 20133006.

Pion, M., J. E. Spangenberg, A. Simon, S. Bindschedler, C. Flury, et al. 2013. "Bacterial Farming by the Fungus *Morchella crassipes*." *Proceedings of the Royal Society B—Biological Sciences* 280(1773): 20132242.

Poulsen, M., and J. J. Boomsma. 2005. "Mutualistic Fungi Control Crop Diversity in Fungus-Growing Ants." *Science* 307(5710): 741–744.

Poulsen, M., H. Hu, C. Li, Z. Chen, L. Xu, et al. 2014. "Complementary Symbiont Contributions to Plant Decomposition in a Fungus-Farming Termite." *Proceedings of the National Academy of Sciences* 111(40): 14500–14505.

Scheuring, I., and D. W. Yu. 2012. "How to Assemble a Beneficial Microbiome in Three Easy Steps." *Ecology Letters* 15(11): 1300–1307.

Schroeder, J. I., E. Delhaize, W. B. Frommer, M. L. Guerinot, M. J. Harrison, et al. 2013. "Using Membrane Transporters to Improve Crops for Sustainable Food Production." *Nature* 497(7447): 60–66.

Schultz, T. R., and S. G. Brady. 2008. "Major Evolutionary Transitions in Ant Agriculture." *Proceedings of the National Academy of Sciences* 105(14): 5435–5440.

Silliman, B. R., and S. Y. Newell. 2003. "Fungal Farming in a Snail." *Proceedings of the National Academy of Sciences* 100(26): 15643–15648.

Thurber, A. R., W. J. Jones, and K. Schnabel. 2001. "Dancing for Food in the Deep Sea: Bacterial Farming by a New Species of Yeti Crab." *PLoS ONE* 6(11).

Ule, E. H. G. 1901. "Ameisengärten im Amazonasgebiet." *Beiblatt zu den Botanischen Jahrbüchern* 68:45–52.

Van Bael, S. A., H. Fernández-Marín, M. C. Valencia, E. I. Rojas, W. T. Wcislo, et al. 2009. "Two Fungal Symbioses Collide: Endophytic Fungi Are Not Welcome in Leaf-Cutting Ant Gardens." *Proceedings of the Royal Society B—Biological Sciences* 276(1666): 2419–2426.

Vanderpool, D., R. R. Bracewell, and J. P. McCutcheon. 2018. "Know Your Farmer: Ancient Origins and Multiple Independent Domestications of Ambrosia Beetle Fungal Cultivars." *Molecular Ecology* 27(8): 2077–2094.

Weber, M. G., and K. H. Keeler. 2013. "The Phylogenetic Distribution of Extrafloral Nectaries in Plants." *Annals of Botany* 111(6): 1251–1261.

IV PATTERNS OF CONVERGENCE IN AGRICULTURALISTS, DOMESTICATES, AND PARASITES

10 Coevolution in the Arable Battlefield: Pathways to Crop Domestication, Cultural Practices, and Parasitic Domesticoids

Dorian Q. Fuller and Tim Denham

This chapter is about domestication of various plants by *Homo sapiens*, the novel ecosystems that such domestication processes created, and other taxa that benefited from these anthropic environments. The coevolution of symbioses of different phyla is by no means unique to the domesticatory relationships of humans and their crops. Indeed, "agriculture" by ants, termites, and beetles are far more ancient in evolutionary terms and widespread across phyla and habitats (Mueller et al. 2005; McGhee 2011; see also chapters 1, 2, and 14, this volume). The parallels between humans and some insects raise fundamental questions about what is meant by terms such as "cultivation," "domestication," and "agriculture." For anthropologists and archaeologists these terms are often taken to be self-explanatory. Cultivation is something that people do—namely, learned cultural behaviors and labor investments whereby people plant crops in prepared plots of land. Domestication refers to the resultant changes, genetic and/or morphological, in the targeted plant taxa that become adapted to cultivation and provide a yield that is economic, often caloric, for the human cultivators. Agriculture is often distinguished based on scale, whether in terms of the degree of dependence on cultivated food for diet or in terms of the level of investment in agricultural activities, which is associated with the importance of cultivated food both to the human economy and ecology and as an obligate part of sustaining human communities and populations. Agricultural societies require cultivation to persist, whereas cultivation and use of domesticates may be undertaken on a variable scale by economies that could still be largely reliant on fishing, hunting, and/or gathering. Economies with a smaller reliance on cultivation, including many in which domestication traits were evolving in the crops, have often been recognized as distinctive and variously termed "intermediate economies" practicing "predomestication cultivation" (Hillman 1975; Harris 1989, 2012), low-level food production (Smith 2001), or food production with/without domesticates (Harris 1996; Fuller et al. 2018). In the general terms laid out by Mueller et al (2005), agriculture as we define it here requires "nutritional dependency" and the reorientation of social life to the production of food, whereas habitual planting, improvement of growth conditions, and harvesting without nutritional dependency constitutes nonagricultural cultivation.

One of the striking features of agriculture is that it has evolved in parallel in different places and at different times, offering the opportunity to consider commonalities in process and causal variables. Mueller et al (2005) concluded that there were seven separate origins of agriculture among beetles, as well as one each among termites (in the Old World) and ants (in the New World). Similarly, among humans, agriculture developed independently at least 14 times, and perhaps as many as 23 times—in different regions, in different cultural traditions, and based on different crop plants (Purugganan and Fuller 2009; Larson et al. 2014; Denham et al. 2020). Our updated understanding of these geographically and culturally separate origins of crop domestication and agriculture are mapped in figure 10.1, which distinguishes centers focused on the cereal domestication pathway (figure 10.1A) and the vegecultural pathway (figure 10.1B).

In comparative terms three things are striking about the human development of agriculture that we wish to explore here. First, the nature of crop types and practices of reproduction vary and are linked to different forms of domestication. We can therefore break human-plant domesticatory relationships into a series of alternative domestication pathways. There is a coherent set of convergent traits for species within each pathway, but there are differences in the kinds of domestication traits between each pathway. Here we explore these pathways for the evolution of grain crops; the weeds of grain crops, some of which were selected as secondary domesticates; vegecultural crops (roots, tubers, and so on); and fruit trees. Each of these pathways has its own distinctive syndrome of domestication traits and tends to be associated with different kinds of agricultural system.

Second, human agriculture has tended to be diverse and diversifying in the sense that the number of plant species that have coevolved with humans to become domesticates has increased over time. We highlight this in the sections below with regional examples of diversification trajectories. This contrasts with insect agricultures, which tend to be an obligate mutualism between a specific insect species and a specific fungal crop (Mueller et al. 2005; chapter 5, this volume; chapter 14, this volume). In humans the range of crop species has progressively increased, and we suggest that this is driven by processes of cultural evolution that continued to increase productivity in order to support long-term population growth and increasing sociocultural complexity (Ellis et al. 2013, 2018). Even in cases where human agriculture may have begun with just a single crop, which is plausibly the case with early rice farming in the Yangtze basin (Fuller et al. 2014), agriculture subsequently diversified to include many more crops and various domesticate types (e.g., melons, peaches, soybeans, water chestnuts, and so on). This diversifying tendency may be unique to human forms of agriculture.

Third, the nature of evolutionary transmission among humans is primarily through cultural information rather than genetics, and this has allowed for a potentially more rapid process. The character of cultural transmission has probably fostered the many pathways to domestication and the long-term diversification of agricultural systems.

Defining the Arable Habitat: Target Crops and Weedy Taxa

Before exploring some of the archaeological cases of agricultural evolution, we will clarify how we understand the cultivated ecosystem of arable habitat. We take the *arable habitat* to refer to those areas of land that have been prepared and planted. Preparations normally include removal of some or most preexisting vegetation, through practices such as cutting, uprooting, digging, or burning, followed by some preparation of soil (tillage). The target

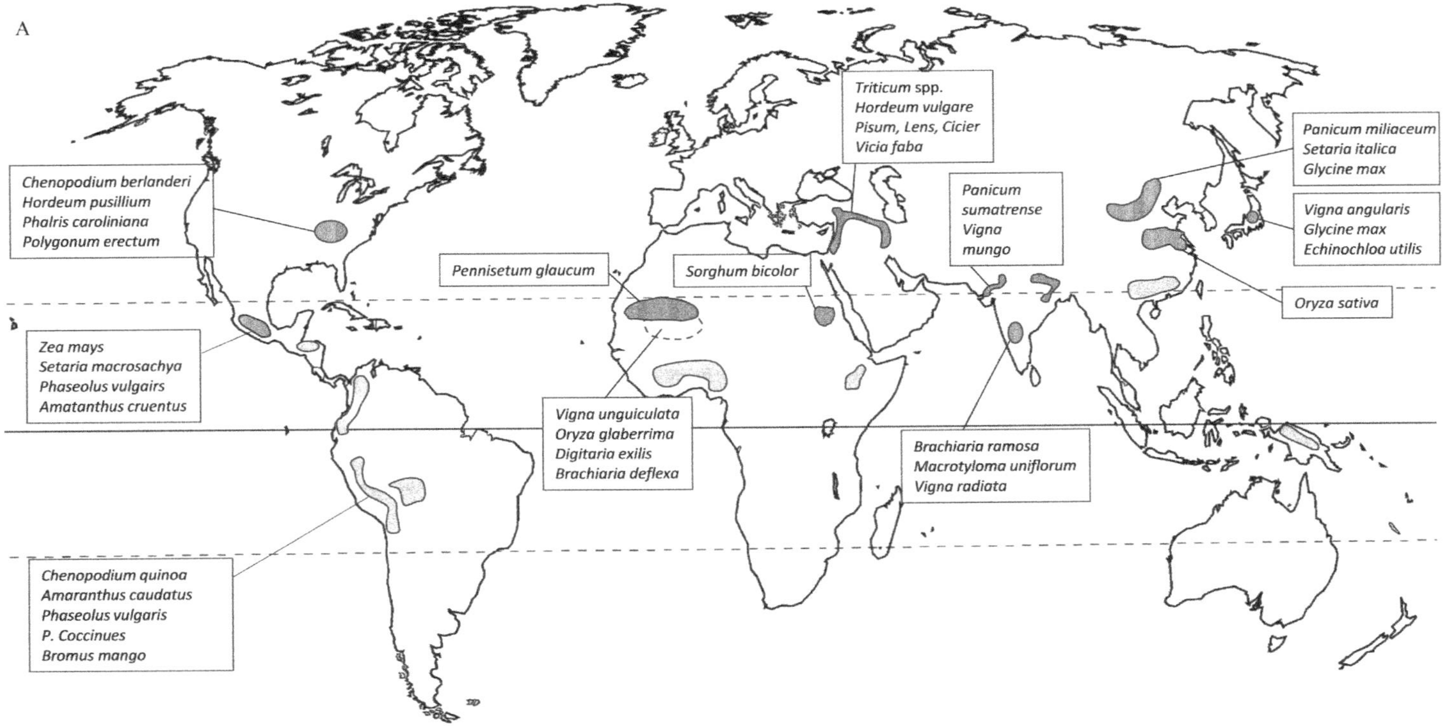

Figure 10.1
Centers of ancient plant domestication: (A) presents selected seed crops in relation to regions of domestication; (B) presents key vegecultural crops in relation to regions of domestication. Adapted from Fuller et al. 2014; Denham et al. 2020.

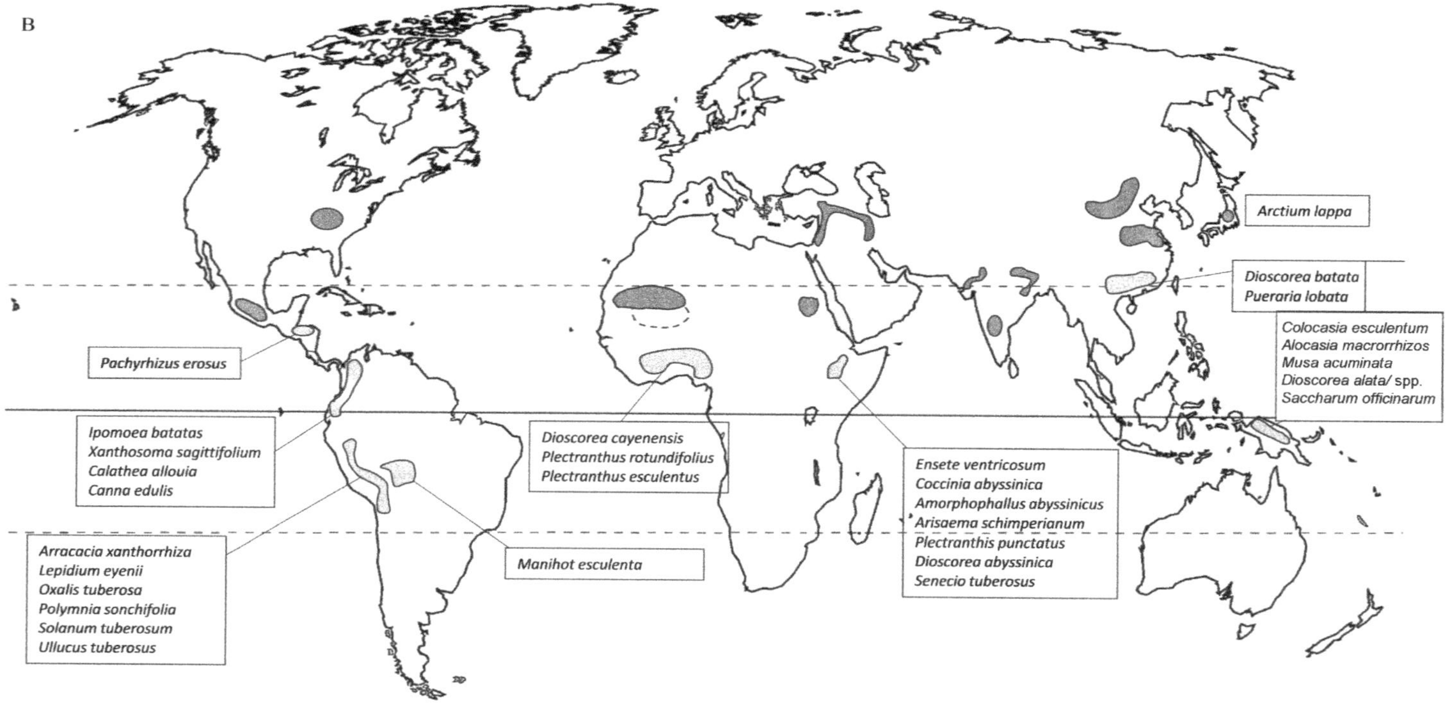

Figure 10.1
(continued)

species of cultivation—namely, the crop—is then planted; planting may be by seed or vegetative organ (cuttings, tubers), and these alternative reproductive systems lead to different patterns in terms of domestication process and domestication traits, as we turn to below. In part, domestication is characterized by convergent evolution, in what botanists have long discussed in terms of a "domestication syndrome" (Harlan, De Wet, and Price 1973; Hammer 1984; Harlan 1995; Fuller 2007; Denham et al. 2020). However, as we explore in this chapter, the symptoms that recur across different crops and that distinguish them from their wild relatives tend to differ based on the mode of reproduction and what we term here *alternative domestication pathways*. In particular we consider four alternative pathways: a seed crop or cereal pathway; a secondary cereal pathway, in which weeds are turned into crops; a vegetative pathway; and, a long-lived perennial, or tree fruit, pathway. Recent years have seen archaeological progress in documenting examples for each of these pathways.

The arable habitat, however, is never restricted to the crop, but also provides a habitat for competing taxa, or what would commonly be termed "agricultural weeds." Because these weeds compete with the crop for resources (soil nutrients, water, light) and success of the weeds tends to decrease the yields of crops, they can be regarded as parasitic upon the symbiotic relationship that is cultivation. Because weedy taxa have also been evolving adaptations to make them more successful as weeds of cultivation, they often possess some adaptations that overlap with those of the target crops. For this reason it has been suggested that they could be considered *parasitic domesticoids* (Fuller and Stevens 2017), inasmuch as they have evolved some traits that are shared with and similar to domesticated crops, but that aid them in parasitizing the resources of the cultivation symbiosis. It has long been noted that some agricultural weeds are today unknown from nonagricultural habitats, which indicates that these so-called obligate weeds have ancestrally evolved from species that originally occupied other habitats and have subsequently been extirpated from such habitats or those habitats have been replaced by anthropogenic habitats. Such obligate weeds are best known from those that evolved early in Southwest Asia as part of cereal agriculture there (Zohary 1950; Willcox 2012; Fuller and Stevens 2019a), but similar cohorts of weeds can doubtless be recognized in other regions too. Many of these weeds spread with agriculture to much greater geographical ranges than they had ever occupied as wild species, with some weeds becoming more globally widespread than many crops (Holm et al. 1977). The adaptations that many of these weeds possess often parallel crops in terms of changes in seed size, dormancy and germination, or growth habit.

The arable field has been regarded as a botanical battleground, where crops and their human cultivators wage an ongoing war with weeds (Jones 1988; Fuller and Stevens 2019a). A key adaptation on the human side of the battle occurs through sociocultural rather than genetic evolution. Human agricultural practices—in terms of various forms of tillage, soil preparation, intercropping, irrigation, weeding, and, in more recent centuries, the use of chemical herbicides—all represent innovations acquired and spread as cultural information in a process that has helped to push some weedy competitors out of the field and increase productivity. However, this battle is never won; as some weeds are pushed out of the segetal (growing within cultivated fields) and relegated to the ruderal (disturbed ground along paths and settlement margins), other weeds take their place. There are no weed-free cultivation systems. Take the example of rainfed rice compared to irrigated rice:

studies in traditional Indian fields recorded an average of 31 weed taxa in dry rice, compared with 13 in wet rice; wet rice has fewer weeds and is much more productive, yet it is not weed-free (Weisskopf et al. 2014). In any given region the long-term history of arable flora is a product of ongoing cultural evolution in agricultural practice, including the introductions of new crops and weeds via trade (Poschlod 2015a, 2015b; Pokorna et al. 2018; Stevens and Fuller 2019).

Other organisms, notably various animals, compete with humans for the calories derived from cultivation by infesting stores of harvested crops. Zoologists often refer to these as "commensal organisms," although they should probably also be classed as parasitic domesticoids. These animal species compete for calories with human cultivators and in essence parasitize the symbiotic relationship of cultivation, while also evolving adaptations that tie them to the human ecology of cultivators, much like domestication traits. This category includes several small animals that dwell in human settlements and feed largely on grain stores or incidental waste from grain stores, such as house mice (*Mus musculus*) (Fuller and Stevens 2017; Weissbrod et al. 2017; Cucchi et al. 2020). In many cases, such as flightless grain weevils (e.g., *Sitophilus granarius, S. oryzae*; see Plarre 2013) and particular mouse subspecies (*Mus musculus domesticus*; see Cucchi et al. 2020), these taxa are reliant on human settlement habitats and human-mediated dispersal. Along with the coevolution of weeds in early fields of cultivation, various animal taxa coevolved to take advantage of human settlements and caloric stores of crops. Thus in addition to the botanical battleground that was the cultivated field, there was an ongoing battlefront in settlements focused on granaries. The diversifications of dogs and domestic cats generated allies for this conflict. The above discussion highlights how human agricultural ecosystems continually acquired and renewed taxonomic diversity.

Archaeological Evidence for Domestication of Seed Annuals: The Cereal Pathway to Agriculture

The best studied cases of agricultural origins involve cereals, or similar grain crops such as pulses. Much of the human world today is reliant on the production of cereal crops, especially a few key species (wheat, rice, maize, barley, and sorghum) that feed the most people. These crops also supported many of the earliest urban human settlements and have attracted the most archaeological attention. In some ways these are also much easier crops to study: the ability to be stored as hard dried seeds results in better archaeological preservation (usually through charring), and seeds are therefore readily recovered archaeologically when sampling for small organics, such as through flotation, is carried out. The greater wealth of evidence for grain crop domestication does not mean that its understanding is without controversy. One can find debate between those who assume that cereal domestication was exceedingly rare and thus posit centric models (e.g., Lev-Yadun, Gopher, and Abbo 2000; Abbo, Lev-Yadun, and Gopher al. 2011, 2014), and those who take a noncentric view that domestication took place numerous times in parallel (e.g., Willcox 2005; Fuller, Willcox, and Allaby 2012; Fuller et al. 2014). It is also the case that those who prefer a centric hypothesis tend to see domestication as a kind of "discovery" driven by human conceptualization of the domesticated as distinct from the wild, whereas multicentric views tend to emphasize the operation of what Darwin (1868) termed "unconscious selection," meaning that domestication features evolved as the unintended evolutionary outcomes of human activities (Zohary 2004). We accept that the multi-centric and unconscious selection processes are more likely and are supported by the majority of archaeologists specializing in the emergence and transformation of early agriculture.

The domestication processes of these cereals result in parallel evolution of a *domestication syndrome* (Harlan, De Wet, and Price 1973; Hammer 1984; Fuller 2007). Domestication represents selection and fixation of genes and morphological adaptations of plants that were either absent or rare variants in preexisting wild populations. These traits evolved over a period of time, which we can define as a *domestication episode*, whereby populations of plants became genetically modified from their wild predecessors and gained adaptations to cultivation and human harvesting. These characteristics can be related to different aspects of cultivation in terms of what causes them to evolve. On the one hand there are traits that offer competitive advantages for the establishment of seedlings in cleared soil conditions within a cultivated field; on the other hand there are features that relate to human harvesting, thereby connecting cereal reproduction to dispersal by the farmer. Current archaeological data suggest that domestication traits evolved in crops over extended periods of time ranging between 2,000 and 4,000 years (see Fuller et al. 2014, 2018). This protracted domestication episode of a few millennia is often referred to as "pre-domestication cultivation" and represents a key period for understanding the origins of agriculture (Harris 1989; Hillman and Davies 1990; Willcox 2005, 2012). Recent work in Southwest Asia suggests that the domestication episode took place between 9500 and 7000 BC for a few species of wheat (*Triticum monococcum, T. dicoccon, T. timopheevi*), barley (*Hordeum vulgare*), lentil (*Lens culinaris*), pea (*Pisum sativum*), chickpea (*Cicer arietinum*), broadbean (*Vicia faba*), and flax (*Linum usitatissimum*) (Zohary, Hopf, and Weiss 2012; Fuller et al. 2018). Recent work in southern China suggests that the domestication of rice (*Oryza sativa*) occurred between ~7000 BC and 4000 BC (Fuller et al. 2014; Stevens and Fuller 2017) and might have also involved *Trapa natans* water chestnuts (Guo et al. 2017). Although there are still relatively few crop domestications that are fully documented through time series of archaeological data, some studies have increased evidence for parts of the sequences of change in several crops, such as indigenous seed crops of North America (Smith and Yarnell 2009; Fuller et al. 2014), sorghum (*Sorghum bicolor*) in Sudan (Winchell et al. 2018; Barron et al. 2020), various Indian Leguminosae (Murphy and Fuller 2017; Fuller et al. 2019), and Andean *Chenopodium* spp. (Langlie et al. 2011; Fritz et al. 2017), to name a few. What these examples have in common are both the protracted time period of change in the morphologies of these crops and the recurrence of morphological changes across species.

Key changes that relate to soil conditions include a trend toward increasing seed size and loss of germination inhibition. Crops tend to germinate as soon as they are wet and planted, in contrast to wild forms that often germinate after certain conditions have passed, such as in response to day length or temperature or after the seed coat is physically damaged. Loss of inhibition can be expected to be selected for under cultivation as seeds that do not germinate will not contribute to the harvest. Seed size increases are thought to relate to increasingly fit seedlings, which are better able to establish themselves quickly in the cleared ground of fields (Fuller 2007). Seed size change is readily studied from archaeological seed remains through the compilation of measurements (figure 10.2).

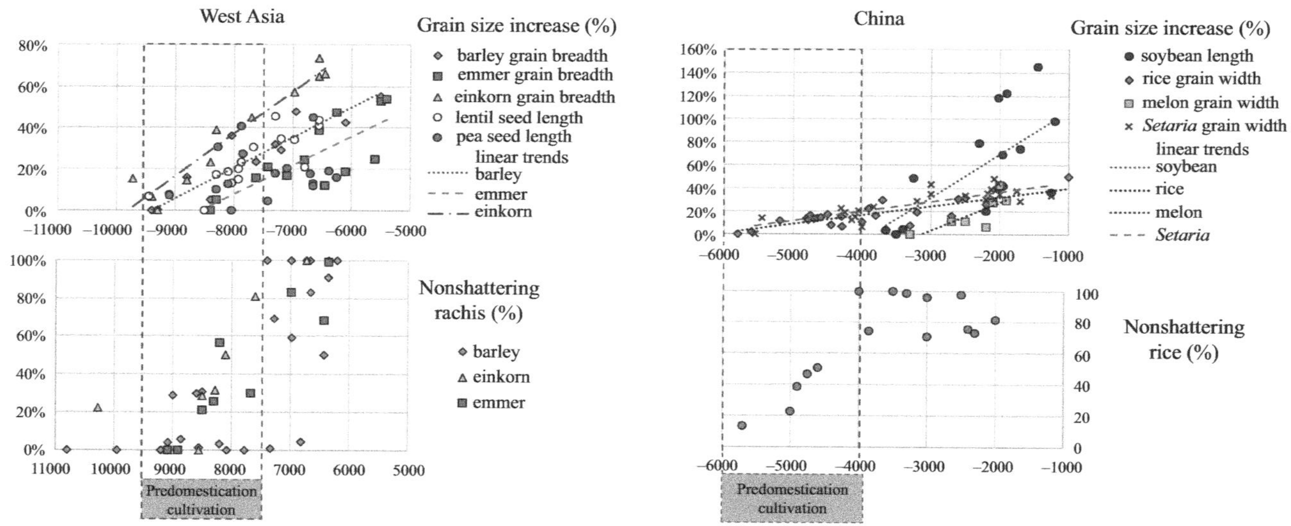

Figure 10.2
Domestication trajectories documented by morphological changes in seed crops: comparing China (right; Stevens and Fuller 2017; Fuller and Stevens 2017) and West Asia (left; Fuller et al. 2018). Archaeological data are plotted against the median calibrated radiocarbon age associated with the seed assemblage. Charts at the top indicate the increase in average seed size expressed as a percentage increase since the earliest/smallest assemblage. The stage defined as pre-domestication cultivation is indicated for each region; timescales are in years BC.

Although preservation through charring may change seed size, usually leading to shrinkage, this is expected to have similar effects across different archaeological samples, allowing comparisons between archaeobotanical assemblages to chart general trends. What is striking is that evidence for grain size increase in cereal crops is focused on the early millennia of domestication, with little significant or directional change afterward. Even modern breeding tends to work within the bounds of the size ranges established during early domestication in grains and grain legumes: the largest of modern wheat or rice grains are not significantly different from those at the end of the Neolithic (Gegas et al. 2010; Okamoto et al. 2012; Lu et al. 2013; Fuller et al. 2014, 2017).

A key change that related to cycles of human harvesting and sowing from stored harvests is the loss of natural seed dispersal, such as through the dehiscence of spikelets, or the "shattering" of cereal ears or panicles. This is often regarded as the single most important domestication trait (Zohary 1969; Harlan, De Wet, and Price 1973; Harlan 1995; Abbo, Lev-Yadun, and Gopher 2014). It makes a species highly dependent upon the human farmer for survival and means that instead of shedding seeds when they are mature, a plant retains them. Those retained seeds must be separated by processing (threshing)—thus, by the addition of human labor. Some of these seeds are further processed for human consumption while others are kept as a seed stock, to be dispersed later by farmers to create further crop generations. Over the era of domestication in Southwest Asia, human groups invested increasing effort in creating bespoke harvesting tools in the form of sickles, and thus technological evolution tracks the evolution of morphological change in wheats and barleys (Maeda et al. 2016). In the case of Chinese rice, specialized harvesting tools—hand-held knives or sickles—appeared around 3000 BC after nonshattering morphologies had already been fixed for some time (Fuller 2007; Stevens and Fuller 2017).

Other changes that are part of the domestication syndrome are either less necessary or harder to recover in the archaeological evidence (Fuller 2007). An increasing reliance on human harvesting also tended to lead to the reduction in seed dispersal aids. Plants often have a range of structures that aid seed dispersal, including hairs, barbs, and awns, and in grasses even the general shape of the spikelet. Thus domesticated wheat spikelets are less hairy, have shorter or no awns, and are plump, whereas in the wild they are heavily haired, barbed, and aerodynamic in shape. Awnless wheat, rice, and other cereals exist under cultivation and are absent from wild populations. While these changes can be regarded as having come about by the removal of natural selection for effective dispersal, allowing for less metabolic "expenditure" on these structures, there is often still much variation in the degree to which some of these structures are retained in some domesticated populations.

Another recurrent change is a shift to a more compact growth habit and increased apical dominance. Harlan (1995, 199) refers to this as the "sunflower effect." Domesticated cereals in general have shifts from their wild progenitors, sometimes involving suppression of tillers (basal secondary stalks) or axillary branches higher up, but also often selecting for even maturation across different branches. For example, in maize, foxtail millet, and pearl millet (*Zea mays*, *Setaria italica, Pennisetum glaucum*), the domesticated form tends to have suppression of any lateral branches or tillers, whereas in wheat, barley, and rice lateral branching is reduced and tillers have been selected to mature more evenly (Doust 2007). These branching patterns are not only determined by genetic combinations, but also respond to environmental cues, conditions of soil, and shading. From the farmer's point of view, being able to pack more erect crops together in a field will increase potential production per unit of land, but it comes at a potential price because each individual plant has less soil space and therefore less "real estate" from which to derive soil nutrients and water (see also, chapters 3 and 4, this volume). Apical dominance may have evolved only after cultivation was established along with fertilizing practices. This in turn means that erect and more productive early crops required more human labor inputs in terms of soil maintenance, the so-called soil nutrient trap (Fuller, Allaby, and Stevens 2010).

While domestication evolved in the context of cultivation, the final shift to agriculture represented an economic reorientation in which time put into cultivation and its products became far more central to human societies. Where detailed data are available, wild foods remained important and even staples alongside cereals undergoing domestication, including in Southwest Asia, where wheat and barley were domesticated (Arranz-Otaegui et al. 2016; Fuller et al. 2018), and in the Yangtze basin in China, where rice was domesticated (e.g., Fuller and Qin 2010). Thus one of the correlates of domestication was a process of investing more in production of crops and divesting from efforts in hunting and gathering. In summarizing our conceptualization of the cereal pathway to agriculture (figure 10.3), we can think of this as a spectrum of stages that tend to move from wild plant exploitation through intermediate economies involving pre-domestication cultivation to agriculture, when the economy (human food web) is focused primarily on the cultivation of genetically transformed taxa. During the intermediate stages we can document the dynamic evolution of the target species in terms of those alleles that confer the adaptations of the domestication syndrome.

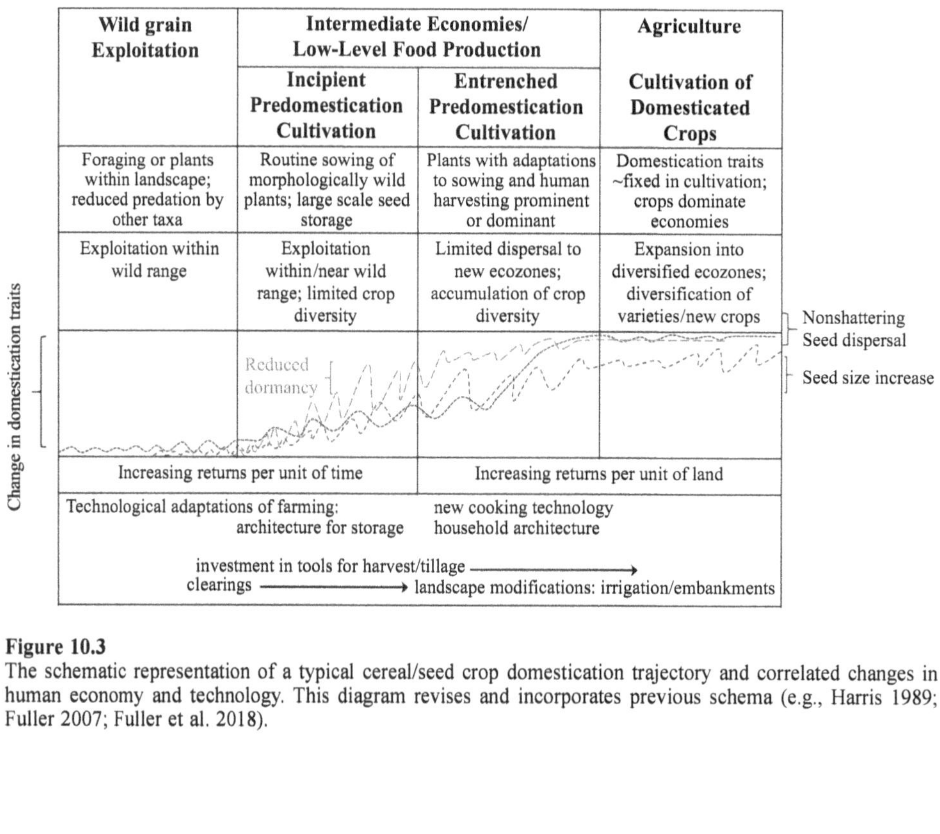

Figure 10.3
The schematic representation of a typical cereal/seed crop domestication trajectory and correlated changes in human economy and technology. This diagram revises and incorporates previous schema (e.g., Harris 1989; Fuller 2007; Fuller et al. 2018).

Diversification in Agriculture: Global and Regional Trends

Once cereals and associated seed crops were domesticated and regional agricultural systems were established, a number of tendencies for change in both the human and crop components of these systems can be recognized. Populations of humans practicing agriculture tended to grow in population density and eventually to disperse outward to additional territory (Bellwood 2005; Shennan 2018). The extent of local population growth and rate at which agriculturalists expanded by migrations varied, depending on factors such as the potential carrying capacity of the agricultural systems in question (Fuller, Champion, and Stevens 2019; Qin and Fuller 2019) and interaction with existing populations in adjacent regions (Donohue and Denham 2010). In some cases the expansion of agriculture into new geographical zones presented challenges, as crops were moved well beyond the ecological parameters that their wild ancestors had been adapted to. Some crops were less adaptable, leading to some loss of crop diversity farther away from places of origin. This was noticeable during the European Neolithic where western and northern areas had much reduced crop diversity, with crops such as lentil, chickpea, and einkorn wheat becoming progressively rarer (McClatchie et al. 2014; Fuller and Lucas 2017). Among the challenges that had to be overcome were changes in seasonal temperature and the distribution of rainfall, which led in many cases to adaptive changes in the sensitivity of photoperiodicity (Lister et al. 2009; Fuller and Lucas 2017; Liu et al. 2017).

Some declines in crop diversity during initial dispersal were, however, ultimately counteracted by large patterns of increasing crop diversity. The increase in crop diversity can be seen at two different scales. At a global scale and over the long term, the total number of grain/seed crops that people around the world domesticated increased, especially as crops were domesticated in more regions (figure 10.4). Thus while domestications of the Early Holocene, from 10,000 to 7,000 years BP, were dominated by Southwest Asian grain domesticates, those between 5,000 and 2,000 years BP involved many more geographical regions, from Africa and the Americas to South and East Asia. But another aspect of diversification is that the number of crops grown in any particular region and documented on a typical site from that region tended to increase over time. Based on a large database of crop presence and absence across the Old World (Stevens et al. 2016), it is possible to see how the average number of crops on a site and the maximum number of crops known in a region tend to increase over time (figure 10.5). Southwest Asia is an exception because there seems to have been a loss of diversity during early agriculture, after which diversity again increased. By contrast, data from China and sub-Saharan Africa show clear rising trends of crop taxa counts after 5000 BC. These differences highlight how the evolution of agriculture was never a single fixed adaptation; rather, it has continually evolved.

Secondary Cereal Domesticates: Crops from Weeds

Another source of new diversity was the expansion of the crop repertoire through additional, secondary crops evolved from weeds. As long as there has been cultivation, there have been weeds of cultivation. Some of those weeds have been extremely successful at adapting to anthropic environments, especially as agriculture was dispersed to an ever-greater

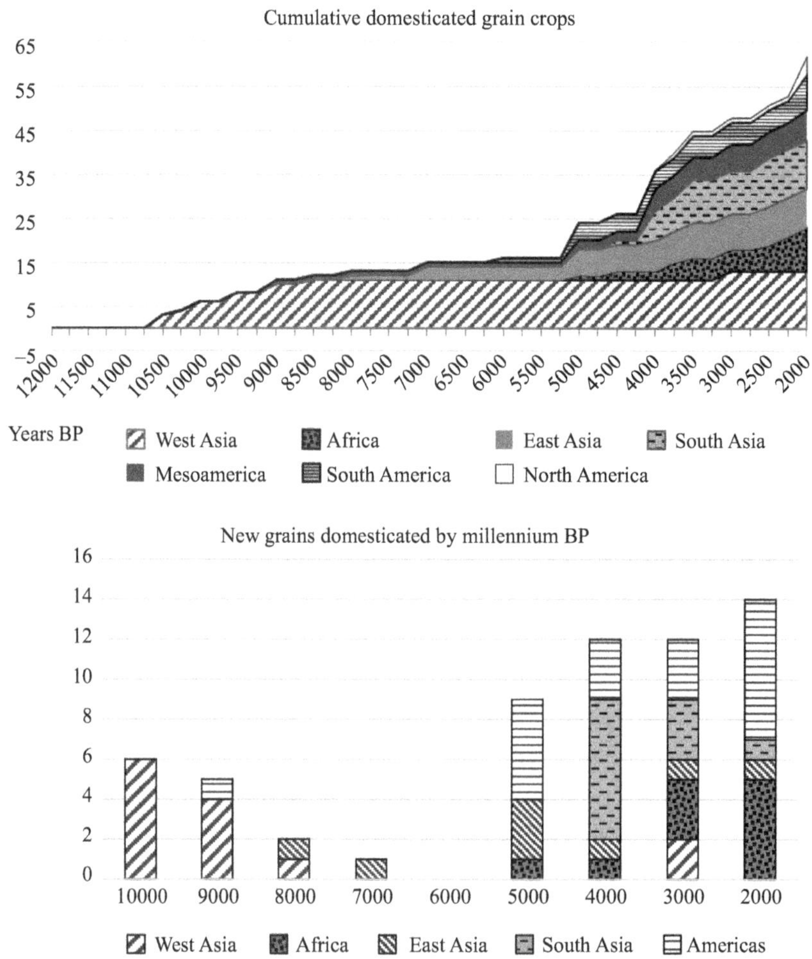

Figure 10.4
Long-term diversification in global seed crop repertoire (Old World data after Fuller, Champion, and Stevens 2019; New World data after Larson et al. 2014).

geographical range, and they increasingly challenged preexisting crops with variations in climate and soil. In some circumstances weeds outcompeted crops, and it was from among those highly successful weeds that some additional crops were domesticated. Following Vavilov (1926 [1992]), these can be defined as secondary cereals and include well-known species such as rye (*Secale cereale*) and various cultivated oats (*Avena sativa, A. byzantina, A. chinensis, A. abyssinica*). To this list can be added some of the millets of India, such as kodo millet (*Paspalum scrobiculatum*) (Fuller 2020) and probably some of the small millets (fonios) of western Africa (*Digitaria exilis, D. iburua, Brachiaria deflexa*) (Champion and Fuller 2018). Some pseudo-cereals (*Chenopodium* spp, *Amaranthus* spp.) and buckwheats (*Fagopyrum* spp.) may have evolved by this pathway, but in general these taxa are much less well documented. This is likely the major pathway for many fodder crops—species grown to provide food for domesticated livestock, such as alfalfa (*Medicago sativa*), common vetch (*Vicia sativa*), and red clover (*Trifolium pratense* var. *sativa*);

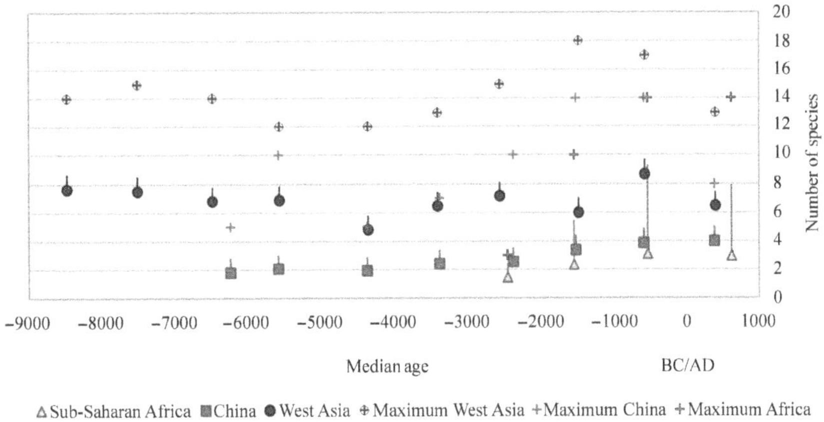

Figure 10.5
Trends toward increased crop diversity by region. Counts of crop species are averaged across sites within each region in millennium bins. Error bars indicate the standard deviation above the mean; maximum outlier values are also indicated. Total site counts: Western Asia = 143, China = 73, sub-Saharan Africa = 41.

many domesticated oats are also grown mainly as fodder (e.g., *Avena strigosa, A. byzantina*). All of these domesticates come from species that had already become well-adapted segetal weeds in arable habitats (figure 10.6). Where genetic evidence is available, it suggests multiple domestications from geographically differentiated populations across the wild or weedy range, as reported for *M. sativa* (Muller et al. 2003, 2006) and likely for *V. sativa* (Erskine, Smartt, and Muehlbauer 1994; Potokina et al. 2002; Zohary, Hopf, and Weiss 2012).

Weeds can be thought of as having undergone partial domestication, as they became adapted to life in the cultivated field (as segetal weeds) (Baucom and Holt 2009). In some cases weeds have evolved crop mimicry in which their vegetative growth makes them more or less indistinguishable from crops (Barrett 1983; Fuller and Stevens 2019a). One source of particularly troublesome weeds are crops themselves, which have undergone *de-domestication* or *feralization* by re-evolving some aspects of wild adaptations (such as seed dispersal), but otherwise retaining adaptations of cultivation (e.g., Sukopp and Sukopp 1993; Londo and Schaal 2007; Thurber et al. 2010; Xia et al. 2011; Qi et al. 2015; Huang et al. 2017). Most major cereal crops, such as rice, barley, and sorghum, have weedy forms that are widespread in global agriculture, creating "crop-weed complexes" (Harlan and De Wet 1965; see also Barnaud et al. 2009; Viguiera, Olsen, and Caicedo 2013). In these cases segetal weeds have evolved as the feral forms of crops.

A simplified schema of thinking about how crops and weeds have mutually evolved is given in figure 10.6. Arrows indicate potential evolutionary (phylogenetic) relationships between taxa, and these are distributed across a grid demarcated by degrees of anthropic habitat modification and by degree of genomic divergence from ancestral, preagricultural, wild forms. Weeds can be divided between those that persist within active cultivation systems (i.e., segetals) versus those that persist in human disturbed ground, or "weeds of waste" (i.e., ruderals). Ruderals and segetals are closely entwined because species may switch between these kinds of habitats depending on the nature of cultivation practices. It

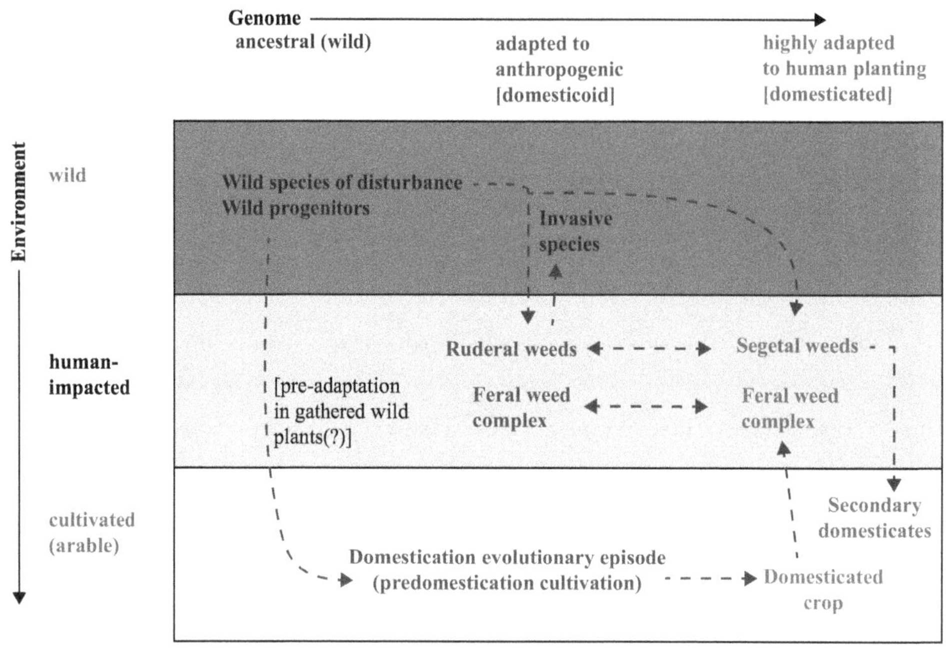

Figure 10.6
A schema of the potential evolutionary trajectories and phylogenetic relationships between crops, weeds, and wild relatives. These are charted in terms of degrees of habitat transformation by humans and in terms of degrees of genomic change (number of adaptative alleles) from hypothetical preagricultural wild populations.

is important to note that many species encountered as ruderals today were segetals in the past. This is a consequence of the ongoing coevolutionary battle between cultivators and weeds (Jones 1988; Poschlod 2015a, 2015b; Pokorna et al. 2018; Stevens and Fuller 2019).

The domestication syndrome in most secondary domesticates is the same as that described for the cereal pathway above, including increased seed size, reduced dormancy, changes in growth habit, and loss of natural seed dispersal. Some of these changes, such as in dormancy and growth habit, can be expected to have evolved in weeds. While minor changes in seed size and even a minor presence of nonshattering could arise in some weeds, a major increase in grain size and the loss of natural seed dispersal evolved rapidly among secondary domesticates once they became target cultivars. The domestications of many weed taxa are not as well documented as those of primary cereals. Nevertheless, with the oats, rye, and common vetch in Europe, domestication occurred just before or during the Iron Age (1000–1 BC) (Zohary, Hopf, and Weiss 2012). Cultivation of fodder crops of alfalfa date back to perhaps 400 BC in the Mediterranean (Rovira and Alonso 2018) and at least AD 100 in Chinese Central Asia (Chen et al. 2020), while *Vicia sativa* was certainly grown in Roman times 2,000 years ago (Erskine, Smartt, and Muehlbauer 1994; Zohary, Hopf, and Weiss 2012). Kodo millet (*Paspalum scrobiculatum*) domestication in India can be attributed to a similar period (Fuller 2020). The domestication episodes for these species took place during periods of well-developed and diversified agriculture, involving farmers with long-established cultural traditions of agronomy, in which crop diversity was an established strategy for coping with environmental variability. Consequently, these secondary taxa evolved domestication traits much more rapidly than earlier cereal domesticates. It is also possible that conscious selection akin to artificial selection in modern times was involved in these processes (artificial selection as distinct from unconscious selection was defined by Darwin [1868]).

Vegetative Domestication of Root Crops

Although not as well documented as for cereals, the domestication of root crops has been almost as significant for subsistence in the past, especially in the wet tropics, mountainous regions, wetlands, and some deserts (Harris 1972). Early farming emerged in several regions based, at least in part, on the vegetative propagation of root crops that today are globally significant, most notably manioc (*Manihot esculenta*), potato (*S. tuberosum*), and sweet potato (*Ipomoea batatas*) in the Americas (Ugent and Peterson 1988; Piperno and Pearsall 1998); and aroids (Araceae) and yams (*Dioscorea* spp.) in the Indo-Pacific region extending from eastern India to New Guinea (Burkill 1935; Li 1970; Yen 1973). Thus, understanding vegetative propagation provides a truly global perspective on the long-term history of human selection and domestication of plants, practices of cultivation, and forms of agriculture.

The vegetative domestication pathway is primarily, although often not exclusively, asexual; that is, it is clonal and occurs through the planting of reproductively viable plant parts, such as fragments of underground storage organs, divided root mass, and vine/stem cuttings. In most root crops, the vegetative plant part harvested, stored, and then replanted is also the part eaten—whether corms in taro (*Colocasia esculenta*), rhizomes in ginger (*Zingiber officinale*), root tubers in manioc (*Cassava manihot*), or stolon tubers in potato (*Solanum tuberosum*) and yams (*Dioscorea* spp.). In other cases, vegetative reproduction can be enabled by another plant part, such as stem/vine slips in sweet potato (*Ipomoea batatas*).

Sexual reproduction still occurs in some cultivars, whereby new, spontaneously emergent sexually reproduced phenotypes are adventitiously incorporated into vegetatively propagated stock. Such practices were important generators of high varietal differentiation in manioc and sweet potato, as well as other root crops. Even in some vegetatively propagated crops that are considered almost exclusively asexual today, sexual reproduction is implicated in the early stages of domestication, such as in greater yam (*Dioscorea alata*; Lebot 2009). For most root crops, with the exception of some tap roots (e.g., carrots, *Daucus carota*) and bulbs (e.g., onions, *Allium* spp.), the primary practice of intervening in plant reproduction is vegetative, and prolonged clonal cultivation seems to have reduced sexual reproductive capacity in some crops (Denham et al. 2020).

Long-term domesticatory relationships can be envisaged for root crops that encompass plant management at the least intensive end of the spectrum to mechanized monoculture at the most intensive end (Harris 1972; Hather 1996). For instance, many nonfarming, foraging groups around the world practice vegetative propagation behaviors such as tending favored plants and replanting viable plant parts during gathering of tubers (Barton and Denham 2018). Over time, human-mediated selective pressures would have gradually increased as a result of preferential plant management and resource intensification through burning, disturbance, and replanting, as well as adventitious growth from discarded and misplaced plant-parts around camps. These practices include both conscious and unconscious selective pressures on plants, yet they are conceptually distinct from other forms of cultivation because they do not involve planting in prepared plots or fields.

At some point in the domestication history of a crop, people started to transplant viable plant-parts of preferred phenotypes into prepared plots to cultivate them and eventually began to disperse them under cultivation (figure 10.7). A general loss of sex in clonal plants under cultivation, namely, a general lowering of sexual reproductive capacity, was stimulated through various mechanisms: some likely inadvertent (unconscious selection), such as movement into new environments (i.e., altitudes and latitudes) where sexual reproduction was not possible; some more conscious, such as preferential selection of sterile forms; and some highly deliberate, such as harvesting before flowering and seed-set. Thus, numerous practices worked to suppress sexual reproduction, with subsequent selection being driven primarily by clonal variation under cultivation.

Clonal propagation exerted greater selective pressure on field crops and allowed a higher degree of control over preferred phenotypes than sexual reproduction. However, in most vegetative crops the degrees to which domestication represents "morphogenetic" fixation, or a plastic response to growth environment, including cultivation practices, are unclear (Denham et al. 2020). It can be argued that plastic adaptation to growth environments in vegetatively propagated crops drives phenotypic variation, in contrast to sexually reproduced crops in which phenotypes are less plastic and more tightly regulated genetically. Nonetheless, even though plasticity may be more visible in clonally reproduced crops there

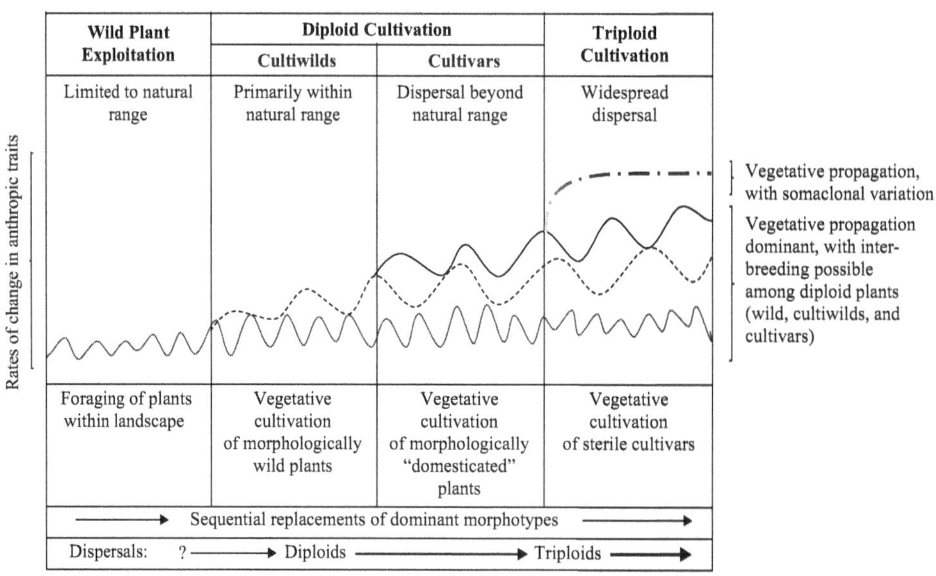

Figure 10.7
The schematic representation of a typical vegeculture domestication trajectory for a root crop with triploid cultivars, such as taro (*Colocasia esculenta*). Exploitation of wild plants (dotted line) eventually led to the cultivation of morphologically wild plants, called *cultiwilds* (dashed line), with the subsequent generation and cultivation of diploid cultivars (solid line), from which triploid cultivars emerged, dispersed, and became the dominant cultivar groups (dotted-dashed line).

is still underlying genetic variation resulting from somaclonal mutations, introgression for sexually viable plants, and other phenomena.

Cultivation practices are intended to create growth environments that favor larger edible underground storage organs by increasing the friability of soils, maintaining soil water content, and reducing competition from other vegetation. Other practices are designed to foster specific phenotypic responses, whether sequential harvesting to prolong timing of production (e.g., many root crops), density of planting for apical dominance (e.g., manioc and some yams), or watering to reduce toxins (e.g., manioc). For many vegetatively propagated plants, though, it is hard to clearly determine domestication traits because either the wild progenitor is unknown, such as for greater yam (*D. alata*) and sweet potato, or little is known about the plant ecology, genetics, and cultivation history of related wild species.

Furthermore, most vegetative crops have low archaeobotanical visibility, principally because the key plant-parts exhibiting potential domestication traits are either soft tissues that preserve only rarely in burned, desiccated, or waterlogged form or are preserved as less diagnostic phytoliths and starch grains. Furthermore, much of the phenotypic variation in plant macrofossils, such as archaeological parenchyma, and in microfossils, such as phytoliths and starch grains, are poorly understood in terms of establishing domestication traits relative to wild populations. To exacerbate matters, these techniques are not routinely applied during archaeological investigations, aside from some noteworthy exceptions (for example, Piperno and Pearsall 1998; Mindzie et al. 2001; Denham et al. 2003; Iriarte et al. 2004).

Furthermore, in comparison with many cereals, the phylohistories and geodomestication pathways for most vegetatively propagated root crops are poorly characterized or are only beginning to be clarified genetically, as is the case for manioc (Wang et al. 2014), potato (Hardigan et al. 2017), sweet potato (Roullier et al. 2013a; Muñoz-Rodríguez et al. 2018), taro (Chaïr et al. 2016), and some yams (Malapa et al. 2005; Scarcelli et al. 2019). However, caution is needed when relying on genetic data from modern populations to infer plant domestication and dispersal histories. In recent times, as in the deep past, specific lineages have come to dominate cultivated assemblages following successive waves of dispersal and widespread adoption, resulting in the near-wholesale replacement or marginalization of preexisting genetic lineages (e.g., Roullier et al. 2013b). Limitations of the archaeobotanical and genetic data hinder a detailed reconstruction of the domestication histories for most root crops.

Taro (*Colocasia esculenta*) is a widespread aroid, exploited primarily for an underground corm, although its leaves are also cooked as a vegetable. Its natural range is poorly determined and potentially extends from the Indian subcontinent, mainland and island Southeast Asia, northward to East Asia and eastward to New Guinea–northern Australia, with multiple domestications potentially occurring within this range. Yams (*Dioscorea* spp.) are a pantropical genus utilized primarily for underground, edible tubers. Yam species were independently domesticated in tropical climates of the Americas, Africa, India, and Southeast Asia–New Guinea, where they are regionally significant subsistence crops. In the Island Southeast Asian–New Guinea region, archaeobotanical finds of several yam species (*D. alata, D. esculenta, D. bulbifera*) and taro, as well as other aroids (i.e, *Alocasia* spp. and *Amorphophallus* spp.), have been documented at Pleistocene- and Holocene-aged sites, yet few of these finds occur in contexts associated with agriculture. Exceptions include starch grain residues of yam and taro, as well as aroid seeds, from early and mid-Holocene contexts at Kuk Swamp in the highlands of New Guinea (Fullagar et al. 2006).

These findings do not indicate that these root crops were domesticated at Kuk; rather, archaeobotanical association with archaeological evidence of cultivation, pedological indications of digging, and palaeoecological evidence of montane forest disturbance suggest that they were cultivated there. Much more recent finds of taro corms, along with banana (*Musa* sp.) skin peelings and sugarcane (*Saccharum officinarum*) stem sections from archaeological contexts associated with the Islamic port at Quseir al-Qadim in Egypt, indicate westward movement of these plants to the Red Sea by at least AD 1040–1160 (Van der Veen and Morales 2011), although all three crops were likely introduced to Africa and the Mediterranean much earlier (Murdock 1960; Perrier et al. 2011; Grimaldi et al. 2018).

In a similar vein, the domestication histories of several major root crops in the Americas are relatively poorly documented archaeobotanically. Although mixed, food-producing economies have been proposed for the Americas during the early and mid-Holocene (Iriarte 2007), there are no clear signatures of domestication in the archaeobotanical record of vegetatively propagated crops for this period. Two globally significant crops, potato (*Solanum tuberosum*) and sweet potato (*Ipomoea batatas*), are discussed below as exemplars.

Potato (*Solanum tuberosum*) originally formed part of regionally important root crop assemblages in the Central Andes, where they were all vegetatively propagated and of variable importance at different altitudes, including oca (*Oxalis tuberosa*), ulluco (*Ullucus tuberosus*), and mashua (*Tropaeolum tuberosum*) (National Research Council 1989). Potato spread significantly only after its post-Columbian introduction to Europe. Desiccated macro-remains of potato (*S. tuberosum*) were excavated from 10,000-year-old contexts in Chilca Canyon, coastal Peru (Engel 1970), as well as from multiple sites dating from 4,250 to 3,500 years ago in the Casma Valley, coastal Peru (Ugent, Pozorski, and Pozorski 1982; Ugent and Peterson 1988). Although these finds are at a remove from the inferred loci of domestication in the Peruvian Andes, their significance varies: the early Holocene remains may not constitute dispersal under cultivation, especially since there are no comparable finds for several thousand years, whereas those dating to ~4,250–3,500 years ago are clearly part of a mixed farming economy.

Sweet potato (*Ipomoea batatas*) may have undergone two domestications in the Americas: a northern domestication in Central America/the Caribbean generated the *camote* and *batata* cultivar groups, and a southern domestication centered on Peru and Ecuador generated the *kumara* cultivar groups (Roullier et al. 2013a). It is an orphaned crop, which has no known wild progenitor. Sweet potato was widely cultivated up to mid-latitudes in North, Central, and South America, as well as in the Caribbean during pre-Columbian times. Sweet potato had also dispersed widely through Eastern and Central Polynesia from at least ~AD 1200–1300, eventually reaching the Hawaiian archipelago, Easter Island, and New Zealand—most probably due to human agency (Yen 1974; Roullier et al. 2013b), although natural seed dispersal cannot be completely discounted (Muñoz-Rodríguez et al. 2018). Archaeobotanical evidence for sweet potato is rare; the earliest comprises desiccated tubers dating to ~4,250 years ago in the Casma Valley, coastal Peru (Ugent, Pozorski, and Pozorski 1981).

The practices, locations, and timings for the domestication of most vegetatively propagated root crops are poorly known. Comparable geodomestication pathways for other types of vegetatively propagated field crops, such as bananas (*Musa* cvs.; Perrier et al. 2011) and sugarcane (*Saccharum officinarum*) (Grivet et al. 2004), are variable in terms of the coverage and consilience of archaeobotanical and genetic evidence. Furthermore, there appears to be considerable variation among clonal plants in the degree to which observed phenotypic change may be considered plastic (i.e., that which is environmentally induced) or nonplastic (i.e., that which is under direct genetic control).

As in grain-based agriculture, vegetative systems also attract weeds, which have generated parasitic domesticoids, although these are less well studied. In vegecultural systems in Southern Ethiopia, for example, weeding is a routine practice to reduce competition from perennial monocots such as *Cynodon* grasses and *Cyperus rotundatus* sedges, which compete with younger *Ensete ventricosum* (Ethiopian banana) cultivars (Tsegaye and Struik 2002). Among the common weeds in Southern Ethiopian fields and gardens are several Araceae species, which are tolerated as their tubers can be processed for edible starch, and in some cases they have come to be cultivated as crops in their own right, as in the case of *Arisaema schimperianum*, *Amorphophallus abyssinicus*, and *Sauromatum nubicum* (Westphal 1975; Engels and Goettsch 1991). Vegetative agriculture in Island Southeast Asia and the Pacific also features several species that span the spectrum from weed to crop. Examples include the kudzu vine (*Pueraria montana*, syn. *P. lobata*) and Polynesian arrowroot (*Tacca leontopetaloides*), both of which are widespread as weeds (Holm et al. 1979), although occasionally they are intentionally planted. Nevertheless these taxa may sometimes be processed for the starchy tubers or for medicine (Barrau 1958; Kirch 1978; Spenneman 1994) and were translocated with the spread of agriculture in the Pacific. It has been speculated that kudzu was originally a major crop and later became a feral weed (Barrau 1965). Kudzu is native to parts of mainland East and Southeast Asia, as well as some islands (Philippines), but it is now ubiquitous in anthropogenic habitats in Melanesia and is known as an invasive clonal weed elsewhere in Polynesia, North America, and Australia (Van der Maesen 2002; Bodner and Hymowitz 2002). Polynesian arrowroot is thought to be native to sandy coastal habitats in Island Southeast Asia and was introduced to the isolated island of Palau around 3,000–2,000 years ago (Farley et al. 2018).

Despite limitations of record and the confounding effects of plasticity, some general trends in the character of vegetative crop plant domestication have been elicited (Denham et al. 2020). There seem to be convergent tendencies to lessen sexual reproductive capacity and increase the size of the edible portion. Other traits seem to be more plastic, including yield, ease of harvesting, timing of fruit production, some aspects of plant architecture, and some defensive adaptations.

Whereas in sexually reproduced crops phenotypic and genotypic transformations associated with early domestication are often portrayed as occurring in lockstep, the considerable variation exhibited by vegetatively propagated plants probably represents phenotypic plasticity resulting from active and recurrent practical management of the plant and its growth environment by people. In some ways, the domestication of sexually reproduced cereals and legumes is articulated primarily in terms of Darwinian adaptation to anthropic environments and human selection under cultivation, whereas vegetatively propagated plants exhibit a greater degree of plastic adaptation to growth environments, in a Lamarckian-type mechanism to drive phenotypic variation.

Archaeobotanical Evidence for Domestication of Long-Lived Perennials

A fourth category of domestication pathway is represented by most fruit and nut trees and long-lived vines. These taxa are woody perennials that have long generation times (5, 10, or 15 years) and lower productivity per year than most annual crops. This is thought to make it harder for them to undergo selection, in contrast to large populations of each annual generation (Zohary, Hopf, and Weiss 2012). An added factor affecting productivity is that most fruit trees have high interannual variability in production, with off years of low productivity (Goldschmidt 2013), which fall on the spectrum of mast-fruiting patterns known for many nuts (Vander Wall 2001). Added complications are that most of these species have insect pollination and therefore regular cross-pollination, in contrast to the high selfing rate in most cereals or other annual crops. Selfing should make it easier to fix favored traits in cereals than would be the case in tree crops (Zohary 2004). Since ancient times most reproduction has been by vegetative means, that is, cloning from cuttings, which is a common practice in orchards and vineyards today. Vegetative cuttings have the advantage of preserving favored traits and reproducing varieties (Zohary and Spiegel-Roy 1975; Zohary, Hopf, and Weiss 2012) and involve a high degree of conscious selection because people deliberately favor phenotypes. The diversity represented by varieties must have originated from cycles of sexual reproduction, as well as the occasional case of somatic mutation—which is well-known in citrus fruits, for example. Our current understanding is that after initial domestication involving sexual reproduction and planting from seed, cloning and vegetative plant propagation became the predominant mode of cultivation; grafting developed later (figure 10.8).

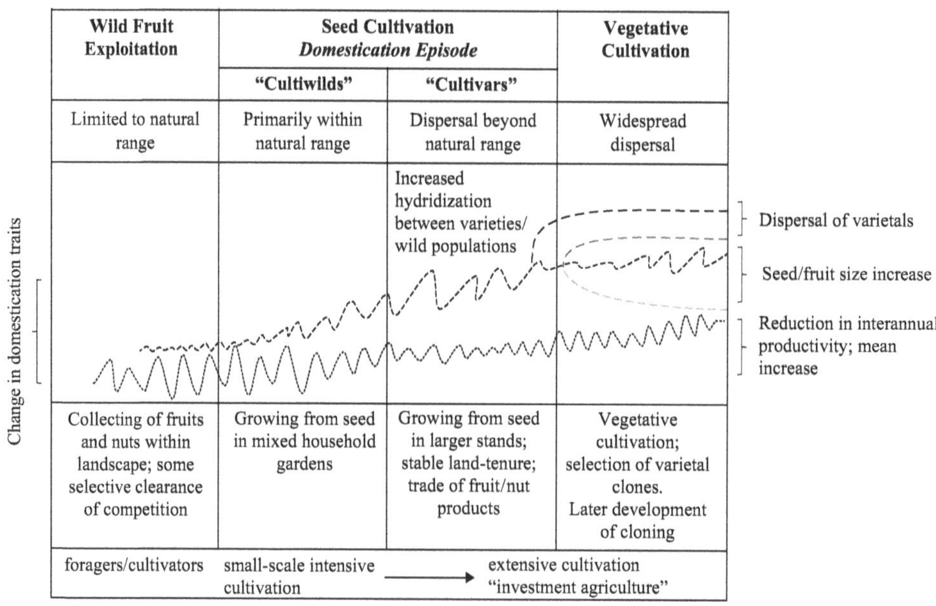

Figure 10.8
The schematic representation of a typical tree fruit crop domestication trajectory indicating how phenotypic changes are expected to correlate with changes in human economy and technology.

Archaeobotanical evidence, however, has made clear that initial cultural uses and domestication must have involved a predominance of reproduction for seed for some millennia (Weiss 2015; Fuller 2018). This is indicated by evidence across several crops, from several different world regions, for directional change in seed size, in particular increasing seed/endocarp length, which is linked to larger fruit sizes with a higher proportion of edible flesh (Liphschitz and Bonani 2001; Zheng, Crawford, and Chen 2014; Dighton et al. 2017; Fuller 2018; Bonhomme et al. 2020). This is well documented in several Mediterranean fruits (grapes, olives, dates), as well as in East Asian peaches, Japanese chestnuts, Pacific nuts such as *Canarium*, Indian mangos, and Mesoamerican avocados, which demonstrates convergent evolution across different cultural traditions and phylogenetically distant taxa (Fuller 2018). It is unclear to what extent this selection for larger fruits might have involved a conscious element on the part of cultivators, but it can be estimated that the rate of phenotypic change, when corrected for generation length, was at least as fast and often faster than evolutionary rates recorded in early seed crops (Fuller 2018). Alongside selection for increased seed size is contemporaneous evidence for the translocation of cultivated fruit trees beyond their regions of origin, with a major period of range expansion coinciding with the era of most marked morphological change in seed/fruit size. This can be illustrated with current data from peaches domesticated in China (figure 10.9) and date palms from western Asia (figure 10.10). Of interest is that after a period of largely directional change in seed morphology, such changes stop and then further variation can be documented between regions and assemblages. This is likely to coincide with the dominance of vegetative propagation of particular varieties of these fruits.

Some generalizations can be offered about how tree fruit impacted human ecology and the organization of land use. Where such crops were added to agricultural traditions that

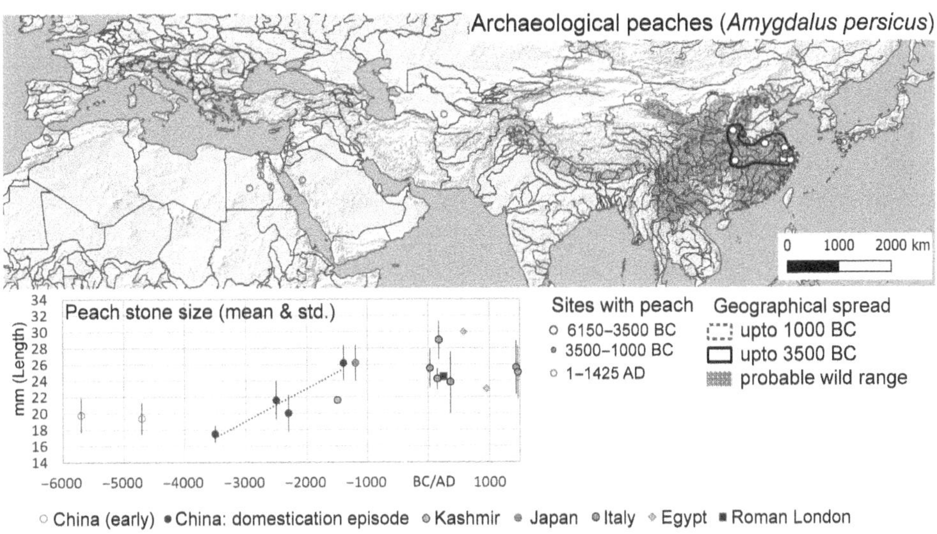

Figure 10.9

Peach (*Amygdalus persicus*) as an example of perennial tree fruit domestication. The map above shows the distribution of archaeological evidence of peaches, illustrating the gradual geographical expansion. Graph at lower left illustrates average and standard deviations of measured stone length plotted against median age estimate, highlighting a domestication episode of morphological change (after Fuller 2018; Fuller and Stevens 2019b).

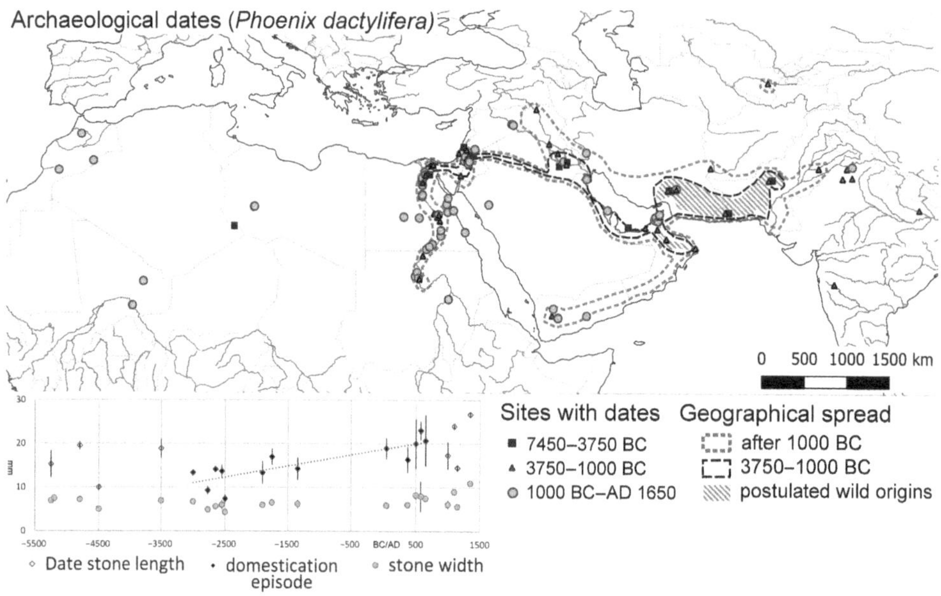

Figure 10.10
Date palm (*Phoenix dactylifera*) as an example of perennial tree fruit domestication. The map above shows the distribution of archaeological evidence of dates, illustrating the gradual geographical expansion. Graph at lower left illustrates average and standard deviations of measured stone length plotted against median age estimate, highlighting a domestication episode of morphological change (after Fuller 2018; Flowers et al. 2019, dataset S1; Fuller and Stevens 2019b).

were already based on seed crops, the addition of tree fruits expanded agriculture not just in diversity but in the nature of temporal cycles, regional flows in foodstuffs, and specialization in production (Renfrew 1972; Sherratt 1999; McCorriston 2009). Unlike annual cereals, these tree-fruits required many years of labor investment before providing returns. This implies changes in temporality—namely, longer time horizons in cycles of planting and return—as well as in social systems—namely, establishment of land tenure over longer periods (of decades or generations). Taken together these have been characterized as moving from earlier "sustainability agriculture" of the Neolithic toward an "investment agriculture" of the Metal Ages (Fuller and Stevens 2019b). With large scale production, surpluses of tree fruit production contributed to increasing trade, because they became commodities of dried fruits, nuts, wine, or olive oils, and as such played key roles in expanding the geographical range over which agricultural producers and consumers were interlinked.

Discussion: Temporalities of Practice and Transmission

If domestication is defined as the management of captive plants and animals, regardless of whether those plants and animals are modified due to domestication by humans, then the earliest detectable changes in the archaeological record provide only minimum dates of origin. In other words, the historical origin of a particular human-domesticate symbiosis is necessarily older than the earliest detectable morphological markers detected in the archaeological record. (Schultz et al. 2005, 164)

Archaeology enables reconstruction of several different pathways of plant domestication, including for cereals, secondary cereals (derived from weeds), vegetatively propagated root crops, and fruit trees. Other possible domestication pathways include ecosystem engineering—sometimes referred to as "landscape domestication" (Yen 1989; Terrell et al. 2003)—and fiber crops, selected for their raw material use, such as cotton (Viot 2019), jute, and several others. Similarly, multiple pathways to domestication have been proposed for domestic animals, principally commensal, directed, and prey pathways (Zeder 2012; Larson and Fuller 2014).

Despite inherent limitations, archaeology has enabled the domestication of several plant species, especially several cereals and legumes, to be reconstructed fairly accurately in terms of the relative timings for the emergence and fixation of domestication traits such as nonshattering and larger grain sizes (Fuller et al. 2014, 2018). For most other types of plant enmeshed in different domestication pathways, archaeological traces are scant. Instead, archaeological inferences often rely on a combination of archaeological, pedological, and palaeoecological evidence for inferring the character of cultivation systems in the past; these provide often indirect evidence for inferring plant domestication drawing on botanical, ecological, and genetic information about present-day domesticates and their wild progenitors, but without the direct evidence of preserved morphological change. There are, though, several key aspects of human-plant domesticatory relationship that emerge.

Foremost, early forms of plant exploitation, movement, and cultivation were designed to favor specific phenotypes, which gradually and inadvertently led to the accumulation of specific genetic lineages at the expense of others. Plants that were adapted to human environments benefited, whether disturbed environments in and around settlements, within managed landscapes, or within small-scale cultivated plots. Through time, selection for favored phenotypes inadvertently led to the accumulation of domestication traits within managed subpopulations relative to the wild population as a whole. Human suppression of other herbivores may also have shifted selection away from some plant defenses, especially where human harvesting led to increased seed dispersal.

However, a stepped change in selective pressure can be envisaged in which people cultivated increasingly curated propagules, whereby seed and vegetative parts were increasingly derived only from cultivated stock within anthropic landscapes that were increasingly denuded of wild plants. The genetic isolation of cultivated stock due to a reduction or absence of cultivated–wild plant interbreeding—as well as through vegetative propagation and movement beyond the natural range—would have greatly accelerated the genotypic divergence of cultivated plants from the wild progenitor population. In other words, these conditions would have greatly accelerated the emergence of domestication traits.

The period of pre-domestication cultivation, during which domestication traits emerged and rose toward fixation, can be estimated to take anywhere from 1,000 to 4,000 years in studied seed crops. Recent estimates of the strength of selection for domestication traits near the start of these domestication pathways suggest that there was an even earlier lead-in with some very weak symbiotic selection (Allaby et al. 2017). Indeed, it is estimated that human-plant symbioses began anywhere from 4,000 to 10,000 years prior to clear evidence for cultivation. These time spans can be contrasted with the period of ~30 million years between the time when attine ants evolved cultivation behaviors and when higher attines coevolved with obligate domesticated fungi (McGhee 2011, 225; Schultz et al.

2005). Unlike among insect agriculturalists, no speciation took place among the cultivating hominins (chapters 12 and 14, this volume), although cultivation may have played a key role in the social and cultural differentiation of human groups. In addition, many of the same human groups also domesticated livestock during the same periods as plant domestication (Larson et al. 2014), and while this is also paralleled in ant-aphid mutualisms, these involved very divergent Formicinae ants (McGhee 2011, 226; chapter 8, this volume).

Although early forms of seed-based cultivation (for cereals and legumes) and vegetative propagation (root crops, as well as bananas and cane grasses) were deliberate interventions in plant reproduction and growth, the domesticatory consequences of these practices were, at least initially, cumulative and unintentional. However, later forms of secondary cereal domestications, and some fruit tree domestications, were more rapid and undertaken by communities already practicing other forms of cultivation; here, people would have been more aware of the longer-term consequences of their actions in terms of making plants more amenable to exploitation and use. In the case of woody fruiting species, these were initially cultivated sexually with some subsequent clonal propagation and later grafting of preferred fruit varieties onto hardy root stock. Only beginning with these later, diversification domestications (secondary seed crops, fruit trees) does anything approaching *artificial selection* (as opposed to unconscious selection) begin to play a role in human agriculture.

Sauer (1952), among others, speculated that vegetative forms of propagation preceded seed-based cultivation. Currently, there is no robust archaeological evidence to support this claim. Rather, people in different parts of the world engaged in different forms of cultivation, which led to seed-based and vegetative domestication pathways. Currently this appears to relate more to environmental differences between biomes and available plant foods: vegetative cultivation is more associated with wetter, tropical environments or mountain zones, whereas early seed-based cultivation is found in semi-arid open ecosystems—including tropical savannahs and temperate steppes.

Certainly, in parts of the world where vegetative propagation is the dominant mode of cultivation, people incorporate into their plots cuttings from varieties grown in neighboring communities as well as those from wild/feral plants (Denham 2018). Over time, these processes can lead to the wholesale replacement of cultivated lineages, as documented for the sweet potato in Polynesia (Roullier et al. 2013b) and postulated for taro in New Guinea (Denham 2018). Furthermore, flowering, pollination, and seed-set are often deliberately suppressed under vegetative cultivation to increase yield, maintain starch or sugar contents, obtain preferred taste, and so on, as well as to restrict sexual reproduction and retain preferred phenotypes (Denham 2018). These human practices are cognate with many of those undertaken by fungus-growing ants (Mueller et al. 2005; Schultz et al. 2005; chapter 14, this volume). Ants vegetatively propagate fungus and periodically translocate and incorporate wild fungi into cultivated stock within their nests, as well as domesticated fungi from other nests. These practices lead to the wholesale replacement of fungal lineages over time. Also comparably to human practices, ants engage in a range of cultivation and harvesting practices to suppress fruiting and sexual reproduction in fungi in order to maintain preferred fungal production and limit gene flow among cultivated and wild fungal populations.

Early forms of human-plant domesticatory relationship can be conceived as mutualistic: they benefited both species. Over time, as environments of cultivation became increasingly anthropic and wild populations became increasingly sparse, this relationship became one of asymmetrical control with people as proactive partner and plants as reactive partner.

Cultivated plants became increasingly reliant on people for reproduction (i.e., in regard to gathering of nonshattering seed heads, processing, storage, and sowing) and growth (i.e., in specially prepared ground within demarcated plots or fields, often improved in terms of moisture and nutrient availability and protected from pests and competitor species). Only with modern agribusiness and genetically engineered seed stock has the relationship changed further toward enslavement—a characteristic most starkly seen in modern factory farming of animals.

The coevolution of humans and domesticates, however, was achieved through two very different forms of evolutionary transmission. Changes in the plants ultimately occurred in their genomes as they acquired adaptations either fixed through vegetative cloning or through the fixation of recessive alleles for many seed crops. Human adaptation mainly took the form of cultural information, practices, technologies, and traditions—memes in the broad sense, which were both handed down through human generations and had the potential to be transmitted laterally between unrelated human individuals and across cultures (see, e.g., Shennan 2002). As noted by Schultz (chapter 14, this volume), this makes humans facultative farmers, not obligate ones. The lateral transmission process of culture—of farming practices—contributed to regional trends towards increasing diversity of domesticated species tied into the human agricultural systems. A parallel in insects is the widespread evidence of lateral transmission of fungal cultivars across species (Mueller et al. 2005), but a contrast is the fungal monocultures that characterize many ant, termite, and beetle crops. The evolution of farming in insects took millions of years and involved adaptive speciation, whereas human facultative farmers could adapt much more rapidly in part through testing and choosing among a diversity of crops at any one time.

The increasing biodiversity of human agriculture over the long term is evident on local, continental, and global scales. However, a reversal away from agricultural diversity has occurred in many regions, mostly over the past century, as the human-domesticate relationship has shifted towards high-throughput monoculture at the expense of species or varietal diversity (Perfecto, Vandermeer, and Wright 2009). This is paralleled by the move to enslavement that can be taken to characterize many industrial animal farms, and perhaps commercial crop varieties with "genetic use restriction technologies" (Lombardo 2014). It is in this more recent phase, since the twentieth century, that humans have begun to use chemical herbicides to control weeds among their crops. By contrast many ants evolved the use of chemical control of weeds among their farmed fungi as an essential part of their cultivation symbioses (Mueller et al. 2005; chapter 11, this volume). Thus despite remarkable parallels, the various pathways of evolution of human agricultural systems and those of insects have been quite different. In this chapter we have outlined four alternative pathways of crop domestication by people, all of which have shared a tendency to contribute to increasing crop diversity over time and increasing lateral transmission across cultures, although these tendencies may have been reversed in the late industrial era.

References

Abbo, S., S. Lev-Yadun, and A. Gopher. 2011. "Origin of Near Eastern Plant Domestication: Homage to Claude Levi-Strauss and 'La Pensée Sauvage.'" *Genetic Resources and Crop Evolution* 58(2): 175–179.

Abbo, S., S. Lev-Yadun, and A. Gopher. 2014. "The 'Human Mind' as a Common Denominator in Plant Domestication." *Journal of Experimental Botany* 65(8): 1917–1920.

Allaby, R. G., C. Stevens, L. Lucas, O. Maeda, and D. Q. Fuller. 2017. "Geographic Mosaics and Changing Rates of Cereal Domestication." *Philosophical Transactions of the Royal Society B—Biological Sciences* 372(1735): 20160429.

Arranz-Otaegui, A., S. Colledge, L. Zapata, L. C. Teira-Mayolini, and J. J. Ibáñez. 2016. "Regional Diversity on the Timing for the Initial Appearance of Cereal Cultivation and Domestication in Southwest Asia." *Proceedings of the National Academy of Sciences* 113(49): 14001–14006.

Barnaud, A., M. Deu, E. Garine, J. Chantereau, J. Bolteu, et al. 2009. "A Weed-Crop Complex in Sorghum: The Dynamics of Genetic Diversity in a Traditional Farming System." *American Journal of Botany* 96(10): 1869–1879.

Barrett, S. H. 1983. "Crop Mimicry in Weeds." *Economic Botany* 37(3): 255–282.

Barrau, J. 1958. *Subsistence Agriculture on Melanesia*, Bishop Museum Bulletin 219. Honolulu: Berenice P. Bishop Museum.

Barrau, J. 1965. "Witnesses of the Past: Notes on Some Food Plants of Oceania." *Ethnology* 4(3): 282–294.

Barron, A., D. Q. Fuller, C. Stevens, L. Champion, F. Winchell, and T.P. Denham. 2020. "Snapshots in Time: MicroCT scanning of Pottery Sherds Determines Early Domestication of Sorghum (*Sorghum bicolor*) in East Africa." *Journal of Archaeological Science* 123:105259.

Barton, H., and T. P. Denham. 2018. "Vegecultures and the Social-Biological Transformations of Plants and People." *Quaternary International* 489:17–25.

Baucom, R. S., and J. S. Holt 2009. "Weeds of Agricultural Importance: Bridging the gap between Evolutionary Ecology and Crop and Weed Science." *New Phytologist* 184(4): 741–743.

Bellwood, P. 2005. *First Farmers. The Origins of Agricultural Societies*. Oxford: Blackwell.

Bodner, C. C., and T. Hymowitz. 2002. "Ethnobotany of *Pueraria* Species." In *Pueraria: The Genus Pueraria*, edited by W. M. Keung, 50–86. London: Taylor and Francis.

Bonhomme, V., S. Picq, S. Ivorra, A. Evin, T. Pastor, et al. 2020. "Eco-Evo-Devo Implications and Archaeobiological Perspectives of Wild and Domesticated Grapevines Fruits Covariating Traits." *bioRxiv*. https://doi.org/10.1101/513036.

Burkill, I. H. 1935. *A Dictionary of the Economic Products of the Malay Peninsula*, 2 vols. London: Crown Agents.

Chaïr, H., R. E. Traore, M. F. Duval, R. Rivallan, A. Mukherjee, et al. 2016. "Genetic Diversification and Dispersal of Taro (*Colocasia esculenta* (L.) Schott)." *PLoS ONE* 11(6): e0157712. doi:10.1371/journal.pone.0157712.

Champion, L., and D. Q. Fuller. 2018. "New Evidence on the Development of Millet and Rice Economies in the Niger River Basin: Archaeobotanical Results from Benin." In *Plants and People in the African Past*, edited by A. Mercuri, A. D'Andrea, R. Fornaciari, and A. Höhn, 529–547. Cham: Springer.

Chen, T., B. Wang, R. C. Power, and H. Jiang. 2020. "The First Archaeobotanical Evidence of *Medicago sativa* L. in China: Hay Fodder for Livestock." *Archaeological and Anthropological Sciences* 12:2. https://doi.org/10.1007/s12520-019-00957-7.

Cucchi, T., K. Papayianni, S. Cersoy, L. Aznar-Cormano, A. Zazzo, et al. 2020. "Tracking the Near Eastern Origins and European Dispersal of the Western House Mouse." *Scientific Reports* 10(1): 1–12.

Darwin, C. 1868. *The Variation of Animals and Plants under Domestication*. London: Appleton and Company.

Denham, T. P. 2018. *Tracing Early Agriculture in the Highlands of New Guinea: Plot, Mound and Ditch*. Oxford: Routledge.

Denham, T., H. Barton, C. Castillo, A. Crowther, E. Dotte-Sarout, et al. 2020. "The Domestication Syndrome in Vegetatively Propagated Field Crops." *Annals of Botany* 125:581–597.

Denham, T. P., S. G. Haberle, C. Lentfer, R. Fullagar, J. Field, et al. 2003. "Origins of Agriculture at Kuk Swamp in the Highlands of New Guinea." *Science* 301:189–193.

Dighton, A., A. Fairbairn, S. Bourke, J. T. Faith, and P. Habgood. 2017. "Bronze Age Olive Domestication in the North Jordan valley: New Morphological Evidence for Regional Complexity in Early Arboricultural Practice from Pella in Jordan." *Vegetation History and Archaeobotany* 26(4): 403–413.

Donohue, M., and T. P. Denham. 2010. "Farming and Language in Island Southeast Asia: Reframing Austronesian History." *Current Anthropology* 51(2): 223–256.

Doust, A. 2007. "Architectural Evolution and Its Implications for Domestication in Grasses." *Annals of Botany* 100(5): 941–950.

Ellis, E. C., J. O. Kaplan, D. Q. Fuller, S. Vavrus, K. K. Goldewijk, et al. 2013. "Used Planet: A Global History." *Proceedings of the National Academy of Sciences* 110(20): 7978–7985.

Ellis, E. C., N. R. Magliocca, C. J. Stevens, and D. Q. Fuller. 2018. "Evolving the Anthropocene: Linking Multi-Level Selection with Long-Term Social-Ecological Change." *Sustainability Science* 13(1): 119–128.

Engel, F. A. 1970. "Explorations of the Chilca Canyon, Peru." *Current Anthropology* 11:55–58.

Engels, J. M. M., and E. Goettsch. 1991. "Konso Agriculture and Its Plant Genetic Resources." In *Plant Genetic Resources of Ethiopia*, edited by J. M. M. Engels, J. G. Hawkes, and M. Worede, 169–196. Cambridge: Cambridge University Press.

Erskine, W., J. Smartt, and F. J. Muehlbauer. 1994. "Mimicry of Lentil and the Domestication of Common Vetch and Grass Pea." *Economic Botany* 48(3): 326–332.

Farley, G., L. Schneider, G. Clark, and S. G. Haberle. 2018. "A Late Holocene Palaeoenvironmental Reconstruction of Ulong Island, Palau, from Starch Grain, Charcoal, and Geochemistry Analyses." *Journal of Archaeological Science: Reports* 22:248–256.

Flowers, J. M., K. M. Hazzouri, M. Gros-Balthazard, Z. Mo, K. Koureoumpa, et al. 2019. "Cross-Species Hybridization and the Origin of North African Date Palms." *Proceedings of the National Academy of Sciences* 116(5): 1651–1658. doi:10.1073/Pnas.1817453116.

Fritz, G. J., M. C. Bruno, B. S. Langlie, B. D. Smith, and L. Kistler. 2017. "Cultigen Chenopods in the Americas: A Hemispherical Perspective." In *Social Perspectives on Ancient Lives from Paleoethnobotanical Data*, edited by M. P. Sayre and M. C. Bruno, 55–75. Cham: Springer.

Fullagar R., J. Field, T. P. Denham, and C. Lentfer. 2006. "Early and Mid Holocene Processing of Taro (*Colocasia esculenta*) and Yam (*Dioscorea* sp.) at Kuk Swamp in the Highlands of Papua New Guinea." *Journal of Archaeological Science* 33:595–614.

Fuller, D. Q. 2007. "Contrasting Patterns in Crop Domestication and Domestication Rates: Recent Archaeobotanical Insights from the Old World." *Annals of Botany* 100(5): 903–924.

Fuller, D. Q. 2018. "Long and Attenuated: Comparative Trends in the Domestication of Tree Fruits." *Vegetation History and Archaeobotany* 27(1): 165–176.

Fuller, D. Q. 2020. "Archaeobotany." In *Excavations at Paithan, Maharashtra. Beyond Boundaries: Religion, Region, Language and the State*, edited by D. Kennet, J. V. Rao, and M. K. Bai, 3:245–280. Berlin: De Gruyter GmbH.

Fuller, D. Q., R. G. Allaby, and C. Stevens. 2010. "Domestication as Innovation: The Entanglement of Techniques, Technology and Chance in the Domestication of Cereal Crops." *World Archaeology* 42(1): 13–28.

Fuller, D. Q., L. Champion, and C. Stevens. 2019. "Comparing the Tempo of Cereal Dispersal and the Agricultural Transition: Two African and One West Asian Trajectory." In *Trees, Grasses and Crops—People and Plants in Sub-Saharan Africa and Beyond*, edited by B. Einchornn and A. Höhn, 119–140. Bonn: Verlag Dr. Rudolf Habelt GmbH.

Fuller, D. Q., S. Colledge, C.Murphy, and C. J. Stevens. 2017. "Sizing Up Cereal Variation: Patterns in Grain Evolution Revealed in Chronological and Geographical Comparisons." In *Miscelánea en homenaje a Lydia Zapata Peña (1965–2015)*, edited by J. Fernández Eraso, J. A. Mujika Alustiza, Á. A. Valbuena, and Marcos García Díez, 131–149. Bilbao: Servicio Editorial Universidad Del País Vasco.

Fuller, D. Q., T. Denham, M. Arroyo-Kalin, L. Lucas, C. J. Stevens, et al. 2014. "Convergent Evolution and Parallelism in Plant Domestication Revealed by an Expanding Archaeological Record." *Proceedings of the National Academy of Sciences* 111(17): 6147–6152.

Fuller, D. Q., and L. Lucas. 2017. "Adapting Crops, Landscapes and Food Choices: Patterns in the Dispersal of Domesticated Plants across Eurasia." In *Human Dispersal and Species Movement: From Prehistory to the Present*, edited by M. Petraglia, N. Boivin, and R. Crassard, 304–331. Cambridge: Cambridge University Press.

Fuller, D. Q., L. Lucas, L. Gonzalez Carretero, and C. Stevens. 2018. "From Intermediate Economies to Agriculture: Trends in Wild Food Use, Domestication and Cultivation among Early Villages in Southwest Asia." *Paléorient* 44:61–76.

Fuller, D. Q., C. Murphy, E. Kingwell-Banham, C. C. Castillo, and S. Naik. 2019. "*Cajanus cajan* (L.) Millsp. Origins and Domestication: The South and Southeast Asian Archaeobotanical Evidence." *Genetic Resources and Crop Evolution* 66(6): 1175–1188.

Fuller, D. Q., and L. Qin. 2010. "Declining Oaks, Increasing Artistry, and Cultivating Rice: The Environmental and Social Context of the Emergence of Farming in the Lower Yangtze Region." *Environmental Archaeology* 15(2): 139–159.

Fuller, D. Q., and C. J. Stevens. 2017. "Open for Competition: Domesticates, Parasitic Domesticoids and the Agricultural Niche." *Archaeology International* 20:112–123.

Fuller, D. Q., and C. J. Stevens. 2019a. "The Making of the Botanical Battleground: Domestication and the Origins of the World's Weed Floras." In *Far from the Hearth. Essays in Honour of Martin K. Jones*, edited by E. Lightfoot, X. Liu, and D. Q. Fuller, 9–22. Cambridge: McDonald Institute of Archaeology.

Fuller, D. Q., and C. J. Stevens. 2019b. "Between Domestication and Civilization: The Role of Agriculture and Arboriculture in the Emergence of the First Urban Societies." *Vegetation History and Archaeobotany* 28(3): 263–282.

Fuller, D. Q., G. Willcox, and R. G. Allaby. 2012. "Early Agricultural Pathways: Moving Outside the 'Core Area' Hypothesis in Southwest Asia." *Journal of Experimental Botany* 63(2): 617–633.

Gegas, V. C., A. Nazari, S. Griffiths, J. Simmonds, L. Fish, et al. 2010. "A Genetic Framework for Grain Size and Shape Variation in Wheat." *Plant Cell* 22(4): 1046–1056.

Goldschmidt, E. E. 2013. "The Evolution of Fruit Tree Productivity: A Review." *Economic Botany* 67(1): 51-62

Grimaldi, I. M., S. Muthukumaran, G. Tozzi, A. Nastasi, N. Boivin, et al. 2018. "Literary Evidence for Taro in the Ancient Mediterranean: A Chronology of Names and Uses in a Multilingual World." *PLoS ONE* 13(6): e0198333. https://doi.org/10.1371/journal.pone.0198333.

Grivet, L., C. Daniels, J. C. Glaszman, and A. D'Hont. 2004. "A review of Recent Molecular Genetics Evidence for Sugarcane Evolution and Domestication." *Ethnobotany Research and Applications* 2:9–17.

Guo, Y., R. Wu, G. Sun, Y. Zheng, and B. T. Fuller. 2017. "Neolithic Cultivation of Water Chestnuts (*Trapa* L.) at Tianluoshan (7000–6300 cal BP), Zhejiang Province, China." *Scientific Reports* 7:16206. https://doi.org/10.1038/s41598-017-15881-w.

Hammer, K. 1984. "Das domestikationssyndrom." *Die Kulturpflanze* 32(1): 11–34.

Hardigan, M. A., F. P. E. Laimbeer, L. Newton, E. Crisovan, J. P. Hamilton, et al. 2017. "Genome Diversity of Tuber-Bearing Solanum Uncovers Complex Evolutionary History and Targets of Domestication in the Cultivated Potato." *Proceedings of the National Academy of Sciences* 114(46): E9999–E10008.

Harlan, J. R. 1995. *The Living Fields: Our Agricultural Heritage*. Cambridge: Cambridge University Press.

Harlan, J. R., and J. M. J. De Wet. 1965. "Some Thoughts about Weeds." *Economic Botany* 19(1): 16–24.

Harlan J. R., J. M. J. De Wet, and E. G. Price. 1973. Comparative Evolution of Cereals." *Evolution* 27:311–325.

Harris, D. R. 1972. "The Origins of Agriculture in the Tropics." *American Scientist* 60:180–193.

Harris, D. R. 1989. "An Evolutionary Continuum of People-Plant Interaction." In *Foraging and Farming. The Evolution of Plant Exploitation*, edited by D. R. Harris and G. C. Hillman, 11–26. London: Routledge.

Harris, D. R. 1996. "Introduction: Themes and Concepts in the Study of Early Agriculture." In *The Origins and Spread of Agriculture and Pastoralism in Eurasia*," edited by D. R. Harris, 1–9. London: University College London Press.

Harris, D. R. 2012. "Evolution of Agroecosystems: Biodiversity, Origins, and Differential Development." In *Biodiversity in Agriculture. Domestication, Evolution, and Sustainability*, edited by P. Gepts, T. R. Famula, R. I. Bettinger, S. B. Brush, A. B. Damania, et al., 21–56. Cambridge: Cambridge University Press.

Hather, J. G. 1996. "The Origins of Tropical Vegeculture: Zingiberaceae, Araceae and Dioscoreaceae in Southeast Asia." In *The Origins and Spread of Agriculture and Pastoralism in Eurasia*, edited by D. R. Harris, 538–550. London: University College London Press.

Hillman, G. 1975. "The Plant Remains from Tell Abu Hureyra: A Preliminary Report." *Proceedings of the Prehistoric Society* 41(1): 70–73.

Hillman, G. C., and M. S. Davies, 1990. "Domestication Rates in Wild-Type Wheats and Barley under Primitive Cultivation." *Biological Journal of the Linnean Society* 39(1): 39–78.

Holm, L., J. V. Pancho, J. P. Herberger, and D. L. Plucknett. 1979. *A Geographical Atlas of World Weeds*. New York: Wiley.

Holm, L. G., D. L. Plucknett, J. V. Pancho, and J. P. Herberger. 1977. *The World's Worst Weeds*. Honolulu: University of Hawaii Press.

Huang, Z., N. D. Young, M. Reagon, K. E. Hyma, K. Olsen, et al. 2017. "All Roads Lead to Weediness: Patterns of Genomic Divergence Reveal Extensive Recurrent Weedy Rice Origins from South Asian Oryza." *Molecular Ecology* 26(12): 3151–3167.

Iriarte, J. 2007. "New Perspectives on Plant Domestication and the Development of Agriculture in the New World." In *Rethinking Agriculture: Archaeological and Ethnoarchaeological Perspectives*, edited by T. P. Denham, J. Iriarte, and L. Vrydaghs, 167–188. Walnut Creek, CA: Left Coast Press.

Iriarte, J., I. Holst, O. Marozzi, C. Listopad, E. Alonso, et al. 2004. "Evidence for Cultivar Adoption and Emerging Complexity during the Mid-Holocene in the La Plata Basin." *Nature* 432:614–17.

Jones, M. K. 1988. "The Arable Field: A Botanical Battleground." In *Archaeology and the Flora of the British Isles: Human Influence on the Evolution of Plant Communities*, edited by M. K. Jones, 86–92. Oxford: Oxford University Committee for Archaeology.

Kirch, P. V. 1978. "Indigenous Agriculture on Uvea (Western Polynesia)." *Economic Botany* 32(2): 157–181.

Langlie, B. S., C. A. Hastorf, M. C. Bruno, M. Bermann, R. M. Bonzani, et al. 2011. "Diversity in Andean Chenopodium Domestication: Describing a New Morphological Type from La Barca, Bolivia 1300–1250 BC." *Journal of Ethnobiology* 31(1): 72–88.

Larson, G., and D. Q. Fuller. 2014. "The Evolution of Animal Domestication." *Annual Review of Ecology, Evolution, and Systematics* 45:115–136.

Larson, G., D. R. Piperno, R. G. Allaby, M. D. Purugganan, L. Andersson, et al. 2014. "Current Perspectives and the Future of Domestication Studies." *Proceedings of the National Academy of Sciences* 111(17): 6139–6146.

Lebot, V. 2009. *Root and Tuber Crops: Cassava, Sweet Potato, Yams and Aroids. Crop Production Science in Horticulture,* vol. 17. Wallingford, Oxfordshire: CABI Publishing.

Lev-Yadun, S., A. Gopher, and S. Abbo. 2000. "The Cradle of Agriculture." *Science* 288(5471): 1602–1603.

Li, H.-L. 1970. "The Origin of Cultivated Plants in Southeast Asia." *Economic Botany* 24: 3–19.

Liphschitz, N., and G. Bonani. 2001. "Wild and Cultivated Date Palm (*Phoenix dactylifera*) from Qumran Cave 24." *Tel Aviv* 28:305–309.

Lister, D. L., S. Thaw, M. A. Bower, H. Jones, M. P. Charles, et al. 2009. "Latitudinal Variation in a Photoperiod Response Gene in European Barley: Insight into the Dynamics of Agricultural Spread from 'Historic' Specimens." *Journal of Archaeological Science* 36(4): 1092–1098.

Liu, X., D. L. Lister, Z. Zhao, C. A. Petrie, X. Zeng, et al. 2017. "Journey to the East: Diverse Routes and Variable Flowering Times for Wheat and Barley en Route to Prehistoric China." *PLoS ONE* 12(11): e0187405.

Lombardo, L., 2014. "Genetic Use Restriction Technologies: A Review." *Plant Biotechnology Journal* 12(8): 995–1005.

Londo, J. P., and B. A. Schaal. 2007. "Origins and Population Genetics of Weedy Red Rice in the USA." *Molecular Ecology* 16(21): 4523–4535.

Lu, L., D. Shao, X. Qiu, L. Sun, W. Yan, et al. 2013. "Natural Variation and Artificial Selection in Four Genes Determine Grain Shape in Rice." *New Phytologist* 200(4): 1269–1280.

Maeda, O., L. Lucas, F. Silva, K. I. Tanno, and D. Q. Fuller. 2016. "Narrowing the Harvest: Increasing Sickle Investment and the Rise of Domesticated Cereal Agriculture in the Fertile Crescent." *Quaternary Science Reviews* 145:226–237.

Malapa, R., G. Arnau, J. L. Noyer, and V. Lebot. 2005. "Genetic Diversity of the Greater Yam (*Dioscorea alata* L.) and Relatedness to *D. nummularia* Lam. and *D. transversa* Br. as Revealed with AFLP Markers." *Genetic Resources and Crop Evolution* 52(7): 919–929.

McClatchie, M., A. Bogaard, S. Colledge, N. J. Whitehouse, R. J. Schulting, et al. 2014. "Neolithic Farming in North-Western Europe: Archaeobotanical Evidence from Ireland." *Journal of Archaeological Science* 51:206–215.

McCorriston, J. 2009. "Domestication and the Dialectic: Archaeobotany and the Future of the Neolithic Revolution in the Near East." In *From Foragers to Farmers*, edited by A. Fairbairn and E. Weiss, 27–36. Oxford: Oxbow Books.

McGhee, G. 2011. *Convergent Evolution. Limited Forms Most Beautiful.* Cambridge, MA: MIT Press.

Mindzie, C. M., H. Doutrelepont, L. Vrydaghs, R. L. Swennen, R. J. Swennen, et al. 2001. "First Archaeological Evidence of Banana Cultivation in Central Africa during the Third Millennium before Present." *Vegetation History and Archaeobotany* 10:1–6.

Mueller, U. G., N. M. Gerardo, D. K. Aanen, D. L. Six, and T. R. Schultz. 2005. "The Evolution of Agriculture in Insects." *Annual Review of Ecology, Evolution and Systematics* 36:563–595.

Muller, M. H., Poncet, C., Prosperi, J. M., Santoni, S., and Ronfort, J. 2006. "Domestication History in the *Medicago sativa* Species Complex: Inferences from Nuclear Sequence Polymorphism." *Molecular Ecology* 15(6): 1589–1602.

Muller, M. H., J. M. Prosperi, S. Santoni, and J. Ronfort. 2003. "Inferences from Mitochondrial DNA Patterns on the Domestication History of Alfalfa (*Medicago sativa*)." *Molecular Ecology* 12(8): 2187–2199.

Muñoz-Rodríguez, P., T. Carruthers, J. R. Wood, B. R. Williams, K. Weitemier, B. Kronmiller, D. Ellis, N. L. Anglin, L. Longway, S. A. Harris, M. D. Rausher, S. Kelly, A. Liston, and R. W. Scotland. 2018. "Reconciling Conflicting Phylogenies in the Origin of Sweet Potato and Dispersal to Polynesia." *Current Biology* 28(8):1246–1256.

Murdock, G. P. 1960. "Staple Subsistence Crops of Africa." *Geographical Review* 50(4): 523–540.

Murphy, C., and D. Q. Fuller. 2017. "Seed Coat Thinning during Horsegram (*Macrotyloma uniflorum*) Domestication Documented through Synchrotron Tomography of Archaeological Seeds." *Scientific Reports* 7:5369. doi:10.1038/s41598-017-05244-w.

National Research Council. 1989. *Lost Crops of the Incas.* Washington, DC: National Academy Press.

Okamoto, Y., T. Kajimura, T. M. Ikeda, and S. Takumi. 2012. "Evidence from Principal Component Analysis for Improvement of Grain Shape and Spikelet Morphology-Related Traits after Hexaploid Wheat Speciation." *Genes & Genetic Systems* 87(5): 299–310.

Perfecto, I., J. Vandermeer, and A. Wright. 2009. *Nature's Matrix. Linking Agriculture, Conservation and Food Sovereignty*. London: Earthscan

Perrier, X., E. De Langhe, M. Donohue, C. Lentfer, L. Vrydaghs, et al. 2011. "Multidisciplinary Perspectives on Banana (*Musa* spp.) Domestication." *Proceedings of the National Academy of Sciences* 108:11311–11318.

Piperno D. R., and D. M. Pearsall. 1998. *The Origins of Agriculture in the Lowland Neotropics*. San Diego: Academic Press.

Plarre, R. 2013. "An Attempt to Reconstruct the Natural and Cultural History of the Granary Weevil, *Sitophilus granarius* (Coleoptera: Curculionidae)." *European Journal of Entymology* 107(1): 1–11.

Pokorna, A., P. Kočár, J. Novak, T. Šálková, P. Žáčková, et al. 2018. "Ancient and Early Medieval Man-Made Habitats in Czech Republic: Colonization History and Vegetation Changes." *Preslia* 90:171–193.

Poschlod, P. 2015a. *Geschichte der Kulturlandschaft*. Stuttgart: Eugen Ulmer KG.

Poschlod, P. 2015b. "The Origin and Development of the Central European Man-Made Landscape, Habitat and Species Diversity as Affected by Climate and Its Changes—A Review." *Interdisciplinaria Archaeologica. Natural Sciences in Archaeology* 6(2): 197–221

Potokina, E., F. Blattner, T. Alexandrova, and K. Bachmann. 2002. "AFLP Diversity in the Common Vetch (*Vicia sativa* L.) on the World Scale." *Theoretical and Applied Genetics* 105(1): 58–67.

Purugganan, M. D., and D. Q. Fuller. 2009. "The Nature of Selection during Plant Domestication." *Nature* 457(7231): 843–848.

Qi, X., Y. Liu, C. C. Vigueira, N. D. Young, A. L. Caicedo, et al. 2015. "More Than One Way to Evolve a Weed: Parallel Evolution of US Weedy Rice through Independent Genetic Mechanisms." *Molecular Ecology* 24(13): 3329–3344.

Qin, L., and D. Q. Fuller 2019. "Why Rice Farmers Don't Sail: Coastal Subsistence Traditions and Maritime Trends in Early China." In *Prehistoric Maritime Cultures and Seafaring in East Asia*, edited by C. Wu and B. Rolett, 159–191. Singapore: Springer.

Renfrew, C. 1972 *The Emergence of Civilisation: The Cyclades and the Aegean in the Third Millennium BC*. London: Methuen.

Roullier, C., L. Benoit, D. B. McKey, and V. Lebot. 2013a. "Historical Collections Reveal Patterns of Diffusion of Sweet Potato in Oceania Obscured by Modern Plant Movements and Recombination." *Proceedings of the National Academy of Sciences* 110:2205–2210.

Roullier, C., A. Duputié, P. Wennekes, L. Benoit, V. M. F. Bringas, et al. 2013b. "Disentangling the Origins of Cultivated Sweet Potato (*Ipomoea batatas* (L.) Lam.)." *PLoS ONE* 8(5): e62707. https://doi.org/10.1371/journal.pone.0062707.

Rovira, N., and N. Alonso. 2018. "Crop Growing and Plant Consumption in Coastal Languedoc (France) in the Second Iron Age: New Data from Pech Maho (Aude), Lattara (Hérault) and Le Cailar (Gard)." *Vegetation History and Archaeobotany* 27(1): 85–97.

Sauer, C. O. 1952. *Agricultural Origins and Their Dispersals*. New York: American Geographical Society.

Scarcelli, N., P. Cubry, R. Akakpo, A. C. Thuillet, J. Obidiegwu, et al. 2019. "Yam Genomics Supports West Africa as a Major Cradle of Crop Domestication." *Science Advances* 5:eaaw1947.

Schultz, T. R., U. G. Mueller, C. R. Currie, and S. A. Rehner. 2005. "Reciprocal Illumination: A Comparison of Agriculture in Humans and Fungus-Growing Ants." In *Insect–Fungal Associations: Ecology and Evolution*, edited by F. Vega and M. Blackwell, 149–190. Oxford: Oxford University Press.

Shennan, S. 2002. *Memes, Genes and Human History*. London: Thames and Hudson.

Shennan, S. 2018. *The First Farmers of Europe. An Evolutionary Perspective*. Cambridge: Cambridge University Press.

Sherratt, A. 1999 "Cash-Crops before Cash: Organic Consumables and Trade." In *The Prehistory of Food: Appetites for Change*, edited C. Gosden and J. Hather, 13–34. London: Routledge.

Smith, B. D. 2001. "Low-Level Food Production." *Journal of Archaeological Research* 9(1): 1–43.

Smith, B. D., and R. A. Yarnell. 2009. "Initial Formation of an Indigenous Crop Complex in Eastern North America at 3800 BP." *Proceedings of the National Academy of Sciences* 106(16): 6561–6566.

Spennemann, D. H. 1994. "Traditional Arrowroot Production and Utilization in the Marshall Islands." *Journal of Ethnobiology* 14(2): 211–34.

Stevens, C. J., and D. Q. Fuller. 2017. "The Spread of Agriculture in Eastern Asia: Archaeological Bases for Hypothetical Farmer/Language Dispersals." *Language Dynamics and Change* 7(2):152–186.

Stevens, C. J., and D. Q. Fuller. 2019. "The Fighting Flora: An Examination of the Origins and Changing Composition of the Weed Flora of the British Isles." In *Far from the Hearth. Essays in Honour of Martin K. Jones*, edited by E. Lightfoot, X. Liu, and D. Q. Fuller, 23–36. Cambridge: McDonald Institute of Archaeology.

Stevens, C. J., C. Murphy, R. Roberts, L. Lucas, F. Silva, et al. 2016. "Between China and South Asia: A Middle Asian Corridor of Crop Dispersal and Agricultural Innovation in the Bronze Age." *The Holocene* 26(10): 1541–1555.

Sukopp, H., and U. Sukopp. 1993. "Ecological Long-Term Effects of Cultigens Becoming Feral and of Naturalization of Non-Native Species." *Experientia* 49(3): 210–218.

Terrell, J. E., J. P. Hart, S. Barut, N. Cellinese, A. Curet, et al. 2003. "Domesticated Landscapes: The Subsistence Ecology of Plant and Animal Domestication." *Journal of Archaeological Method and Theory* 10:323–368.

Thurber, C. S., M. Reagon, B. L. Gross, K. M. Olsen, Y. Jia, et al. 2010. "Molecular Evolution of Shattering Loci in US Weedy Rice." *Molecular Ecology* 19(16): 3271–3284.

Tsegaye, A., and P. C. Struik. 2002. "Analysis of Enset (*Ensete ventricosum*) Indigenous Production Methods and Farm-Based Biodiversity in Major Enset-Growing Regions of Southern Ethiopia." *Experimental Agriculture* 38(3): 291–315.

Ugent, D., and L. W. Peterson. 1988. "Archaeological Remains of Potato and Sweet Potato in Peru." *Circular (International Potato Center)* 16:1–10.

Ugent, D., S. Pozorski, and T. Pozorski. 1981. "Prehistoric Remains of the Sweet Potato from the Casma Valley of Peru." *Phytologia* 49:401–415.

Ugent, D., S. Pozorski, and T. Pozorski. 1982. "Archaeological Potato Tuber Remains from the Casma Valley of Peru." *Economic Botany* 36:182–192.

Van der Maesen, L. J. G. 2002. "Pueraria: Botanical Characteristics." In *Pueraria. The Genus Pueraria*, edited by W. M. Keung, 21–49. London: Taylor and Francis.

Van der Veen, M., and J. Morales. 2011. "Summer Crops—From Trade to Innovation." In *Consumption, Trade and Innovation. Exploring the Botanical Remains from the Roman and Islamic Ports at Quseir al-Qadim, Egypt*, edited by M. Van der Veen, 75–119. Frankfurt: Africa Magna Verlag.

Vander Wall, S. B. 2001. "The Evolutionary Ecology of Nut Dispersal." *Botanical Review* 67(1): 74–117.

Vavilov, N. I. 1992 [1926]. *Origin and Geography of Cultivated Plants*, translated by D. Love. Cambridge: Cambridge University Press.

Vigueira, C. C., K. M. Olsen, and A. L. Caicedo. 2013. "The Red Queen in the Corn: Agricultural Weeds as Models of Rapid Adaptive Evolution." *Heredity* 110(4): 303–311.

Viot, C. 2019. "Domestication and Varietal Diversification of Old World Cultivated Cottons (*Gossypium* sp.) in the Antiquity." *Revue d'Ethnoécologie* 15. doi:10.4000/Ethnoecologie.4404.

Wang, W., B. Feng, J. Xiao, Z. Xia, X. Zhou, et al. 2014. "Cassava Genome from a Wild Ancestor to Cultivated Varieties." *Nature Communications* 5(1): 5110.

Weiss, E. 2015. "Beginnings of Fruit Growing in the Old World—Two Generations Later." *Israel Journal of Plant Sciences* 62(1–2): 75–85.

Weissbrod, L., F. B. Marshall, F. R. Valla, H. Khalaily, G. Bar-Oz, et al. 2017. "Origins of House Mice in Ecological Niches Created by Settled Hunter-Gatherers in the Levant 15,000 Y Ago." *Proceedings of the National Academy of Sciences* 114(16): 4099–4104.

Weisskopf, A., E. Harvey, E. Kingwell-Banham, M. Kajale, R. Mohanty, et al. 2014. "Archaeobotanical Implications of Phytolith Assemblages from Cultivated Rice Systems, Wild Rice Stands and Macro-Regional Patterns." *Journal of Archaeological Science* 51: 43–53.

Westphal, E. 1975. *Agricultural Systems in Ethiopia*. Wageningen, The Netherlands: Centre for Agricultural Publishing and Documentation.

Willcox, G., 2005. "The Distribution, Natural Habitats and Availability of Wild Cereals in Relation to Their Domestication in the Near East: Multiple Events, Multiple Centres." *Vegetation History and Archaeobotany* 14(4): 534–541.

Willcox, G. 2012. "Searching for the Origins of Arable Weeds in the Near East." *Vegetation History and Archaeobotany* 21(2): 163–167.

Winchell, F., M. Brass, A. Manzo, A. Beldados, V. Perna, et al. 2018. "On the Origins and Dissemination of Domesticated Sorghum and Pearl Millet across Africa and into India: A View from the Butana Group of the Far Eastern Sahel." *African Archaeological Review* 35(4): 483–505.

Xia, H. B., H. Xia, N. C. Ellstrand, C. Yang, and B. R. Lu. 2011. "Rapid Evolutionary Divergence and Ecotypic Diversification of Germination Behavior in Weedy Rice Populations." *New Phytologist* 191(4):1119–1127.

Yen, D. E. 1973. "The Origins of Oceanic Agriculture." *Archaeology and Physical Anthropology in Oceania* 8:68–85.

Yen, D. E. 1974. *The Sweet Potato and Oceania: An Essay in Ethnobotany*. Honolulu: Bishop Museum Bulletin 236.

Yen, D. E. 1989. “The Domestication of Environment.” In *Foraging and Farming: The Evolution of Plant Exploitation*, edited by D. R. Harris and G. C. Hillman, 55–75. London: Routledge.

Zeder, M. 2012. “Pathways to Animal Domestication.” In *Biodiversity in Agriculture: Domestication, Evolution and Sustainability*, edited by P. Gepts, T. R. Famula, R. L. Bettinger, S. B. Brush, A. B. Damania, et al., 227–259. Cambridge: Cambridge University Press.

Zheng, Y., G. W. Crawford, and X. Chen. 2014. “Archaeological Evidence for Peach (*Prunus persica*) Cultivation and Domestication in China.” *PloS ONE* 9(9): e106595.

Zohary, D. 1969. “The Progenitors of Wheat and Barley in Relation to Domestication and Agriculture Dispersal in the Old World.” In *The Domestication and Exploitation of Animals and Plants*, edited by P. J. Ucko and G. W. Dimbleby, 47–66. London: Duckworth.

Zohary, D. 2004. “Unconscious Selection and the Evolution of Domesticated Plants.” *Economic Botany* 58(1): 5–10.

Zohary, D., M. Hopf, and E. Weiss. 2012. *Domestication of Plants in the Old World*, 4th ed. Oxford: Oxford University Press.

Zohary, D., and P. Spiegel-Roy, P. 1975. “Beginnings of Fruit Growing in the Old World.” *Science* 187(4174): 319–327.

Zohary, M. 1950. “The Segetal Plant Communities of Palestine.” *Vegetatio* 2:387–411.

11 Convergent Adaptation and Specialization of Eukaryotic Pathogens across Agricultural Systems

Nicole M. Gerardo

Domestication of crops provides an opportunity for pathogens to exploit novel, abundant hosts. Pathogens specializing on human crops have repeatedly emerged over the history of human agriculture. Similarly, pathogens have tracked domestication of the fungal crops of ants (Caldera et al. 2009). Like pathogens in nondomesticated systems, agricultural pathogens often face selection pressures to overcome host defenses, and such adaptation may lead to increased specialization (Flor 1956; Burdon 1987; Thompson and Burdon 1992). Specialization may lock pathogens into narrow host ranges, such that as host crops diverge from one another, either through intentional or unintentional selection by their domesticators, pathogens also diverge, potentially leading to patterns of co-divergence between the pathogens and their hosts. Alternatively, despite specialization, pathogens may make occasional host jumps, switching to hosts distantly related to the original host, which can lead to the emergence of novel disease.

Here, I highlight four key issues in the study of crop pathogen evolution and specialization and the approaches used to address those questions, focusing on fungal and oomycete pathogens in human and ant agriculture. These include (1) using experimental and molecular approaches to compare pathogens in domesticated and nondomesticated systems in order to determine the origins of crop pathogens; (2) elucidating the relationships between pathogen strains that utilize different crops to gain insights into the degree and evolution of pathogen specialization; (3) leveraging genomics and genetics to elucidate the mechanisms underlying pathogen specialization; and (4) studying how, like their crop hosts, pathogens evolve in response to agricultural practices. In many cases, we have more advanced understanding of human crop pathogens, but the study of ant crop pathogens provides a unique opportunity to study crop pathogen evolution over a much longer time scale.

The Diversity of Eukaryotic Pathogens Attacking Human and Ant Crops

Over the last 12,000 years, humans have domesticated a broad array of plant and fungal crops. All of these crops are parasitized by pathogens. The impact of pathogens on crop yield and crop quality has profoundly shaped human agriculture. Over the history of agriculture, humans have abandoned some crop varieties because of high pathogen burdens

(Heslop-Harrison and Schwarzacher 2007). Humans also have minimized disease through selection for crop resistance and through use of management strategies and chemical fungicides to minimize pathogen spread and pathogen damage.

The pathogens of human crops are as diverse as the crops themselves. In many agricultural systems, fungal and oomycete pathogens cause the most disease burden. Some of these pathogen species are described as being specialized (e.g., *Blumeria* spp., *Puccinia* spp., *Ustilago* spp.; table 11.1) (León-Ramírez et al. 2004; Dracatos et al. 2018), with each species attacking only one or a few host species, or, at a finer scale, with strains within a species being able to attack only certain genotypes of a given host species. Other pathogen species are considered to have broad host ranges (e.g., *Botrytis cinerea*, *Fusarium oxysporum*, *Phakopsora pachyrhizi*, *Rhizoctonia solani*; table 11.1) (Goellner et al. 2010; Dean et al. 2012; Hane et al. 2014), though within these species there may be fine-scale specificity due to genotypic interactions and local adaptation to both abiotic and biotic factors as well. Pathogens also differ in their level of virulence and economic impact.

Humans are not the only farmers. Agriculture, defined here as large-scale dispersal, seeding, and management of food propagules that are sustainably harvested, is a feature of other animal systems as well (introduction, this volume). Nonhuman agriculture is dominated by domestication of microbial crops (though see chapter 9, this volume). Slime molds cultivate bacteria (Brock et al. 2011). Snails cultivate fungi (Silliman and Newell 2003), as do ants, termites, and beetles (Mueller et al. 2005). While crop pathogens have not been identified in many of these systems, identification of a specialized fungal pathogen in the fungus-farming ant system has made the system a model for studying host-pathogen evolution. The primary pathogens in the ant system are fungi in the genus *Escovopsis*.

Fungus-growing ants have evolved for 50 million years into more than 245 species in over 19 genera of obligate farmers that propagate their cultivated fungi primarily vertically between the nests of mothers and daughters (Schultz and Brady 2008). Like human populations specialized at growing particular crops, most species of ants grow a narrow range of fungal cultivars (Mueller et al. 2018). *Escovopsis* spp. parasites attack and consume the fungal cultivars, which can lead to rapid colony decline (Reynolds and Currie 2004; Currie 2001). Different *Escovopsis* spp. attack different fungal crops (Currie et al. 2003; Gerardo et al. 2004; Gerardo, Mueller, and Currie 2006; Pagnocca, Masiulionis, and Rodrigues 2012; Custodio and Rodrigues 2019). To manage *Escovopsis* infections, ants actively eliminate infected portions of garden (Currie and Stuart 2001) and use antibiotic-producing bacteria that suppress fungal infection (Currie et al. 1999; Poulsen et al. 2009).

Human and ant agricultural systems exhibit many similarities as they relate to disease transmission and management (figure 11.1). For example, in both human and ant agriculture, crops are typically grown as monocultures of single species (Mueller et al. 2010; McDonald and Stukenbrock 2016), which leads to high host density relative to density of nondomesticated hosts (chapter 4, this volume). This may facilitate transmission within fields (human agriculture) and within colonies (ant agriculture). At the same time, these field and colony patches may be spatially separated from one another, which could lead to pathogen extinction at local scales or pathogen adaptation to local hosts and conditions (i.e., local adaptation) (Croll and McDonald 2017). Furthermore, pathogens of both human and ant agriculture must contend with an array of defense strategies, including utilization of behavioral and chemical defenses. Crop rotation, for example, is a behavior employed

Table 11.1
Key fungal and oomycete pathogens of human crops

Pathogen	Host crop(s)	Disease	Notes on specialization
Blumeria graminis[F]	cereal crops and other grasses	powdery mildew	Genetic analyses indicate that strains cluster according to host range, suggesting specialization of groups within the species (Wyand and Brown 2003).
Botrytis cinerea[F]	broad host range, including grapes	gray mold, noble rot	While there is some evidence of adaptation to alternative agricultural host plants, strains have very broad host ranges and show little evidence of specificity (Bardin et al. 2018).
Colletotrichum spp.[F]	broad host range, including bananas, cassava, and many stored fruits	anthracnose spots and blights, post harvest blights	The phenomenon of cultivar specificity was first recognized in *C. lindemuthianum*.
Fusarium graminearum[F]	cereals, maize	fusarium head blight	Infection across host species indicates little evidence for specificity (Kuhnem et al. 2015).
Fusarium oxysporum[F]	broad host range, including bananas	fusarium wilt	While the species is considered to have a broad host range, individual isolates are highly specialized.
Melampsora lini[F]	flax, linseed	flax rust	*M. lini* serves as a model for studying gene-for-gene specificity in pathogen-plant relationships (Flor 1956).
Mycosphaerella graminicola (syn. *Zymoseptoria tritici*)[F]	wheat	*Septoria tritici* blotch (STB)	While *M. graminicola* is specific to wheat, other *Mycosphaerella* spp. infect other grasses. Comparative genomics and genetic manipulations have elucidated some aspects of host specialization (Poppe et al. 2015; Stukenbrock et al. 2011).
Phakopsora pachyrhizi[F]	broad host range, including soybean and many legumes	Asian soybean rust	It is not clear whether this species' broad host range is due to unusual plasticity for a rust fungus or whether there are underlying intraspecific populations that differ in virulence and host specificity (Ono, Buriticá, and Hennen 1992).
Phytophthora infestans[O]	potato, tomato	late blight, tomato blight	*P. infestans* is closely related to other species with different host ranges.
Puccinia spp.[F]	cereals	rust	Strains are host specific, though they can infect alternative hosts at certain stages.
Pyricularia graminis-tritici[F]	wheat	wheat blast	*P. graminis-tritici* is part of a species complex that can cause disease on a variety of grasses. There is debate as to whether this should or should not be split from *P. oryzae* (Gladieux et al. 2018; Valent et al. 2019).

(*continued*)

Table 11.1
(continued)

Pathogen	Host crop(s)	Disease	Notes on specialization
Pyricularia oryzae (syn. *Magnaporthe oryzae*)[F]	rice	rice blast	*P. oryzae* is part of a species complex that can cause disease on a variety of grasses, though the name *P. oryzae* is sometimes used only for those that infect rice. Several genes have been identified that may underlie host specificity (Yoshida et al. 2016).
Rhizoctonia solani[F]	broad host range, including cereals, beans, and brassica	damping off, seed rot, root rot, black scurf and other diseases	Some groups of strains may be more specific than others (Keijer et al. 1997).
Ustilago maydis[F]	corn	corn smut	While host species-specific in nature, *U. maydis* can infect a broad range of hosts under lab conditions (León-Ramírez et al. 2004).

Notes: List of eukaryotic pathogens based on previous surveys of plant mycologists to identify fungal pathogens of plants that are of economic concern or that serve as models for studying fungal disease (Dean et al. 2012). Additional fungal and oomycete pathogens of human crops mentioned in the text have been added.

[F]fungal pathogen, [O]oomycete pathogen.

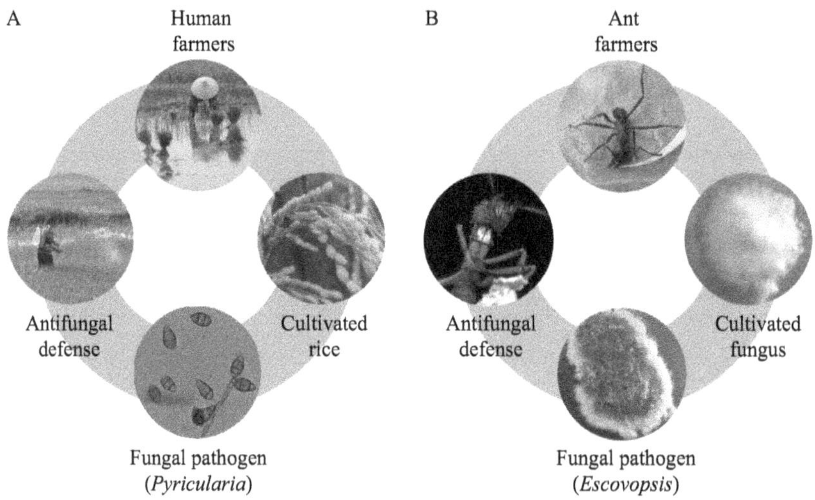

Figure 11.1
Parallels between farmer-crop-pathogen interactions. (A) The role of pathogens in shaping human agriculture is exemplified by infection of rice by the fungal pathogen *Pyricularia oryzae* (syn. *Magnaporthe oryzae*). Human farmers (top) actively cultivate rice (right) in large monoculture fields. *P. oryzae* (bottom) infects rice, causing rice blast. To combat the disease, humans use an array of defense strategies, including chemical fungicides (left). (B) Leaf-cutting ants (top) cut fresh vegetation to feed to their cultivated fungus (right), which is grown in monoculture within ant colonies. The cultivated fungus is attacked by fungal pathogens in the genus *Escovopsis* (bottom). To combat infection, the ants deploy an array of defenses, including the use of antimicrobials produced by bacteria that grows on the ants' cuticles (white patch under the ant's head, left). Ant photographs: Alex Wild.

by humans to minimize pathogen build-up in fields. While ants do not switch crops readily (Mueller, Poulin, and Adams 2004), they can invest heavily in weeding their gardens to remove pathogens (Currie and Stuart 2001). Humans utilize a suite of defensive chemicals (e.g., fungicides) to combat pathogens at great expense, and ants utilize bacteria that produce antifungals to inhibit their gardens' pathogens (Kelly and Ward 2018). Humans also both consciously and unconsciously select for crops with higher resistance. While there is no direct evidence that ants do the same, there has been ample time over the millions of years of ant-cultivar-*Escovopsis* association for the ants to at least occasionally propagate more resistant crop strains. These parallels could lead to convergent processes of pathogen evolution in these systems.

Studying the Origins of Crop Pathogens

There are several potential mechanisms by which pathogens may begin to utilize crops as hosts. These include host range expansion (i.e., beginning to utilize a new host while retaining the ability to utilize current hosts), host shifts (i.e., switching to utilize a new host closely related to the current host), host jumps (i.e., switching to utilize a new host distantly related to the current host), and adaptation to a pathogenic lifestyle from a free-living, commensal, or mutualistic form (figure 11.2). Host range expansion, host shifts, and host jumps require contacting novel hosts, which can be facilitated by migration. In human agriculture, for example, migration and colonization have played an important role in the establishment of novel pathogens (Goodwin, Cohen, and Fry 2006). Whether a pathogen begins to associate with a crop due to range expansion, host shift, or host jump, the ancestral host could be either a nondomesticated or a domesticated species.

Determination of the origins of a pathogen relies on comparative analyses across pathogen species or strains isolated from alternative hosts. In some cases, patterns will be clear, highlighting relationships between pathogens that suggest the origins of novel infections. Dating methods can provide insight into when a pathogen emerged in an agricultural system. Phylogenetic analyses of *Pyricularia* spp. (syn. *Magnaporthe* spp.) strains isolated from a variety of grass species, for example, indicate that *P. oryzae*, a virulent pathogen of rice, originated from a host jump from millet to rice. It then made several host shifts from rice to grassy weeds associated with farming (Couch et al. 2005). Similarly, phylogenetic relationships of isolates of *Phytophthora infestans*, the causal agent of the Irish Potato Famine, are consistent with a hypothesis that the virulent pathogen spread from Mexico, where diverse potato varieties are grown. Of note is that there is substantially more genetic variation in the Mexican *Phytophthora* population, which suggests that the pathogen went through a series of bottlenecks as it spread globally (Goodwin, Cohen, and Fry 2006).

Experimental infections can also suggest something about pathogen origins. *Pyricularia graminis-tritici* is an emerging pathogen of global concern. In Brazil and surrounding countries in South America, it has consistently caused wheat losses of 40–100% since first being reported in 1985 (McDonald and Stukenbrock 2016; Ceresini et al. 2018). It has also been reported in Bangladesh, where it has also caused devastating crop losses (Islam et al. 2016). While it was suspected that the pathogen originated by a host shift of *Pyricularia* found on rice to wheat, experiments demonstrated that *Pyricularia graminis-tritici* strains

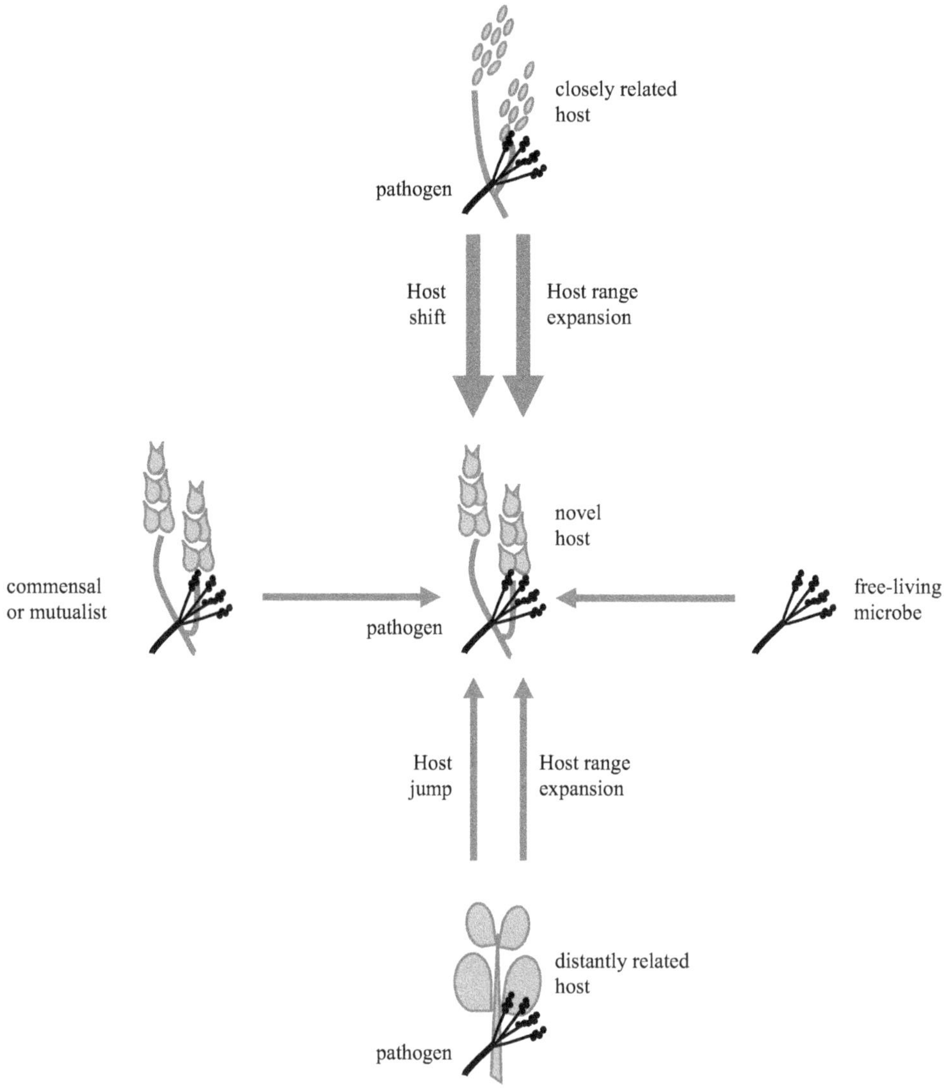

Figure 11.2
Origins of crop pathogens. Novel agricultural pathogens can arise from a number of sources. Common routes (top) include pathogens that evolve to use a novel host that is closely related to the original host. This can occur through host shifts, when populations or variants of the pathogen evolve to use the novel host and lose the ability to infect the original host, or host range expansion, when pathogens evolve to become more generalist, being able to infect both the original and novel host. Less common are host jumps (bottom), where pathogens evolve to infect novel hosts distantly related to the original host, or host range expansion, where pathogen host range evolves to include distantly related hosts. While not commonly observed, there is the potential for free-living (non-host associated) (right) and commensal and mutualistic microbes (left) to evolve into host-associated pathogens.

are not virulent on rice; this suggests that they instead originated from strains infecting a different grass host (Urashima, Igarashi, and Kato 1993; Castroagudín et al. 2016).

For *Escovopsis* spp., the only known pathogens of ant-cultivated fungi, the first phylogenetic analyses based on samples of *Escovopsis* and cultivars associated with a subset of ant species demonstrated phylogenetic congruence between the ants, their cultivars, and *Escovopsis* (Currie et al. 2003). This is consistent with a process of strict specialization and host fidelity such that pathogen species, once they emerge, rarely switch to new hosts through host jumps. Deeper sampling, however, indicates that the evolutionary history of the pathogen has included at least occasional jumps to distantly related hosts (Gerardo, Mueller, and Currie 2006; Meirelles et al. 2015).

For many human crop pathogens, such as *P. oryzae* mentioned above, researchers have identified closely related pathogen strains living on free-living hosts; this finding has facilitated determination of how and when the pathogens of domesticated crops diverged from free-living associates. No putative free-living relatives of *Escovopsis* have been identified. Assuming that the ancestor was a pathogen itself, further sampling of fungal pathogens of nondomesticated, free-living fungi in the Neotropics could potentially identify such a relative. Limited dating does suggest that the origin of the *Escovopsis* genus may be ancient and coincident with the evolution of ant agriculture (de Man et al. 2016). Such dating methods should be extended to comprehensively investigate the origins of clades within the *Escovopsis* genus.

Experimental Approaches, Phylogenetics, and Population Genetics Elucidate Patterns of Specialization

A defining feature of many host-pathogen systems is arms race coevolution, where there is strong selection on hosts to resist pathogens and strong selection on pathogens to overcome resistance (Haldane 1949). These processes can lead to the establishment and maintenance of specialization by pathogens because selection for the ability to infect and overcome resistance in some hosts leads to loss of the ability to effectively utilize other hosts (Flor 1956; Burdon 1987; Thompson and Burdon 1992), though this depends on the genetic basis of host resistance and pathogen virulence (Parker 1994).

In some respects, the degree to which a pathogen species is considered specialized is dictated by how we define a species. If ecological isolation based on utilization of alternative hosts precludes genetic exchange between two or more pathogen populations, then, by the biological species concept, researchers may define these two populations as separate species (Restrepo et al. 2014). In tandem or alternatively, phylogenetic species concepts may be utilized to name species based on the formation of monophyletic groups. Regardless of how a species is defined, specialization can be investigated by comparing the ability of different species or strains within species to infect a set of hosts. Experimental investigation can be coupled with phylogenetic and population genetic investigations to elucidate the relationships between strains found on different hosts in nature.

A common approach to study pathogen specialization is to screen the ability of strains isolated from different environments to utilize different hosts. These experiments can provide insight into the degree of host specialization (how many hosts a pathogen can use) and the breadth of host range (how closely related useable hosts are to one another). Cross inoculations of potato and tomato with strains of *P. infestans* isolated from both crops, for example, indicates that potato strains are specialized on potato, having very low fitness when put on tomato. Tomato strains are more generalist, though they do perform better on tomato than potato (Kröner et al. 2017) (figure 11.3A). Similar cross inoculations have been conducted utilizing *Escovopsis* isolated from colonies of different ant species that grow distinct fungal crops. Like *P. infestans*, these *Escovopsis* strains are more likely to successfully infect their typical than their atypical host (Gerardo et al. 2004, 2006; Birnbaum and Gerardo 2016) (figure 11.3B). Also like *P. infestans*, some *Escovopsis* strains appear to be more generalist than others (Birnbaum and Gerardo 2016).

Phylogenetic and population genetic analyses of many strains from different hosts can also provide insights into the degree of host specialization. Experimental results suggesting that *P. graminis-tritici* is specialized on wheat (Urashima, Igarashi, and Kato 1993; Castroagudín et al. 2016), for example, are consistent with population genetic analyses that demonstrated that rice- and wheat-infecting strains form distinct genetic clusters with little evidence for gene flow between them (Chiapello et al. 2015), though with evidence of some gene flow between strains infecting wheat and other hosts (e.g., oats) (Gladieux et al. 2018). Furthermore, phylogenetic analyses corroborated the host specificity of lineages and revealed that wheat-infecting *P. graminis-tritici* are more closely related to strains found in association with oats than they are to strains found in association with rice (Castroagudín et al. 2016; Yoshida et al. 2016). Thus, these analyses together elucidated both the origins and specialization of this fungal pathogen.

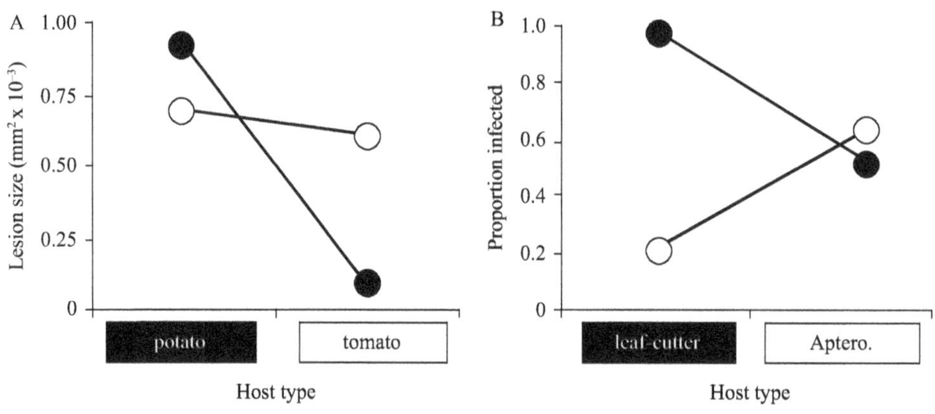

Figure 11.3
Parallel investigation of pathogen host range. The host ranges of pathogen strains are often investigated by doing cross-inoculations of alternative hosts with pathogens isolated from those hosts. Specificity is exemplified by pathogens being able to better infect their native host than an alternative host. This is exemplified by (A) cross inoculations of potato and tomato plants with strains of *Phytophthora infestans* (black indicates strains from potato, and white indicates strains from tomato; figure adapted from (Kröner et al. 2017)); and (B) cross inoculations of cultivated fungi from leaf-cutter and *Apterostigma* ant colonies with strains of *Escovopsis* spp. (black indicates strains from leaf-cutter colonies [*Atta* and *Acromyrmex* spp. combined], and white indicates strains from *Apterostigma* colonies); figure adapted from Birnbaum and Gerardo (2016). Note that, in the former case (A), infection success is measured by the virulence of the pathogen, as measured by the size of lesions created by the infection, while in the latter case (B) infection success is measured by establishment of the pathogen on the hosts, as measured by the proportion of hosts that are infected.

Phylogenetic investigations of *Escovopsis* have demonstrated that, at the broad scale, *Escovopsis* spp. are specialized on particular cultivar hosts (Currie et al. 2003; Gerardo, Mueller, and Currie 2006). At finer scales, population genetic analyses of *Escovopsis* strains isolated from different colonies of the same ant species that are growing closely related cultivars often indicate that genetic distances are not correlated with differences in host utilization (Gerardo and Caldera 2007; Taerum et al. 2007). This contradicts both genetic and experimental studies with other *Escovopsis* strains that highlight that closely related strains of the same *Escovopsis* spp. may differ in which cultivar strains they can infect, which suggests that *Escovopsis*-cultivar dynamics can exhibit genotype-genotype specificity (Birnbaum and Gerardo 2016; Kellner et al. 2018; Custodio and Rodrigues 2019). These differences suggest that some *Escovopsis* spp. are more generalist than others.

The Use of Genetics and Genomics to Reveal Mechanisms of Host Utilization and Specialization

Several fungal and oomycete pathogens of human crops have been developed as tractable genetic models to study the evolutionary dynamics of interactions between species. For example, the model for gene-for-gene matching between pathogen genes underlying virulence (known as "avirulence" genes) and host genes underlying resistance was first described for *Melampsora lini*, a fungal pathogen of flax and linseed (Flor 1956). While not of huge economic concern, the genetic tractability of this plant-pathogen system is a model for studying plant resistance, plant immunity, and mechanisms of pathogen virulence (Chisholm et al. 2006; Dean et al. 2012). Other fungal pathogens of agricultural crops have demonstrated how both multigenic, quantitative traits and ecological conditions can shape the outcomes of host-pathogen interactions (Croll and McDonald 2017).

While genetic technologies for some crop pathogens have been available for decades, the utilization of genomics to investigate eukaryotic pathogens has provided novel insights into the genetic underpinnings of host utilization and host specialization in many systems without tools for genetic manipulation. Crop pathogen genome sequencing efforts have identified genes that may underlie host utilization and virulence, including genes involved in secondary metabolism, carbohydrate utilization, and manipulation of host metabolism (Dean et al. 2005; Kämper et al. 2006; Chiapello et al. 2015). Genomes of crop pathogens have also revealed that eukaryotic crop pathogens differ markedly in their genomic structure and organization. For example, while some fungal pathogens have streamlined genomes (de Man et al. 2016; Benevenuto et al. 2018), others are packed with repetitive elements and have large genomes (Raffaele and Kamoun 2012), suggesting that these organisms adapt to hosts under fundamentally different genomic constraints.

Comparative genomic projects have moved beyond the annotation of single genomes to investigate differences between strains or closely related species that utilize different hosts. For example, comparison of the genomes of eight *Pyricularia oryzae* (syn. *Magnaporthe oryzae*) strains that differ in host specificity revealed that the gene content and genomic features of these genomes are highly similar. One notable signature of host specialization, however, is that strains adapted to rice share a specific set of gene families that are not found in the strains attacking other hosts (and vice versa). The rice-specific gene families include known avirulence genes and some genes involved in secondary metabolism (Chiapello et al. 2015). Beyond identifying differences in the presence and absence of genes, comparative genomic approaches can also assess signatures of selection across genomes, potentially revealing those regions that may be under selection because of their role in host-pathogen interactions. For example, many human crop pathogens exhibit diversifying selection at loci underlying virulence (van de Wouw et al. 2010; Kelly and Ward 2018). Other studies assessing patterns of positive selection across pathogen genomes adapted to different hosts have revealed complex patterns of selection across many loci; this finding suggests that adaptation to hosts may involve multigenic selection on genes underlying interaction with both abiotic and biotic stressors (Benevenuto et al. 2018; Mohd-Assaad, McDonald, and Croll 2018). Results from these genome-wide studies suggest that the prevailing view of pathogen evolution being defined by a few loci underlying pathogen-host gene-for-gene interactions may be missing key elements of pathogen adaptation to the environment.

Molecular investigations of the mechanisms underlying *Escovopsis* specialization in the ant agricultural system are in their infancy. Sequencing and annotation of the first *Escovopsis* genome provided some insights into the basis of *Escovopsis*'s host utilization (de Man et al. 2016). As with human crop pathogens, *Escovopsis* genomes encode for the capacity to produce a rich array of secondary metabolites (de Man et al. 2016; Heine et al. 2018) and to utilize carbohydrates abundant in their fungal crop hosts. Of note is that the genome annotation revealed that *Escovopsis* cannot produce key enzymes necessary for utilizing plant material that can be abundant in ant gardens—a finding that highlights that it is indeed a specialized pathogen of the ants' fungal crop (de Man et al. 2016). Comparative genomics across *Escovopsis* spp. associated with different hosts may reveal how these features change with host range. As with human crop pathogens, development of tools for genetic manipulation could greatly facilitate study of the evolution of *Escovopsis* specialization.

Consideration of How Agricultural Practices Shape Pathogen Evolution

Due to their large population sizes, short generation times, and ability to disperse, eukaryotic pathogens have amazing potential to adapt to abiotic and biotic selection pressures, including agricultural practices. Because of its applied importance and experimental tractability, one of the most common phenotypes studied in this regard is the evolution of resistance to fungicides. Resistance in fungal pathogens appears to often arise via *de novo* mutations in the target sites of the fungicides (Hawkins et al. 2019). The mutations can then spread rapidly, shaping pathogen population structure in agricultural settings (Kozhar et al. 2020).

The fungus-growing ant system provides a particularly interesting case study in terms of how pathogens adapt to host defenses. Unlike humans, who typically use chemical fungicides in agriculture, ants use fungicides produced by bacteria (Currie et al. 1999). Evolution of fungicide resistance in the pathogens could select for the ants to switch to novel bacterial partners that produce alternative fungicides or could select for the bacteria to modify the fungicides that they produce. While *Escovopsis* spp. and strains within species vary in terms of their susceptibility to these bacteria-derived fungicides (Poulsen et al. 2009), nothing is known about the mechanisms of resistance. Elucidation of these mechanisms could open up avenues to explore how fungicide resistance evolves in this ancient agricultural system and whether the pathogens and fungicide-producing bacteria evolve in response to one another in an arms-race–like fashion.

Another feature of agriculture that can shape pathogen evolution is the spatial structure of the agricultural landscape (Papaïx et al. 2015). Large, uniform monocultures, for example, may facilitate the evolution of pathogen specialization (Débarre and Gandon 2010), whereas landscape heterogeneity, providing potential reservoirs of noncrop host species, could hinder the evolution of specialization. Crop harvest and rotation practices, both of which influence host density and availability, may also influence pathogen evolution, particularly evolution of traits that influence transmission and virulence (van den Berg et al. 2011). A proxy for spatial structure in ant agriculture is colony density and proximity to colonies of the same or other species. These community features have not yet been considered in terms of their impacts on cultivar-*Escovopsis* disease dynamics over ecological time or evolution over longer time scales.

Conclusions and Implications

Pathogens place strong evolutionary pressures on their hosts and have shaped domestication processes. Because pathogens can be spread with their hosts and can diversify along with their hosts, studying pathogen diversification may inform our understanding of ancient domestication processes. Understanding the origins and adaptive processes of current pathogens in agricultural systems can also have important applied implications. The degree of specialization of crop pathogens and the mechanisms underlying this specialization can provide insight into the likelihood of disease spread and can therefore inform management strategies. Furthermore, if studies across agricultural pathogen systems indicate similar origins for many pathogens, as they do with zoonotic diseases, then the routes of transmission that facilitate these origins may be more carefully controlled to minimize the emergence of novel diseases. Beyond these applied reasons for studying agricultural pathogens, these systems can serve as models for studying evolutionary processes. These experimentally tractable systems, for example, allow researchers to explore the proximate and ultimate mechanisms underlying coevolution and co-speciation.

Acknowledgments

Thank you to the GerDer lab for feedback and to the attine community for their continued support. Writing was supported by grants from the National Science Foundation (DEB 1754595, DEB 1821533).

References

Bardin, M., C. Leyronas, C. Troulet, and C. E. Morris. 2018. "Striking Similarities between *Botrytis cinerea* from Non-Agricultural and from Agricultural Habitats." *Frontiers in Plant Science* 9:1820. https://doi.org/10.3389/fpls.2018.01820.

Benevenuto, J., N. S. Teixeira-Silva, E. E. Kuramae, D. Croll, and C. B. Monteiro-Vitorello. 2018. "Comparative Genomics of Smut Pathogens: Insights from Orphans and Positively Selected Genes into Host Specialization." *Frontiers in Microbiology* 9:660. https://doi.org/10.3389/fmicb.2018.00660.

Birnbaum, S. S. L., and N. M. Gerardo. 2016. "Patterns of Specificity of the Pathogen *Escovopsis* across the Fungus-Growing Ant Symbiosis." *American Naturalist* 188(1): 52–65. https://doi.org/10.5061/dryad.5b0s8.

Brock, D. A., T. E. Douglas, D. C. Queller, and J. E. Strassmann. 2011. "Primitive Agriculture in a Social Amoeba." *Nature* 469(7330): 393–396. https://doi.org/10.1038/nature09668.

Burdon, J. J. 1987. *Diseases and Plant Population Biology*. Cambridge: Cambridge University Press.

Caldera, E. J., M. Poulsen, G. Suen, and C. R. Currie. 2009. "Insect Symbioses: A Case Study of Past, Present, and Future Fungus-Growing Ant Research." *Environmental Entomology* 38(1): 78–92. https://doi.org/10.1603/022.038.0110.

Castroagudín, V. L., S. I. Moreira, D. A. S. Pereira, S. S. Moreira, P. C. Brunner, et al. 2016. "*Pyricularia graminis-tritici*, a New *Pyricularia* Species Causing Wheat Blast." *Persoonia: Molecular Phylogeny and Evolution of Fungi* 37:199–216. https://doi.org/10.3767/003158516X692149.

Ceresini, P. C., V. L. Castroagudín, F. Á. Rodrigues, J. Rios, C. Eduardo Aucique-Pérez, et al. 2018. "Wheat Blast: Past, Present, and Future." *Annual Review of Phytopathology* 56:427–456. https://doi.org/10.1146/annurev-phyto-080417-050036.

Chiapello, H., L. Mallet, C. Guérin, G. Aguileta, J. Amselem, et al. 2015. "Deciphering Genome Content and Evolutionary Relationships of Isolates from the Fungus *Magnaporthe oryzae* Attacking Different Host Plants." *Genome Biology and Evolution* 7(10): 2896–2912. https://doi.org/10.1093/gbe/evv187.

Chisholm, S. T., G. Coaker, B. Day, and B. J. Staskawicz. 2006. "Host-Microbe Interactions: Shaping the Evolution of the Plant Immune Response." *Cell* 124(4): 803–814. https://doi.org/10.1016/j.cell.2006.02.008.

Couch, B. C., I. Fudal, M.-H. Lebrun, D. Tharreau, B. Valent, et al. 2005. "Origins of Host-Specific Populations of the Blast Pathogen *Magnaporthe oryzae* in Crop Domestication with Subsequent Expansion of Pandemic Clones on Rice and Weeds of Rice." *Genetics* 170:613–630. https://doi.org/10.1534/genetics.105.041780.

Croll, D., and B. A. McDonald. 2017. "The Genetic Basis of Local Adaptation for Pathogenic Fungi in Agricultural Ecosystems." *Molecular Ecology* 26(7): 2027–2040. https://doi.org/10.1111/mec.13870.

Currie, C. R. 2001. "Prevalence and Impact of a Virulent Parasite on a Tripartite Mutualism." *Oecologia* 128(1): 99–106. https://doi.org/10.1007/s004420100630.

Currie, C. R., J. A. Scott, R. C. Summerbell, and D. Malloch. 1999. "Fungus-Growing Ants Use Antibiotic-Producing Bacteria to Control Garden Parasites." *Nature* 398:701–704. https://doi.org/10.1038/19519.

Currie, C. R., and A. E. Stuart. 2001. "Weeding and Grooming of Pathogens in Agriculture by Ants." *Proceedings of the Royal Society of London B—Biological Sciences* 268(1471): 1033–1039. https://doi.org/10.1098/rspb.2001.1605.

Currie, C. R., B. Wong, A. E. Stuart, T. R. Schultz, S. A. Rehner, et al. 2003. "Ancient Tripartite Coevolution in the Attine Ant-Microbe Symbiosis." *Science* 299(5605): 386–388. https://doi.org/10.1126/science.1078155.

Custodio, B. C., and A. Rodrigues. 2019. "*Escovopsis kreiselii* Specialization to Its Native Hosts in the Fungiculture of the Lower Attine Ant *Mycetophylax morschi. Antonie van Leeuwenhoek* 112(2): 305–317. https://doi.org/10.1007/s10482-018-1158-x.

Dean, R. A., J. A. L. Van Kan, Z. A. Pretorius, K. E. Hammond-Kosack, A. Di Pietro, et al. 2012. "The Top 10 Fungal Pathogens in Molecular Plant Pathology." *Molecular Plant Pathology* 13(4): 414–430. https://doi.org/10.1111/j.1364-3703.2011.00783.x.

Dean, R. A., N. J. Talbot, D. J. Ebbole, M. L. Farman, T. K. Mitchell, et al. 2005. "The Genome Sequence of the Rice Blast Fungus *Magnaporthe grisea*." *Nature* 434(7036): 980–986. https://doi.org/10.1038/nature03449.

Débarre, F., and S. Gandon. 2010. "Evolution of Specialization in a Spatially Continuous Environment." *Journal of Evolutionary Biology* 23(5): 1090–1099. https://doi.org/10.1111/j.1420-9101.2010.01966.x.

De Man, T. J. B., J. E. Stajich, C. P. Kubicek, C. Teiling, K. Chenthamara, et al. 2016. "Small Genome of the Fungus *Escovopsis weberi*, a Specialized Disease Agent of Ant Agriculture." *Proceedings of the National Academy of Sciences* 113(13): 3567–3572. https://doi.org/10.1073/pnas.1518501113.

Dracatos, P. M., R. Haghdoust, D. Singh, and R. F. Park. 2018. "Exploring and Exploiting the Boundaries of Host Specificity Using the Cereal Rust and Mildew Models." *New Phytologist* 218(2): 453–462. https://doi.org/10.1111/nph.15044.

Flor, H. H. 1956. "The Complementary Genic Systems in Flax and Flax Rust." *Advances in Genetics* 8:29–54. https://doi.org/10.1016/S0065-2660(08)60498-8.

Gerardo, N. M., and E. J. Caldera. 2007. "Labile Associations between Fungus-Growing Ant Cultivars and Their Garden Pathogens." *ISME Journal* 1(5): 373–384. https://doi.org/10.1038/ismej.2007.57.

Gerardo, N. M., S. R. Jacobs, C. R. Currie, and U. G. Mueller. 2006. "Ancient Host-Pathogen Associations Maintained by Specificity of Chemotaxis and Antibiosis." *PLoS Biology* 4(8): e235–236. https://doi.org/10.1371/journal.pbio.0040235.

Gerardo, N. M., U. G. Mueller, and C. R. Currie. 2006. "Complex Host-Pathogen Coevolution in the *Apterostigma* Fungus-Growing Ant-Microbe Symbiosis." *BMC Evolutionary Biology* 6:88. https://doi.org/10.1186/1471-2148-6-88.

Gerardo, N. M., U. G. Mueller, S. L. Price, and C. R. Currie. 2004. "Exploiting a Mutualism: Parasite Specialization on Cultivars within the Fungus-Growing Ant Symbiosis." *Proceedings of the Royal Society of London B—Biological Sciences* 271(1550): 1791–1798. https://doi.org/10.1098/rspb.2004.2792.

Gladieux, P., B. Condon, S. Ravel, D. Soanes, J. L. N. Maciel, et al. 2018. "Gene Flow between Divergent Cereal- and Grass-Specific Lineages of the Rice Blast Fungus *Magnaporthe oryzae*." *MBio* 9(1): e01219–17. https://doi.org/10.1128/mBio.01219-17.

Goellner, K., M. Loehrer, C. Langenbach, U. Conrath, E. Koch, and U. Schaffrath. 2010. "*Phakopsora pachyrhizi*, the Causal Agent of Asian Soybean Rust: Pathogen Profile." *Molecular Plant Pathology* 11(2): 169–177. https://doi.org/10.1111/j.1364-3703.2009.00589.x.

Goodwin, S. B., B. A. Cohen, and W. E. Fry. 2006. "Panglobal Distribution of a Single Clonal Lineage of the Irish Potato Famine Fungus." *Proceedings of the National Academy of Sciences* 91(24):11591–11595. https://doi.org/10.1073/pnas.91.24.11591.

Haldane, J. B. S. 1949. "Disease and Evolution." *Ricerca Scientifica* 19:68–76.

Hane, J. K., J. P. Anderson, A. H. Williams, J. Sperschneider, and K. B. Singh. 2014. "Genome Sequencing and Comparative Genomics of the Broad Host-Range Pathogen *Rhizoctonia solani* AG8." *PLoS Genetics* 10(5): 11–12. https://doi.org/10.1371/journal.pgen.1004281.

Hawkins, N. J., C. Bass, A. Dixon, and P. Neve. 2019. "The Evolutionary Origins of Pesticide Resistance." *Biological Reviews* 94(1): 135–155. https://doi.org/10.1111/brv.12440.

Heine, D., N. A. Holmes, S. F. Worsley, A. C. A. Santos, T. M. Innocent, et al. 2018. "Chemical Warfare between Leafcutter Ant Symbionts and a Co-Evolved Pathogen." *Nature Communications* 9(1): 2208. https://doi.org/10.1038/s41467-018-04520-1.

Heslop-Harrison, J. S., and T. Schwarzacher. 2007. "Domestication, Genomics and the Future for Banana." *Annals of Botany* 100(5): 1073–1084. https://doi.org/10.1093/aob/mcm191.

Islam, M. T., D. Croll, P. Gladieux, D. M. Soanes, A. Persoons, et al. 2016. "Emergence of Wheat Blast in Bangladesh Was Caused by a South American lineage of *Magnaporthe oryzae*." *BMC Biology* 14(1): 84. https://doi.org/10.1186/s12915-016-0309-7.

Kämper, J., R. Kahmann, M. Bölker, L.-J. Ma, T. Brefort, et al. 2006. "Insights from the Genome of the Biotrophic Fungal Plant Pathogen *Ustilago maydis*." *Nature* 444(7115): 97–101. https://doi.org/10.1038/nature05248.

Keijer, J., M. G. Korsman, A. M. Dullemans, P. M. Houterman, J. De Bree, et al. 1997. "*In vitro* Analysis of Host Plant Specificity in *Rhizoctonia solani*." *Plant Pathology* 46(5): 659–669. https://doi.org/10.1046/j.1365-3059.1997.d01-61.x.

Kellner, K., M. R. Kardish, J. N. Seal, T. A. Linksvayer, and U. G. Mueller. 2018. "Symbiont-Mediated Host-Parasite Dynamics in a Fungus-Gardening Ant." *Microbial Ecology* 76(2): 530–543. https://doi.org/10.1007/s00248-017-1124-6.

Kelly, A. C., and T. J. Ward. 2018. "Population Genomics of *Fusarium graminearum* Reveals Signatures of Divergent Evolution within a Major Cereal Pathogen." *PLoS ONE* 13(3): e0194616. https://doi.org/10.1371/journal.pone.0194616.

Kozhar, O., M. M. Larsen, N. J. Grünwald, and T. L. Peever. 2020. "Fungal Evolution in Anthropogenic Environments: *Botrytis cinerea* Populations Infecting Small Fruit Hosts in the Pacific Northwest Rapidly Adapt to Human-Induced Selection Pressures." *Applied and Environmental Microbiology* 86(9): e02908–19. https://doi.org/10.1128/aem.02908-19.

Kröner, A., R. Mabon, R. Corbière, J. Montarry, and D. Andrivon. 2017. "The Coexistence of Generalist and Specialist Clonal Lineages in Natural Populations of the Irish Famine Pathogen *Phytophthora infestans* Explains Local Adaptation to Potato and Tomato." *Molecular Ecology* 26(7): 1891–1901. https://doi.org/10.1111/mec.14004.

Kuhnem, P. R., E. M. Del Ponte, Y. Dong, and G. C. Bergstrom. 2015. "*Fusarium graminearum* Isolates from Wheat and Maize in New York Show Similar Range of Aggressiveness and Toxigenicity in Cross-Species Pathogenicity Tests." *Phytopathology* 105(4): 441–448. https://doi.org/10.1094/PHYTO-07-14-0208-R.

León-Ramírez, C. G., J. L. Cabrera-Ponce, A. D. Martínez-Espinoza, L. Herrera-Estrella, L. Méndez, et al. 2004. "Infection of Alternative Host Plant Species by *Ustilago maydis*." *New Phytologist* 164(2): 337–346. https://doi.org/10.1111/j.1469-8137.2004.01171.x.

McDonald, B. A., and E. H. Stukenbrock. 2016." Rapid Emergence of Pathogens in Agro-Ecosystems: Global Threats to Agricultural Sustainability and Food Security." *Philosophical Transactions of the Royal Society B—Biological Sciences* 371(1709): 20160026. https://doi.org/10.1098/rstb.2016.0026.

Meirelles, L. A., S. E. Solomon, M. Bacci Jr., A. M. Wright, U. G. Mueller, et al. 2015. "Shared *Escovopsis* Parasites between Leaf-Cutting and Non-Leaf-Cutting Ants in the Higher Attine Fungus-Growing Ant Symbiosis." *Royal Society Open Science* 2(9): 150257. https://doi.org/10.1098/rsos.150257.

Mohd-Assaad, N., B. A. McDonald, and D. Croll. 2018. "Genome-Wide Detection of Genes under Positive Selection in Worldwide Populations of the Barley Scald Pathogen." *Genome Biology and Evolution* 10(5): 1315–1332. https://doi.org/10.1093/gbe/evy087.

Mueller, U. G., N. M. Gerardo, D. K. Aanen, D. L. Six, and T. R. Schultz. 2005. "The Evolution of Agriculture in Insects." *Annual Review of Ecology, Evolution, and Systematics* 36:563–595. https://doi.org/10.1146/annurev.ecolsys.36.102003.152626.

Mueller, U. G., M. R. Kardish, H. D. Ishak, A. M. Wright, S. E. Solomon, et al. 2018. "Phylogenetic Patterns of Ant-Fungus Associations Indicate That Farming Strategies, Not Only a Superior Fungal Cultivar, Explain the Ecological Success of Leafcutter Ants." *Molecular Ecology* 27(10): 2414–2434. https://doi.org/10.1111/mec.14588.

Mueller, U. G., J. Poulin, and R. M. M. Adams. 2004. "Symbiont Choice in a Fungus-Growing Ant (Attini, Formicidae)." *Behavioral Ecology* 15(2): 357–364. https://doi.org/10.1093/beheco/arh020.

Mueller, U. G., J .J. Scott, H. D. Ishak, M. Cooper, and A. Rodrigues. 2010. "Monoculture of Leafcutter Ant Gardens." *PLoS ONE* 5(9): e12668. https://doi.org/10.1371/journal.pone.0012668.

Ono, Y., P. Buriticá, and J. F. Hennen. 1992. "Delimitation of *Phakopsora, Physopella* and *Cerotelium* and Their Species on Leguminosae." *Mycological Research* 96(10): 825–850. https://doi.org/10.1016/S0953-7562(09)81029-0.

Pagnocca, F. C., V. E. Masiulionis, and A. Rodrigues. 2012. "Specialized Fungal Parasites and Opportunistic Fungi in Gardens of Attine Ants." *Psyche* 2012:905109. https://doi.org/10.1155/2012/905109.

Papaïx, J., J. J. Burdon, J. Zhan, and P. H. Thrall. 2015. "Crop Pathogen Emergence and Evolution in Agro-Ecological Landscapes." *Evolutionary Applications* 8(4): 385–402. https://doi.org/10.1111/eva.12251.

Parker, M. A. 1994. "Pathogen and Sex in Plants." *Evolutionary Ecology* 8(5): 560–584.

Poppe, S., L. Dorsheimer, P. Happel, and E. H. Stukenbrock. 2015. Rapidly Evolving Genes Are Key Players in Host Specialization and Virulence of the Fungal Wheat Pathogen *Zymoseptoria tritici (Mycosphaerella graminicola)*." *PLoS Pathogens* 11(7): e1005055. https://doi.org/10.1371/journal.ppat.1005055.

Poulsen, M., M. J. Cafaro, D. P. Erhardt, A. E. F. Little, N. M. Gerardo, et al. 2009. "Variation in *Pseudonocardia* Antibiotic Defence Helps Govern Parasite-Induced Morbidity in *Acromyrmex* Leaf-Cutting Ants." *Environmental Microbiology Reports* 2(4): 534–540. https://doi.org/10.1111/j.1758-2229.2009.00098.x.

Raffaele, S., and S. Kamoun. 2012. "Genome Evolution in Filamentous Plant Pathogens: Why Bigger Can Be Better." *Nature Reviews Microbiology* 10(6): 417–430. https://doi.org/10.1038/nrmicro2790.

Restrepo, S., J. F. Tabima, M. F. Mideros, N. J. Grünwald, and D. R. Matute. 2014. "Speciation in Fungal and Oomycete Plant Pathogens." *Annual Review of Phytopathology* 52(1): 289–316. https://doi.org/10.1146/annurev-phyto-102313-050056.

Reynolds, H. T., and C. R. Currie. 2004. "Pathogenicity of *Escovopsis weberi*: The Parasite of the Attine Ant-Microbe Symbiosis Directly Consumes the Ant-Cultivated Fungus." *Mycologia* 96(5): 955–959. https://doi.org/10.1080/15572536.2005.11832895.

Schultz, T. R., and S. G. Brady. 2008. "Major Evolutionary Transitions in Ant Agriculture." *Proceedings of the National Academy of Sciences* 105(14): 5435–5440. https://doi.org/10.1073/pnas.0711024105.

Silliman, B. R., and S. Y. Newell. 2003. "Fungal Farming in a Snail." *Proceedings of the National Academy of Sciences* 100(26): 15643–15648. https://doi.org/10.1073/pnas.2535227100.

Stukenbrock, E. H., T. Bataillon, J. Y. Dutheil, T. T. Hansen, R. Li, et al. 2011. "The Making of a New Pathogen: Insights from Comparative Population Genomics of the Domesticated Wheat Pathogen *Mycosphaerella graminicola* and Its Wild Sister Species." *Genome Research* 21(12): 2157–2166. https://doi.org/10.1101/gr.118851.110.

Taerum, S. J., M. J. Cafaro, A. E. F. Little, T. R. Schultz, and C. R. Currie. 2007. "Low Host-Pathogen Specificity in the Leaf-Cutting Ant-Microbe Symbiosis." *Proceedings of the Royal Society B—Biological Sciences* 274(1621): 1971–1978. https://doi.org/10.1006/mpev.1997.0466.

Thompson, J. N., and J. J. Burdon. 1992. "Gene-for-Gene Coevolution between Plants and Parasites." *Nature* 360:121–125.

Urashima, A. S., S. Igarashi, and H. Kato. 1993. "Host Range, Mating Type, and Fertility of *Pyricularia grisea* from Wheat in Brazil." *Plant Disease* 77(12): 1211–1216. https://doi.org/10.1094/PD-77-1211.

Valent, B., M. Farman, Y. Tosa, D. Begerow, E. Fournier, et al. 2019. "*Pyricularia graminis-tritici* Is Not the Correct Species Name for the Wheat Blast Fungus: Response to Ceresini et al. (MPP 20:2)." *Molecular Plant Pathology* 20(2): 173–179. https://doi.org/10.1111/mpp.12778.

Van den Berg, F., N. Bacaer, J. A. J. Metz, C. Lannou, and F. van den Bosch. 2011. "Periodic Host Absence Can Select for Higher or Lower Parasite Transmission Rates." *Evolutionary Ecology* 25(1): 121–137. https://doi.org/10.1007/s10682-010-9387-0.

Van de Wouw, A. P., A. J. Cozijnsen, J. K. Hane, P. C. Brunner, B. A. McDonald, et al. 2010. "Evolution of Linked Avirulence Effectors in *Leptosphaeria maculans* Is Affected by Genomic Environment and Exposure to Resistance Genes in Host Plants." *PLoS Pathogens* 6(11): e1001180. https://doi.org/10.1371/journal.ppat.1001180.

Wyand, R. A., and J. K. M. Brown. 2003. "Genetic and Forma Specialis Diversity in *Blumeria graminis* of Cereals and Its Implications for Host-Pathogen Co-Evolution." *Molecular Plant Pathology* 4(3): 187–198. https://doi.org/10.1046/j.1364-3703.2003.00167.x.

Yoshida, K., D. G. O. Saunders, C. Mitsuoka, S. Natsume, S. Kosugi, et al. 2016. "Host Specialization of the Blast Fungus *Magnaporthe oryzae* Is Associated with Dynamic Gain and Loss of Genes Linked to Transposable Elements." *BMC Genomics* 17:370. https://doi.org/10.1186/s12864-016-2690-6.

12 Evaluating Potential Proximate and Ultimate Causes of Phenotypic Change in the Human Skeleton over the Agricultural Transition

Lumila P. Menéndez and Laura T. Buck

Gordon Childe coined the term "Neolithic Revolution" to characterize the fundamental changes in the human ecological niche driven by the domestication of plants and animals (Childe 1936). From Childe's viewpoint, this revolution resulted in technological innovations and trends toward sedentism that brought improvements in health, food quality, and a lessening of labor demands (Childe 1936). It is now understood that the adoption of agriculture was in fact a complex and gradual process that occurred independently among humans and in different species of insects (i.e., ants, termites, beetles) at different times and in many geographic areas of the world (Bar-Yosef 1998; Schultz et al. 2005; Stock and Pinhasi 2011; Armelagos and Cohen 2013; Biedermann and Vega 2020). Regarding humans, despite shifts in understanding since Childe's work, it can still be argued that agriculture resulted in pivotal changes in the relationship between human behavior and the natural environment, albeit with varied impacts in different populations.

There are ongoing discussions regarding the areas in which early agriculture originated independently and those in which it was secondarily adopted due to the diffusion of people or ideas (e.g., Bellwood 2005). Current consensus, however, points to at least the following eight regions of agricultural innovation: the Levant (10,000 BP), India (5,000–4,000 BP), the Yangtze and Yellow River basins in south China (8,000 BP), the New Guinea Highlands (7,000 BP), west sub-Saharan Africa (4,000–2,000 BP), Central Mexico (10,000–4,000 BP), western South America (8,000–5,000 BP), and eastern North America (5,000–4,000 BP) (Diamond and Bellwood 2003; Bellwood 2005; Price and Bar-Yosef 2011; chapter 10, this volume).

Over the past several decades, researchers have questioned the initial assumptions of the Childean school of thought, which suggested that the transition to agriculture was of unadulterated benefit to humanity. This has led to much discussion of the effects of different subsistence practices on human skeletal morphology in workshops and projects leading to influential books (Cohen and Armelagos 1984; Steckel and Rose 2002; Ungar 2006; Pinhasi and Stock 2011). More recently, dietary influences on skeletal morphology have been well documented in comparative analyses of recent populations from all around the world (e.g., von Cramon-Taubadel 2011; Noback and Harvati 2015; Katz, Grote, and Weaver 2017) and/or specific subcontinents (e.g., Pinhasi and Pluciennik 2004; Paschetta et al. 2010; Perez et al. 2011; Cheronet, Finarelli, and Pinhasi 2016). However, since morphological variation between human populations results from several evolutionary and ecological processes, it becomes difficult to disentangle the impact of different subsistence practices on the skeleton. Additionally, the effect that diverse diets have on the skeleton differs when comparing populations at distinct geographical scales. For instance, on a worldwide scale, genetic drift is expected to have the strongest effect (von Cramon-Taubadel 2014; Katz, Grote, and Weaver 2017), while on smaller subcontinental or regional scales other factors such as diet or climate play a stronger role (Paschetta et al. 2010; Perez et al. 2011; Evteev et al. 2014).

In order to address the complexity of causative factors where possible, the association between subsistence and skeletal variation should be evaluated in diachronic sequences in areas where agriculture either originated locally or was acquired early via knowledge transfer. More specifically, geographic areas where agriculture was introduced as the result of migrations from other areas should be excluded from analysis in order to avoid biological variation due to the influx of a migrant population. Skeletal studies on the morphological changes resulting from the adoption of agriculture are available from only about half of the regions and time periods of agricultural innovation listed above. These include the Levant (Smith, Bar-Yosef, and Sillen 1984; Eshed, Gopher, and Hershkovitz 2006; Pinhasi, Eshed, and Shaw 2008; Pinhasi, Eshed, and von Cramon-Taubadel 2015; May et al. 2018; Pokhojaev et al. 2019), eastern North America (Illinois and Ohio; Buikstra 1984; Cassidy 1984; Cook 1984; Goodman et al. 1984; Perzigian, Tench, and Braun 1984; Paschetta et al. 2010), and the central South American Andes (~Peruvian Andes; Gehlert 1979; Allison 1984; Benfer 1984, 1990) (figure 12.1). In other areas such as India, the New Guinea Highlands, west sub-Saharan Africa, China (but see Li, Zhang, and Zhu 2012), and Central Mexico (but see Marquez-Morfin and Storey 2007), the skeletal changes over the time in question have not been explored so far, at least in the English- or Spanish-language literature. This could be due to a lack of available skeletal samples covering the whole diachronic sequence or to biases in the distribution of bioarchaeological investigation.

There are, however, other relevant skeletal studies in areas neighboring those regions of agricultural innovation, owing to systematic archaeological excavations and good bone preservation. Despite known population turnover in at least some of these regions (e.g., Japan, see Hudson, Nakagome, and Whitman 2020), we include them here to increase our ability to detect global trends, always relying more heavily on those populations where there is a good argument for biological continuity. Some examples of these neighboring regions include Nubia (Greene, Ewing, and Armelagos 1967; Carlson and Van Gerven 1977; Hinton and Carlson 1979; Martin et al. 1984; Calcagno 1986; Galland et al. 2016), Japan (Kaifu 1997; Kanazawa and Kasai 1998; Fukase and Suwa 2008; Fujita and Ogura 2009; Temple 2011; Hoover and Williams 2015), South Africa (Ginter 2011), the Maghreb (Sardi, Ramirez-Rozzi, and Pucciarelli 2004), Iran and Iraq (Rathbun 1984), western and southern North America (Larsen 1981, 1984; Dickel, Schulz, and McHenry 1984; Rose et al. 1984), South Asia (Kennedy 1984), and southeastern Europe (y'Edynak and Fleisch 1983; Pinhasi and Meiklejohn 2011; Cheronet, Finarelli, and Pinhasi 2016; Macintosh, Pinhasi, and Stock 2016) (figure 12.1).

By synthesizing results from studies documenting morphological changes across the agricultural transition from a global sample of research, we hope to better understand the impact that dietary change had on the skeleton and the evolutionary mechanisms underlying it, as detailed below. Despite the substantial body of research, the causes of agriculture-

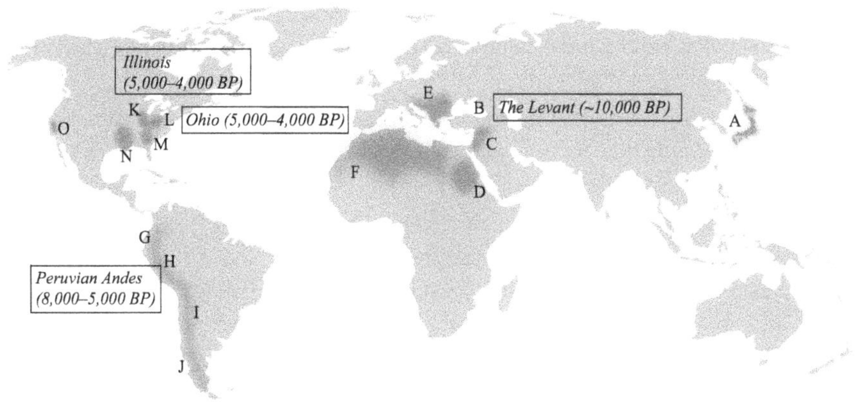

Figure 12.1
Map showing approximate regions referred to in text. (A) Japan, (B) the Levant, (C) Anatolia, (D) Nubia, (E) the Balkan Peninsula, (F) the Maghreb, (G) the Ecuadorean Andes, (H) the Peruvian Andes, (I) the Chilean Andes, (J) the southern Argentinian Andes, (K) Illinois, (L) the Ohio River Valley, (M) Georgia, (N) the Mississippi River Valley, (O) Central California. The centers of agricultural innovation that are mentioned in this chapter are labeled together with the approximate origin dates. Global map: Public domain.

related morphological changes remain unclear and are much debated. One way to organize the currently disparate interpretations of cause and effect is by applying Mayr's definition of proximate and ultimate causes (Mayr 1961). We understand the limitations of their application; notably, this framework does not allow us to invoke developmental mechanisms to explain evolutionary novelties (Baedke 2020); nor does it recognize that biological systems have a reciprocal relationship with their environments (Laland et al. 2013). However, in this chapter, we depart from an evo-devo perspective (Hall 2012) and consider developmental plasticity as an ultimate cause, because morphological changes occurring through developmental plasticity can have long-term evolutionary effects, just as selection does (Paaby and Testa 2018; Brown 2020). The proximate causes of the morphological changes associated with the adoption of agriculture may be a decrease in masticatory loading, the impact of dental pathology, or nutritional deficiency. The ultimate causes proposed for these morphological changes are developmental plasticity and/or generational responses to directional selection. It should be noted, however, that these definitions are working concepts that we apply as a means of grouping and understanding the different types of explanations current in the literature. In this chapter, we present a synthesis summarizing the morphological changes that have been linked to the transition to agriculture and discuss the main evolutionary explanations given for those changes.

Morphological Changes Associated with the Transition to Agriculture

The skeletal effects of the agricultural transition have been described mainly in the teeth, mandible, skull, and long bones (Cohen and Armelagos 1984; Larsen 1995, 2006; Pinhasi and Stock 2011; Macintosh, Pinhasi, and Stock 2016). This representation could be due either to a bias in research designs toward the study of those skeletal structures or to their being more affected due to their function and its relationship with diet and subsistence practices. The main morphological changes described as resulting from a shift to cultivated foods can be summarized as a decrease in overall skeletal size, concomitant with allometric and shape changes that can be interpreted as a trend toward gracilization (i.e., less robust morphology) (Carlson and Van Gerven 1977; Larsen 2006). Additionally, across the studied populations, there are often morphological changes associated with the deterioration of health status, either as a consequence of a more sedentary life and/or of a less diverse diet (Diamond 1987; Ulijaszek 1991; Larsen 1995, 2006). In the following section we compile the evidence for morphological changes in the mandible, cranium, dentition, and long bones from studies with a wide geographic range.

The Mandible

Due to its primarily masticatory function, the mandible has been described as the bone most influenced by changes in subsistence (Lieberman 2011; von Cramon-Taubadel 2011). The global pattern of mandibular variation in recent humans reflects a dichotomous distinction between hunter-gatherer (HG) and agriculturalist/pastoralist subsistence economies (von Cramon-Taubadel 2011; Katz, Grote, and Weaver 2017). Both globally and regionally, the change in mandibular morphology from an HG economy to one based on animal and/or plant domesticates can be summarized as a trend of decreasing mandibular size and robusticity (Kaifu 1997; Fukase and Suwa 2008; Holmes and Ruff 2011; von Cramon-Taubadel 2011; Galland et al. 2016; Katz, Grote, and Weaver 2017; Pokhojaev et al. 2019) (figure 12.2).

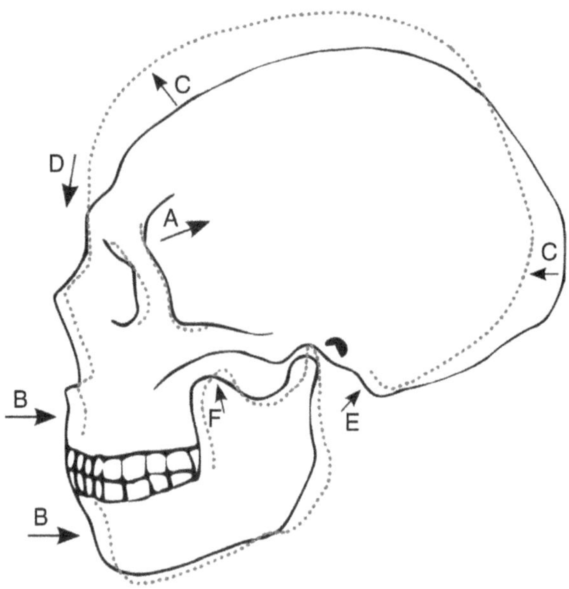

Figure 12.2
Summary of morphological changes in the skull and mandible observed in hunter-gatherers (black solid line) when compared with agriculturalists (dashed grey line). The main morphological changes are (A) a reduction in the size of the muscles of mastication and a relatively more posterior site of origin; (B) a reduced anteroposterior growth of the maxillomandibular complex; (C) a reduction in the relative length and increase in the relative height of the cranial vault, which becomes more globular; (D) a reduction in the size of the face, which becomes more inferoposteriorly located; (E) an overall reduction, expressed in some parts such as the mastoid; and (F) an increase in mandibular ramus height. Reduction in teeth size is not shown here. Adapted from Carlson and Van Gerven (1977).

Given that the Levant is one of the earliest centers of agricultural innovation, it is perhaps not surprising that much research on the skeletal effects of this dietary shift have focused there (Diamond and Bellwood 2003; Price and Bar-Yosef 2011). In the Levant, a reduction in mandibular size and robusticity, coinciding with the arrival of a domesticated diet, can be seen by comparing the HGs of the Natufian period to the later agriculturalist Neolithic populations. Natufian mandibles are short and heavy, and over time there is an increase in maximum length, corpus length, and ramus height, but a decrease in ramus width (Smith, Bar-Yosef, and Sillen 1984; Pokhojaev et al. 2019) (figure 12.2). The wider ramus in Natufian mandibles is accompanied by a shorter, wider coronoid process and wide mandibular notch, which gives way over time to a taller, narrower coronoid and a narrow mandibular notch in the Neolithic (May et al. 2018; Pokhojaev et al. 2019) (figure 12.2). Anterior symphyseal height reduces significantly in southern Levantine populations over time (Pinhasi, Eshed, and Shaw 2008; May et al. 2018), while the projection of the mental eminence increases (Pokhojaev et al. 2019). Natufian mandibles show a more lingual (horizontal) inclination of the posterior part of the body of the mandible, whereas later farming populations have a more buccal (vertical) orientation in this region (May et al. 2018).

In Nubian populations from northeastern Africa, Mesolithic individuals present larger mandibles with wider and more robust corpuses, shorter, wider, and more upright rami and coronoid processes, and longer mandibular condyles than later Neolithic agriculturalists (Galland et al. 2016) (figure 12.2). This reduction in mandible size over the course of a shift in subsistence corresponds to the smaller dimensions of the temporomandibular joint (TMJ) in the Nubian Neolithic (Greene, Ewing, and Armelagos 1967; Hinton and Carlson 1979). In general, during the transition to agricultural subsistence in Nubia, there is a reduction of the sexual dimorphism in TMJ size and more posteriorly positioned masticatory muscles (Hinton and Carlson 1979).

Several studies have compared different aspects of the mandibular morphology of prehistoric Japanese foragers (Jomon) with the agriculturalist Yayoi who succeeded them (Kaifu 1997; Kanazawa and Kasai 1998; Fukase and Suwa 2008). The latter are smaller and lighter with thinner cortical bone (Kaifu 1997; Fukase and Suwa 2008).

The Cranium

On a worldwide scale, diet-related morphological differences between recent populations of farmers and HGs are seen most strongly in temporalis muscle size and shape variation (Sardi, Ramirez-Rozzi, and Pucciarelli 2004; Noback and Harvati 2015; Cheronet, Finarelli, and Pinhasi 2016) (figure 12.2). Compared with crania from farmers whose diets are largely based on crops, crania from populations relying heavily on hunting and/or fishing show larger and more anteriorly placed masticatory muscles accompanied by widened alveolar processes, larger nuchal planes, larger mastoid processes, and wider faces (Noback and Harvati 2015). In general terms, the main craniofacial change in populations across the adoption of agriculture is an increase in cranial breadth, or a brachycephalization of the cranial vault (e.g., Sardi, Ramirez-Rozzi, and Pucciarelli 2004; Cheronet, Finarelli, and Pinhasi 2016).

On a regional level, in the Levant, the crania of Neolithic farmers are shorter than those of preceding Natufian HGs, and their zygomatic breadths are narrower, resulting in narrower faces and more globular (i.e., spherical) neurocrania (Smith, Bar-Yosef, and Sillen

1984). As the Levant is a widely accepted center of agricultural innovation, it is likely that there is biological continuity between HG and agriculturalist groups. Therefore, since there is no archaeological evidence for population replacement or migrations associated with agriculture (Smith, Bar-Yosef, and Sillen 1984; Bar-Yosef 1998; Cheronet, Finarelli, and Pinhasi 2016), diachronic comparisons in this region strengthen the evidence that these changes are diet-related.

Sardi and collaborators (2004) compared individuals from northwest Africa belonging to the Upper Paleolithic (Iberomaurisian Complex; Afalou and Taforalt archaeological sites in Morocco and Algeria) with Neolithic ones (Guanches from the Canary Islands, Phoenicians and Arabs from Tunis, and modern Berbers from Algeria), and found that overall cranial size is larger in the first group. More specifically, the most divergent variables between HG and agricultural groups in prehistoric northwest Africa are the rates of change in the sizes of the neural and facial parts of the cranium. In the Upper Paleolithic, the facial region is relatively larger than it is in the Neolithic, while in the Neolithic the neural region is relatively larger than it is in the Upper Paleolithic (Sardi, Ramirez-Rozzi, and Pucciarelli 2004) (figure 12.2). In Nubia, from the Mesolithic to Neolithic there is a decrease in the size and robusticity of the whole craniofacial complex (Carlson and Van Gerven 1977; Martin et al. 1984; Galland et al. 2016), and later agricultural populations show an increase in cranial height with a concomitant decrease in cranial length, effectively increasing cranial globularity (Carlson and Van Gerven 1977; Martin et al. 1984; Galland et al. 2016) (figure 12.2). Overall, there is a reorganization of the craniofacial complex such that the vault becomes more anterosuperiorly located, while the mid- and lower face becomes more inferoposteriorly located (Carlson and Van Gerven 1977; Martin et al. 1984) (figure 12.2). Neolithic Nubian faces are smaller, lower, and more retracted, compared with their Mesolithic forerunners. They present less deep, narrower zygomatics, less pronounced alveolar prognathism, less projecting glabella regions, and shorter mastoid processes (Galland et al. 2016). These changes result in the Neolithic crania being overall more gracile (i.e., less robust) and more globular, as with the Levantine agriculturalists when compared with preceding foragers (figure 12.2).

In South America, farmers present a reduction in the size of the masticatory component of the cranium when compared with HG groups (Gonzalez-José et al. 2005). Most studies in this region have been conducted using variables identified following the craniofunctional method (Pucciarelli 2008; Sardi 2017), linking cranial components to specific functions such as mastication and allowing functional interpretations of the observed anatomical changes. In the southern Andes (central-west Argentina), in addition to the reduction in the masticatory component, the transition to agriculture is accompanied by a reduction in the posteroneuronal part of the cranium (Sardi, Novellino, and Pucciarelli 2006) (figure 12.2). This pattern of masticatory and posteroneuronal reduction is found throughout the South American continent, where there is a clinal pattern of size and allometric cranial shape variation from southeast to northwest. Dietary variation is a plausible explanation for the majority of this pattern, which is characterized by smaller sizes in the northwest and increasing robusticity toward the southeast of the continent. The most influential variables in this size variation are cranial width, height of the masticatory functional component, facial width, and neurocranial length (Perez et al. 2011; Menéndez et al. 2014). Generally, in South America there is sexual dimorphism in the extent of masticatory reduction across the shift to agriculture, with greater size reduction in females than in males (Sardi and Béguelin 2011).

In North America, particularly in individuals from the Ohio Valley, there are significant differences between HGs and farmers in the size of the temporal fossa and masticatory complex (including the TMJ), as well as allometric differences in the alveolar region (Paschetta et al. 2010) (figure 12.2). Similar trends of reduction in the size of the face and jaw are seen in populations from the American state of Georgia across 4,000 years of dietary changes (Larsen 1981, 1984). Despite localized skeletal changes, which fit with the wider global picture of responses to dietary change, HGs from the Ohio Valley actually present overall smaller cranial dimensions than farmers from the same region, combined with less robusticity or smaller facial size relative to total size (Paschetta et al. 2010), showing regional variation in the patterns of craniofacial response to changing diets. The evidence from the Ohio Valley is of particular importance, as it is one of those rare regions where biological continuity can be inferred with reasonable certainty, thus avoiding confounding factors such as population history that might also affect shape (Smith 1989; Paschetta et al. 2010).

The Dentition

There is a pattern of dental reduction in many populations after the adoption of agriculture as their main subsistence practice. This has been found in such disparate geographic regions as the Levant (Pinhasi, Eshed, and Shaw 2008), Nubia (Martin et al. 1984; Calcagno 1986; Calcagno and Gibson 1988), the eastern Mediterranean and Balkans (y'Edynak and Fleisch 1983; y'Edynak 1989), the American state of Georgia (Larsen 1981), Peru (Benfer 1990), and South Asia (Kennedy 1984). In addition to trends of decreasing dental size, there are changes in dental wear and the frequencies of dental and oral pathologies associated with dietary shifts.

In the Levant, Neolithic teeth are smaller than those of earlier groups, but dental size reduction appears to be restricted to buccolingual dimensions (Pinhasi, Eshed, and Shaw 2008). In Georgia (US), after the transition to agriculture there is a reduction in dental size that affects females to a greater extent than males, perhaps due to women being more responsible for agriculture-related activity and thus having a diet with a greater domesticated component (Larsen 1981). In Nubia, there is a strong decrease in dental length, breadth, and occlusal area in both males and females, starting in the Mesolithic and continuing across agricultural periods. This trend is followed by an ongoing, albeit diminished, trend of reduction for only the molar teeth between later agriculturalist groups (Calcagno 1986). Finally, dental asymmetry in buccolingual and mesiodistal diameters decreases, although the differences are not statistically significant, over the transition to agriculture in Peru, as seen from the diachronic sequence at Paloma (Gehlert 1979; Benfer 1984).

Alongside trends of dental reduction associated with dietary change, there are concomitant changes in dental wear. In some geographic areas, such as Japan (Fujita and Ogura 2009) and Peru (Benfer 1984), there is a reduction in the magnitude of dental wear associated with a transition to agriculture, due to the consumption of less abrasive foods. Yet in others, such as Ecuador (Buikstra 1984), dental wear increases from HG to agriculturalist populations, which could be explained by increased grit incorporated into food as a result of grinding grains (Hartnady and Rose 1991). Patterns of dental wear also tend to change over a dietary shift from HG to domesticated foods, from flat to angled and cupped; this pattern is seen in the Levant, Nubia, and Japan (Smith, Bar-Yosef, and Sillen 1984; Kasai and Kawamura 2001; Pinhasi, Eshed, and Shaw 2008). That these changes are consistently associated with change in subsistence in such widely separated geographic regions suggests an effect of diet on dental wear patterns.

There is an extensive literature describing changes in dental health as a result of adopting agricultural subsistence strategies. In general, foragers are described as presenting higher degrees of dental wear, but less pathology such as caries, tooth rotation, crowding, alveolar lipping, and abscesses than agriculturalists, and fewer problems with bone remodeling and maintenance. Examples of this trend come from Nubia (Greene, Ewing, and Armelagos 1967; Martin et al. 1984; Martin and Armelagos 1984), Japan (Fujita and Ogura 2009), the Levant (Smith, Bar-Yosef, and Sillen 1984), Greece (Papathanasiou 2011), Peru (Allison 1984; Benfer 1990), Ecuador (Buikstra 1984), Chile (Allison 1984), South Asia (Kennedy 1984), Central California (Dickel, Schulz, and McHenry 1984), Dickson Mounds, Illinois (Goodman et al. 1984), and the Ohio valley (Perzigian, Tench, and Braun 1984). In one example, in the former Yugoslavia, y'Edynak and Fleisch (1983) found that Mesolithic individuals had more chipped and cracked teeth than Neolithic individuals, whereas the Neolithic group had a higher prevalence of hypoplasia and higher grades of alveolar resorption. However, this widespread trend is not universal. In Japan, Jomon and Yayoi individuals did not have significantly different carious tooth frequencies (Temple and Larsen 2007). This has been interpreted as a result of the Jomon reliance on cariogenic plants such as acorns, which have similar cariogenic properties to the wet rice that the Yayoi consumed (Temple and Larsen 2007).

The Postcranial Skeleton

Regarding variation in postcranial skeletal morphology related to the transition to agriculture, researchers have described changes in cortical thickness, body size, stature, degree of sexual dimorphism, and the frequency of degenerative joint diseases. Reduction in size is perhaps the most consistent change associated with a shift in subsistence and is seen in many disparate populations. This relates particularly to shorter stature in farmers, but also in some cases to reduced limb length and body mass. Compared with these fairly consistent shifts in size, changes in cortical thickness, which reflect bone strength, and sexual dimorphism show a more complex pattern that differs among the regions considered here.

There are numerous examples of stature reduction associated with shifts in subsistence practices in South America (Chile, Argentina, Ecuador) (Allison 1984; Buikstra 1984; Sardi and Béguelin 2011), the Levant (Smith, Bar-Yosef, and Sillen 1984), South Asia (Kennedy 1984), Nubia (Martin et al. 1984), the Ohio River Valley, Georgia (US), the lower Mississippi Valley, central California, and Dickson Mounds, Illinois, in North America (Dickel, Schulz, and McHenry 1984; Goodman et al. 1984; Larsen 1984; Perzigian, Tench, and Braun 1984; Rose et al. 1984). Despite being widespread, the trend of stature reduction does not seem to be universal, however. At the site of Paloma in Peru, where the earlier levels are HG and the later ones are agriculturalist, stature actually increases over time (Benfer 1984). This may be a sampling artifact, however, rather than an actual increase in height in farming populations, as the more recent chronological periods when agriculture was fully established are not well represented (Benfer 1984).

Stature also slightly increases in females from the lower Illinois Valley over the transition to agriculture (Cook 1984), and, in a comparative study, Auerbach (2011) argues that overall, stature is higher among southwestern agriculturalists (Arkansas, Louisiana) than southeastern foragers. These last results should be considered with caution, since the southwestern agriculturalists have not been shown to be direct descendants of the comparative sample of southeastern foragers, meaning that factors other than subsistence might explain such morphological differences, as mentioned above.

Decrease in body mass is recorded among Neolithic agriculturalists, when compared with previous HG populations from Nubia (Stock et al. 2011), with similar trends in the lower Mississippi (Rose et al. 1984), and Georgia (US) (Larsen 1984). As with changes in stature, however, this trend is not globally consistent; in the southern part of North America there are increases in southwestern agriculturalists' body masses when compared with southeastern foragers, with the former also presenting wider body breadths (Auerbach 2011).

Cortical thickness is a determinant of long bone strength. Changes in cortical thickness, calculated as the percentage of cortical area in relation to the total area, are associated with nutritional and mechanical factors (Larsen 1995). Despite being developmentally plastic (e.g., Ruff, Walker, and Trinkaus 1994), cortical thickness does not appear to respond consistently to changes in lifestyle associated with the transition to agriculture across different geographic regions, although some general trends can be established. In general, farmers present a decrease in long bone shaft cross-sectional dimensions and an increase in circularity, suggesting reduced strength and activity (figure 12.3). There is a reduction in cortical thickness when comparing long bones from HGs with those of later agriculturalists in Nubia (Martin et al. 1984), and Dickson Mounds, Illinois (Goodman et al. 1984), but no significant changes in cortical thickness when comparing HGs with later agriculturalists in the Illinois Valley (Cook 1984). Long bone diameter and rigidity, additional

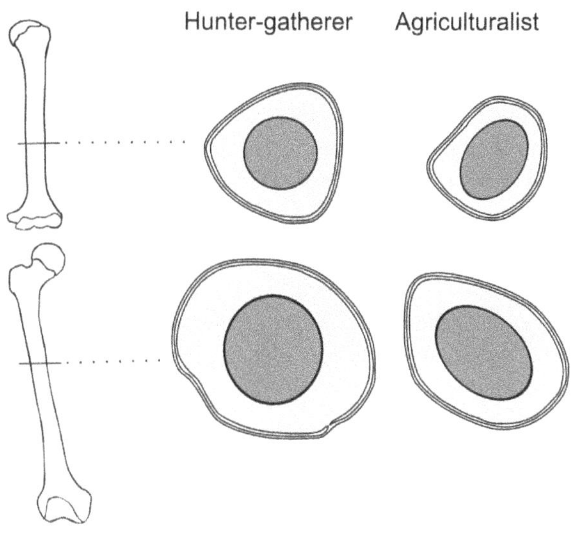

Figure 12.3
Long bone cross-sections showing the main changes in shape, size, and distribution of subperiosteal bone in the humerus and femur among hunter-gatherers and agriculturalists.

proxies for bone strength, are reduced in agriculturalists when compared with previous HGs in Illinois (Cook 1984), Paloma, Peru (Benfer 1984), the Danube Gorges in southeast Europe (Pinhasi et al. 2011), and the Nile Valley (Stock et al. 2011). However, again these are not ubiquitous trends: Larsen and Ruff (2011) and Rose and collaborators (1984) describe an increase in femoral diameter across the transition to agriculture in the lower Mississippi Valley, while Sardi and Béguelin (2011) describe an increase in femoral mid-shaft shape index (also denoting an increase in bone strength) among farmer females, compared with HG females, from the southern Andes.

In addition to the divergent diachronic changes in female height in North America and the change in female femoral strength in the southern Andes described above, there is other evidence of sexual dimorphism in skeletal responses to a change in diet. In Georgia (US) females show a greater diet-related decrease in skeletal size, robusticity, and stature than males (Larsen 1984). In fact, there is often an increase in sexual dimorphism accompanying a change in subsistence from HGs to agriculturalists. This is the case among central-west Argentinian farmers compared with earlier HGs (Sardi and Béguelin 2011), in agricultural populations from Ohio River Valley compared with previous HGs (Perzigian, Tench, and Braun 1984), in Georgia (Larsen 1984), in most archaeological sites from the Mississippi Valley (Rose et al. 1984), and overall in North America (Auerbach 2011). However, as with most of the postcranial trends associated with the transition to agriculture, these patterns are not the same in all populations. At the archaeological site of Caddoan in the Mississippi Valley, as well as at sites in Peru and the Nile Valley, sexual dimorphism decreases, rather than increases, in agricultural groups when compared with earlier HGs from the same regions (Benfer 1984; Rose et al. 1984; Stock et al. 2011). Finally, there are no changes in the degree of sexual dimorphism when comparing HGs with later agriculturalists from Dickson Mounds, Illinois (Goodman et al. 1984).

As with the dental evidence, frequencies of postcranial skeletal pathologies change after the transition to agriculture in many populations (Cook 1984; Goodman et al. 1984; Larsen 1984, 1995, 2006; Martin et al. 1984; Rose et al. 1984). In Georgia (US), there is an increase in the frequency of postcranial periosteal reactions, reflecting a decline in skeletal health, but a decrease in pathology related to mechanical stress (degenerative joint disease; Larsen 1981). A decrease in degenerative joint disease is also present among agriculturalists from Nubia (Martin et al. 1984) and the lower and central Mississippi Valley (Cook 1984; Rose et al. 1984). In general, both nonspecific markers of skeletal health and degenerative diseases associated with high levels of activity decline with the adoption of agriculture (Cook 1984; Larsen 1981; Martin et al. 1984; Rose et al. 1984). These results are congruent with previous findings that hunter-gatherers present a higher frequency of degenerative joint disease (osteoarthritis) than agriculturalists (Jurmain 1977; Larsen 1981). However, this trend may be population-dependent to an extent, since after studying a large sample of postcranial remains from individuals from across southern North America, Auerbach (2011) claims there is no general decline in health despite previous evidence to the contrary (Larsen 1995, 2006), and an increase in health is actually described among agriculturalists from Illinois in comparison with their HG ancestors (Goodman et al. 1984).

Discussion

Convergent and Divergent Trends Worldwide

The research we present above shows that the skeletal effects of the transition to agriculture are modest, directional, and most pronounced in craniofacial and mandibular features directly involved in mastication. There are some consistent trends seen in many geographically disparate regions (table 12.1). The mandible presents a size reduction and gracilization trend and a move toward a less upright ramus (von Cramon-Taubadel et al. 2011; Galland et al. 2016; Katz, Grote, and Weaver 2017). Over the course of the transition to agriculture, farmers tend to present more globular neurocranial vaults relative to lower faces, shorter and more inferiorly placed temporalis muscles, reduction in the maxillomandibular complex, smaller faces, modestly taller palates, shorter toothrows, and posterior displacement of the dentition due to the reduced length of the maxilla (Katz, Grote, and Weaver 2017). There is a generalized trend toward dental reduction in many regional studies conducted on teeth in North America, the Levant, North Africa, and South America, although the specific pattern of reduction is somewhat variable (y'Eynak and Fleisch 1983; Kennedy 1984; Larsen 1984; Martin et al. 1984; Calcagno 1986; Benfer 1990; Pinhasi, Eshed, and Shaw 2008). These changes are accompanied by an increase in dental caries (Larsen 1995, 2006), except in some areas such as East Asia, where agriculture was based on rice and caries frequency does not follow this trend (Temple and Larsen 2007).

Conversely, some anatomical structures appear to change as a result of the transition to agriculture but do not follow a widespread trend, instead varying with population or geographic region (table 12.1). These include most of the postcranial changes associated with shifts to domesticated diets, such as changes in stature, body mass, cortical thickness, and especially patterns of sexual dimorphism. In the following section we evaluate the potential proximal and ultimate causes for trends in morphological change across the transition to agriculture.

Table 12.1
Convergent and divergent phenotypic trends across geographic regions after the adoption of agriculture

	Near East	North Africa	North America	South America
Gracilization and mandibular size reduction	X	X	?	?
Trend toward less upright mandibular ramus	X	X	?	?
More globular neurocrania	X	X	X	X
Lower and smaller facial skeleton	X	X	X	X
Shorter temporalis muscle	X	X	X	X
Dental size reduction	X	X	X	X
Stature reduction	X	X	X	X
Body mass reduction	?	X	X	?
Postcranial bone cortical thickness reduction	?	X	X	X
Sexual dimorphism	?	Reduction	Variable	Increase

Notes: X = presence in region; ? = uncertainty due to insufficient data.

Proximate and Ultimate Causes of Morphological Change across the Agricultural Transition

As suggested at the beginning of this chapter, one way to organize the currently disparate interpretations of the morphological effects of agriculture is to apply Mayr's (1961) definition of proximate and ultimate causes. This approach allows us to distinguish between different levels of explanation that are not mutually exclusive but refer to different aspects of the phenomenon. In his influential essay, Mayr wrote that "proximate causes govern the responses of the individual (and his organs) to immediate factors of the environment, while ultimate causes are responsible for the evolution of the particular DNA code of information with which every individual of every species is endowed" (Mayr 1961, 1503). According to more recent perspectives, proximate causation can be equated with immediate factors (for example, physiology) and ultimate causation with evolutionary explanations (for example, selection) (Laland et al. 2011). Here we use Mayr's framework to discuss the following potential proximate causes for the morphological differences between HGs and farmers: (1) direct biomechanical changes due to alterations in the intensity of masticatory loading or physical activity and (2) systematic changes due to adjustments in dietary nutrition or activity levels resulting in alterations in hormonal circulation levels. We then examine possible ultimate causes (i.e., the evolutionary processes behind the morphological changes) and evaluate the potential roles of (3) phenotypic plasticity as a result of environmental influences on individuals during development and (4) directional selection, whereby morphological adaptation involves changes to DNA sequences over many generations. Since developmental plasticity, the ability of an individual to alter its phenotype in response to the environment (Fusco and Minelli 2010), has a relevant role in evolutionary change (Nijhout 1990; Gilbert 2001; Pigliucci 2001; West-Eberhard 2003), we will here consider it as an ultimate cause, despite the fact that this is a highly debatable point (see Brown 2020).

Proximate Causes of Morphological Changes over the Agricultural Transition

When evaluating explanations provided for proximate causes behind morphological changes across the transition to agriculture, most interpretations in the literature can be attributed to one of two alternative mechanisms: (1) direct biomechanical changes due to alterations in loading or (2) systematic changes due to adjustments in the diet or changes in physical activity.

Direct Biomechanical Causes

Biomechanical explanations for diet-related morphological changes in the skull are very widespread (Lieberman 2011). They usually invoke a decrease in masticatory stress due to changes in food preparation techniques (cooking, grinding, and so on), which were critical in producing softer food, and/or the reduction or complete replacement of meat and fiber resulting in a diet with a large proportion of softer, grain-based foods (food consistency). As these factors are usually entangled and could be causally related, they will be considered together.

The suggestion that reduced masticatory stress seen across the transition to agriculture led to the morphological changes in the skull has been well developed by Carlson and Van Gerven (1977) as a set of premises that compose the masticatory-functional hypothesis

(MFH). The MFH states that shifts from HG to agricultural diets resulted in decreasing functional demands on the masticatory complex, bringing about alterations in skull morphology, including a reduction in size of the masticatory muscles, reduction in the growth of the maxillomandibular complex, and compensatory changes in the shape of the cranial vault. As a consequence, individuals from agricultural populations tend to present allometric shape changes, such as a more posterior origin of masticatory muscles, a more inferiorly placed and retrognathic face, a more globular vault, and concomitant size reductions in teeth and alveolar region (Carlson and Van Gerven 1977). The first assumption behind the MFH is known as *Wolff's law*, which states that bone is deposited in presence of functional demand and resorbed in its absence in such a way that a decrease in musculoskeletal stress results in size reductions (Wolff 1892, 1986). Though the nuances of this relationship have been much questioned, the basic relationship between stress and bone deposition is supported (Ruff, Holt, and Trinkaus 2006). The second, implicit, assumption is that an HG diet results in greater stresses on the masticatory complex than an agricultural one (Carlson and Van Gerven 1977).

A reduction in the magnitude and change in direction of masticatory forces resulting from changes in subsistence and food-producing techniques has been claimed as the main mechanism behind morphological change in many disparate geographic regions. The MFH has been invoked to account for the decrease in cranial, mandibular, and TMJ size in populations from Nubia (Hinton and Carlson 1979; Galland et al. 2016), the former Yugoslavia (y'Edynak & Fleisch 1983), Japan (Kaifu 1997; Hoover and Williams 2015), the Maghreb (Sardi, Ramirez-Rozzi, and Pucciarelli 2004), and the Levant (Pinhasi , Eshed, and Shaw 2008; Pinhasi, Eshed and von Cramon-Taubadel 2015; May et al. 2018; Pokhojaev et al. 2019). Despite morphological integration between the mandible and maxillary regions, they can evolve independently (von Cramon-Taubadel 2011), and studies comparing responses to dietary change from HG to farming across the whole skull provide a complex scenario. Some describe diet-related morphological changes in the shape of the whole skull (Carlson and Van Gerven 1977; Larsen 1995; Gonzalez-José et al. 2005; Sardi, Ramirez-Rozzi, and Pucciarelli 2004; Perez et al. 2011), while others describe more localized changes, such as the more geographically widespread changes in the attachment of the temporalis muscle area (Paschetta et al. 2010; von Cramon-Taubadel 2011; Noback and Harvati 2015). The size reduction of the masticatory component and posteroneuronal region in crania from the southern Andes has been interpreted as a direct and localized change resulting from changing masticatory loading due to reduced dietary hardness in agriculturalists (Sardi, Novellino, and Pucciarelli 2006). An increase in the vertical height of the mandible, combined with a reduction in the mandibular condyle and coronoid process, as described from Levantine and Nubian sites, are all associated with the reduction in stresses resulting from anterior dental loading (Hinton and Carlson 1979), which is known to be reduced in agriculturalists compared with foragers (Hinton and Carlson 1979; y'Edynak and Fleisch, 1983; Galland et al. 2016; May et al. 2018). Similarly, the changes in orientation in the posterior part of the mandibular body from a more lingual inclination to a more buccal one in later Levantine populations may suggest a reduction in torsional forces on the mandible, resulting from a diet that was easier to chew (May et al. 2018).

Support for a causal relationship between the morphological changes described across the transition to agriculture and reduced masticatory stress comes from experimental studies on nonhuman models. Studies comparing animals raised on foods varying in consistency (hardness or toughness) support the conclusions drawn from the bioarchaeological literature, as described above, by showing morphological differentiation in the same parts of the cranium and mandible (Bouvier and Hylander 1982; Corruchini and Beecher 1982, 1984; Hinton 1983, 1992; Yamada and Kimmel 1991; Tuominen, Kantomaa, and Pirttiniemi 1993; Ciochon, Nisbett, and Corruccini 1997; Liberman et al. 2004). Experimental studies on nonhuman primates fed tough foods also indicate that high strains are generated in the same regions as those seen to undergo change in the human cranium during the course of the agricultural transition (Hylander, Johnson, and Picq 1991; Ross and Hylander 1996; Ravosa et al. 2000), suggesting these regions would be where adaptation would be most likely, according to Wolff's law.

Dental size reduction in agriculturalists, compared with forager groups, has most often been interpreted as a secondary consequence of facial reduction driven by biomechanical shifts (Carlson and Van Gerven 1977; Smith, Bar-Yosef, and Sillen 1984; Lieberman 2011), yet there is no uniform association between facial and dental reduction. In the Levant, for example, teeth are reduced in buccolingual dimensions, but facial size reduction appears to be restricted to the mandible, with specific changes to the corpus height at the mandibular symphysis and a reduction in ramus breadth (Pinhasi, Eshed, and Shaw 2008). Since it is difficult to see how these specific mandibular changes would affect buccolingual dental dimensions, this may indicate that, at least in some populations, the reduction in masticatory stresses is acting directly on dental size.

Diachronic changes in dental wear patterns in the Levant, Nubia, and Japan have been interpreted as evidence of reduction in dietary toughness from an HG diet to an agricultural one consisting of cereal-based foods (Smith, Bar-Yosef, and Sillen 1984; Kasai and Kawamura 2001; Pinhasi, Eshed, and Shaw 2008). The cooking practices required to make such foods edible may further reduce toughness and dental wear (Eshed, Gopher, and Hershkovitz 2006; Fujita and Ogura 2009). Where the reverse is seen—namely, an increase in dental wear from earlier forager to more recent farmer populations—it is interpreted as a side-effect of non-cooking food processing techniques, such as grinding, which can introduce grit into food and lead to considerable wear (Watson 2008). The increase in dental pathologies seen in adopters of agriculture may be a side-effect of facial reduction resulting from biomechanical changes. Y'Edynak and Fleisch (1983) argue that softer food produces less stress on the mandible and that in turn the mandible does not achieve its maximum phenotypic development. The shortened mandible then produces crowded and rotated teeth resulting in possible maladjustments. Maladjustments such as crowded and rotated teeth produce uneven stresses on the periodontal membrane, resulting in inflammation of the gingival tissue, diseases of the periodontium, and eventually tooth loss and alveolar resorption, as seen at higher levels in many agriculturalist populations compared with foragers (y'Edynak and Fleisch 1983). Additionally, most agricultural diets are richer in carbohydrates than those of foragers, and these foodstuffs are more cariogenic, a situation worsened by maladjustments that lead to food sticking to teeth and contribute to the greater frequencies of caries in agriculturalist populations (Hillson 1986; Larsen 1997; Ogden 2008). In contrast, HGs tend to use their anterior dentition much more than agriculturalists, which involves greater loading on the temporalis and masseter muscles, and results in an increased incidence of dental chipping in HG groups (y'Edynak and Fleisch 1983).

The widely seen reduction in postcranial skeletal size and robusticity across the agricultural transition has been linked to biomechanical factors resulting from a less active lifestyle. The exact mechanism thought to underpin changes is rarely specified explicitly, however. A lifeway based on hunting and gathering appears to involve more functional demand on the body than one in which agriculture is the primary mode of subsistence (Larsen 1984; Ruff, Larsen, and Hayes 1984; Stock and Pfeiffer 2001; Macintosh, Pinhasi, and Stock 2016). This reduction in functional demand in farmers could be at least in part the result of a decrease in mobility (Larsen and Ruff 1991). Ruff, Larsen, and Hayes (1984) have shown that the cross-sectional geometric properties of limb bones reflect a reduction in biomechanical demands from HG to farmers in Georgia (US), following Wolff's law. They propose that HG activity is characterized by climbing and running, while farmers perform more walking, lifting, and carrying (Ruff, Larsen, and Hayes 1984). These results are supported by longitudinal and comparative studies in contemporary populations that show increases in bone size, cross-sectional geometric measures of bone strength, bone mineral density, and body mass when performing moderate to high exercise (Taaffe et al. 1997; Bradney et al. 1998; Daly et al. 2004; Guadalupe-Grau et al. 2009; Shaw and Stock 2009; Suominen 2012).

Theories about changes in levels of sexual dimorphism over the agricultural transition have often used biomechanical processes as their implicit underpinning. The increase in sexual dimorphism in postcranial strength accompanying a change in subsistence from HG to agriculture in many geographic regions has been interpreted as resulting from strong sexual division of labor in farmers, as opposed to both sexes being involved more evenly in foraging activities (Larsen 1984; Perzigian, Tench, and Braun 1984; Rose et al. 1984; Auerbach 2011; Sardi and Béguelin 2011). In contrast, a decrease in size sexual dimorphism over the transition to agriculture has been reported for some populations. In those cases, the decrease results from a relative increase in female size, which may be due either to an increasing physical demand in females (Cook 1984; Rose et al. 1984) or to lessening sexual division of labor (Benfer 1990). The lack of a global pattern in sexual dimorphism likely relates to the types of task being carried out by both sexes and how differentiated they are, factors that differ with local diet, culture, and ecology.

Systematic Explanations

Some scholars argue that many size reductions in populations transitioning to an agricultural diet may be related to a lack of protein and other nutrients, at the expense of an increase in the consumption of carbohydrates, experienced during development. Environmental influences during growth and development have been shown to have profound consequences on the subsequent phenotypic expression of biological traits during adulthood (Bogin 1999; Frisancho 2009). This mechanism has been suggested for interpreting morphometric patterns in crania and postcrania particularly among South American populations (Sardi, Novellino, and Pucciarelli 2006; Perez et al. 2011; Menéndez et al. 2014), but also for North American ones (Larsen 1981, 1984, 1995; Martin et al. 1984; Perzigian, Tench, and Braun 1984). In South America, smaller and more gracile crania from the northwest of the continent have been interpreted as a result of diets based mostly on processed carbohydrates, which, being less nutritious than varied forager diets, would have induced systematic changes leading to a general decrease in cranial and body size, together with allometric changes in shape (Menéndez 2015). Both South American and North American populations traditionally shared a reliance on maize, a cereal low in two essential amino acids: lysine and tryptophan. Amino acids are essential for cell multiplication in the growth and differentiation of structural tissues, such as muscle and bone (Stini 1971). While humans synthesize some amino acids, others must be ingested. If any of them is lacking in the diet, however, this will preclude the utilization of the rest (Stini 1971). The cessation of amino acid supply reduces the maintenance and growth processes of skeletal muscle, producing a slow and delayed growth that results in the size reduction of the skeleton (Stini 1975).

In an experiment conducted on squirrel monkeys, Cónsole and collaborators (2001) found that a low-protein diet induced a decrease in growth hormone (GH) and prolactin cell populations, resulting in changes of craniofacial morphology, especially in the masticatory module. Low protein intake impairs both the production and action of insulin-like growth factor-I, which is essential for longitudinal bone growth and bone formation, as it stimulates proliferation and differentiation of chondrocytes in the epiphyseal plate (Bonjour et al. 2001). Cranial and mandibular size changes associated with the adoption of a diet overly reliant on carbohydrates and with reduced protein are further supported by experimental studies on rats and squirrel monkeys (Pucciarelli 1980, 1981; Dressino and Pucciarelli 1999; Miller and German 1999). Decrease in skull size associated with dietary shifts also could be interpreted as the result of changes in hormonal pathways due to the under-availability of nonprotein nutrients required for growth (Nijhout 2003). Ginter (2011) proposed a reduced-growth model for South African populations after the adoption of agriculture. She suggests that decrease in skeletal size resulted from stunted development, mitigating nutritional insufficiency, and that this explains the later rebound in stature after improvements in agricultural practices (Ginter 2011). Direct evidence of malnutrition in the archaeological populations where size reduction has been recorded comes from palaeopathology analyses; studies have linked the presence of pathologies such as *cribra orbitalia* in prehistoric populations with nutritional deficiencies in proteins, calcium, and iron (Angel 1984; Rathbun 1984; Smith, Bar-Yosef, and Sillen 1984; Ulijaszek 1991). These theories are supported by comparative studies in human populations undergoing protein malnutrition, in which individuals have reduced body size and stature (Frisancho, Garn, and Ascoli 1970; Stini 1972; Bogin and MacVean 1981).

Changes in hormonal pathways as a result of reduced physical activity have also been proposed for interpreting changes in the skull (both mandible and cranium) and postcranial skeleton across the agricultural transition (Smith, Bar-Yosef, and Sillen 1984; Sardi, Ramirez-Rozzi, and Pucciarelli 2004). Regarding the systematic effects of reduced activity on bone growth, some studies have found a direct association between growth hormone circulation and intensity of physical activity, in which GH circulation increases following moderate to intense exercise, promoting the incremental growth of skeletal and muscle mass (Vogl et al. 1993; Kalu, Banu, and Wang 2000; Weltman et al. 2001). Variation in GH circulation during development is linked to variation in the length of the limbs (Ohlsson et al. 1998), while in adults it is associated with variation in muscle strength and bone mineral density (Johansson et al. 1994). A more sedentary lifestyle and changes in activity type might have reduced GH circulation in farmers, which contributes to lower skeletal mass and accounts for many of the size and strength differences in the postcranial skeleton observed when they are compared with HGs (Pfeiffer and Sealy 2006). Sardi, Ramirez-Rozzi, and Pucciarelli (2004) suggest that the relative reduction in the face and midneural component in Neolithic northwest Africans, as a result of overall skull reduction, could be explained by the differences in subsistence and mobility in comparison with Upper Paleolithic HGs. Smith, Bar-Yosef, and Sillen (1984) suggested that a combination of reduced game hunting and the more complex tool kit, including for example projectile points, which characterized most agriculturalists, could have led to smaller, more gracile bodies. This hypothesis originally was proposed for a Neolithic to Upper Paleolithic comparison in Europe (Frayer 1980), and it has been argued that there is insufficient evidence to support it in the Levantine context (Smith, Bar-Yosef, and Sillen 1984), thus it cannot be invoked as a general mechanism across populations.

Ultimate Causes of Morphological Changes over the Agricultural Transition

The long-term patterns of morphological variation, such as those seen across the agricultural transition, could be interpreted as a result of developmental plasticity in each generation and/or directional selection. Distinguishing between plasticity and selection is not an easy task due to the complexity of their interactions within and between populations (Gilbert and Epel 2009). Some expectations can be drawn, however, to infer the prevalence of one over the other in different circumstances. In this section we explore how the contribution of these processes are interpreted in the literature with regard to morphological changes in populations that shifted from HG to agricultural subsistence.

Developmental Plasticity

Developmental plasticity is a long-term process by which an individual adjusts to environmental conditions during growth and development (West-Eberhard 2003). Plasticity-induced variants have been interpreted as being adaptive or as accelerating the pace of genetic adaptation by providing a source of raw variability upon which natural selection can act to shape subsequent genetic adaptation (Baldwin 1896; Schmalhausen 1949; Waddington 1953; West-Eberhard 2003). Although the long-term effects of plasticity have not yet been fully considered as an alternative ultimate explanation for explaining morphological change, work by Waddington (1953), as recently explained by Fabris (2018, 2019), emphasized the role that developmental plasticity has in evolution.

Through developmental plasticity, human populations can evolve relatively rapidly when confronted with environmental change (Perez et al. 2011). The phenotype may adjust to recent changes or prevailing conditions by reorganizing life history strategies and altering the balance of energy allocation among the areas of growth, maintenance, defense, and reproduction (Wells and Stock 2020; chapter 13, this volume). Most of the morphological changes interpreted as resulting from the transition to agriculture are inferred to be the outcomes of developmental plasticity exclusively. This is especially the case for research on craniofacial data (Larsen 1984; Gonzalez-José et al. 2005; Holmes and Ruff 2011; Perez et al. 2011; Katz, Grote, and Weaver 2017), but also applies to dental and postcranial differences between groups (Kaifu 1997; Macintosh, Pinhasi, and Stock 2016).

These plastic interpretations of morphological change resulting from the adoption of agriculture are supported by a wealth of evidence. Holmes and Ruff (2011) compared ontogenetic changes in mandibular morphology in two human populations with contrasting diets: late prehistoric Inuit HG from Alaska, who are characterized by a very demanding masticatory regime, and proto-historic Arikara farmers from South Dakota. Although they found large differences in mandibular morphology between the Inuit and Arikara adults, those differences were very subtle between the youngest juveniles of both series. These authors suggest that mandibular shape differences between these populations developed gradually during growth due to increasing exposure to dietary loading, suggesting plasticity as the main evolutionary process behind them (*sensu* Lieberman 2011). This argument is reinforced by *in silico* biomechanical modeling analyses of Paleolithic, Mesolithic, and recent human mandibles that show that differences in mandibular size and shape between populations are consistent with plasticity acting on variation in mandibular loading during ontogeny (Stansfield, Evteev, and O'Higgins 2018). Regarding the craniofacial complex, Gonzalez-José and collaborators (2005) compared phenotypic distances for different cranial modules between populations from South America with different subsistence practices. They stated that, since the level of phenotypic differentiation between subsistence groups is lower than the level of intrapopulation variation, the morphological changes observed are not genetically fixed and concluded that plasticity is the most plausible explanation. The mandible also provides evidence for plasticity. Kaifu (1997) argued that the speed of the rapid changes observed in mandibular size reduction accompanying dietary change in Japan shows that developmental plasticity is a better explanation for this trend than selection. The effects of plastic responses to diet in the skull are demonstrated by bioarchaeological studies on dental health, showing greater malocclusion and dental crowding among farmers (Larsen 2006; Lieberman 2011; von Cramon-Taubadel 2011; Katz, Grote, and Weaver 2017). These conditions result from an inadequate coordination between facial and dental growth, since bone responds directly to biomechanical forces, while teeth do not (von Cramon-Taubadel 2017). As a consequence, in a population that has recently changed to a softer diet, the degree of tooth size reduction is considerably less than the degree of reduction of the supporting jaw and facial skeleton, which produces less room for the dentition. Tooth crown reduction is also much greater among permanent teeth than deciduous ones (Hillson and Trinkaus 2002). The difference is due to the fact that the former grow while the individual is already chewing with the deciduous teeth, allowing changes due to epigenetic mechanisms (i.e., non-DNA-based, developmentally induced regulatory forms of phenotypic modulation; Jablonka and Lamb 2010). According to Lieberman (2011), given the lack of membrane separating the developing teeth from the alveolus, expectations from the MFH (Carlson and Van Gerven 1977) and Wolff's law (Wolff 1892, 1986), which were described above, can be also extended to the permanent teeth. Thus, the chewing of highly processed food may have contributed to the size decline observed in permanent teeth during human evolution.

The importance of the role of developmental plasticity in explaining morphological variation between populations with different subsistence practices is further supported by numerous experimental studies on nonhuman samples. This research shows rapid changes in the size and shape of the maxilla, orbital plane, and mandibular ramus and corpus depending on whether animals were fed hard/high masticatory stress food or soft/low masticatory stress food. In addition to the characteristic bony morphology of animals fed on soft diets, they also exhibit dental malocclusions as adults, as seen in humans with arguably less biomechanically demanding diets (Larsen 2006; Lieberman 2011; von Cramon-Taubadel 2011; Katz,

Grote, and Weaver 2017). These results have been repeated in a variety of phylogenetically and morphologically diverse taxa (e.g., Corruccini and Beecher 1982, 1984; Yamada and Kimmel 1991; Ross and Hylander 1996; Ciochon, Nisbett, and Corruccini 1997; Lieberman et al. 2004; Scott et al. 2014).

Finally, an additional source of support for developmental plasticity responding to dietary specializations and biological differences across populations comes from studies on the gut microbiome (Schnorr et al. 2014; Schnorr 2018). Schnorr and collaborators (2014) studied gut microbiome richness in Hazda HGs in comparison with neighboring farmers and urban groups and found that the Hadza display unique features that enhance their ability to digest and extract valuable nutrition from fibrous wild plant foods. Since the gut microbiome is acquired by each individual during ontogeny, its phylogenetic diversity and taxonomic variation will be shaped by the subsistence economy and the lifetime exposure to a certain natural environment (Schnorr 2018). Therefore, both biological differences between HGs and farmers, and the existence of convergent traits between groups with similar diets could be interpreted as a result of their specific lifestyle and ecological niche, particularly in terms of diet.

Directional Selection

In addition, or as an alternative, to developmental plasticity, directional selection has been proposed as an evolutionary process, or ultimate cause, that might explain some of the morphological changes associated with the transition from an HG to an agricultural diet. Selection has been invoked particularly to account for the trend in dental reduction (Greene, Ewing, and Armelagos 1967; O'Connor, Franciscus, and Holton 2005), but there are also some claims for its influence on the postcranial skeleton (e.g., Ginter 2011).

Specific mandibular traits have been interpreted as resulting from selection, independently of the dentition. These include the thick basal cortical bone of the mandible in Japanese HGs (Kanazawa and Kasai 1998) and the large mandibular condyle size in Nubian foragers (Hinton and Carlson 1979). Pinhasi, Eshed, and Shaw (2008) pointed out that since only very specific mandibular traits, such as mandibular ramus breadth and anterior height, are reduced in combination with dental buccolingual dimensions, simple systematic explanations cannot be invoked, and selective pressures acting locally might be the most plausible explanation. Similarly, May and collaborators (2018) claimed that the complex changes in chin height and mandibular angle associated with the adoption of an agricultural diet cannot be explained solely through developmental plasticity and reduced stresses on the masticatory system. Instead, decreases in the size and robusticity of the mandible are argued to be subject to selective pressures, while the width and robusticity of the facial skeleton might be due to masticatory-induced phenotypic plasticity (Galland et al. 2016).

The idea that selection is the main process behind dietary adaptation in the skull is supported by studies showing that some morphological differences consistent with variation in dietary functional demands are manifested very early in development and therefore likely indicate a genetic component (Fukase and Suwa 2008; Gonzalez, Perez, and Bernal 2010; Katz, Grote, and Weaver 2017). Fukase and Suwa (2008) found greater bone mass and significantly thicker cortical bone in mandibles from HGs (Jomon) than recent Japanese people, which they relate to differences in diets. More importantly, even the youngest

Jomon mandibles in this ontogenetic sample exhibited most of the adult characteristics that differ from recent Japanese mandibles. Considering that the more robust morphology of the Jomon is found even in presumably pre-weaning infants, Fukase and Suwa (2008) suggest that there is at least a genetic component to the overall robusticity of the mandible, although they acknowledge the likely existence of complementary plastic remodeling in response to lifetime diet. Similarly, Gonzalez, Perez, and Bernal (2010) analyzed the ontogenetic development of craniofacial robusticity in human populations from South America by performing morphometric comparisons of populations with different degrees of robusticity. They found that the robust cranial traits of Fueguians, which are consistent with the functional requirements of a HG diet, are already established early in life and that differences in the extension of allometric trajectories contributed to the variation in robust traits that is observed in adults.

Teeth do not remodel as bone does, though epigenetic mechanisms may alter developing teeth to some extent (Lieberman 2011); thus they are far less developmentally plastic to environmental stimuli than the rest of the skeleton. This means that consistent association between dental morphology and diet may provide the strongest evidence of directional selection. Tooth reduction in agriculturalists compared with preceding HGs has been interpreted as a result of directional selection in South Asia (Kennedy 1984), Nubia (Martin et al. 1984; Calcagno and Gibson 1988), and Anatolia (Pinhasi, Eshed, and von Cramon-Taubadel 2015). Pinhasi and Meiklejohn (2011) suggest that the most parsimonious mechanism behind dental reduction is directional selection, since drift alone could not explain a significant diachronic reduction process. Y'Edynak and Fleisch (1983) argue that in the absence of the need for large, robust jaws suited to tough forager diets, by the Neolithic selection had led to populations with reduced posterior teeth and as a result, still smaller, developmentally less costly jaws. In a related hypothesis with a slightly different focus, Greene, Ewing, and Armelagos (1967) postulated that selection for less-complex and more caries-resistant teeth resulted in a smaller and morphologically less complex dentition in farmers from Sudan. Carlson and Van Gerven (1977), however, claimed that although Green's hypothesis had merit, it was not the only causal process involved. They explained the diachronic alteration and gracilization of the craniofacial complex in prehistoric Nubians as the result of two independent processes. Since agricultural foodstuffs resulted in an increased prevalence of caries and related pathologies, selective pressures may have acted to reduce overall size and morphological complexity of the dentition. Simultaneously, reduction in the functional demands on the masticatory complex would have also led to alterations in the growth of the maxillomandibular complex in such a way that the face became smaller, less robust, and oriented more inferoposteriorly (Carlson and Van Gerven 1977; see above). This dual-mechanism theory likely reflects the real-world complexity of integrated skeletal changes resulting from the interplay of multiple, interrelated factors.

In terms of postcranial shifts in morphology, Smith, Bar-Yosef, and Sillen (1984) suggested that reduced selective pressures resulting from a combination of reduced hunting and more sophisticated Neolithic technology could have led to smaller size and reduced sexual dimorphism. As noted above, however, in many populations we see increased, rather than decreased, sexual dimorphism after the transition to agriculture (Larsen 1984; Perzigian, Tench, and Braun 1984; Rose et al. 1984; Auerbach 2011; Sardi and Béguelin

2011), which reduces the strength of this argument in that respect at least. Ginter (2011) offers a more nuanced and complex model for the interaction between nutritional insufficiency and body size. She suggests that developmental plasticity in the first instance, but also longer-term directional selection, could have combined to produce the decrease in skeletal size in South African populations seen after the adoption of agriculture. Selection would have resulted from the energetic advantages of being small-bodied in a resource-poor environment (Ginter 2011)—another important reminder that evolutionary mechanisms in nature rarely act independently. Wells and Stock (2020) provide a life-history framework within which these mechanisms may interact. They propose that, as a result of the shift to farming, size reduction is a product of changes in energy allocation toward immune defense and reproduction at the expense of growth and maintenance. In complementary fashion, Gawne and McKenna (chapter 13, this volume) consider such size changes as morphogenetic trade-offs resulting from high-level developmental processes that govern the production of form during ontogeny. These suggestions bring together insights into the subtle and complex ways in which multiple mechanisms lead to morphological change at the individual and population levels.

A final source of support for the assertion that dietary specializations show discernable patterns of genetic divergence between HGs and farmers, and thus likely directional selection, comes from genetic studies on diet-related pathways (Perry et al. 2007; Tishkoff et al. 2007; Fumagalli et al. 2015; Heath et al. 2016; Raj et al. 2019). A pioneering study by Perry and collaborators (2007) provided an example of selection in populations with high-starch (agricultural) diets. They found a strong positive correlation between the copy number of the salivary amylase gene (AMY1) and salivary amylase protein level, which is present at higher levels in individuals from populations with high-starch diets and improves the digestion of starchy foods. Some adaptive responses to different subsistence practices such as lactase persistence in Eurasians and Africans (Tishkoff et al. 2007) and low iron levels in Europeans (Heath et al. 2016), which both emerged with agriculture, are widespread, while others, such as adaptations to fatty marine HG diets in Greenland Inuit (Fumagalli et al. 2015), are more specialized. With direct relevance to crop-based agricultural diets, Raj and collaborators (2019) studied genetic variation in 29 Asian populations to evaluate if the range of domesticated foods available in different regions has created regionally distinct nutrient intake profiles and deficiencies. They found a correlation between genetic variation in diet-related pathways and dietary differences among Asian populations. They conclude that diet-related selection on genes for salivary glands and cellular processes in the pancreas drove the genetic adaptations of Asian populations. This research demonstrates some of the adaptation via directional selection that has taken place in recent humans to aid adaptation to an agricultural diet. Since such physiological dietary adaptations can be shown to be encoded in the genome, this suggests evidence for selection leading to skeletal adaptations will also be found there.

Conclusion

In this chapter we have provided a geographically wide-ranging synthesis of the phenotypic changes in the human skeleton resulting from the transition to agriculture, as interpreted from morphological analyses, by accounting for the plausible proximate and ultimate explanations within an evolutionary biology framework. We find that, while some of the morphological consequences of transferring from a mixed, wild diet to one based mostly on domesticated crops have been similar across the world (e.g., tooth reduction, gracilization of the skull), others are expressed differently and vary locally (e.g., degree of sexual dimorphism, tooth decay) (table 12.1).

Underlying the various morphological trends observed, there is evidence for several different mechanisms at work alone or in combination. Alternate proximate mechanisms could have produced morphological changes in populations undergoing separate dietary shifts and, due to the modular complexity of the skeleton, each of its parts could be influenced by different factors to varying extents. This being the case, given the current evidence available to us, it is not possible to determine without doubt the causes of any specific morphological change associated with the transition to agriculture. Though it is undeniably complicated, however, there are specific morphological trends in which the balance of evidence points toward a majority role for one proximate cause or another.

Perhaps the strongest argument for a subsistence-related biomechanically driven proximate cause is in changes to mandibular morphology. This claim can be supported using a number of sources of evidence: archaeological, functional, comparative, and computer modeling. The closest relationship between diet and global variation in human skeletal morphology is found in the mandible (von Cramon-Taubadel 2011), and when specifically considering changes over the course of the agricultural transition, some mandibular changes such as the lengthening and narrowing (anteroposteriorly and mediolaterally) of the ramus are repeatedly seen in disparate geographic regions such as the Levant (Smith, Bar-Yosef, and Sillen 1984; May et al. 2018; Pokhojaev et al. 2019), Nubia (Galland et al. 2016), and Japan (Kaifu 1997). At least one of these regions, the Levant, is likely a region of biological continuity over the transition (Bar-Yosef 1998). Biomechanical analysis shows that a short, wide ramus is advantageous for a high-strain diet, as it results in a more vertically oriented (providing greater efficiency) and larger temporalis muscle and reduces regionally high levels of strain (Korioth, Romilly, and Hannam 1992; Nicholson and Harvati 2006; Sella-Tunis et al. 2018). Thus, it seems reasonable to suggest optimization of the mandible to reduce masticatory forces resulting from a shift to agriculture was largely responsible for the changes in ramus shape seen in many populations undergoing this shift.

Systematic hormonal effects as a proximate cause are most plausible in the trend for smaller body size in farmers. Some size reduction in skeletal regions not functionally related to subsistence practices is likely the consequence of stunted growth due to insufficient protein and other nutrients. Reducing growth buffers the maintenance of essential bodily functions under conditions of malnutrition (Ginter 2011; Menéndez 2015). As described above, the results of malnutrition on size are well documented from a wide range of human biomedical and comparative experimental studies in other species. The differential effects of this particular physiological process and other factors leading to smaller size on specific populations will have varied depending on local environments and diets, but it may be possible to identify stunting, as it preferentially affects certain regions of the body (Pomeroy et al. 2012). Stunting may also be accompanied by other nonspecific indicators of stress, such as linear enamel hypoplasia (Temple 2008; Vercellotti et al. 2014), and, taken together, these pathologies enable a better understanding of the mechanisms resulting in smaller-sized agricultural populations.

The co-occurrence of the same proximate mechanisms in different individuals from the same population and the persistence of the subsequent morphological changes over many generations could be interpreted as a result of different ultimate mechanisms acting on the human skeleton. Although distinguishing between these mechanisms is again difficult, there are specific morphological trends in which the action of a particular ultimate cause is more likely.

There is a strong case for developmental plasticity as the ultimate mechanism in the gracilization of specific masticatory regions of the skull, such as those where the masticatory muscles attach, as a result of a decrease in the loading of the masticatory apparatus. This gracilization is seen in numerous populations from different geographic regions and is duplicated both in diachronic sequences across the transition to agriculture and in contemporaneous groups with different subsistence strategies. Ontogenetic series showing that population-specific morphology is not established at birth are particularly convincing (Holmes and Ruff 2011). The relationship between bone deposition, remodeling, and masticatory strain has been shown by both comparative studies using primates (Hylander 1979), hyraxes (Lieberman et al. 2004), and rodents (Yamada and Kimmel 1991) and *in silico* simulations of human morphology (Korioth, Romilly, and Hannam 1992; Sella-Tunis et al. 2018); Wolff's law is a persuasive argument for the mechanism behind these functional changes (Wolff 1892, 1986; Ruff, Holt, and Trinkaus 2006).

The lack of remodeling in teeth, particularly those that develop *in utero*, makes them less subject to developmental plasticity than other parts of the skeleton (Hillson and Trinkaus 2002; Lieberman 2011). Dental changes are therefore the best evidence for the ultimate cause of directional selection resulting from dietary change over the transition to agriculture. At least some of the trends for dental reduction likely result from directional selection, possibly due to selective pressures for smaller, less complex teeth to avoid tooth decay (e.g., Pinhasi, Eshed, and Shaw 2008). The strength of the potential selective pressure is evident in the increasing levels of caries across a globally distributed sample of diachronic series (Allison 1984; Buikstra 1984; Dickel, Schulz, and McHenry 1984; Goodman et al. 1984; Kennedy 1984; Martin et al. 1984; Perzigian, Tench, and Braun 1984; Smith, Bar-Yosef, and Sillen 1984; Papathanasiou 2011), and additional factors such as decreasing jaw size (Larsen 2006; Lieberman 2011; von Cramon-Taubadel 2011; Katz, Grote, and Weaver 2017) may have intensified this pressure by leading to malocclusions, which further increase the risk of caries.

Ultimately the interplay between alternate proximal and ultimate causes is complex and no doubt varies between individuals and populations, yet when considered as a whole, the research on the consequences of adopting agriculture reveals repeated phenotypic trends and suggestive evidence of their causal mechanisms, as detailed in this chapter. Our comparative approach, addressing the underlying mechanisms driving phenotypic changes in human populations that adopted agriculture, is relevant when establishing connections with the evolution of agriculture in other taxa. It particularly allows inferences to be drawn on the impact that such relatively recent dietary change has had on phenotype, the differential responses between populations of a species with a global spread, and the possible proximate and ultimate causes driving some of those changes.

We conclude with some suggestions for investigators working in this area and hopes for future research, drawn from our synthesis of the existing literature. An additional source of complexity that could be somewhat mitigated in future work is morphological covariance. Patterns of integration and modularity within the human skeleton affect to what extent adaptation in one region leads to changes in others. These patterns are as yet incompletely understood, which complicates unpicking which morphological changes associated with the transition to agriculture are the direct consequences of dietary change and which are indirectly affected via covariance. By studying functional cranial modules separately, it might be possible to reduce the complexity of the question and untangle the different evolutionary processes influencing each of them (Gonzalez-José et al. 2005).

Throughout this synthesis we have used, where possible, archaeological evidence from populations with probable biological continuity over the transition to agriculture, such as those from the Levant (Bar-Yosef 1998; May et al. 2018) and Ohio (Smith 1989; Paschetta et al. 2010). In many regions of interest, however, such data are not available, and this impedes the understanding of the morphological changes associated with the transition to agriculture by introducing variation due to population history or adaptation to different local environments. Increasingly the combination of archaeological, bioarchaeological, and genetic approaches may clarify population history in regions of interest and enable the identification of further skeletal series with biological continuity over the transition to agriculture. One example of where this has begun is in the exploration of the population turnover accompanying the arrival of Neolithic culture in Britain and other parts of Western Europe (Brace et al. 2019).

To better assess alternative proximate causes associated with phenotypic change, we suggest future researchers comparatively evaluate levels of intrapopulation variation within collections of skeletal remains. To date there are few accounts of these data, yet they can be extremely informative. If plasticity leads to morphological change, we expect considerable variation within a population for each of the studied traits, as a result of the different genetic backgrounds of each individual. In this case, the most adaptive phenotype will be present only among some individuals. On the other hand, if directional selection for particular traits has been involved, then we expect the resultant intrapopulation variation to be more constrained to the phenotypes that present the most adaptive traits, due to greater underlying genetic similarity achieved over generations of selection (Gonzalez-José et al. 2005). Where possible, the study of ontogenetic series is also extremely informative in building a case for the action of plasticity or selection in causing a particular morphological change (e.g., Fukase and Suwa 2008; Gonzalez, Perez, and Bernal 2010; Holmes and Ruff 2011).

Finally, interdisciplinary studies combining morphometric analysis with techniques such as nitrogen and carbon stable isotopic analysis, dental macro- and microwear analysis, cross-sectional geometry, biomechanics, and bone remodeling patterns from the same set of individuals (Hogue and Melsheimer 2008; Menéndez et al. 2014; May et al. 2018; Stansfield, Evteev, and O'Higgins 2018; Brachetta-Aporta et al. 2019), should be conducted to answer some of the remaining questions regarding the impact of this fundamental change in subsistence practices on the phenotype.

Acknowledgments

We would like to thank Ted R. Schultz, Peter Peregrine, and Rick Gawne for organizing the fascinating workshop "The Convergent Evolution of Agriculture in Humans and Insects" at the KLI, which inspired us to write this manuscript, and for inviting us to publish it as part of this volume. Special thanks to Ted and Rick for the comments that allowed us to improve a former version of this manuscript. We would also like to thank Flavia Fabris for her insightful comments and constructive feedback on the theoretical aspects of this chapter. This article is part of the KLI postdoctoral project by LPM.

References

Allison, M. J. 1984. "Paleopathology in Peruvian and Chilean Populations." In *Paleopathology at the Origins of Agriculture*, edited by M. N. Cohen and G. J. Armelagos, 515–530. Orlando, FL: Academic Press.

Angel, J. L. 1984. "Health as a Crucial Factor in the Changes from Hunting to Developed Farming in the Eastern Mediterranean." In *Paleopathology at the Origins of Agriculture*, edited by M. N. Cohen and G. J. Armelagos, 51–74. Orlando, FL: Academic Press.

Armelagos, G. J., and M. N. Cohen. 2013. "Preface to the 2013 Edition." In *Paleopathology at the Origins of Agriculture*, edited by M. N. Cohen and G. J. Armelagos, xvii–xxxvi. Orlando, FL: Academic Press.

Auerbach, B. M. 2011. "Reaching Great Heights: Changes in Indigenous Stature, Body Size and Body Shape with Agricultural Intensification in North America." In *Human Bioarchaeology of the Transition to Agriculture*, edited by R. Pinhasi and J. T. Stock, 203–233. Chichester, UK: John Wiley & Sons.

Baedke, J. 2020. "Mechanisms in Evo-Devo." In *Evolutionary Developmental Biology*, edited by L. Nuño de la Rosa and G. Müller, 1–14. Cham: Springer Nature.

Baldwin, J. M. 1896. "A New Factor in Evolution." *American Naturalist* 30(354): 441–451.

Bar-Yosef, O. 1998. "The Natufian Culture in the Levant, Threshold to the Origins of Agriculture." *Evolutionary Anthropology* 6(5): 159–177.

Bellwood, P. 2005. *First Farmers: The Origins of Agricultural Societies*. Oxford: Blackwell.

Benfer, R. A. 1984. "The Challenges and Rewards of Sedentism: The Preceramic Village of Paloma, Peru." In *Paleopathology at the Origins of Agriculture*, edited by M. N. Cohen and G. J. Armelagos, 531–558. Orlando, FL: Academic Press.

Benfer, R. A. 1990. "The Preceramic Period Site of Paloma, Peru: Bioindications of Improving Adaptation to Sedentism." *Latin American Antiquity* 1(4): 284–318.

Biedermann, P. H. W., and F. E. Vega. 2020. "Ecology and Evolution of Insect-Fungus Mutualisms." *Annual Review of Entomology* 65:431–455.

Bogin, B. 1999. *Patterns of Human Growth*. Cambridge: Cambridge University Press.

Bogin, B., and R. B. MacVean. 1981. "Body Composition and Nutritional Status of Urban Guatemalan Children of High and Low Socioeconomic Class." *American Journal of Physical Anthropology* 55(4): 543–551.

Bonjour, J. P., P. Ammann, T. Chevalley, and R. Rizzoli. 2001. "Protein Intake and Bone Growth." *Canadian Journal of Applied Physiology* 26(S1): S153–S166.

Bouvier, M., and W. L. Hylander. 1982. "The Effect of Dietary Consistency on Morphology of the Mandibular Condylar Cartilage in Young Macaques (*Macaca mulatta*)." *Progress in Clinical and Biological Research* 101:569–579.

Brace, S., Y. Diekmann, T. J. Booth, L. van Dorp, Z. Faltyskova, et al. 2019. "Ancient Genomes Indicate Population Replacement in Early Neolithic Britain." *Nature Ecology & Evolution* 3(5): 765–771.

Brachetta-Aporta, N., Gonzalez, P. N., and V. Bernal. 2019. "Variation in Facial Bone Growth Remodeling in Prehistoric Populations from Southern South America." *American Journal of Physical Anthropology* 169(3): 422–434.

Bradney, M., G. Pearce, G. Naughton, C. Sullivan, S. Bass, et al. 1998. "Moderate Exercise during Growth in Prepubertal Boys: Changes in Bone Mass, Size, Volumetric Density, and Bone Strength: A Controlled Prospective Study." *Journal of Bone and Mineral Research* 13(12): 1814–1821.

Brown R. L. 2020. "Proximate versus Ultimate Causation and Evo-Devo." *In Evolutionary Developmental Biology*, edited by L. Nuño de la Rosa and G. Müller. Cham: Springer. http://doi-org-443.webvpn.fjmu.edu.cn/10.1007/978-3-319-33038-9_97-1.

Buikstra, J. E. 1984. "The Lower Illinois River Region: A Prehistoric Context for the Study of Ancient Diet and Health." In *Paleopathology at the Origins of Agriculture*, edited by M. N. Cohen and G. J. Armelagos, 215–234. Orlando, FL: Academic Press.

Calcagno, J. M. 1986. "Dental Reduction in Post-Pleistocene Nubia." *American Journal of Physical Anthropology 70*(3): 349–363.

Calcagno, J. M., and K. R. Gibson. 1988. "Human Dental Reduction: Natural Selection or the Probable Mutation Effect." *American Journal of Physical Anthropology* 77(4): 505–517.

Carlson, D. S., and D. P. Van Gerven. 1977. "Masticatory Function and Post-Pleistocene Evolution in Nubia." *American Journal of Physical Anthropology* 46(3): 495–506.

Cassidy, C. M. 1984. "Prehistoric Health in the Ohio River Valley." In *Paleopathology at the Origins of Agriculture*, edited by M. N. Cohen, and G. J. Armelagos, 307–346. Orlando, FL: Academic Press.

Cheronet, O., J. A. Finarelli, and R. Pinhasi. 2016. "Morphological Changes in Cranial Shape Following the Transition to Agriculture across Western Eurasia." *Scientific Reports* 6:33316.

Childe, V. G. 1936. *Man Makes Himself.* London: Watts and Co.

Ciochon, R. L., R. A. Nisbett, and R. S. Corruccini. 1997. "Dietary Consistency and Craniofacial Development Related to Masticatory Function in Minipigs." *Journal of Craniofacial Genetics and Developmental Biology* 17(2): 96–102.

Cohen, M. N., and G. J. Armelagos, eds. 1984. *Paleopathology at the Origins of Agriculture*. Orlando, FL: Academic Press.

Cónsole, G. M., E. E. Oyhenart, S. B. Jurado, F. L. Riccillo, H. M. Pucciarelli, et al. 2001. "Effect of Undernutrition on Cranial Components and Somatotroph-Lactotroph Pituitary Populations in the Squirrel Monkey (*Saimiri sciureus boliviensis*)." *Cells Tissues Organs* 168(4): 272–284.

Cook, D. C. 1984. "Subsistence and Health in the Lower Illinois Valley: Osteological Evidence." In *Paleopathology at the Origins of Agriculture*, edited by M. N. Cohen and G. J. Armelagos, 235–270. Orlando, FL: Academic Press.

Corruccini, R. S., and R. M. Beecher. 1982. "Occlusal Variation Related to Soft Diet in a Nonhuman Primate." *Science* 218(4567): 74–76.

Corruccini, R. S., and R. M. Beecher. 1984. "Occlusofacial Morphological Integration Lowered in Baboons Raised on Soft Diet." *Journal of Craniofacial Genetics and Developmental Biology* 4(2): 135–142.

Daly, R. M., L. Saxon, C. H. Turner, A. G. Robling, and S. L. Bass. 2004. "The Relationship between Muscle Size and Bone Geometry during Growth and in Response to Exercise." *Bone* 34(2): 281–287.

Diamond, J. 1987. "The Worst Mistake in the History of the Human Race." *Discover* 8(5): 64–66.

Diamond, J., and P. Bellwood. 2003. "Farmers and Their Languages: The First Expansions." *Science* 300(5619): 597–603.

Dickel, D. N., P. D. Schulz, and H. M. McHenry. 1984. "Central California: Prehistoric Subsistence Changes and Health." In *Paleopathology at the Origins of Agriculture*, edited by M. N. Cohen and G. J. Armelagos, 439–462. Orlando, FL: Academic Press.

Dressino, V., and H. M. Pucciarelli. 1999. "Growth of Functional Cranial Components in *Saimiri sciureus boliviensis* (Cebidae): A Longitudinal Study." *Growth, Development, and Aging* 63(3): 111–127.

Eshed, V., A. Gopher, and I. Hershkovitz. 2006. "Tooth Wear and Dental Pathology at the Advent of Agriculture: New Evidence from the Levant." *American Journal of Physical Anthropology* 130(2): 145–159.

Evteev, A., A. L. Cardini, I. Morozova, and P. O'Higgins. 2014. "Extreme Climate, Rather Than Population History, Explains Mid-facial Morphology of Northern Asians." *American Journal of Physical Anthropology* 153(3): 449-462.

Fabris, F. 2018. "Waddington's Processual Epigenetics and the Debate over Cryptic Variability." In *Everything Flows: Towards a Processual Philosophy of Biology*, edited by D. J. Nicholson and J. Dupré, 246–263. Oxford: Oxford University Press.

Fabris, F. 2019. "Conrad Hal Waddington (1905–1975)." In *Evolutionary Developmental Biology*, edited by L. Nuño de la Rosa and G. Müller, 1–15. Cham: Springer Nature.

Frayer, D. W. 1980. "Sexual Dimorphism and Cultural Evolution in the Late Pleistocene and Holocene of Europe." *Journal of Human Evolution* 9(5): 399–415.

Frisancho, A. R. 2009. "Developmental Adaptation: Where We Go from Here." *American Journal of Human Biology* 21(5): 694–703.

Frisancho, A. R., S. M. Garn, and W. Ascoli. 1970. "Childhood Retardation Resulting in Reduction of Adult Body Size Due to Lesser Adolescent Skeletal Delay." *American Journal of Physical Anthropology* 33(3): 325–336.

Fujita, H., and M. Ogura. 2009. "Degree of Dental Attrition with Sex and Aging among Jomon and Edo people in Japan." *Journal of Oral Biosciences* 51(3): 165–171.

Fukase, H., and G. Suwa. 2008. "Growth-Related Changes in Prehistoric Jomon and Modern Japanese Mandibles with Emphasis on Cortical Bone Distribution." *American Journal of Physical Anthropology* 136(4): 441–454.

Fumagalli, M., I. Moltke, N. Grarup, F. Racimo, P. Bjerregaard, et al. 2015. "Greenlandic Inuit Show Genetic Signatures of Diet and Climate Adaptation." *Science* 349(6254): 1343–1347.

Fusco, G., and A. Minelli. 2010. "Phenotypic Plasticity in Development and Evolution: Facts and Concepts." *Philosophical Transactions of the Royal Society of London, Series B: Biological Sciences* 365:547–556.

Galland, M., D. P. Van Gerven, N. von Cramon-Taubadel, and R. Pinhasi. 2016. "11,000 Years of Craniofacial and Mandibular Variation in Lower Nubia." *Scientific Reports* 6:31040.

Gehlert, S. 1979. "Dental Asymmetry in Two Peruvian Archaeological Samples." Masters dissertation, University of Missouri, Columbia.

Gilbert, S. F. 2001. "Ecological Developmental Biology: Developmental Biology Meets the Real World." *Developmental Biology* 233(1): 1–12.

Gilbert, S. F., and D. Epel. 2009. *Ecological Developmental Biology: Integrating Epigenetics, Medicine, and Evolution.* Sunderland, MA: Sinauer Associates.

Ginter, J. K. 2011. "Using a Bioarchaeological Approach to Explore Subsistence Transitions in the Eastern Cape, South Africa during the Mid- to Late Holocene." In *Human Bioarchaeology of the Transition to Agriculture*, edited by R. Pinhasi and J. T. Stock, 107–152. Chichester, UK: John Wiley & Sons.

Gonzalez, P. N., S. I. Perez, and V. Bernal. 2010. "Ontogeny of Robusticity of Craniofacial Traits in Modern Humans: A Study of South American Populations." *American Journal of Physical Anthropology* 142(3): 367–379.

González-José, R., F. Ramírez Rozzi, M. Sardi, N. Martínez-Abadías, M. Hernández, et al. 2005. "Functional-Cranial Approach to the Influence of Economic Strategy on Skull Morphology." *American Journal of Physical Anthropology* 128(4): 757–771.

Goodman, A. H., J. Lallo, G. J. Armelagos, and J. C. Rose. 1984. "Health Changes at Dickson Mounds, Illinois (AD 950–1300)." In *Paleopathology at the Origins of Agriculture*, edited by M. N. Cohen and G. J. Armelagos, 271–306. Orlando, FL: Academic Press.

Greene, D. L., G. H. Ewing, and G. J. Armelagos. 1967. "Dentition of a Mesolithic Population from Wadi Halfa, Sudan." *American Journal of Physical Anthropology* 27(1): 41–55.

Guadalupe-Grau, A., T. Fuentes, B. Guerra, and J. A. Calbet. 2009. "Exercise and Bone Mass in Adults." *Sports Medicine* 39(6): 439–468.

Hall, B. K. 2012. "Evolutionary Developmental Biology (Evo-Devo): Past, Present, and Future." *Evolution: Education and Outreach* 5:184–193.

Hartnady, P., and J. E. Rose. 1991. "Abnormal Tooth-Loss Patterns among Archaic-Period Inhabitants of the Lower Pecos Region, Texas." In *Advances in Dental Anthropology*, edited by M. A. Kelley and C. S. Larsen, 267–278. New York: Wiley-Liss.

Heath, K. M., Axton, J. H., McCullough, J. M., and N. Harris. 2016. "The Evolutionary Adaptation of the C282 Y Mutation to Culture and Climate during the European Neolithic." *American Journal of Physical Anthropology* 160(1): 86–101.

Hillson, S. 1986. *Teeth.* Cambridge: Cambridge University Press.

Hillson, S. W., and E. Trinkaus. 2002. "Comparative Dental Crown Metrics." In *Portrait of the Artist as a Child. The Gravettian Human Skeleton from the Abrigo do Lagar Velho and Its Archaeological Context*, edited by J. Zilhao and E. Trinkaus, 356–364. Trabalhos de Arqueologia 22. Lisboa: Instituto Portugues de Arqueologia.

Hinton, R. J. 1983. "Relationships between Mandibular Joint Size and Craniofacial Size in Human Groups." *Archives of Oral Biology* 28(1): 37–43.

Hinton, R. J. 1992. "Alterations in Rat Condylar Cartilage Following Discectomy." *Journal of Dental Research* 71(6): 1292–1297.

Hinton, R. J., and D. S. Carlson. 1979. "Temporal Changes in Human Temporomandibular Joint Size and Shape." *American Journal of Physical Anthropology* 50(3): 325–333.

Hogue, S. H., and R. Melsheimer. 2008. "Integrating Dental Microwear and Isotopic Analyses to Understand Dietary Change in East-Central Mississippi." *Journal of Archaeological Science* 35(2): 228–238.

Holmes, M. A., and C. B. Ruff. 2011. "Dietary Effects on Development of the Human Mandibular Corpus." *American Journal of Physical Anthropology* 145(4): 615–628.

Hoover, K. C., and F. L. E. Williams. 2015. "Variation in Regional Diet and Mandibular Morphology in Prehistoric Japanese Hunter-Gatherer-Fishers." *Quaternary International* 30:1–9.

Hudson, M. J., S. Nakagome, and J. B. Whitman. 2020. "The Evolving Japanese: The Dual Structure Hypothesis at 30." *Evolutionary Human Sciences* 2:1–13.

Hylander, W. L. 1979. "Mandibular Function in *Galago crassicaudatus* and *Macaca fascicularis*: An *in Vivo* Approach to Stress Analysis of the Mandible." *Journal of Morphology* 159:253–296.

Hylander, W. L., K. R. Johnson, and P. G. Picq. 1991. "Masticatory Stress Hypotheses and the Supraorbital Region of Primates." *American Journal of Physical Anthropology* 86(1): 1–36.

Jablonka, E., and M. J. Lamb. 2010. "Transgenerational Epigenetic Inheritance." In *Evolution: The Extended Synthesis*, edited by M. Pigliucci and G. B. Mueller, 137–174. Cambridge, MA: MIT Press.

Johansson, A. G., A. Forslund, L. Hambraeus, W. F. Blum, and S. Ljunghall. 1994. "Growth Hormone-Dependent Insulin-Like Growth Factor Binding Protein Is a Major Determinant of Bone Mineral Density in Healthy Men." *Journal of Bone and Mineral Research* 9(6): 915–921.

Jurmain, R. D. 1977. "Stress and the Etiology of Osteoarthritis." *American Journal of Physical Anthropology* 46(2): 353–365.

Kaifu, Y. 1997. "Changes in Mandibular Morphology from the Jomon to Modern Periods in Eastern Japan." *American Journal of Physical Anthropology* 104(2): 227–243.

Kalu, D. N., J. Banu, and L. Wang. 2000. "How Cancellous and Cortical Bones Adapt to Loading and Growth Hormone." *Journal of Musculoskeletal and Neuronal Interactions* 1(1): 19–23.

Kanazawa, E., and K. Kasai. 1998. "Comparative Study of Vertical Sections of the Jomon and Modern Japanese Mandibles." *Anthropological Science* 106:107–118.

Kasai, K., and A. Kawamura. 2001. "Correlation between Buccolingual Inclination and Wear of Mandibular Teeth in Ancient and Modern Japanese." *Archives of Oral Biology* 46(3): 269–273.

Katz, D. C., M. N. Grote, and T. D. Weaver. 2017. "Changes in Human Skull Morphology across the Agricultural Transition Are Consistent with Softer Diets in Preindustrial Farming Groups." *Proceedings of the National Academy of Sciences* 114(34): 9050–9055.

Kennedy, K. A. R. 1984. "Growth, Nutrition, and Paleopathology in Changing Paleodemographic Settings in South Asia." In *Paleopathology at the Origins of Agriculture*, edited by M. N. Cohen and G. J. Armelagos, 169–192. Orlando, FL: Academic Press.

Korioth, T. W. P., D. P. Romilly, and A. G. Hannam. 1992. "Three-Dimensional Finite Element Stress Analysis of the Dentate Human Mandible." *American Journal of Physical Anthropology* 88:69–96.

Laland, K. N., J. Odling-Smee, W. Hoppitt, and T. Uller. 2013. "More on How and Why: Cause and Effect in Biology Revisited." *Biology & Philosophy* 28(5): 719–745.

Laland, K. N., K. Sterelny, J. Odling-Smee, W. Hoppitt, and T. Uller. 2011. "Cause and Effect in Biology Revisited: Is Mayr's Proximate-Ultimate Dichotomy Still Useful?" *Science* 334(6062): 1512–1516.

Larsen, C. S. 1981. "Skeletal and Dental Adaptations to the Shift to Agriculture on the Georgia Coast." *Current Anthropology* 22(4): 422–423.

Larsen, C. S. 1984. "Health and Disease in Prehistoric Georgia: The Transition to Agriculture." In *Paleopathology at the Origins of Agriculture*, edited by M. N. Cohen and G. J. Armelagos, 367–392. Orlando, FL: Academic Press.

Larsen, C. S. 1995. "Biological Changes in Human Populations with Agriculture." *Annual Review of Anthropology* 24:185–213.

Larsen, C.S. 1997. *Bioarchaeology: Interpreting Behaviour from the Human Skeleton*. Cambridge: Cambridge University Press.

Larsen, C. S. 2006. "The Agricultural Revolution as Environmental Catastrophe: Implications for Health and Lifestyle in the Holocene." *Quaternary International* 150(1): 12–20.

Larsen, C. S., and C. B. Ruff. 1991. "Biomechanical Adaptation and Behavior on the Prehistoric Georgia Coast." In *What Mean These Bones? Studies in Southeastern Bioarchaeology*, edited by M. L. Powell, P. S. Bridges, and A. M. Wagner, 102–113. Tuscaloosa: University of Alabama Press.

Larsen, C. S., and C. B. Ruff. 2011. "An External Agency of Considerable Importance: The Stresses of Agriculture in the Foraging-to-Farming Transition in Eastern North America." In *Human Bioarchaeology of the Transition to Agriculture*, edited by R. Pinhasi and J. T. Stock, 293–316. Chichester, UK: John Wiley & Sons.

Li, H., Q. Zhang, and H. Zhu. 2012. "The Size Variation and Related Implications of Mandibles in Northern China in the Past 7000 Years." *Chinese Science Bulletin* 57(4): 387–394.

Lieberman, D. E. 2011. *The Evolution of the Human Head*. Cambridge, MA: Harvard University Press.

Lieberman, D. E., G. E. Krovitz, F. W. Yates, M. Devlin, and M. St Claire. 2004. "Effects of Food Processing on Masticatory Strain and Craniofacial Growth in a Retrognathic Face." *Journal of Human Evolution* 46:655–677.

Macintosh, A. A., R. Pinhasi, and J. T. Stock. 2016. "Early Life Conditions and Physiological Stress Following the Transition to Farming in Central/Southeast Europe: Skeletal Growth Impairment and 6000 Years of Gradual Recovery." *PloS ONE* 11(2): e0148468.

Marquez-Morfin, L., and R. Storey. 2007. "From Early Village to Regional Center in Mesoamerica: An Investigation of Lifestyles and Health." In *Ancient Health: Skeletal Indicators of Agricultural and Economic Intensification*, edited by M. N. Cohen and G. Crane-Kramer, 80–91. Gainesville: University of Florida Press.

Martin, D. L., and G. J. Armelagos. 1985. Skeletal Remodeling and Mineralization as Indicators of Health: An Example from Prehistoric Sudanese Nubia. *Journal of Human Evolution 14*(5): 527–537.

Martin, D. L., G. J. Armelagos, A. H. Goodman, and D. P. Van Gerven. 1984. "The Effects of Socioeconomic Change in Prehistoric Africa: Sudanese Nubia as a Case Study." In *Paleopathology at the Origins of Agriculture*, edited by M. N. Cohen and G. J. Armelagos, 193–214. Orlando, FL: Academic Press.

May, H., T. Sella-Tunis, A. Pokhojaev, N. Peled, and R. Sarig. 2018. "Changes in Mandible Characteristics during the Terminal Pleistocene to Holocene Levant and Their Association with Dietary Habits." *Journal of Archaeological Science: Reports* 22:413–419.

Mayr, E. 1961. "Cause and Effect in Biology." *Science* 134(3489): 1501–1506.

Menéndez, L. P. 2015. *Diversificación Morfológica Craneofacial y Diversidad en la Dieta. El Caso de la Región Centro-Oeste de Argentina durante el Holoceno tardío.* BAR-S2743. Oxford, UK: Archaeopress.

Menéndez, L. P., V. Bernal, P. Novellino, and S. I. Perez. 2014. "Effect of Bite Force and Diet Composition on Craniofacial Diversification of Southern South American Human Population." *American Journal of Physical Anthropology* 155(1): 114–127.

Miller, J. P., and R. Z. German. 1999. "Protein Malnutrition Affects the Growth Trajectories of the Craniofacial Skeleton in Rats." *Journal of Nutrition* 129(11): 2061–2069.

Nijhout, H. F. 1990. "Problems and Paradigms: Metaphors and the Role of Genes in Development." *BioEssays* 12(9): 441–446.

Nijhout, H. F. 2003. "The Control of Growth." *Development* 130:5863–5867.

Nicholson, E., and K. Harvati. 2006. "Quantitative Analysis of Human Mandibular Shape Using Three-Dimensional Geometric Morphometrics." *American Journal of Physical Anthropology* 131(3): 368–383.

Noback, M. L., and K. Harvati. 2015. "The Contribution of Subsistence to Global Human Cranial Variation." *Journal of Human Evolution* 80:34–50.

O'Connor, C. F., R. G. Franciscus, and N. E. Holton. 2005. "Bite Force Production Capability and Efficiency in Neandertals and Modern Humans." *American Journal of Physical Anthropology* 127(2): 129–151.

Ogden, A. R. 2008. "Advances in the Paleopathology of Teeth and Jaws." In *Advance in Human Paleopathology*, edited by S. Mays and R. Pinhasi, 283–307. Chichester, UK: John Wiley & Sons.

Ohlsson, C., B. A. Bengtsson, O. G. Isaksson, T. T. Andreassen, and M. C. Slootweg. 1998. "Growth Hormone and Bone." *Endocrine Reviews* 19(1): 55–79.

Paaby, A. B., and N. D. Testa. 2018. "Developmental Plasticity and Evolution." In *Evolutionary Developmental Biology*, edited by L. Nuño de la Rosa and G. Müller. Cham: Springer Nature. https://doi.org/10.1007/978-3-319-33038-9_110-1.

Papathanasiou, A. 2011. "Health, Diet and Social Implications in Neolithic Greece from the Study of Human Osteological Material." In *Human Bioarchaeology of the Transition to Agriculture*, edited by R. Pinhasi and J. T. Stock, 87–106. Chichester, UK: John Wiley & Sons.

Paschetta, C., S. de Azevedo, L. Castillo, N. Martinez-Abadias, M. Hernandez, et al. 2010. "The Influence of Masticatory Loading on Craniofacial Morphology: A Test Case across Technological Transitions in the Ohio Valley." *American Journal of Physical Anthropology* 141(2): 297–314.

Perez, S. I., V. Lema, J. A. F. Diniz-Filho, V. Bernal, P. N. Gonzalez, et al. 2011. "The Role of Diet and Temperature in Shaping Cranial Diversification of South American Human Populations: An Approach Based on Spatial Regression and Divergence Rate Tests." *Journal of Biogeography* 38(1): 148–163.

Perry, G. H., N. J. Dominy, K. G. Claw, A. S. Lee, H. Fiegler, et al. 2007. "Diet and the Evolution of Human Amylase Gene Copy Number Variation." *Nature Genetics* 39(10): 1256–1260.

Perzigian, A. J., P. A. Tench, and D. J. Braun. 1984. "Prehistoric Health in the Ohio River Valley." In *Paleopathology at the Origins of Agriculture*, edited by M. N. Cohen and G. J. Armelagos, 347–366. Orlando, FL: Academic Press.

Pfeiffer, S., and J. Sealy. 2006. "Body Size among Holocene Foragers of the Cape Ecozone, Southern Africa." *American Journal of Physical Anthropology* 129(1): 1–11.

Pigliucci, M. 2001. *Phenotypic Plasticity: Beyond Nature and Nurture*. Baltimore, MD: Johns Hopkins University Press.

Pinhasi, R., V. Eshed, and P. Shaw. 2008. "Evolutionary Changes in the Masticatory Complex Following the Transition to Farming in the Southern Levant." *American Journal of Physical Anthropology* 135(2): 136–148.

Pinhasi, R., V. Eshed, and N. von Cramon-Taubadel. 2015. "Incongruity between Affinity Patterns Based on Mandibular and Lower Dental Dimensions Following the Transition to Agriculture in the Near East, Anatolia and Europe." *PLoS ONE* 10(2): e0117301.

Pinhasi, R., and C. Meiklejohn. 2011. "Dental Reduction and the Transition to Agriculture in Europe." In *Human Bioarchaeology of the Transition to Agriculture*, edited by R. Pinhasi and J. T. Stock, 451–474. Chichester, UK: John Wiley & Sons.

Pinhasi, R., and M. Pluciennik. 2004. "A Regional Biological Approach to the Spread of Farming in Europe: Anatolia, the Levant, South Eastern Europe, and the Mediterranean." *Current Anthropology* 45(S4): S59–S82.

Pinhasi, R., S. Stefanovic, A. Papathanasiou, and J. T. Stock. 2011. "Variability in Long Bone Growth and Limb Proportions within and amongst Mesolithic and Neolithic Populations from Southeast Europe." In *Human Bioarchaeology of the Transition to Agriculture*, edited by R. Pinhasi and J. T. Stock, 177–202. Chichester, UK: John Wiley & Sons.

Pinhasi, R., and J. T. Stock, eds. 2011. *Human Bioarchaeology of the Transition to Agriculture*. Chichester, UK: John Wiley & Sons.

Pokhojaev, A., H. Avni, T. Sella-Tunis, R. Sarig, and H. May. 2019. "Changes in Human Mandibular Shape during the Terminal Pleistocene-Holocene Levant." *Scientific Reports* 9:8799.

Pomeroy, E., J. T. Stock, S. Stanojevic, J. J. Miranda, T. J. Cole, et al. 2012. "Trade-Offs in Relative Limb Length among Peruvian Children: Extending the Thrifty Phenotype Hypothesis to Limb Proportions." *PLoS ONE* 7(12): e51795.

Price, T. D., and O. Bar-Yosef. 2011. "The Origins of Agriculture: New Data, New Ideas: An Introduction to Supplement 4." *Current Anthropology* 52(S4): S163–S174.

Pucciarelli, H. M. 1980. "The Effects of Race, Sex, and Nutrition on Craniofacial Differentiation in Rats. A Multivariate Analysis." *American Journal of Physical Anthropology* 53(3): 359–368.

Pucciarelli, H. M. 1981. "Growth of the Functional Components of the Rat Skull and Its Alteration by Nutritional Effects. A Multivariate Analysis." *American Journal of Physical Anthropology* 56(1): 33–41.

Pucciarelli, H. M. 2008. *Evolución y diversificación biológica humana desde la perspectiva craneofuncional*. Mexico City: Universidad Autónoma de México.

Raj, S. M., A. Pei, M. Foll, F. Schlamp, L. Excoffier, et al. 2019. "Reconstruction of Nine Thousand Years of Agriculture-Based Diet and Impact on Human Genetic Diversity in Asia." *bioRxiv*. https://doi.org/10.1101/747709.

Rathbun, T. A. 1984. "Skeletal Pathology from the Paleolithic through the Metal Ages in Iran and Iraq." In *Paleopathology at the Origins of Agriculture*, edited by M. N. Cohen and G. J. Armelagos, 137–168. Orlando, FL: Academic Press.

Ravosa, M. J., V. E. Noble, W. L. Hylander, K. R. Johnson, and E. M. Kowalski. 2000. "Masticatory Stress, Orbital Orientation and the Evolution of the Primate Postorbital Bar." *Journal of Human Evolution* 38(5): 667–693.

Rose, J. C., B. A. Burnett, M. W. Blaeuer, and M. S. Nassaney. 1984. "Paleopathology and the Origins of Maize Agriculture in the Lower Mississippi Valley and Caddoan Culture Areas." In *Paleopathology at the Origins of Agriculture*, edited by M. N. Cohen and G. J. Armelagos, 393–424. Orlando, FL: Academic Press.

Ross, C. F., and W. L. Hylander. 1996. "In Vivo and in Vitro Bone Strain in the Owl Monkey Circumorbital Region and the Function of the Postorbital Septum." *American Journal of Physical Anthropology* 101(2): 183–215.

Ruff, C., B. Holt, and E. Trinkaus. 2006. "Who's Afraid of the Big Bad Wolff?: "Wolff's Law" and Bone Functional Adaptation." *American Journal of Physical Anthropology* 129(4): 484–498.

Ruff, C. B., C. S. Larsen, and W. C. Hayes. 1984. "Structural Changes in the Femur with the Transition to Agriculture on the Georgia Coast." *American Journal of Physical Anthropology* 64(2): 125–136.

Ruff, C. B., A. Walker, and E. Trinkhaus. 1994. "Postcranial Robusticity in Homo. III. Ontogeny." *American Journal of Physical Anthropology* 95(1): 35–54.

Sardi, M. L. 2017. "El método craneofuncional." In *De cómo cruzar fronteras en la ciencia. Homenaje a Héctor Pucciarelli*, edited by C. Paschetta, R. González-José, and J. L. Lanata, 120–135. Puerto Madryn, Argentina: CONICET.

Sardi, M. L., and M. Béguelin. 2011. "Skeletal Differentiation at the Southernmost Frontier of Andean Agriculture." In *Human Bioarchaeology of the Transition to Agriculture*, edited by edited by R. Pinhasi and J. T. Stock, 429–450. Chichester, UK: John Wiley & Sons.

Sardi, M. L., P. S. Novellino, and H. M. Pucciarelli. 2006. "Craniofacial Morphology in the Argentine Center-West: Consequences of the Transition to Food Production." *American Journal of Physical Anthropology* 130(3): 333–343.

Sardi, M. L., F. Ramirez-Rozzi, and H. M. Pucciarelli. 2004. "The Neolithic Transition in Europe and North Africa: The Functional Craneology Contribution." *Anthropologischer Anzeiger* 62(2): 129–145.

Schmalhausen, I. I. 1949. *Factors of Evolution: The Theory of Stabilizing Selection*. Philadelphia: Blakiston Company.

Schnorr, S. L. 2018. "Meanings, Measurements, and Musings on the Significance of Patterns in Human Microbiome Variation." *Current Opinion in Genetics & Development* 53:43–52.

Schnorr, S. L., M. Candela, S. Rampelli, M. Centanni, C. Consolandi, et al. 2014. "Gut Microbiome of the Hadza Hunter-Gatherers." *Nature Communications* 5:3654.

Schultz, T. R., U. G. Mueller, C. R. Currie, and S. A. Rehner. 2005. "Reciprocal Illumination a Comparison of Agriculture in Humans." In *Insect–Fungal Associations: Ecology and Evolution*, edited by F. Vega and M. Blackwell, 149–190. New York: Oxford University Press.

Scott, J. E., McAbee, K. R., Eastman, M. M., and M. J. Ravosa. 2014. "Experimental Perspective on Fallback Foods and Dietary Adaptations in Early Hominins." *Biology Letters* 10(1): 20130789.

Sella-Tunis, T., A. Pokhojaev, R. Sarig, P. O'Higgins, and H. May. 2018. "Human Mandibular Shape Is Associated with Masticatory Muscle Force." *Scientific Reports* 8(1): 1–10.

Shaw, C.N., and J. T. Stock. 2009. "Habitual Throwing and Swimming Correspond with Upper Limb Diaphyseal Strength and Shape in Modern Human Athletes." *American Journal of Physical Anthropology* 140(1): 160–172.

Smith, B. D. 1989. "Origins of Agriculture in Eastern North America." *Science* 246:1566–1571.

Smith, P., O. Bar-Yosef, and A. Sillen. 1984. "Archaeological and Skeletal Evidence for Dietary Change during the Late Pleistocene/Early Holocene in the Levant." In *Paleopathology at the Origins of Agriculture*, edited by M. N. Cohen and G. J. Armelagos, 101–136. Orlando, FL: Academic Press.

Stansfield, E., A. Evteev, and P. O'Higgins. 2018. "Can Diet Be Inferred from the Biomechanical Response to Simulated Biting in Modern and Pre-Historic Human Mandibles?" *Journal of Archaeological Science: Reports* 22:433–443.

Steckel, R. H., and J. C. Rose. 2002. *The Backbone of History: Health and Nutrition in the Western Hemisphere*. Cambridge: Cambridge University Press.

Stini, W. A. 1971. "Evolutionary Implications of Changing Nutritional Patterns in Human Populations." *American Anthropologist* 73(5): 1019–1030.

Stini, W. A. 1972. "Reduced Sexual Dimorphism in Upper Arm Muscle Circumference Associated with Protein-Deficient Diet in a South American Population." *American Journal of Physical Anthropology* 36(3): 341–351.

Stini, W. A. 1975. "Adaptive Strategies of Human Populations under Nutritional Stress." In: *Biosocial Interrelations in Population Adaptation*, edited by E. S. Watts, F. E. Johnston, and G. W. Lasker, 19–42. The Hague: Mouton & Co.

Stock, J., and S. Pfeiffer. 2001. "Linking Structural Variability in Long Bone Diaphyses to Habitual Behaviors: Foragers from the Southern African Later Stone Age and the Andaman Islands." *American Journal of Physical Anthropology* 115(4): 337–348.

Stock, J. T., and R. Pinhasi. 2011. "Introduction: Changing Paradigms in Our Understanding of the Transition to Agriculture: Human Bioarchaeology, Behaviour and Adaptation." In *Human Bioarchaeology of the Transition to Agriculture*, edited by R. Pnihasi and J. T. Stock, 1–13. Chichester, UK: John Wiley & Sons.

Stock, J. T., M. C. O'Neill, C. B. Ruff, M. Zabecki, L. Shackelford, et al. 2011. "Body Size, Skeletal Biomechanics, Mobility and Habitual Activity from the Late Paleolithic to the Mid-Dynastic Nile Valley." In *Human Bioarchaeology of the Transition to Agriculture*, edited by R. Pinhasi and J. T. Stock, 347–370. John Wiley & Sons.

Suominen, H. 2012. "Improving Athletic Performance in Older Adults." *Journal of Aging and Physical Activity* 20:25–26.

Taaffe, D. R., T. L. Robinson, C. M. Snow, and R. Marcus. 1997. "High-Impact Exercise Promotes Bone Gain in Well-Trained Female Athletes." *Journal of Bone and Mineral Research* 12(2): 255–260.

Temple, D. H. 2008. "What Can Variation in Stature Reveal about Environmental Differences between Prehistoric Jomon Foragers? Understanding the Impact of Systemic Stress on Developmental Stability." *American Journal of Human Biology* 20(4): 431–439.

Temple, D. H. 2011. "Evolution of Postcranial Morphology during the Agricultural Transition in Prehistoric Japan." In *Human Bioarchaeology of the Transition to Agriculture*, edited by R. Pinhasi and J. T. Stock, 235–264. Chichester, UK: John Wiley & Sons.

Temple, D. H., and C. S. Larsen. 2007. "Dental Caries Prevalence as Evidence for Agriculture and Subsistence Variation during the Yayoi Period in Prehistoric Japan: Biocultural Interpretations of an Economy in Transition." *American Journal of Physical Anthropology* 134(4): 501–512.

Tishkoff, S. A., F. A. Reed, A. Ranciaro, B. F. Voight, C. C. Babbitt, et al. (2007). "Convergent Adaptation of Human Lactase Persistence in Africa and Europe." *Nature Genetics* 39(1): 31–40.

Tuominen, M., T. Kantomaa, and P. Pirttiniemi. 1993. "Effect of Food Consistency on the Shape of the Articular Eminence and the Mandible: An Experimental Study on the Rabbit." *Acta Odontologica Scandinavica* 51(2): 65–72.

Ulijaszek, S. J. 1991. "Human Dietary Change." *Philosophical Transactions of the Royal Society of London, Series B: Biological Sciences* 334(1270): 271–279.

Ungar, P. S. 2006. *Evolution of the Human Diet: The Known, the Unknown, and the Unknowable*. Oxford: Oxford University Press.

Vercellotti, G., B. A. Piperata, A. M. Agnew, W. M. Wilson, D. L. Dufour, et al. 2014. "Exploring the Multidimensionality of Stature Variation in the Past through Comparisons of Archaeological and Living Populations." *American Journal of Physical Anthropology* 155(2): 229–242.

Vogl, C., W. R. Atchley, D. R. Cowley, P. Crenshaw, J. Murray, et al. 1993. "The Epigenetic Influence of Growth Hormone on Skeletal Development." *Growth Development and Aging* 57:163–182.

Von Cramon-Taubadel, N. 2011. "Global Human Mandibular Variation Reflects Differences in Agricultural and Hunter-Gatherer Subsistence Strategies." *Proceedings of the National Academy of Sciences* 108:19546–19551.

Von Cramon-Taubadel, N. 2014. "Evolutionary Insights into Gobal Patterns of Human Cranial Diversity: Population History, Climatic and Dietary Effects." *Journal of Anthropological Sciences* 92(4): 43–77.

Von Cramon-Taubadel, N. 2017. "Measuring the Effects of Farming on Human Skull Morphology." *Proceedings of the National Academy of Sciences* 114(34): 8917–8919.

Waddington, C. H. 1953. "Genetic Assimilation of an Acquired Character." *Evolution* 7(2): 118–126.

Watson, J. T. 2008. "Changes in Food Processing and Occlusal Dental Wear during the Early Agricultural Period in Northwest Mexico." *American Journal of Physical Anthropology* 135(1): 92–99.

Wells, J. C., and J. T. Stock. 2020. "Life History Transitions at the Origins of Agriculture: A Model for Understanding How Niche Construction Impacts Human Growth, Demography and Health." *Frontiers in Endocrinology* 11:325.

Weltman, A., J. Y. Weltman, J. D. Veldhuis, and M. L. Hartman. 2001. "Body Composition, Physical Exercise, Growth Hormone and Obesity." *Eating and Weight Disorders* 6(3): 28–37.

West-Eberhard, M. J. 2003. *Developmental Plasticity and Evolution*. New York: Oxford University Press.

Wolff, J. 1892. *Das Gesetz der Transformation der Knochen*. Berlin: Hirshwald.

Wolff, J. 1986. "Concept of the Law of Bone Remodelling." In *The Law of Bone Remodelling*. Berlin: Springer.

Yamada, K., and D. B. Kimmel. 1991. "The Effect of Dietary Consistency on Bone Mass and Turnover in the Growing Rat Mandible." *Archives of Oral Biology* 36(2): 129–138.

Y'Edynak, G. 1989. "Yugoslav Mesolithic Dental Reduction." *American Journal of Physical Anthropology* 78(1): 17–36.

Y'Edynak, G., and S. Fleisch. 1983. "Microevolution and Biological Adaptability in the Transition from Food-Collecting to Food-Producing in the Iron Gates of Yugoslavia." *Journal of Human Evolution* 12(3): 279–296.

13 Hammond's Law: A Mechanism Governing the Development and Evolution of Form in Domesticated Organisms

Richard Gawne and Kenneth Z. McKenna

The entire process of the development of living forms is guided by [the Malthusian] principle. The struggle for existence . . . for food and propagation, takes place at all the stages of life between all orders of living units from . . . the elements that are accessible to direct observation, to the cells, and still higher up, to individuals and colonies.
—August Weismann (1896)

I call the balancing between the volume of organic masses, and by shorthand the balancing of organs, [the fourth] law of living nature, by virtue of which a normal or pathological organ never acquires extraordinary prosperity in such a way that another of its system or its relations does not suffer as a result.
—Geoffroy Saint-Hilaire (1822)

As other chapters in this volume have indicated (chapters 1, 2, and 14), agriculture appears to constitute a striking example of evolutionary convergence. Several distantly related lineages of animals, including hominids (Smith 1998; Rindos 1984), insects (Mueller et al. 2005), and snails (Silliman and Newell 2003), have evolved a life-history strategy that involves the farming of other organisms for food. Although the agricultural lifestyle is not universally fitness-enhancing (Mueller et al. 2005; Mummert et al. 2011), there are many contexts in which it has significant advantages over hunting and gathering. As a consequence, it is hardly surprising that a number of different animals have become farmers. Somewhat more counterintuitive is the fact that entering into an agricultural arrangement can be beneficial to the organism being tended (chapter 5, this volume). Over large time scales, plants, animals, and fungi that become domesticates can outnumber their wild ancestors by orders of magnitude (Razifard et al. 2020). However, entering into an agricultural relationship is certainly not a risk-free life-history strategy. Artificial selection often produces "improvements" in crops and livestock that reduce their fitness in natural habitats, thereby making them increasingly dependent upon the services provided by their farmers.

The specific modifications exhibited by a domesticated lineage are likely determined by a complex array of factors that include internal developmental constraints (Alberch 1982; Smith 1998), local environmental conditions (Hammond 1936), and the precise artificially applied selective pressures it experiences (Gregory 2009). However, looking across species, it is possible to identify general patterns of morphological change that seem to recur over and over again, even in widely diverged organisms (Wilkins, Wrangham, and Fitch 2014; Zeder 2006, 2015). Among other things, there is a tendency for domesticates to exhibit extreme enlargements of body parts that are deemed valuable to their farmers (Smith 1998; Fuller et al. 2014). These traits are ecologically and evolutionarily noteworthy because they have the potential to make a domesticate's relationship with its farmer obligate, but their importance does not end there. The highly derived morphologies of domesticates are also of significant interest from a developmental point of view. If a particular body part or the overall form of a domesticated organism differs in some way from that of its known wild ancestor, this gives us two populations with quantifiable morphological variation and immediately invites the question of what developmental mechanisms are responsible for producing the observed differences.

Numerous approaches have been used to address questions of this type. Some researchers have looked for genomic signatures of selective sweeps (Tian, Stevens, and Buckler 2009), while others have focused on the enrichment of proposed candidate genes (Domyan et al. 2016) or attempted to locate hormonal differences between the populations (Haase and Donham 1980), and so on. These studies have undeniably provided important insight into the processes that are responsible for generating domesticated morphologies. Yet, they are limited in the sense that the developmental differences identified between the ancestral and domesticated populations are almost always necessary, rather than sufficient, for generating the observed variation in form (Domyan et al. 2016; Domyan and Shapiro 2017). Jointly, these studies tend to yield an ever-increasing list of causally relevant factors that play a role in shaping the morphologies of the focal character(s), but provide little understanding of their relative significance, and no obvious insight into how they interact.

This emphasis on list-making and lineage-specific pathway minutiae rather than the "big picture" is increasingly being recognized as a significant flaw in the way that developmental biology is practiced (Gawne, McKenna, and Nijhout 2018; Nicholson and Dupré 2018; Bizzarri et al. 2019; DiFrisco and Jaeger 2019, 2020). Contrast the state of affairs in development to those of closely related fields such as ecology and evolution. In the latter, nearly all aspects of research are driven by generalized theoretical hypotheses that apply across both taxa and time. Even when a particular study is narrowly focused on a specific organism, the results are often used to test and refine large-scale theories that matter to the larger community. This back-and-forth dialogue between theory and data is largely absent in developmental biology, in part because the conceptual foundations of the field remain in a somewhat primitive state (Woodger 1929; Minelli and Pradeu 2014). Development clearly lacks a universally applicable theory that is analogous in scope to Darwinian evolution, but to make matters worse, it also has remarkably few context-specific hypotheses that would be similar in breadth to, for example, inclusive fitness theory (West, Griffin, and Gardner 2007), Batesian mimicry (Ruxton et al. 2019), the neutral theory of ecology (Hubbell 2001), or sexual selection (Andersson 1994). This translates to an open opportunity for future research dedicated to the construction of developmental theories.

It is well known that early evolutionary biologists were able to formulate sweeping generalizations about the nature of natural selection by studying how artificial selection alters the morphology of domesticates (Evans 1984; Gregory 2009). There is nothing stopping developmental biologists from taking a page from this playbook and using the same basic dataset to identify the ontogenetic principles that help to facilitate these phenotypic changes. Indeed, that is precisely what this chapter aims to do. In the sections that follow, we reintroduce and refine a developmental theory formulated by the British agricultural scientist John Hammond (1889–1964), which predicts how, when, and why an organism's form can be quantitatively modified to suit the needs of farmers. We refer to this hypothesis as "Hammond's law" and show that it (1) has the potential to explain how past selection on physiological mechanisms helped to produce the exaggerated morphologies observed in certain domesticates and (2) provides insight into the conditions under which the phenotypes of tended plants, animals, and fungi can be further altered to increase their productivity. In short, Hammond's law provides a generalized account of the processes that facilitate and constrain an organism's ability to move through morphospace in response to artificial selection. More importantly, perhaps, we predict that the theory is equally applicable to wild organisms that are subjected to naturally occurring selective pressures, making it a rare instance of a truly broad-brush developmental hypothesis.

Developmental Constraints on Organismal Form

The domesticated organisms that fill our fields and refrigerators often bear only a vague resemblance to their wild ancestors. Many fruits and vegetables, for example, have undergone monstrous changes of proportion (figure 13.1) and, in some cases, these alterations have taken place in a remarkably short amount of time (Meyer, DuVal, and Jensen 2012). This is shown from scientific data, including findings from the archaeological record (Zeder 2006; Zeder et al. 2006; chapter 10, this volume), as well as human cultural artifacts such as paintings (Zoltán et al. 2007) and descriptions found in ancient texts (Andrews 1949). Dramatic phenotypic changes have also been produced in pet animals (Drake and Klingenberg 2010; Komiyama et al. 2009) (figure 13.2) and prey species such as cattle (Bovine HapMap Consortium 2009) and sheep (Hammond 1932). Surveying the vast array of quantitative morphological change that has been induced in these organisms, it is easy to start thinking of domesticates as being almost infinitely labile. If the need arises, and we initiate the right artificial selection regime, it seems that the phenotypes of these plants and animals can effectively be altered at will.

There is a long history of conceiving of domesticated organisms in this way. Darwin commented on this fact in the *Origin of Species*, noting that "breeders habitually speak of an animal's organisation as something quite plastic, which they can model almost as they please" (Darwin 1859, 31). To support this point, he quotes the agriculturalist/politician John Sebright, who once boasted that with pigeons, he could "produce any given feather in three years" but would need "six years to obtain head and beak" (Darwin 1859, 31). In some ways, this optimism is warranted—just look at all the previous success we've had in reshaping the morphology of our domesticates. More importantly, think about how organisms evolve in the wild. Given enough time and the right sort of variation to work with, natural selection can produce absolutely stunning morphological changes. Artificial selection is not fundamentally different from natural selection, so it seems logical that we should be able to accomplish something similar—and presumably in a much shorter amount of time—using selective pressures that we apply ourselves.

Figure 13.1
Domesticated organisms exhibiting exaggerated morphological traits. Over time, artificial selection by humans has produced significant changes in the anatomical features of domesticates. (A) The fruits of the modern domesticated pumpkin *Cucurbita pepo* are significantly larger than those of its wild ancestor, and their sizes have continued to increase in recent years. Left: a "large" pumpkin photographed in 1928. Right: a large pumpkin by current standards. The weight of this particular fruit is not documented, but specimens are known to exceed 2,000 lbs. (B) Left: Like the ancestors of other domesticated fruits, wild strawberries (*Fragaria sp.*) tend to be small in size. Right: Modern domesticated varieties (*Fragaria sp.*) are often larger by several orders of magnitude. (C) Left: The domesticated chicken is descended from the red jungle fowl (*Gallus gallus*). Artificial selection has dramatically increased the body size of domesticated breeds, with some such as the Brahma (right) weighing in at over 15 lbs. All photographs shown are courtesy of Wikimedia Commons and have been edited by the authors to produce the final figure. (For author and license credits, see Acknowledgments.)

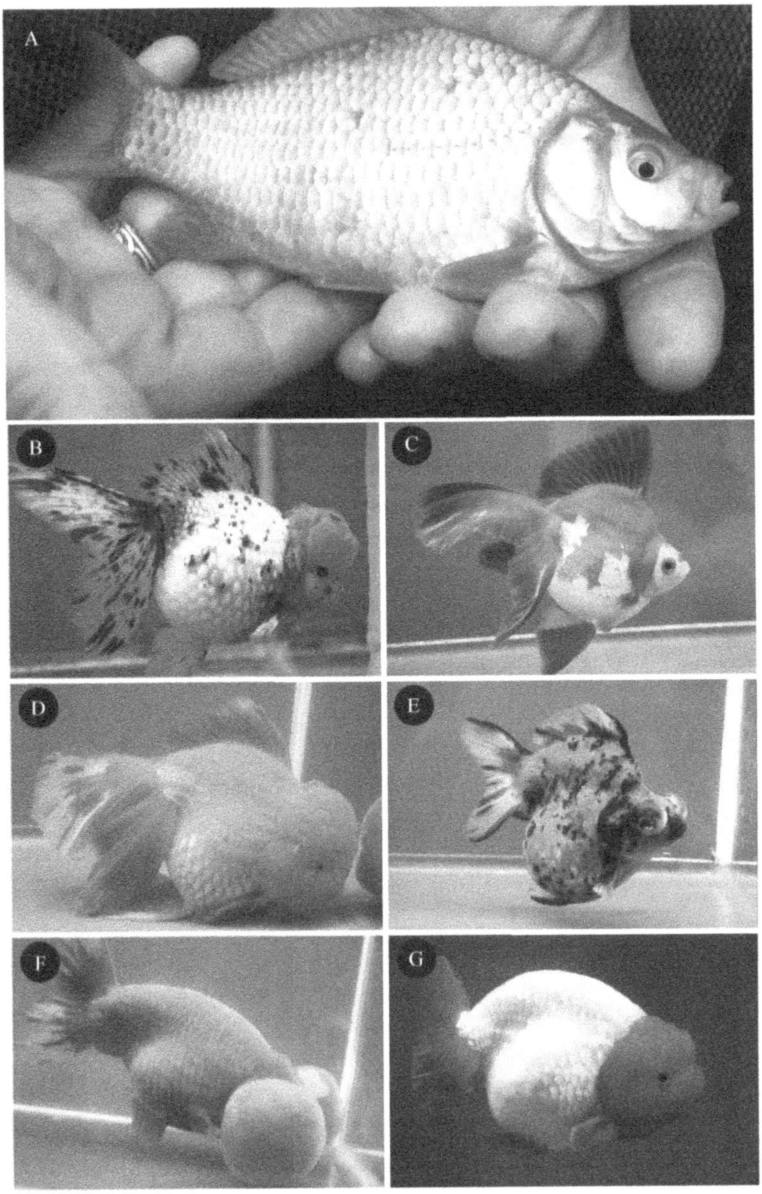

Figure 13.2
Morphological change produced by artificial selection in the domesticated goldfish (*Carassius auratus*). When attempting to highlight the extreme morphological changes humans have produced through artificial selection, researchers often appeal to case studies involving food plants such as maize. However, equally impressive (and more rapid) changes have been elicited in companion domesticates such as the goldfish. *C. auratus* is thought to be descended from the crucian carp (*Carassius sp.*), shown at the top of the figure (A). The following domesticated varieties are displayed below: (B) crowned pearlscale, (C) ryukin, (D) oranda, (E) telescope cross ryukin, (F) bubble eye, and (G) ranchu. Many phenotypic alterations, including changes in the patterns of pigmentation, head and tail morphology, scaling of the abdomen, and the complete loss of the dorsal fin can be observed. These morphologies might be favored in the tightly controlled environments provided by humans, but would almost certainly carry significant fitness costs in the wild. All photos are courtesy of Wikimedia Commons and have been edited by the authors to produce the final figure. (For author and license credits, see Acknowledgments.)

It certainly would be nice if our crops and livestock were this malleable. There is always an incentive to improve the health of our domesticates, increase yields, and, in general, provide the highest nutritional returns at the lowest possible financial cost (Timmusk et al. 2017; Tubiello et al. 2008). However, despite having ample motivation, improving established domesticates is often extremely difficult, even with modern techniques (Sinclair, Purcell, and Sneller 2004; Denison 2012). This could be because we have not found the right candidate genes or have been prevented by ethical panels from doing the appropriate trans-species knock-ins. It could also be that we just have not been diligent enough in our application of artificial selection. A more pessimistic possibility is that none of these techniques will yield the sort of aggressive morphological changes we would like to see because the phenotypic flexibility of domesticated organisms has already been more or less exhausted.

To understand what is being suggested here, consider the effects of natural selection acting on wild populations. Selection is an incredibly powerful creative force, yet there are countless morphological forms that have not been produced in the past 3.7 billion years even though they would likely be fitness enhancing. This is partly because various features of an organism's development can limit the efficacy of selection. These "developmental constraints" (Alberch 1982; Smith 1998) have an important role in determining which regions of morphospace a species can and cannot occupy (McGhee 2015; chapter 1, this volume). Since the mid-1900s (Woodger 1945), discussion of internal constraints has become increasingly common, and they are widely acknowledged to exist in principle. Nevertheless, it remains unclear what they actually are empirically. Even something as simple as deciding when and where to start looking for constraints is a challenge. Using data from free-living populations, it is often unclear what we should infer from the absence of a particular morphology. When a certain form does not occur, this could be due to checks imposed by internal constraints, but it is also possible that the appropriate selective pressures have never been applied. This makes it extremely difficult to identify the areas of morphospace that are truly "off limits" developmentally, and under such circumstances trying to identify constraint-inducing processes is destined to be an exercise in futility.

The study of domesticated organisms can help us get through this quagmire because we can be confident that various selective pressures have been applied but have not been completely successful (Denison 2012; Zeder 2018). Under such circumstances, it is reasonable to infer that some sort of internal developmental constraints really are at work. The next step is identifying the specific morphogenetic processes that are imposing these restrictions on the focal species' form. This is what Hammond's law does.

Hammond's Law

In simple terms, Hammond's law is a hypothesis that specifies how the availability and subsequent use of growth factors during ontogeny produces variation in adult phenotypes (Hammond 1947; Hammond and Zuckerman 1950). Hammond developed his ideas over the course of several decades, with many of the relevant studies being completed from the 1940s to the 1960s. His original papers contain a wealth of data and verifiable predictions that we cannot hope to fully summarize in a single book chapter. Our aim in this section is to provide a broad overview of Hammond's ideas, sacrificing some depth for breadth in order to make his work accessible to a wide scientific audience. We begin by trimming his logic down to a pair of core principles and then selectively focus on a small number of supporting case studies.

Principle 1: Body Parts Compete for Resources during Development

The idea that growth is energetically costly lies at the heart of Hammond's theory. Over the course of development, the majority of the body's parts increase in size. This growth is generally brought about by mitotic divisions that require metabolically combustible "fuels" such as amino acids, lipids, and other circulating macronutrients (Masumura et al. 2000; Géminard, Rulifson, and Léopold 2009; McCue 2010; Koyama, Mendes, and Mirth 2013). These internal developmental resources are often acquired or indirectly synthesized from dietary factors that are sequestered from the local environment (Koyama, Mendes, and Mirth 2013; Koyama, Syropyatova, and Riddiford 2008; Nijhout and McKenna 2018). The fact that the availability of growth factors is linked to diet is important because it implies that external environmental conditions can dictate how much an animal is able to grow over the course of its development. By the 1920s and 1930s, it was widely known that organisms reared under optimal nutritional conditions generally grow to a larger final size than those that experience nutritional stress (Hammond 1936). Hammond's contribution was recognizing that even in times of abundance, the growth of certain body parts can be constrained through a type of developmental interaction that is now referred to as "character-character competition" (Gawne, McKenna, and Levin 2020). These competitive interactions are intensified when nutrition is suboptimal, sometimes leading to significant changes in the organism's overall form.

Competition between characters takes place when multiple body parts rely on the same limited supply of metabolic fuel to drive their growth. Consider a situation in which two body parts concurrently grow in a nonisometric manner, meaning that one of the traits (A) increases in size more rapidly than some other trait (B). All else being equal, A will achieve a larger final size than B due to its higher relative growth rate. The differential growth rate of A and B translates into a nonidentical rate of resource consumption. To be more specific, we would expect A to consume more fuel than B due to its relatively higher rate of growth. If the resource usage differential between A and B is especially high, or internal growth factors are in unusually short supply due to poor dietary conditions, the growth of character A can actively constrain the growth of B by depleting a disproportional percentage of the available resources. The idea that body parts can compete in this way during morphogenesis is implicit in much of Hammond's work on character-specific growth rates in domesticates (Hammond 1947; Hammond and Zuckerman 1950), and recent studies have confirmed that these interactions regularly occur in both domesticated and free-living organisms (Nijhout and Emlen 1998; Klingenberg and Nijhout 1998; Denison 2012; Gawne, McKenna, and Levin 2020).

Principle 2: Resource Availability and the Timing/Rate of Growth Lead to Character-Character Competition

The experimental studies Hammond conducted on domesticated farm animals revealed that body parts tend to reach their maximal growth rate in a highly stereotyped sequence.

Early in development, (1) the bones, cranium, and gut fat attain their maximal rate of growth, followed by (2) the major muscle groups and subcutaneous fat, and (3) the intramuscular fat and muscle tissue around the vertebrae (McMeekan 1940; Hammond 1947; Hammond and Zuckerman 1950). This temporal partitioning of maximal growth rates plays a crucial role in the establishment of organismal form, especially when nutritional intake is varied over the course of ontogeny. For example, Hammond found that pigs will obtain different body proportions—not just sizes—when they are fed a low-nutrient versus a high-nutrient diet during development, and something similar was documented when the nutritional treatments were switched before the animals reached maturity (figures 13.3 and 13.4). Pigs that received a high-nutrient diet throughout ontogeny tended to develop large frames and a substantial amount of body fat. However, Hammond discovered that if the animals were initially fed a high-nutrient diet and were later switched to a low-nutrient diet, they often failed to produce adipose tissue. Conversely, he found that pigs that were fed a low-nutrient diet immediately after birth displayed stunted skeletal growth and possessed little fat, yet if the animals were switched to a high-nutrient diet later in ontogeny, they eventually produced relatively large quantities of fat (McMeekan 1940; Hammond 1947; Hammond and Zuckerman 1950). These seemingly mundane observations are important because they indicate that developing organisms are able to alter the way growth factors are partitioned in response to dietary intake, leading to changes in form. To put it another way, a developing animal makes "decisions" about when and where to allocate internal resources, and different body parts are prioritized when the nutritional regime is altered.

Hammond suggested that the allocation of internal resources during development is determined by two factors: (1) the timing of growth, or the order in which body parts attain their maximum growth rate, and (2) character metabolic rate, understood as the speed at which a body part increases in size due to an underlying rate of resource consumption. He hypothesized that the brain and peripheral nervous system are given priority early in development and achieve their maximal rate of growth before other parts of the body. In addition, Hammond inferred that these tissues must be especially costly to construct and maintain compared with other characters. The fact that the nervous system begins to grow early and is energetically expensive to produce means that it will sequester the greatest proportion of the internal growth factors that the animal has available. Ultimately, this entails that the body parts that reach their maximal rate of growth later in development will be vying for a reduced supply of metabolically combustible fuels, especially when dietary conditions are suboptimal.

After the nervous system, the growth of the bones is prioritized. These characters represent the next most significant energetic investment during development, followed by muscle and fat tissue (figure 13.5). In pregnant females that are themselves still growing, the situation becomes slightly more complicated because the mammary glands and unborn fetus sequester resources that the mother could otherwise use to construct her own bones, fat, and muscle. Describing his version of the diagram depicted in figure 13.4, which illustrates high-priority structures with multiple arrows and low-priority structures with fewer arrows (depicted, respectively, as thicker and thinner arrows, in figure 13.4), Hammond summarized the situation as follows:

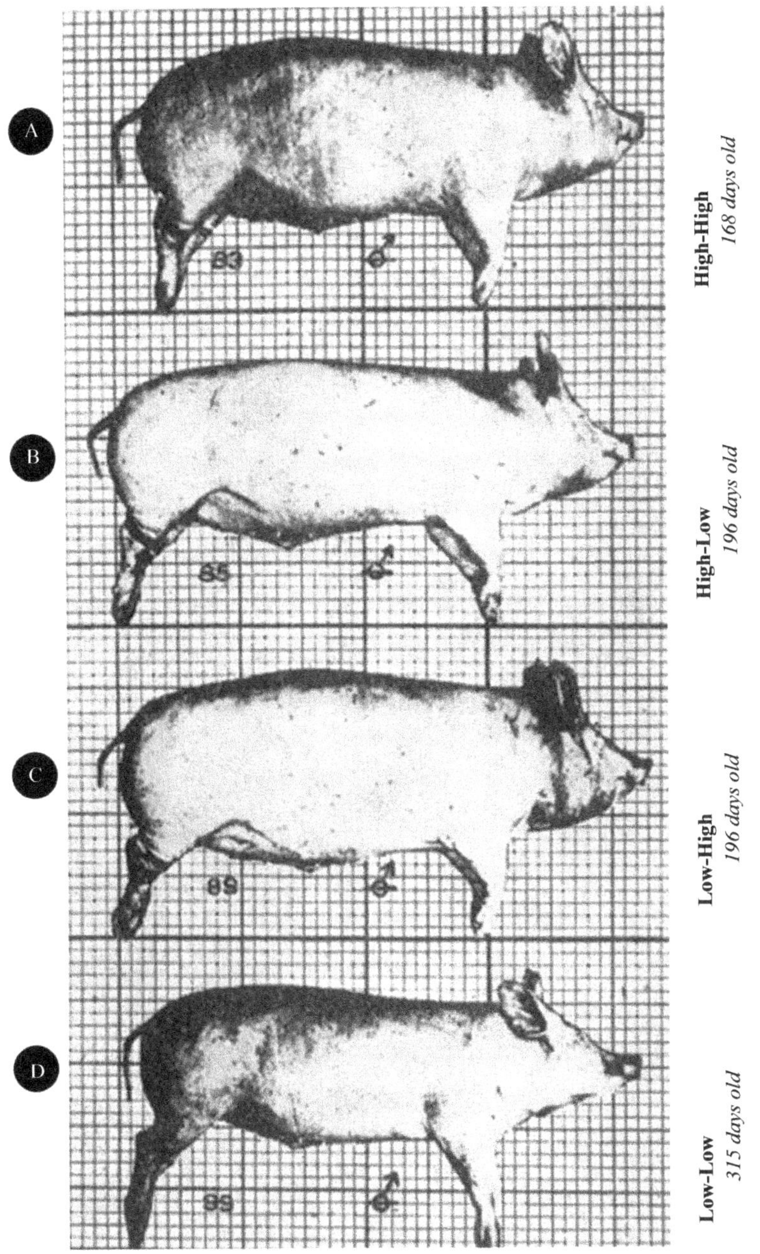

Figure 13.3
Differences in the body proportions of pigs of the same relative age obtained by changing nutrition during postembryonic development. The pigs are scaled to the same shoulder height. Each animal is a representative sample from the dietary treatment group described on the right-hand side of the figure. The pig shown in (A) was fed on a high-nutrition diet throughout development (high-high). These animals grew larger and reached their final adult size more quickly (168 days) than those in other treatments. The pig shown in (B) belongs to a treatment group that began development on a high-nutrition diet, but was switched to a lower plane of nutrition at 16 weeks of age (high-low). Prior to the change in diet, pigs in the high-low treatment group grew rapidly. When transferred to the poor-nutrient diet, growth slowed substantially, and the animals attained their final adult size at approximately 196 days. (C) When pigs were initially fed a low-nutrient diet and switched to a high-nutrient diet at 16 weeks of age (low-high) an effect opposite of that observed in high-low group was documented. Growth was initially slow, but ramped up significantly following the transition to a diet with optimal nutrition. Pigs in this treatment reached their final size at approximately 196 days, similar to what was observed in the high-low treatment. However, although they shared similarities in the timing of their development, the body proportions of the high-low and low-high groups differ. (D) Pigs that were fed on a low plane of nutrition throughout ontogeny grew the slowest and reached a smaller final size than those in the other experimental treatments. These pigs took 315 days to reach adulthood and developed almost no fat deposits in their bodies. The results described are noteworthy because they indicate that character-specific growth rates are altered in response to nutritional intake. Figure adapted from Hammond (1950) with permission from the publisher.

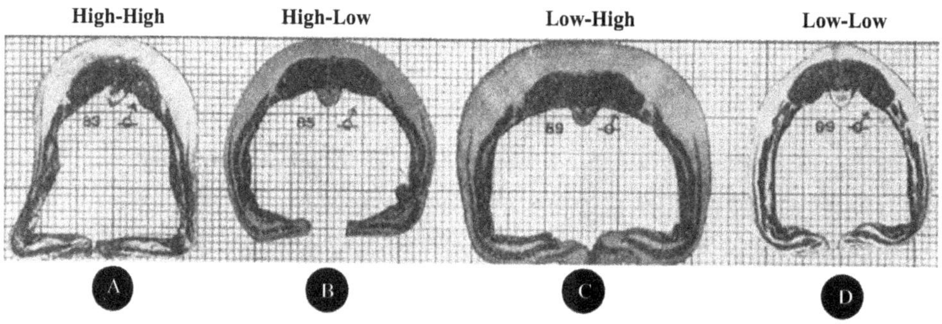

Figure 13.4
The proportion of muscle and fat in the experimental treatment groups of pigs described in figure 13.3. Temporal variation in nutrition has a strong influence on the partitioning of nutrients during postembryonic development. (A) The high-high treatment produced pigs with a large frame and substantial amounts of muscle and fat. (B) Pigs reared on the high-low dietary regime developed a large frame, but failed to grow fat. (C) Conversely, on the low-high regime, skeletal growth was reduced. However, optimal nutrition later in development facilitated the deposition of large quantities of fat on this small frame. (D) Pigs fed on a poor diet throughout development displayed stunted skeletal growth and almost no fat deposition. The differences between (B) and (C) are especially noteworthy. These pigs are the same age and similar in weight, yet they show considerable differences in body composition. This suggests that the timing of nutritional uptake, and not just the quality of the materials ingested, is an important determinant of organismal form in some domesticated organisms. Figure adapted from Hammond (1950) with permission from the publisher.

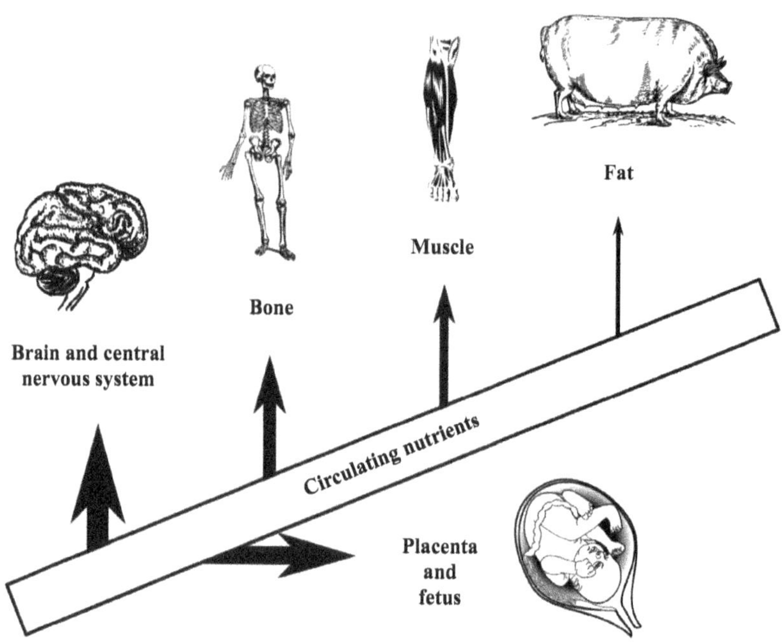

Figure 13.5
A visual representation of Hammond's theory of nutrient partitioning during development. Hammond believed that the nutrients circulating in the bloodstream during development (i.e., growth factors) are allocated in a controlled manner over the course of ontogeny. The thickness of the arrows is proportional to the amount of nutrients each part of the body sequesters, and their order from left to right represents the order in which they are prioritized under nutritional stress (left: high priority; right: low priority). In developing vertebrates, the brain and nervous system consume most of the available growth factors, followed by the placenta and fetus in pregnant females. Skeletal components are the next most costly bodily components, followed by muscle and fat. Reading this figure from right to left, one sees the order in which various features of the body will be diminished in size under suboptimal nutritional conditions. This figure is based on an illustration from Hammond (1950).

If nutrition is on a very high plane all tissues are served to their maximum requirements. When the plane of nutrition is lowered one arrow is deducted from each, so that fat ceases to be put on, but the other tissues continue growing but at a slower rate. When the plane of nutrition is lowered still further and two arrows are taken from each, brain and bone still continue to grow but at a much slower rate, muscle ceases to grow and fat is withdrawn to supply energy for growth of the brain etc. (Hammond 1952, 222)

To reiterate a point made earlier, this means that if a character grows later in development, or grows at a time when many other body parts are also rapidly increasing in size, it will have less fuel at its disposal and, as a consequence, could be reduced in size. In addition, this implies that if a costly structure grows early in ontogeny, its resource consumption patterns could inhibit the maturation of later-developing body parts.

Developing characters might be modular in the sense that distinct batteries of genes could be expressed in their tissues (Arnone and Davidson 1997; Raff and Sly 2000; Wagner 2018), but when the type of interactions described by Hammond occur, their growth can still be tightly correlated. Many classically trained embryologists were aware that developing body parts sometimes compete for internal resources (Roux 1881; Weismann 1896; Saint-Hilaire 1822). However, because these interactions are fundamentally physiological in nature, they cannot be identified or studied from the genetically based reductionist perspective that has dominated developmental biology since the 1950s. This is a case where modern "advances" have impeded scientific progress by causing researchers to forget something that was formerly known to the community.

Putting It All Together: Formalization, Predictions, and Evidence for Hammond's Law

Hammond's law can be more formally stated as follows: (1) all growth is energetically costly, (2) many body parts rely on the same finite supply of internal resources, and, therefore, (3) the growth of one body part can constrain the growth of others. This implies that (4) the overall availability of internal resources, in conjunction with the timing and rate at which body parts grow, can be an important determinant of the organism's final adult form. Hammond suggested that previous work on pigs (McMeekan 1940) and sheep (Verges 1939) supported his hypothesis, but it was also validated by his own work on milk production in cattle.

First, Hammond showed that pregnant cows grow to different weights based on their plane of nutrition (Hammond, 1936, 1947; Hammond, 1952). The nutritional intake of female cattle also influenced their ability to develop mammary glands and subsequently produce milk. More specifically, cows fed on a low plane of nutrition were unable to produce the same quantity of milk as siblings reared under more nutrient-rich control conditions (Wallace 1944; Hammond 1947). Hammond followed this study with a historical account of selection for milk production in zebu cattle (Hammond 1947). For several generations, farmers in India selected for improved milk yields in their cows but saw no increases in output until the government adopted more regimented feeding practices that were used in other countries. This change in the rearing environment was crucial because it eliminated variation in diet across the population and allowed the farmers to select for increased milk production under a single environmental condition. In 1912, cows could produce an average daily yield of 5.8 pounds of milk. By 1922, this had nearly doubled to 10.8 pounds and, roughly a decade later, the cattle produced approximately 18.5 pounds of milk per day on average (Hammond 1947).

From the farmers' perspective, the exaggeration of the udder is often highly desirable when nutrition is ample; however, selection for enlarged udders under a high plane of nutrition can also come at a significant cost. Over time, artificially selecting for enlarged udders under optimal dietary conditions sometimes results in organs that are so monstrously outsized (see, e.g., figure 13.6) that they reduce the viability of the animals when they are reared on a lower-quality diet. The udders are prioritized during development, and when animals are reared under nutrient-poor conditions, they sequester a substantial proportion of the body's internal supply of nutrients, which would otherwise be allocated elsewhere. As a consequence, cows that experience nutritional stress can exhibit a host of undesirable defects, including a higher susceptibility to disease (Caeneiro and Rhoad 1936; van der Waaij 2004). The consequences of this unwanted character-character competition are so serious that agriculturalists now select for increased milk production under less-than optimal nutritional conditions (van der Waaij 2004). As farmers themselves have occasionally put it, this ensures that females will not "milk the flesh off [their] bones" (Hammond 1947, 199) due to the intense character-character competition that an extremely enlarged udder initiates.

Figure 13.6
Artificial selection for increased milk production in cattle breaks the wild-type scaling relationship between the udders and other body parts. The udders of domesticated cattle are sometimes so monstrously enlarged that they hinder the animal's movement and must be supported by a harness in order to prevent injury. Increased udder size improves milk yields, but as Hammond (1947) noted, the enlargement of this body part can lead to unwanted character-character competition during development. In particular, the udder can consume growth factors that are normally allocated to the bones and other body parts, leading to a feeble animal with low fitness, even in the highly controlled agricultural environment. The photograph shown is courtesy of Wikimedia Commons and has been edited by the authors to produce the final figure. (For author and license credits, see Acknowledgments.)

Nutritional Variation and Character-Specific Reaction Norms: Adaptation to Controlled Agricultural Environments Breaks Functional Constraints on Development

One of the things that makes agricultural niches unique is the fact that environmental variables tend to be tightly controlled in order to promote an optimal level of growth in the domesticate. The fact that external conditions are stabilized in this way means that selection on phenotypic traits tends to occur in a more-or-less homogenous environment, wherein phenotypic variation is almost entirely the result of underlying genetic variation. These conditions differ substantially from those experienced in the wild, where temperature, photoperiod, and nutrition are always in flux. The fact that environmental conditions are seldom constant in natural settings means that the development of most organisms is adapted to respond to a wide range of external variables. When a feature of the environment varies continuously, it often produces an associated range of phenotypic outcomes (Shingleton et al. 2007). One of the first researchers to document this trend was the German biologist Richard Woltereck (1909; Nanjundiah 2020), who noted that quantitative traits are generally tightly distributed around a mean in any given environment. Crucially, Woltereck also noted that the value of the mean tends to differ significantly from environment to environment, and he found that when nutrition varies continuously, certain quantitative characters vary in an analogous manner. He referred to the range of phenotypic variance produced under these circumstances as a "reaction norm" and, in coining this term, was the first of many biologists to realize that a single genotype can give rise to a host of different phenotypic outcomes.

Woltereck hypothesized that the mean phenotypic value obtained for any given environmental condition results from the fact that organisms in a population rely on common regulatory mechanisms to control their development. Variance around the mean was then attributed to individual genetic variation that determines how an organism responds to changes in the focal environmental condition(s) (Woltereck 1909). Conceived in this way, the particular norm of reaction an organism exhibits is specific to its genotype. Woltereck's studies are important for our purposes because they suggest that phenotypic plasticity—that is, the ability to produce a range of phenotypic outcomes in response to changes in environmental variables—has a genetic basis and can itself be an adaptation if a single, nonplastic phenotype would have different fitness coefficients in different environments.

In recent years, reaction norms and developmental plasticity have been widely studied (Shingleton et al. 2007; Koyama, Mendes, and Mirth 2013; Thompson 2019; Frankino et al. 2019; Suzuki, McKenna, and Nijhout 2020), but much of this research overlooks the fact that each character in the organism has a unique norm of reaction that results from internal developmental mechanisms that function specifically to control growth. One of the best examples of character-specific reaction norms comes from the literature on *Drosophila melanogaster*. In these animals, nutritional variation affects growth rate via insulin-like peptides (ILPs). The concentration of ILPs is directly correlated with nutritional intake (Géminard, Rulifson, and Léopold 2009), meaning that, all else being equal, organisms that develop under optimal dietary conditions will have more ILPs circulating in their bodies than those that experience nutritional stress. This is important because, in most insects, the availability of ILPs helps to determine the final sizes of the body's various parts (Koyama, Syropyatova, and Riddiford 2008; Nijhout and McKenna 2018). The wings and legs, for example, exhibit dramatically reduced growth when ILP concentrations drop, but organs such as the gonads maintain a similar rate of growth across all ILP concentrations (Tang et al. 2011). These findings suggest that each organ or appendage has a unique dose-response curve for ILPs that determines the norm of reaction it will exhibit when it is exposed to different nutritional regimes.

Numerous studies have shown that character-specific scaling of growth in *Drosophila* is facilitated by differential expression of intracellular components involved in the insulin/TOR cascade, which transduces the ILP signal and thereby produces cellular proliferation. In the gonads, this intracellular cascade is set up in a way that ensures these tissues are always attempting to grow, regardless of the ILP availability. This makes the reproductive organs ultrasensitive to the insulin signal at all concentrations (Tang et al. 2011). In contrast, the wings and legs respond to the presence of ILPs in a concentration-dependent manner, meaning their growth rate increases and decreases according to ILP availability (Tang et al. 2011). The fact that the characters in question respond differently to the ILP signal is noteworthy because it demonstrates that the scaling of growth in response to environmental variation can be controlled on a character-by-character basis throughout the organism. Perhaps even more importantly, the work in question suggests that the range of phenotypes produced must be heritable because genetic factors play a role in determining how the signal transduction network in any given body part responds to ILP concentration.

The fact that the scaling of growth is partially controlled by genetic interactions that are modularly tuned within each body part is what allows the reaction norms of characters to vary independently of one another. Support for this hypothesis can be found in a series of studies examining how nutritional status affects patterns of correlated growth in the limb bones of the common rat (*Rattus norvegicus*) (Miller and German 1999). Male rats reared on a protein-deficient diet tend to grow more slowly and reach a significantly smaller final size than littermates fed a protein-rich control diet. In addition, the growth of long bones in animals reared under the low-protein condition displays a strong signature of integration, or correlated growth, that is not present in the control treatment (Zelditch 1988; Miller and German 1999; Reichling and German 2000; Hallgrímsson et al. 2004) (figure 13.7). This suggests that the genetic variation in the developmental mechanisms associated with each body part can manifest differently across nutritional environments. More specifically, the growth kinetics of bones tend to be nonidentical when protein is abundant, but when nutrition is restricted, this variation collapses, and the characters begin to grow more or less isometrically.

For our purposes, the above findings are of interest because they add further emphasis to the point that farmers wishing to induce directional changes in the morphology of domesticates need to be conscious of the conditions under which artificial selective pressures are applied. With the zebu cattle mentioned by Hammond, rearing animals under optimal nutrition allowed for selection on the genetic variants that produced a maximal udder size. However, a selective regime of this type can turn out to be problematic when no consideration is given to how the enlarged udder will scale with the rest of the body under suboptimal dietary conditions. As discussed above, when developing cattle experience nutritional stress, their udders often continue to grow at a maximal rate, which further

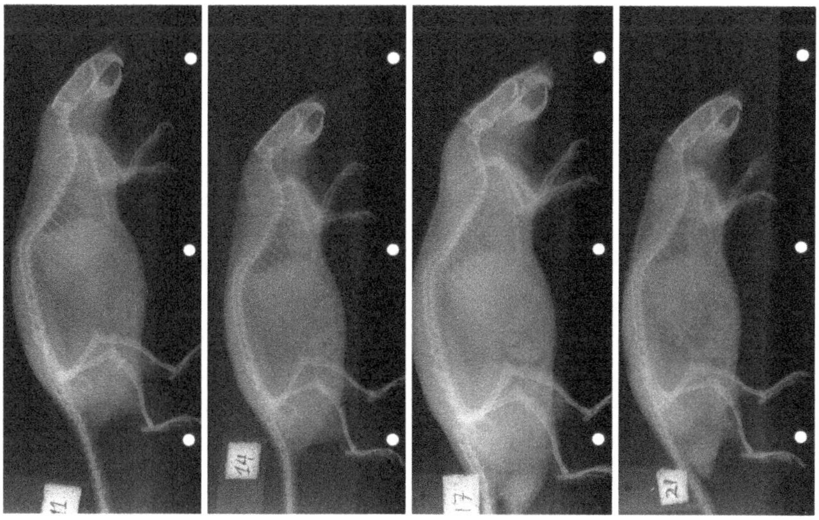

Figure 13.7
Adult male and female rats fed on different dietary regimes during development. The images shown above are of littermates that received either an optimal diet or a low-protein diet. Animals that were fed the low-protein diet stopped growing at substantially smaller body sizes than littermates from the control group. All images were taken once growth had ceased and the scaling is constant across the four treatments. Similar to what Hammond observed in domesticated farm animals (figures 13.3 and 13.4), this suggests that organismal form is highly sensitive to nutritional status. Images were graciously provided by Rebecca German and Fred Nijhout.

decreases the amount of resources available for use in the development of other body parts and leads to health problems. In this case, a reaction norm that is clearly deleterious under suboptimal environmental conditions might not be selected against in the agricultural setting, as it likely would be in the wild. Under natural conditions, nutritional availability frequently oscillates over the lifetimes of individuals and across generations, which means that reaction norms and their associated genotypes are consistently subjected to purifying selection.

Most exaggerated characters produced by natural processes, including sexually selected body parts in insects and vertebrates, display hyperallometric (slope >1) reaction norms, wherein the focal character is exaggerated in large males and more normally sized in smaller males (O'Brien et al. 2018; Rodríguez and Eberhard 2019). Males with larger body sizes are generally able to sequester more resources from the local environment than smaller males (Emlen and Nijhout 2000; Emlen et al. 2012), which means that there will likely be enough internal resources available for all body parts to reach their maximal genetically prescribed size. Under these circumstances, selection can favor the enlargement of certain body parts, because the development of other characters will not be significantly affected. Conversely, in males that fail to attain their maximal size, selection for extreme enlargement of a particular character can be deleterious. This is because smaller males sequester comparatively fewer resources from the local environment and thus have a more limited supply of internal macronutrients available to fuel their growth. With smaller individuals, enlargement of, say, a horn could constrain the development of other important characters, leading to decreases in fitness. For this reason, in animals with smaller body sizes selection would be expected to favor the maintenance of existing body proportions.

The idea that there is little selection against reaction norms in agricultural environments may have been first stated by Helen Spurway (Spurway 1955; Nanjundiah 2020). In an interesting 1955 paper that has been almost completely lost to history, she suggested that, prior to human intervention, domesticates experienced more uniform selective pressures that eliminated genotypes that disproportionately affect the way developmental events are patterned. The agricultural environments created by humans destabilize the selective pressures that normally act on developmental processes because these niches are tightly controlled, thereby removing organisms from many of the environmental factors they would experience in the wild. In this respect, domesticated organisms are of interest to basic research because they allow us to see certain potentials of developmental systems that are generally hidden/suppressed by natural selection. The fact that numerous organisms have been domesticated and are readily obtainable provides biologists with an opportunity to study how internal resources are allocated during development.

Potential Applications of Hammond's Law to Insect Agricultural Systems

In terms of our more immediate task of identifying parallels between the evolution of agriculture in humans and insects, if Hammond is correct and competitive interactions between characters have shaped the morphologies of our domesticates, it is possible that they could do the same in organisms farmed by insects. For example, several recent studies have shown that defensive compounds are costly for plants to produce and tend to come at the expense of somatic tissue growth (Huot et al. 2014). Many of the plants domesticated by humans exhibit decreased levels of protective secondary metabolites in their tissues, which is thought to create a surplus of internal resources that allows other features of the organism to increase in size (Rosenthal and Dirzo 1997; Massei and Hartley 2000; Moreira et al. 2018). Extending this logic to nonhuman agricultural systems, it would be interesting to determine whether the loss of ancestral chemical defenses in insect-tended fungi or plants has facilitated an analogous upregulation in growth. The fact that the lower- and higher-attine fungi are cultivated on different substrates and have different morphological features (chapters 8 and 14, this volume) is also of interest from the perspective of developmental energetics. The nutrient-dense gongylidia present in the fungi of the higher-attine ants (chapters 8 and 14, this volume) are likely metabolically costly to produce (Henrik, Boomsma, and Tunlid 2014). If the fungal substrate is reduced in quality, it would be worth examining whether the gongylidia decrease in size/nutritional quality, or remain more or less unaffected (Masiulionis et al. 2014). The former would indicate that the development of these structures is nutrition-sensitive, while the latter would be indicative of developmental prioritization, similar to what Hammond documented in the udders of cattle.

The principles identified by Hammond could also potentially help us to make sense of the growth dynamics of the fungal mycelium in insect agricultural systems. Ant-tended fungi tend to remain in a vegetative, mycelial state as they grow and rarely produce the fruiting bodies that are needed for sexual reproduction (Mueller et al. 2001). The fact that the fungi exist in this vegetative form is sometimes attributed to the activities of their ant farmers, which are constantly dividing and pruning the mycelial connections (Fisher, Stradling, and Pegler 1994a, 1994b; Mueller 2002). It is also possible, however, that the constant supply of high-quality nutritional substrate provided by the ants encourages the fungus to grow asexually, in its mycelial state. The basic idea is that when resources are abundant, and/or the ratio of proteins to carbohydrates in the substrate is optimized (Shik et al. 2016), the fungus will take advantage of this situation as long as possible by increasing in size, rather than attempting to reproduce sexually. There is, in fact, some evidence to suggest that life history progression in wild leucocoprineaceous fungi is tied to nutrition in this manner (Vellinga 2004), at least in tropical, wet-forest species. This finding is intriguing because it provides insight into how insects might have first been able to domesticate their fungal cultivars. If the fungi had an ancestral tendency to stay in their asexual state when the nutritional environment was conducive to growth, insect farmers could have exploited this fact by providing fungi brought to the nest with a surplus of decomposable resources. This would serve the dual function of ensuring that edible mycelium is continually produced, while also preventing the fungus from "escaping" by entering the sexual stage of its life cycle. Under this scenario, the ants could effectively force fungi into becoming their crops by exploiting their existing developmental tendencies. Once the fungi reach a certain size, environmental factors such as rainfall and temperature would be predicted to promote fruiting, even when nutrition is abundant. At this stage, the hypothesis would be that active pruning of mycelial connections would then become the dominant means by which ants deter sexual reproduction in their crops.

These are just a handful of examples of how taking a developmental-energetics point of view could be of use in understanding nonhuman agricultural systems. If the evolution of agriculture in humans and insects is truly a case of convergent evolution, it seems plausible that Hammond's law will be broadly applicable to all domesticated organisms, regardless of what species takes on the role of the farmer.

Hammond's Law as a Generalized Developmental Theory

Development has long been regarded as one of the least understood and poorly synthesized branches of biology (Woodger 1929; von Bertalanffy 1933). Not even a century ago, a handful of embryologists were still speculating that it might be an impenetrable subject, which is either mostly or entirely screened off from scientific understanding. When confronted with laboratory results that were difficult to explain, vitalists such as Hans Driesch (1867–1941) turned tail and retreated to mystical explanations involving immaterial forces (Driesch 1908). Emboldened by the molecular revolution of the 1950s, researchers eventually stopped running from the problems of development and, instead, began to ignore them entirely. Given its simplicity and seemingly unparalleled explanatory power, an implicit form of genetic determinism has slowly but surely become the guiding principle of developmental biology. Indeed, reading the contemporary literature, one can easily get the impression that morphogenesis is nothing more than a complex series of interactions that are both initiated by, and completely reducible to, the activities of gene products (Nijhout 1990; Larsen 2005; Gawne, McKenna, and Nijhout 2018).

Part of the allure of genetic determinism is the fact that it allows us to get a lot done experimentally. Environmental influences on phenotypes are backgrounded, and physiological processes that are difficult to control are treated as inconsequential downstream events or dismissed outright as a type of developmental noise. This leaves researchers free to focus on things such as candidate genes and other genetic factors that are readily identifiable and easy to manipulate in the lab. Although it has utility in certain contexts, it is important to recognize that the stripped-down form of genetic determinism that dominates modern developmental biology is not a true predictive theory. Instead, it is a sort of baseline assumption that provides a methodological script for doing experimental work. For any given aspect of character formation, we are assured that there are primitive genetic causes and, as biologists, we are told that it is our job to identify these factors.

Over the years, research born out of this tradition has generated an enormous amount of genetic, genomic, and transcriptomic data for nearly every phenotype imaginable. The assumption seems to be that if we keep our heads down and continue to amass more and more empirical data, theoreticians will eventually be able to construct some sort of generalized hypothesis that will "effectively enable one to compute the adult organism from the genetic information in the egg" (Wolpert and Lewis 1975, 21). If future generations of scientists can pull something like this off, it would certainly be an astonishing intellectual achievement. But of course, there is no guarantee that this day will ever come. And more importantly, we do not need big-picture developmental theories 50, 100, or 1,000 years in the future. We need them right now. The idea that we should continue to hoard genetic data simply because there is a chance that someone might eventually be able to use it to construct large-scale predictive models is little more than a procrastinator's excuse that allows us to avoid doing the messy theoretical work ourselves.

If we hope to formulate a developmental theory with wide applicability, it is necessary to abandon genetic predetermination and acknowledge the fact that organisms are hierarchically arranged systems whose final adult phenotypes are produced and maintained by nonlinear causal interactions (Gawne, McKenna, and Nijhout 2018; DiFrisco and Jaeger 2020; Suzuki, McKenna, and Nijhout 2020). The second thing we need to do is reorient our understanding of what developmental biology is ultimately trying to accomplish. The chief objective of the field is not necessarily the identification of one-to-one causal mappings between molecular processes and specific phenotypic characters, as the genetic paradigm suggests. Prior to the mid-twentieth century, classically trained embryologists tended to regard developmental biology as the science of form (Holmes 1904; Zuckerman 1950; Medawar and Zuckerman 1950). Accordingly, their goal was to determine how scaling relationships and the overall shape of organisms arise over the course of ontogeny (Huxley 1932; Huxley and Teissier 1936; Thompson 1942). The form of a plant or animal is the sum total of the sizes and shapes of all its individual body parts, and as a consequence it cannot be explained by simple causes such as a mutation, single nucleotide polymorphism, or duplication event.

Understanding how form is produced requires that we shift our attention to the higher-level interactions that take place between developing body parts. As Hammond recognized nearly a century ago, the development of organs, appendages, and other characters is seldom fully independent or "modular" (Hammond 1947; Gawne, McKenna, and Levin 2020). The growth rate and final size of most traits can be affected by morphogenetic events that take place in other parts of the body. Hammond identified one of the primary means by which these interactions take place—namely, character-character competition. In modern terms, Hammond's law is valuable because it helps to explain the proximate causes that constrain organismal form: when the growth of body parts is driven by the same macronutrients, increases in the size of one part often restricts the development of others. Growth rate is correlated with metabolic rate, and, if structures are growing at the same time, those with a higher rate of growth that grow earlier than others will consume a disproportionate percentage of the body's internal resources. When this occurs, the traits that are competing will exhibit correlated growth and their final scaling relationship will be allometric, because the larger, more mitotically active body parts constrain the growth of their smaller counterparts that grow more slowly.

Hammond's law explains how scaling relationships can be produced, making it a theory of organismal form. It is a generalized theory of form because the predictions it makes are not taxon-specific. Any organism that undergoes size increases though mitotic divisions has the potential to experience the type of character-character competition his hypothesis describes. More work must be done to determine how often character-character competition occurs and what body parts are affected, but, in principle, Hammond's law has an extraordinarily wide domain of applicability. When developmental biologists attempt to construct theoretical hypotheses, there is a tendency to reach for mathematical tools that describe the phenomena in technical terms that only a small fraction of the field can understand, let alone critique. However, there is no reason to think that every generalized hypothesis needs to be chromed-out with flashy formulas, especially when it is first conceived. Indeed, anyone who opens *The Origin of Species* will find persuasive evidence to the contrary. Like Darwin, Hammond supported his hypothesis using careful observations of common domesticated animals. This approach might seem "unsophisticated" to developmental biologists trained in the molecular era, but as more supporting studies are conducted and the theoretical predictions of Hammond's law are further refined, it could very well be transformed into the sort of truly generalized developmental theory that modern methods have systematically failed to produce.

Acknowledgments

We thank the participants in the 38th Altenberg Workshop in Theoretical Biology for helpful comments on the ideas presented in this chapter, and many other stimulating discussions on the convergent evolution of agriculture in humans and insects. Special thanks are due to R. Ford Denison, whose book *Darwinian Agriculture* first encouraged us to start thinking about competitive interactions in the context of domesticated organisms. Ted Schultz and Jon Shik provided important insight into how Hammond's law could potentially be applied to insect domesticates. Workshop funding from the Konrad Lorenz Institute and a postdoctoral fellowship awarded to R. G. made this research possible. Contributions by R. G. and K. Z. M. are equal.

Author and license information for figures that incorporate Wikimedia Commons source material: Figure 13.1A: (left) author, unknown; license, public domain; (right) author, David. politzer; license, Creative Commons Attribution-Share Alike 4.0 International. Figure 13.1B:

(left) author, Dmitry Makeev; license: Creative Commons Attribution-Share Alike 4.0 International; (right) author, Mauro Cateb; license: Creative Commons Attribution-Share Alike 4.0 International. Figure 13.1C: (left) author, Subramanya C K; license, Attribution-Share Alike 3.0 Unported; (right) author, Hansige; license: Creative Commons Attribution-Share Alike 3.0 Unported, 2.5 Generic, 2.0 Generic and 1.0 Generic. Figure 13.2A: author: Zimpenfish; license: Creative Commons Attribution 2.0 Generic; figures 13.2B–G: author: Melanochromis; license: public domain. Figure 13.6: author, Christian Bickel; license, Attribution-Share Alike 2.0 Germany.

References

Alberch, P. 1982. "Developmental Constraints in Evolutionary Processes." In *Evolution and Development*, edited by J. T. Bonner, 313–32. Berlin: Springer.

Andersson, M. 1994. *Sexual Selection*. Monographs in Behavior and Ecology. Princeton, NJ: Princeton University Press.

Andrews, Alfred C. 1949. "The Carrot as a Food in the Classical Era." *Classical Philology* 44(3) 182–196. doi:10.1086/363204.

Arnone, M. I., and E. H. Davidson. 1997. "The Hardwiring of Development: Organization and Function of Genomic Regulatory Systems." *Development* 124(10): 1851.

Bizzarri, Mariano, Douglas E. Brash, James Briscoe, Verônica A. Grieneisen, Claudio D. Stern, and Michael Levin. 2019. "A Call for a Better Understanding of Causation in Cell Biology." *Nature Reviews Molecular Cell Biology* 20(5): 261–262. doi:10.1038/s41580-019-0127-1.

Bovine HapMap Consortium. 2009. "Genome-Wide Survey of SNP Variation Uncovers the Genetic Structure of Cattle Breeds." *Science* 324(5926): 528–532.

Caeneiro, G. G., and A. O. Rhoad. 1936. "The Development of Calves in the Tropics." *Tropical Agriculture, Trinidad and Tobago* 13:177–180.

Darwin, Charles. 1859. *On the Origin of Species by Means of Natural Selection*. London: John Murray.

Denison, R. F. 2012. *Darwinian Agriculture: How Understanding Evolution Can Improve Agriculture*. Princeton, NJ: Princeton University Press.

DiFrisco, James, and Johannes Jaeger. 2019. "Beyond Networks: Mechanism and Process in Evo-Devo." *Biology & Philosophy* 34(6): 54. doi:10.1007/s10539-019-9716-9.

DiFrisco, James, and Johannes Jaeger. 2020. "Genetic Causation in Complex Regulatory Systems: An Integrative Dynamic Perspective." *BioEssays* 42(6): 1900226. doi:10.1002/bies.201900226.

Domyan, Eric T., Zev Kronenberg, Carlos R. Infante, Anna I. Vickrey, Sydney A. Stringham, et al. 2016. "Molecular Shifts in Limb Identity Underlie Development of Feathered Feet in Two Domestic Avian Species," edited by Marianne E Bronner. *ELife* 5(March): e12115. doi:10.7554/eLife.12115.

Domyan, Eric T., and Michael D. Shapiro. 2017. "Pigeonetics Takes Flight: Evolution, Development, and Genetics of Intraspecific Variation." *Changes in Developmental Controls Underlying the Evolution of Animal Body Plans* 427(2): 241–250. doi:10.1016/j.ydbio.2016.11.008.

Drake, Abby Grace, and Christian Peter Klingenberg. 2010. "Large-Scale Diversification of Skull Shape in Domestic Dogs: Disparity and Modularity." *American Naturalist* 175 (3): 289–301. doi:10.1086/650372.

Driesch, H. 1908. *The Science and Philosophy of the Organism: The Gifford Lectures Delivered before the University of Aberdeen in the Year 1907[–08]*. London: Adam and Charles Black.

Emlen, Douglas J., and H. F. Nijhout. 2000. "The Development and Evolution of Exaggerated Morphologies in Insects." *Annual Review of Entomology* 45(1): 661–708. doi:10.1146/annurev.ento.45.1.661.

Emlen, Douglas J., Ian A. Warren, Annika Johns, Ian Dworkin, and Laura Corley Lavine. 2012. "A Mechanism of Extreme Growth and Reliable Signaling in Sexually Selected Ornaments and Weapons." *Science* 337(6096): 860. doi:10.1126/science.1224286.

Evans, L. T. 1984. "Darwin's Use of the Analogy between Artificial and Natural Selection." *Journal of the History of Biology* 17(1): 113–140. doi:10.1007/BF00397504.

Fisher, P. J., D. J. Stradling, and D. N. Pegler. 1994a. "Leaf Cutting Ants, Their Fungus Gardens and the Formation of Basidiomata of Leucoagaricus Gongylophorus." *Mycologist* 8(3): 128–131.

Fisher, P. J., D. J. Stradling, and D. N. Pegler. 1994b. "Leucoagaricus Basidiomata from a Live Nest of the Leaf-Cutting Ant Atta Cephalotes." *Mycological Research* 98(8): 884–888.

Frankino, W. Anthony, Eric Bakota, Ian Dworkin, Gerald S. Wilkinson, Jason B. Wolf, and Alexander W. Shingleton. 2019. "Individual Cryptic Scaling Relationships and the Evolution of Animal Form." *Integrative and Comparative Biology* 59(5): 1411–1428. doi:10.1093/icb/icz135.

Fuller, Dorian Q., Tim Denham, Manuel Arroyo-Kalin, Leilani Lucas, Chris J. Stevens, et al. 2014. "Convergent Evolution and Parallelism in Plant Domestication Revealed by an Expanding Archaeological Record." *Proceedings of the National Academy of Sciences* 111(17): 6147–6152.

Gawne, Richard, Kenneth Z. McKenna, and Michael Levin. 2020. "Competitive and Coordinative Interactions between Body Parts Produce Adaptive Developmental Outcomes." *BioEssays* 42(8): 1900245. doi:10.1002/bies.201900245.

Gawne, Richard, Kenneth Z. McKenna, and H. F. Nijhout. 2018. "Unmodern Synthesis: Developmental Hierarchies and the Origin of Phenotypes." *BioEssays* 40(1): 1600265. doi:10.1002/bies.201600265.

Géminard, Charles, Eric J. Rulifson, and Pierre Léopold. 2009. "Remote Control of Insulin Secretion by Fat Cells in Drosophila." *Cell Metabolism* 10(3): 199–207. doi:10.1016/j.cmet.2009.08.002.

Geoffroy-Saint-Hilaire, E. 1822. *Philosophie Anatomique*. Paris: J. B. Baillière.

Gregory, T. Ryan. 2009. "Artificial Selection and Domestication: Modern Lessons from Darwin's Enduring Analogy." *Evolution: Education and Outreach* 2(1): 5–27. doi:10.1007/s12052-008-0114-z.

Haase, E., and R. S. Donham. 1980. "Hormones and Domestication." In *Hormones and Domestication*, edited by August Epple, Milton H. Stetson, and Barbara J. Nibbio, 549–565. London: Academic Press Inc.

Hallgrímsson, Benedikt, Curtis J. Dorval, Miriam Leah Zelditch, and Rebecca Z. German. 2004. "Craniofacial Variability and Morphological Integration in Mice Susceptible to Cleft Lip and Palate." *Journal of Anatomy* 205(6): 501–517. doi:10.1111/j.0021–8782.2004.00356.x.

Hammond, John. 1932. *Growth and the Development of Mutton Qualities in the Sheep. A Survey of the Problems Involved in Meat Production*. London: Oliver and Boyd.

Hammond, John. 1936. "Environmental Conditions and Livestock Breeding." *Probl. Zivotn* 1936: 101–112.

Hammond, John. 1947. "Animal Breeding in Relation to Nutrition and Environmental Conditions." *Biological Reviews* 22(3): 195–213. doi:10.1111/j.1469–185X.1947.tb00330.x.

Hammond, John. 1952. "Physiological Limits to Intensive Production in Animals." *British Agricultural Bulletin* 4:222–224.

Hammond, John, and Solly Zuckerman. 1950. "Measuring Growth in Farm Animals." *Proceedings of the Royal Society of London B—Biological Sciences* 137(889): 452–461. doi:10.1098/rspb.1950.0053.

Henrik, H., Jacobus J. Boomsma, and Anders Tunlid. 2014. "Symbiotic Adaptations in the Fungal Cultivar of Leaf-Cutting Ants." *Nature Communications* 5(1): 1–10.

Holmes, Samuel J. 1904. "The Problem of Form Regulation." *Archiv Für Entwicklungsmechanik Der Organismen* 17(2–3): 265–305.

Hubbell, S. P. 2001. *The Unified Neutral Theory of Biodiversity and Biogeography (MPB-32)*. Monographs in Population Biology. Princeton, NJ: Princeton University Press.

Huot, Bethany, Jian Yao, Beronda L. Montgomery, and Sheng Yang He. 2014. "Growth–Defense Tradeoffs in Plants: A Balancing Act to Optimize Fitness." *Molecular Plant* 7(8): 1267–1287.

Huxley, J. 1932. *Problems of Relative Growth*. London: Methuen & Co.

Huxley, J. S., and G. Teissier. 1936. "Terminology of Relative Growth." *Nature* 137(3471): 780–781. doi:10.1038/137780b0.

Klingenberg, Christian Peter, and H. F. Nijhout. 1998. "Competition among Growing Organs and Developmental Control of Morphological Asymmetry." *Proceedings of the Royal Society of London B—Biological Sciences* 265(1401): 1135–1139. doi:10.1098/rspb.1998.0409.

Komiyama, Tomoyoshi, Hiroyuki Kobayashi, Yoshio Tateno, Hidetoshi Inoko, Takashi Gojobori, and Kazuho Ikeo. 2009. "An Evolutionary Origin and Selection Process of Goldfish." *Gene* 430(1): 5–11. doi:10.1016/j.gene.2008.10.019.

Koyama, Takashi, Cláudia Mendes, and Christen Mirth. 2013. "Mechanisms Regulating Nutrition-Dependent Developmental Plasticity through Organ-Specific Effects in Insects." *Frontiers in Physiology* 4:263. doi:10.3389/fphys.2013.00263.

Koyama, Takashi, Maria O. Syropyatova, and Lynn M. Riddiford. 2008. "Insulin/IGF Signaling Regulates the Change in Commitment in Imaginal Discs and Primordia by Overriding the Effect of Juvenile Hormone." *Developmental Biology* 324(2): 258–265. doi:10.1016/j.ydbio.2008.09.017.

Larsen, Ellen. 2005. "Developmental Origins of Variation." In *Variation*, edited by Benedikt Hallgrímsson and Brian K. Hall, 113–129. Burlington, MA: Academic Press. doi:10.1016/B978-012088777-4/50009–0.

Masiulionis, Virginia E., Christian Rabeling, H. Henrik, Ted Schultz, Mauricio Bacci Jr., et al. "A Brazilian Population of the Asexual Fungus-Growing Ant *Mycocepurus smithii* (Formicidae, Myrmicinae, Attini) Cultivates Fungal Symbionts with Gongylidia-Like Structures." *PLoS ONE* 9(8): e103800.

Massei, Giovanna, and Susan E. Hartley. 2000. "Disarmed by Domestication? Induced Responses to Browsing in Wild and Cultivated Olive." *Oecologia* 122(2): 225–31.

Masumura, Makoto, Shin'Ichiro Satake, Hironao Saegusa, and Akira Mizoguchi. 2000. "Glucose Stimulates the Release of Bombyxin, an Insulin-Related Peptide of the Silkworm Bombyx Mori." *General and Comparative Endocrinology* 118(3): 393–399. doi:10.1006/gcen.1999.7438.

McCue, Marshall D. 2010. "Starvation Physiology: Reviewing the Different Strategies Animals Use to Survive a Common Challenge." *Comparative Biochemistry and Physiology Part A: Molecular & Integrative Physiology* 15 (1): 1–18. doi:10.1016/j.cbpa.2010.01.002.

McGhee, George R. 2015. "Limits in the Evolution of Biological Form: A Theoretical Morphologic Perspective." *Interface Focus* 5(6): 20150034.

McMeekan, C. P. 1940. "Growth and Development in the Pig, with Special Reference to Carcass Quality Characters. I." *Journal of Agricultural Science* 30(2): 276–343. doi:10.1017/S0021859600048024.

Medawar, Peter Brian, and Solly Zuckerman. 1950. "Transformation of Shape." *Proceedings of the Royal Society of London B—Biological Sciences* 137(889): 474–479. doi:10.1098/rspb.1950.0057.

Meyer, Rachel S., Ashley E. DuVal, and Helen R. Jensen. 2012. "Patterns and Processes in Crop Domestication: An Historical Review and Quantitative Analysis of 203 Global Food Crops." *New Phytologist* 196(1): 29–48. doi:10.1111/j.1469–8137.2012.04253.x.

Miller, Jeffrey P., and Rebecca Z. German. 1999. "Protein Malnutrition Affects the Growth Trajectories of the Craniofacial Skeleton in Rats." *Journal of Nutrition* 129(11): 2061–2069. doi:10.1093/jn/129.11.2061.

Minelli, A., and T. Pradeu. 2014. *Towards a Theory of Development*. New York: Oxford University Press.

Moreira, Xoaquín, Luis Abdala-Roberts, Rieta Gols, and Marta Francisco. 2018. "Plant Domestication Decreases Both Constitutive and Induced Chemical Defences by Direct Selection against Defensive Traits." *Scientific Reports* 8(1): 1–11.

Mueller, Ulrich G. 2002. "Ant versus Fungus versus Mutualism: Ant-Cultivar Conflict and the Deconstruction of the Attine Ant-Fungus Symbiosis." *American Naturalist* 160(S4): S67–S98.

Mueller, Ulrich G., Nicole M. Gerardo, Duur K. Aanen, Diana L. Six, and Ted R. Schultz. 2005. "The Evolution of Agriculture in Insects." *Annual Review of Ecology, Evolution, and Systematics* 36(1): 563–595. doi:10.1146/annurev.ecolsys.36.102003.152626.

Mueller, Ulrich G., Ted R. Schultz, Cameron R. Currie, Rachelle M. M. Adams, and David Malloch. 2001. "The Origin of the Attine Ant-Fungus Mutualism." *Quarterly Review of Biology* 76(2): 169–197.

Mummert, Amanda, Emily Esche, Joshua Robinson, and George J. Armelagos. 2011. "Stature and Robusticity during the Agricultural Transition: Evidence from the Bioarchaeological Record." *Economics & Human Biology* 9(3): 284–301. doi:10.1016/j.ehb.2011.03.004.

Nanjundiah, Vidyanand. 2020. "Preface: Early Thoughts on Phenotypic Plasticity in Development and Evolution." In *Phenotypic Switching*, edited by Herbert Levine, Mohit Kumar Jolly, Prakash Kulkarni, and Vidyanand Nanjundiah, xxi–xxxix. London: Academic Press. doi:10.1016/B978-0-12-817996-3.00035–9.

Nicholson, Daniel J., and John Dupré, eds. 2018. *Everything Flows: Towards a Processual Philosophy of Biology*. Oxford: Oxford University Press. doi:10.26530/OAPEN_1000196.

Nijhout, H. F. 1990. "Problems and Paradigms: Metaphors and the Role of Genes in Development." *BioEssays* 12(9): 441–446. doi:10.1002/bies.950120908.

Nijhout, H. F., and D. J. Emlen. 1998. "Competition among Body Parts in the Development and Evolution of Insect Morphology." *Proceedings of the National Academy of Sciences* 95(7): 3685. doi:10.1073/pnas.95.7.3685.

Nijhout, H. F., and Kenneth Z. McKenna. 2018. "The Distinct Roles of Insulin Signaling in Polyphenic Development." *Current Opinion in Insect Science* 25(February): 58–64. doi:10.1016/j.cois.2017.11.011.

O'Brien, Devin M., Cerisse E. Allen, Melissa J. Van Kleeck, David Hone, Robert Knell, et al. 2018. "On the Evolution of Extreme Structures: Static Scaling and the Function of Sexually Selected Signals." *Animal Behaviour* 144 (October): 95–108. doi:10.1016/j.anbehav.2018.08.005.

Raff, Rudolf A., and Belinda J. Sly. 2000. "Modularity and Dissociation in the Evolution of Gene Expression Territories in Development." *Evolution & Development* 2(2): 102–113. doi:10.1046/j.1525–142x.2000.00035.x.

Razifard, Hamid, Alexis Ramos, Audrey L. Della Valle, Cooper Bodary, Erika Goetz, et al. 2020. "Genomic Evidence for Complex Domestication History of the Cultivated Tomato in Latin America." *Molecular Biology and Evolution* 37(4): 1118–1132. doi:10.1093/molbev/msz297.

Reichling, Tim D., and Rebecca Z. German. 2000. "Bones, Muscles and Visceral Organs of Protein-Malnourished Rats (Rattus Norvegicus) Grow More Slowly but for Longer Durations to Reach Normal Final Size." *Journal of Nutrition* 130(9): 2326–2332. doi:10.1093/jn/130.9.2326.

Rindos, D. 1984. *The Origins of Agriculture: An Evolutionary Perspective*. London: Academic Press.

Rodríguez, Rafael Lucas, and William G. Eberhard. 2019. "Why the Static Allometry of Sexually-Selected Traits Is So Variable: The Importance of Function." *Integrative and Comparative Biology* 59(5): 1290–1302. doi:10.1093/icb/icz039.

Rosenthal, Joshua P., and Rodolfo Dirzo. 1997. "Effects of Life History, Domestication and Agronomic Selection on Plant Defence against Insects: Evidence from Maizes and Wild Relatives." *Evolutionary Ecology* 11(3): 337–355. doi:10.1023/A:1018420504439.

Roux, W. 1881. *Der Zuchtende Kampf Der Teile, Oder Die 'Teilauslese' Im Organismus (Theorie Der 'Funktionellen Anpassung')*. Leipzig: Wilhelm Engelmann.

Ruxton, G. D., W. L. Allen, T. N. Sherratt, and M. P. Speed. 2019. *Avoiding Attack: The Evolutionary Ecology of Crypsis, Aposematism, and Mimicry*. Oxford: Oxford University Press.

Shik, Jonathan Z., Ernesto B. Gomez, Pepijn W. Kooij, Juan C. Santos, William T. Wcislo, et al. 2016. "Nutrition Mediates the Expression of Cultivar–Farmer Conflict in a Fungus-Growing Ant." *Proceedings of the National Academy of Sciences* 113(36): 10121. doi:10.1073/pnas.1606128113.

Shingleton, Alexander W., W. Anthony Frankino, Thomas Flatt, H. Frederik Nijhout, and Douglas. J. Emlen. 2007. "Size and Shape: The Developmental Regulation of Static Allometry in Insects." *BioEssays* 29(6): 536–548. doi:10.1002/bies.20584.

Silliman, Brian R., and Steven Y. Newell. 2003. "Fungal Farming in a Snail." *Proceedings of the National Academy of Sciences* 100(26): 15643. doi:10.1073/pnas.2535227100.

Sinclair, Thomas R., Larry C. Purcell, and Clay H. Sneller. 2004. "Crop Transformation and the Challenge to Increase Yield Potential." *Trends in Plant Science* 9(2): 70–75. doi:10.1016/j.tplants.2003.12.008.

Smith, B. D. 1998. *The Emergence of Agriculture*. Scientific American Library Series. New York: Scientific American Library.

Spurway, Helen. 1955. "The Causes of Domestication: An Attempt to Integrate Some Ideas of Konrad Lorenz with Evolution Theory." *Journal of Genetics* 53(2): 325–62.

Suzuki, Yuichiro, Kenneth Z. McKenna, and H. Frederik Nijhout. 2020. "Regulation of Phenotypic Plasticity from the Perspective of Evolutionary Developmental Biology." In *Phenotypic Switching*, edited by Herbert Levine, Mohit Kumar Jolly, Prakash Kulkarni, and Vidyanand Nanjundiah, 403–442. London: Academic Press. doi:10.1016/B978-0-12-817996-3.00012-8.

Tang, Hui Yuan, Martha S. B. Smith-Caldas, Michael V. Driscoll, Samy Salhadar, and Alexander W. Shingleton. 2011. "FOXO Regulates Organ-Specific Phenotypic Plasticity in Drosophila." *PLoS Genetics* 7(11): e1002373. doi:10.1371/journal.pgen.1002373.

Thompson, Daniel. 2019. "Diet-Induced Plasticity of Linear Static Allometry Is Not So Simple for Grasshoppers: Genotype–Environment Interaction in Ontogeny Is Masked by Convergent Growth." *Integrative and Comparative Biology* 59(5): 1382–1398. doi:10.1093/icb/icz137.

Thompson, Darcy Wentworth. 1942. *On Growth and Form*. Cambridge: Cambridge University Press.

Tian, Feng, Natalie M. Stevens, and Edward S. Buckler. 2009. "Tracking Footprints of Maize Domestication and Evidence for a Massive Selective Sweep on Chromosome 10." *Proceedings of the National Academy of Sciences* 106(Supplement 1): 9979. doi:10.1073/pnas.0901122106.

Timmusk, Salme, Lawrence Behers, Julia Muthoni, Anthony Muraya, and Anne-Charlotte Aronsson. 2017. "Perspectives and Challenges of Microbial Application for Crop Improvement." *Frontiers in Plant Science* 8:49. doi:10.3389/fpls.2017.00049.

Tubiello, Francesco, Josef Schmidhuber, Mark Howden, Peter G. Neofotis, Sarah Park, et al. 2008. *Climate Change Response Strategies for Agriculture: Challenges and Opportunities for the 21st Century*. Agriculture and Rural Development Discussion Paper 42.Washington, DC: The World Bank.

Vellinga, Else C. 2004. "Ecology and Distribution of Lepiotaceous Fungi (Agaricaceae)–A Review." *Nova Hedwigia* 78(3–4): 273–299.

Verges, J. B. 1939. "Effect of the Plane of Nutrition on the Carcase Quality of Suffolk Cross Lambs." *Suffolk Sheep Society Yearbook*, 27–37.

Van der Waaij, E. H. 2004. "A Resource Allocation Model Describing Consequences of Artificial Selection under Metabolic Stress." *Journal of Animal Science* 82(4): 973–981. doi:10.1093/ansci/82.4.973.

Von Bertalanffy, Ludwig. 1933. *Modern Theories of Development: An Introduction to Theoretical Biology*. Translated by J. H. Woodger. Oxford: Oxford University Press.

Wagner, G. P. 2018. *Homology, Genes, and Evolutionary Innovation*. Princeton, NJ: Princeton University Press.

Wallace, L. R. 1944. "Feeding during Pregnancy." *Report of Proceedings of British Society of Animal Production* 1944(2): 58–61. doi:10.1017/S0369852100000075.

Weismann, August. 1896. "Germinal Selection." *The Monist* 6(2): 250–293.

West, Stuart A., Ashleigh S. Griffin, and Andy Gardner. 2007. "Evolutionary Explanations for Cooperation." *Current Biology* 17(16): R661–R672. doi:10.1016/j.cub.2007.06.004.

Wilkins, Adam S., Richard W. Wrangham, and W. Tecumseh Fitch. 2014. "The 'Domestication Syndrome' in Mammals: A Unified Explanation Based on Neural Crest Cell Behavior and Genetics." *Genetics* 197(3): 795. doi:10.1534/genetics.114.165423.

Wolpert, L., and J. H. Lewis. 1975. "Towards a Theory of Development." In *Biology of Aging and Development*, edited by G. Jeanette Thorbecke, 21–34. Boston: Springer. doi:10.1007/978-1-4684-2631-1_4.

Woltereck, Richard. 1909. "Weitere Experimentelle Unter-Suchungen Uber Artveranderung, Speziell Uber Das Wesen Quantitativer Artunterschiede Bei Daphniden." *Versuschung Deutsch Zool. Gesell.* 19:110–172.

Woodger, J. H. 1929. *Biological Principles: A Critical Study*. International Library of Philosophy. London: Taylor & Francis.

Woodger, J. H. 1945. "On Biological Transformations." In *Essays on Growth and Form*, 95–120. Oxford: Clarendon.

Zeder, Melinda A. 2006. "Central Questions in the Domestication of Plants and Animals." *Evolutionary Anthropology: Issues, News, and Reviews* 15(3) 105–117. doi:10.1002/evan.20101.

Zeder, Melinda A. 2015. "Core Questions in Domestication Research." *Proceedings of the National Academy of Sciences* 112(11): 3191. doi:10.1073/pnas.1501711112.

Zeder, Melinda A. 2018. "Why Evolutionary Biology Needs Anthropology: Evaluating Core Assumptions of the Extended Evolutionary Synthesis." *Evolutionary Anthropology: Issues, News, and Reviews* 27(6): 267–284.

Zeder, Melinda A., Eve Emshwiller, Bruce D. Smith, and Daniel G. Bradley. 2006. "Documenting Domestication: The Intersection of Genetics and Archaeology." *Trends in Genetics* 22(3): 139–155. doi:10.1016/j.tig.2006.01.007.

Zelditch, Miriam Leah. 1988. "Ontogenetic Variation in Patterns of Phenotypic Integration in the Laboratory Rat." *Evolution* 42(1): 28–41. doi:10.1111/j.1558-5646.1988.tb04105.x.

Zoltán, Tóth, Gábor Gyulai, Zoltán Szabó, Lajos Horváth, and László Heszky. 2007. "Watermelon (*Citrullus l. Lanatus*) Production in Hungary from the Middle Ages (13th Century)." *Hungarian Agricultural Research* 4:14–19.

Zuckerman, S. 1950. "A Discussion on the Measurement of Growth and Form." *Proceedings of the Royal Society B—Biological Sciences* 137(889):433–523.

14 The Convergent Evolution of Agriculture in Humans and Fungus-Farming Ants

Ted R. Schultz

Leaf-cutter ants are the preeminent herbivores of the New World tropics. Below ground, their colonies contain millions of worker ants that occupy thousands of chambers. Above ground, the soil they excavate forms huge mounds that dominate the forests and grasslands of Central and South America. Hundreds of thousands of ant foragers stream outward from these mounds, sharing the radiating, well-worn paths with equal numbers of inward-streaming sisters, the latter returning with cut leaf fragments held aloft in their mandibles like miniature pennants. Marveling at this spectacle, generations of humans have reasoned that leaf-cutter ants were vegetarians, consuming the cuttings in subterranean dining halls. It remained for the nineteenth-century naturalist Thomas Belt to discover and record the unexpected truth about the purpose of the leaves: "I believe the real use [the ants] make of [the leaf fragments] is as a manure, on which grows a minute species of fungus, on which they feed;—that they are, in reality, mushroom growers and eaters" (Belt 1874, 79). Belt's description of the ants as "mushroom growers" so concisely communicated the behavior he observed that ant researchers have continued to use similar agricultural terminology ever since.

The attribution of agriculture to a nonhuman animal was paradigm-shifting and has been the subject of occasional skepticism. It has been challenged as anthropomorphic (Kermarrec, Decharme, and Febvay 1986; Rodgers 2008). It has also been challenged on the basis that it is more likely that the fungi are cultivating the ants than vice versa (Ridley, Howse, and Jackson 1996; North, Jackson, and Howse 1997; see also below). In this book, we regard agriculture as a special subset of the more general phenomenon of symbiosis. Under this view, it is appropriate to extend the aforementioned skepticism to the unique status traditionally afforded to human agriculture, which, like fungus-farming ant agriculture, is also a symbiosis and about which we might also reasonably question which of the symbionts is in control. Although it has been accelerated by intent, learning, and cultural transmission of information, human agriculture is at its core a biological phenomenon, and, as such, it is subject to the rules of Darwinian natural selection (Rindos 1984, 1989; Allaby 2010; chapter 10, this volume). All agricultural symbioses may therefore prove to share analogous elements, and those elements may in some cases be the results of convergent evolution. If so, then the comparative study of multiple agricultural symbioses may reveal common evolutionary mechanisms that influence the origins, population-genetic and coevolutionary dynamics, and genomic architectures underlying agricultural evolution. The purpose of this chapter is to summarize what is broadly known about human agricultural evolution and to explore parallels, if any, with the evolution of fungus-farming ant agriculture.

Agriculture is a form of symbiosis in which one symbiont (the "farmer") benefits by cultivating the other symbiont (the "crop" or, if the crop is genetically modified due to the symbiosis, the "domesticate"). Symbioses between two organisms can be broadly classified as commensalism (in which neither party benefits or incurs a cost), parasitism (in which one party benefits at the expense of the other), and mutualism (in which both parties benefit). The seeming straightforwardness of these categories is complicated by their context-dependency—that is, the nature of a symbiotic interaction can change depending on multiple factors, one of which is number of participating symbionts. As knowledge about symbioses increases, we are realizing that multipartite symbioses are common and are likely the rule, so that at some point symbiology grades into community ecology. This is proving true even for human agriculture (Fuller and Stevens 2017; chapter 10, this volume).

In this chapter, agricultural terminology will be used to refer to the full range of human cultivation and domestication of plants, animals, fungi, eukaryotic microorganisms, and bacteria. It will focus mostly on nutritional agricultural symbioses, or those that benefit the farmer by providing nutrition, but it is important to bear in mind the many agricultural symbioses that serve nonnutritional needs such as protection from other organisms (guard dogs, *Streptomyces*), production of materials (trees, bottle gourds, cotton, silkworms, alpaca), generation of physical labor (horses, oxen, camels, elephants), induction of pleasure and enlightenment (yeast, coffee, kava, tobacco, *Cannabis*), aesthetics (diverse ornamental plants, goldfish), and so on.

In the introduction to this book, *agriculture* is defined as large-scale cultivation upon which the farmers have become economically (i.e., obligately) dependent. In the first two sections below I will briefly introduce the fungus-farming ant agricultural symbiosis and summarize the differences between human and ant agriculture. In the remaining sections I will take advantage of research on the convergent features of multiple, diverse human agricultural systems to show that many of the evolutionary stages shared by human agricultural symbioses are also shared by the fungus-farming ant agricultural symbosis. These stages include (1) preagricultural hunting-gathering and niche construction by central-place foraging social groups; (2) preagricultural incidental and, subsequently, intentional cultivation; (3) domestication; (4) the parallel rise of biodiverse agroecosystems; and (5) the postagricultural rise of highly complex societies.

The Fungus-Farming Ants

Fungus-farming, or "attine," ants (subfamily Myrmicinae, tribe Attini, subtribe Attina) are a clade of (as of this writing) 246 species, including five fossil species, in 20 genera. They are descended from a common ancestor that lived in South America around 55–65 Mya, shortly after the end-of-Cretaceous mass-extinction event that briefly shut down photosynthesis (Janzen 1995; Kaiho et al. 2016). All attine ants are native to the New World and all obligately depend on fungus-farming for food. When departing from the maternal nest on her mating flight, a daughter queen carries a pellet of fungus from her mother's garden to use as the starting culture for her new garden. The conspicuous leaf-cutting ants, which make up 20% of attine ants, have received the great majority of scientific study because they are the primary pests of human agriculture in Central and South America (Cherrett 1989; Della Lucia 2011). The remaining 80% of fungus-farming ants are mostly cryptic species with small colony sizes that are rarely encountered by humans and about which, with few exceptions, we know very little (Weber 1972; Mehdiabadi and Schultz 2009; Hölldobler and Wilson 2010).

The Five Ant Agricultural Systems

As mentioned above, the definition of "agriculture" used in this book is large-scale cultivation upon which the farmers have become economically (i.e., obligately) dependent. Because all fungus-farming ants must cultivate fungi in order to survive, the term "agriculture" is used here to refer to the entire diversity of attine-ant fungus farming. However, as will be discussed below, the scale on which such cultivation takes place varies dramatically across attine-ant species. All fungus-farming ants grow fungi in the order Agraricales. Most grow fungi in the family Agaricaceae, which also includes the fungus most cultivated by humans for food, the button mushroom *Agaricus bisporus* (Miles and Chang 2004). Based largely on ant-fungal associations, ant agriculture is divided into five systems (figure 14.1): (i) lower agriculture, the ancestral agricultural system, (ii) coral-fungus agriculture, (iii) yeast agriculture, (iv) higher agriculture, and (v) leaf-cutter agriculture, a derived subset of higher agriculture (Schultz and Brady 2008; Mehdiabadi and Schultz 2009; Hölldobler and Wilson 2010; Della Lucia 2011; Branstetter et al. 2017).

In lower agriculture, 84 species of ants in 11 genera cultivate multiple species of parasol mushrooms currently assigned to the genera *Leucocoprinus, Leucoagaricus*, and *Lepiota* in the tribe Leucocoprineae (family Agaricaceae) (figure 14.1A). In coral-fungus agriculture, a derived clade of 29 species in the attine-ant genus *Apterostigma* are the only fungus-farming ants that cultivate fungi outside of the Agaricaceae, multiple species of the genus *Myrmecopterula* in the coral-fungus family Pterulaceae (figure 14.1B). In yeast agriculture, a derived subset of 17 species in the attine-ant genus *Cyphomyrmex* cultivates a small clade of fungal species descended from a lower fungal ancestor that, when associated with ants, grow in a unicellular "yeast-like" phase otherwise unknown in the order Agaricales (figure 14.1C). In higher agriculture, 111 ant species (59 non-leaf-cutting and 52 leaf-cutting higher-attine ants) largely cultivate a clade of multiple fungal species descended from a lower fungal ancestor that became an obligate symbiont (figure 14.1D). Unlike their lower fungal ancestors, higher-attine fungi are never found living apart from ants, they are polyploid, and they consistently produce "gongylidia," nutritious swollen hyphal tips that are preferentially harvested and eaten by the ants. Within higher agriculture, leaf-cutter agriculture is the product of a major evolutionary transition in which a non-leaf-cutting, higher-attine ant ancestor acquired the ability to cut, process, and utilize fresh vegetation (leaves, grasses, and flowers) as the substrate for its fungus gardens (figure 14.1E) (Hölldobler and Wilson 1990; Mehdiabadi and Schultz 2009; Hölldobler and Wilson 2010; Leal-Dutra et al. 2020).

Agricultural System Boundaries

As mentioned above, garden fungi are typically transmitted vertically from one generation to the next, carried by daughter queens when founding new colonies. Genetic data indicate

Figure 14.1
Representative ants and gardens of the five fungus-farming ant agricultural systems: (A) lower agriculture: *Mycetophylax asper*; (B) coral-fungus agriculture: *Apterostigma dentigerum;* (C) yeast agriculture: *Cyphomyrmex* sp. (*rimosus* group); (D) higher agriculture: *Trachymyrmex septentrionalis;* and (E) leaf-cutter agriculture: *Atta cephalotes*. (F) A minor worker (left) and a soldier (right), sisters from the same colony of *Atta cephalotes*, illustrating the extreme worker polymorphism of leaf-cutting ants. Reproduced with the permission of the photographers: Don Parsons (A); Alex Wild (B, C, D, F); Karolyn Darrow (E).

that horizontal transmission occurs frequently over evolutionary time periods, however, so that closely related ants may cultivate distantly related fungi and the same fungal species may be cultivated by distantly related ants. Horizontal transfer can occur when ants acquire fungal cultivars from each other (Adams 2000a; Green, Mueller, and Adams 2002; Howe, Schiøtt, and Boomsma 2018). It can also occur when ants acquire free-living conspecifics of cultivated fungi, which are known to occur in lower agriculture and yeast agriculture (Mueller, Rehner, and Schultz 1998; Vo, Mueller, and Mikheyev 2009). When cultivated by ants, fungi are propagated clonally, but when free-living, they reproduce sexually, so at least in lower and yeast agriculture there is occasional sexual recombination due to genetic continuity between cultivated and free-living populations. Importantly, horizontal transfer events are constrained by the boundaries of the five agricultural systems, i.e., the cultivation of a fungus from one agricultural system by an ant species from a different agricultural system is rare. Violations of the boundaries defining yeast agriculture and coral-fungus agriculture are entirely unknown, i.e., all observed associations occur solely between ants and fungi within those groups. Only one lower-attine ant species is known to cultivate a higher-attine fungus (figure 14.2) (Schultz et al. 2015), whereas a few non-leaf-cutting higher-attine ant species have been found cultivating lower-attine fungi (Mueller, Rehner, and Schultz 1998; Mueller et al. 2018; Solomon et al. 2019). Finally,

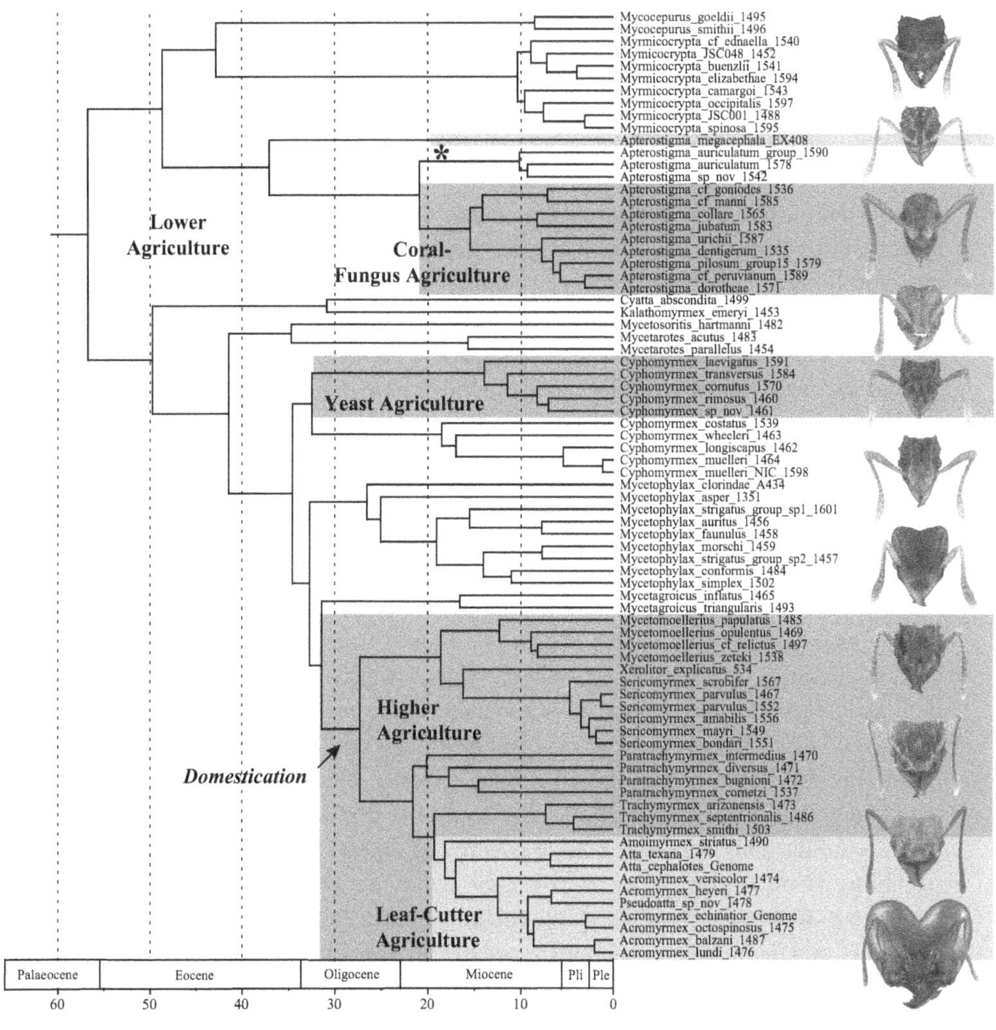

Figure 14.2
Time-dated chronogram of fungus-farming ants. Numbers at bottom indicate millions of years before present. The major agricultural systems (labeled) are characterized by the nearly monolithic fidelity of their component ants and fungi. The asterisk (*) indicates the only lower-attine ant species, *Apterostigma megacephala*, known to cultivate a higher-attine fungus. Adapted from Branstetter et al. (2017).

most leaf-cutting ants cultivate *Leucoagaricus gongylophorus*, a single species of higher-attine fungus, but a few leaf-cutting species have been found cultivating other species of higher-attine fungi (Mueller et al. 2018). The biological mechanisms constraining the agricultural-system boundaries remain unknown, but it has been suggested that they may be physiological, with ants in each agricultural system dependent on that system's fungi to supply an essential, system-specific combination of nutrients and/or micronutrients or, alternatively, that the mechanisms may be microbial-symbiotic, with each system hosting a system-specific, non-interchangeable consortium of microbes (Mueller et al. 2005; Seal and Mueller 2014; Schultz et al. 2015; but see chapters 3 and 4, this volume).

Crop Diseases and Biocontrol

Fungus-farming ant agriculture has crop diseases. Most notably, mycoparasitic ascomycete fungi in the closely related genera *Escovopsis* and *Escovopsioides* have been part of the symbiosis since its origin and have coevolved and codiversified along with the ants and their fungal cultivars (Currie et al. 1999a, 1999b, 2003; Gerardo et al. 2004; Meirelleset al. 2015b; chapter 11, this volume). The ants control them by weeding and by the application of antibiotics (i.e., "herbicides"), some of which originate in the ants, notably in their metapleural glands, antibiotic-producing glands that occur in almost all ants. In addition, most fungus-farming ants host actinomycete bacteria primarily in the genus *Pseudonocardia*, closely related to *Streptomyces*, the genus from which the majority of human antibiotics are derived. Housed in specialized crypts on the ants' integuments and nourished by glandular secretions, mutualistic *Pseudonocardia* produce antibiotics that are effective against *Escovopsis* (Currie 1999b, 2006; Li et al. 2018). The attine fungal cultivars also produce defensive antifungal chemicals (Worsley et al. 2018).

Differences between Human and Ant Agriculture

Before discussing possibly convergent traits in ant and human agriculture, it is important to point out the many differences between the two. Human agriculture is the result of diffuse coevolutionary processes that began with or even prior to the origin of *Homo sapiens* around 200 thousand years ago and domestication processes that began over ten thousand years ago. In contrast, ant agriculture is the result of processes that began as many as 65 Mya. Whereas human farmers represent a single species, fungus-farming ants comprise 246 known extant and five fossil species. Although much of human proto-agriculture (i.e., niche construction, ecosystem engineering, incidental cultivation, and early domestication, discussed below) occurred unintentionally (i.e., without conscious planning), intent has played an increasingly significant role in the evolution of human agriculture. Within the span of a single lifetime, humans observe, learn, and modify their behaviors in order to manipulate nature and obtain desired outcomes. They then transmit that information to the next generation. Fungus-farming ants have also "learned," but changes in their hard-wired behaviors have occurred over evolutionary time spans through the trial-and-error process of natural selection. In the course of that evolution ants have become obligate farmers, unable to survive without their fungal crops. On a biological level humans are, strictly speaking, facultative farmers, individually able to return to hunting-gathering or another nonagricultural way of obtaining food. On economic and cultural levels, however, modern humans are perhaps not quite so different from ants in their dependence on industrial-scale agriculture for maintaining current global population levels and infrastructure.

Humans cultivate thousands of species of plants, animals, fungi, and bacteria. Fungus-farming ants grow only fungi, and only fungi drawn from two subclades of two families in the order Agaricales, although they simultaneously host diverse microbes, a subset of which occur more or less consistently in their gardens (Pinto-Tomás et al. 2009; Aylward et al. 2012, 2014; Worsley et al. 2018). Ants grow their fungus gardens in climate-controlled chambers, usually underground, in a way that is similar to human greenhouse agriculture. Depending on the cultivar, humans use diverse cultivation techniques, although their highest-biomass crops are grown in open-plot, arable fields, which they may intensively prepare and to which they may add water, fertilizers, and pesticides. In contrast to human photosynthetic cultivars that derive nourishment from soil and sunlight, the heterotrophic attine fungal cultivars require living or dead organic matter for food, so, like their hunting-gathering ancestors, fungus-farming ants continue to forage outside the nest but, unlike those ancestors, they forage for food for their fungus gardens.

Hunting-Gathering, Niche Construction, and Cultivation

Humans

The end of the Younger Dryas near-glacial period around 12,000 years ago coincided with generally warmer and wetter conditions worldwide. Within a few thousand years, human agriculture had originated independently in the Near East, Mesoamerica, and South China, and by the mid-Holocene it had arisen in parallel in as many as twenty different locations (Smith 1998b; Diamond 2002; Fuller et al. 2014; Zeder 2018; chapter 10, this volume). Full-blown agriculture was not suddenly "invented" in each of these locations at a particular time in the past, however. It was preceded by hundreds of thousands of years of dietary evolution, during which humans broadened the plant-based portions of their diets, and by coevolution with plants, some of which would eventually become human domesticates (Rindos 1984; Hillman and Wollstonecroft 2014; Allaby et al. 2015). This protracted adaptive process required human genetic evolution (Mathieson et al. 2015) as well as the development of technologies such as fire and stone tools, the latter for grinding seeds and digging up roots (Hillman 1989; Hillman and Wollstonecroft 2014). Humans were cooking tubers at least 170,000 years ago (Wadley et al. 2020) and grass seeds (including the future domesticate sorghum) at least 105,000 years ago (Mercader 2009). Fifty thousand years ago, Neanderthals were cooking and consuming plants that would subsequently be domesticated by *Homo sapiens* (Henry, Brooks, and Piperno 2011, 2014).

Human Niche Construction

Paleolithic hunting-gathering humans lived in social groups and had strong impacts on their local biotic environments (Smith 2011a, 2011b; Petraglia 2017). Foraging individually or in groups, they transported desirable plants and animals back to their settlements to prepare, eat, and perhaps store for future use. As a result, without conscious planning or intent, hunting-gathering humans engineered ecological niches to which communities of plants and animals adapted (Smith 2011b, 2011a; Allaby et al. 2015; Zeder 2018). At some point, beginning more than 45,000 years ago, this niche construction became a conscious goal and humans began to intentionally "domesticate" their environments (Roberts et al. 2017). They cleared the land around and otherwise defended favored plants such as fruit trees (Ames 1939; Rindos 1984; Anderson and Wohlgemuth 2012). They diverted water to enhance the growth of favored wild plants (Anderson and Wohlgemuth 2012). Perhaps most universally, they employed fire to alter their landscapes. They burned vegetation to encourage the growth of early-succession grasses and other food plants and to eliminate unfavored, post-fire-succession species (Smith 1998b; Anderson and Wohlgemuth 2012; Roberts et al. 2017). They also burned to drive out small game, which they hunted, and to increase small-animal hunting productivity (Bliege Bird et al. 2008). So-called "firestick agriculture" is still practiced by the Aboriginal people of Australia (Bliege Bird et al. 2008). Repeated burning coupled with deposition of organic matter over long periods of time created the nutrient-rich *terra preta* (black soils) of the Amazon (Denevan 2001).

Camp Followers

In addition to larger-scale niche construction, the immediate areas around human settlements were ideal habitats for a range of disturbance-adapted plant and animal species, which became increasingly associated with humans. Humans also incidentally imported plant species into their settlements when they returned with gathered food plants, dispersing the seeds (or other propagules) into the settlement area, particularly into refuse piles and waste middens. Camp-following species thus included a mixture of species useful, nonuseful, or even detrimental to humans. Among the useful species, future domesticates such as chenopods (e.g., quinoa), cucurbits (melons, squash, pumpkin, cucumber), sunflower, tomato, and chile pepper are thought to have arisen as camp-followers (Rindos 1984; Harlan 1992). Examples of ultimately useful camp-following animals include dogs, cats, and spiders, whereas examples of nonuseful camp-following animals include house mice, weevils, and fruit flies (Fuller and Stevens 2017; chapter 10, this volume). As will be discussed below for ants, camp-following species are not unique to humans; the curcurbit *Cucumis humifructus*, the "aardvark pumpkin," is an obligate camp-follower of the African aardvark (Kirkbride 1993).

Interpreted retrospectively, in their daily encounters with thousands of species of plants and animals, including both camp-following and forage/prey species, preagricultural humans were unintentionally "auditioning" them as possible future domesticates, and, to avoid anthropomorphism, it should be said that those species were likewise auditioning humans as possible future domesticators. In evolutionary terms, some plants and animals were preadapted for agricultural symbioses whereas others were decidedly maladapted for that role (Diamond 1997, 2002, 2012; Smith 1998b; Zeder 2012). Interestingly, guilds of plants and animals not directly useful or sometimes even detrimental to humans have nonetheless persisted in human agroecosystems, in many cases carried along by the same initially unconscious artificial selection that humans applied to their cultivars (Harlan 1992; Spahillari et al. 1999; Howard, Archer, and Turley 2016; chapter 10, this volume). In an archaeobotanical study of a 23,000-year-old preagricultural site in Israel, Snir et al. (2015) report over 140 plant species gathered by humans, including the future domesticates wheat and barley and 13 well-known current weeds of human agriculture. In another study of three horizons spanning a 300-year period beginning 6,900 years ago and documenting the early stages of rice domestication, Fuller et al. (2009) similarly report more than 50 species of plants, including what appear to be small-seeded future weeds of rice cultivation. One of the most widespread invasive plants on the planet, Johnsongrass, originated as a human food plant closely related to domesticated sorghum (Paterson et al. 2020).

Cultivation and Mixed Food-Acquisition Strategies

At various points in time and in different places, humans began to intentionally cultivate plants. For some vegetatively propagated plants, such as manioc (*Manihot* spp.), cultivation is no more complicated than sticking a cutting into the ground during the rainy season (Harlan 1992; chapter 10, this volume). The cultivation of yams and other root crops similarly requires little more than reburying the tuber, as is often done by hunter-gatherers as a method of storing surplus forage, or of cutting off and discarding the woody heads, which frequently resprout without further effort (Harlan 1992; chapter 10, this volume). For annual plants, rudimentary cultivation requires only the knowledge that seeds gathered in one year can sprout into plants in the following year and cultivation can consist of no more than broadcasting seeds into a suitable habitat without additional care. The Cocopa societies of the lower Colorado River, for example, broadcast panic-grass (*Panicum* sp.) seeds harvested in the previous year in muddy river bottoms recently exposed by receding flood waters, giving them no further attention until harvest (Smith 2001a).

Using such simple and non-time-consuming methods, hunter-gatherers added cultivation to their repertoires of mixed food-acquisition strategies. Tudge (1998) describes Paleolithic proto-farmers as "hobby farmers," cultivating small plots of plants as a back-up while continuing to hunt for animals and gather wild plants for food. He points out the ecological advantage of nonreliance on a single food source, especially with regard to local overhunting. Indeed, the historical record suggests that mixed food-acquisition strategies represent stable states in a complex continuum that do not necessarily replace pure hunting-gathering or inevitably lead to domestication and agriculture (Harris 1989; Smith 1998a, 1998b, 2001a, 2001b; chapter 10, this volume).

Ants

Niche Construction

The common ancestor of all ants was a hunter-gatherer, taking prey as food primarily for the colony's larvae and taking plant carbohydrates as food primarily for the adults. It lived in eusocial family groups consisting of the mother queen and her nonreproductive (worker) daughters, cooperatively rearing the next generation(s) of workers as well as reproductive (virgin queen) daughters/sisters and reproductive sons/brothers. (There are no worker males in ant colonies.) The familial colony occupied a semipermanent nest and, as a central-place forager, influenced the ecologies of the organisms living within its territory. This strategy, which originated ~103–124 Mya (Borowiec et al. 2017), was overwhelmingly successful, resulting in the estimated >25,000 extant species of ants (Ward 2014). Today, ants are keystone species in nearly every ecosystem on the planet and are universally recognized as premier niche constructors/ecosystem engineers (Folgarait 1998; Schultz and McGlynn 2000; Jouquet et al. 2006; Vandermeer and Perfecto 2007; Sanders and Veen 2011; Meyer et al. 2013; chapter 8, this volume). Their core hunting-gathering life-history strategy has repeatedly given rise to increasingly complex, mixed food-acquisition strategies in which ants supplement hunting-gathering with, for example, honeydew "milked" from aphids, lipid-rich elaiosomes (food rewards) attached to ant-dispersed seeds, or carbohydrate and protein rewards supplied by plants that ants protect from herbivory (Beattie 1985; Hölldobler and Wilson 1990; Jolivet 1996; Rico-Gray and Oliveira 2007; chapter 8, this volume). Other ant lineages became specialized carnivores, in some cases adapted to very specific prey items such as spider eggs, polyxenid millipedes, termites, or other ants. The army ants have become both carnivorous and nomadic, managing arthropod prey populations across territories equivalent to those occupied by medium-to-large-sized vertebrate predators (Tobin 1994; Gotwald 1995; Brown 2000).

Camp-Following Fungi

Around 55–65 Mya in South America, the core hunting-gathering life-history strategy gave rise to fungus-farming ant agriculture. Although all extant attine ants obligately grow fungi for food, their ancestors almost certainly passed through a period of facultative fungivory—that is, they were hunter-gatherers with a diet that sometimes included fungi. Although previously thought to be relatively rare in ants (Tobin 1994), multiple new examples of fungivory have been discovered in the past two decades (Mueller et al. 2001; Witte and Maschwitz 2008; Blatrix et al. 2012; Beeren, Mair, and Witte 2014; Mayer et al. 2018; chapter 8, this volume). Based on the phylogenetics, behavior, and ecology of extant species, the ancestral attine ant lived and foraged in the leaf litter, where, like its modern descendants, it frequently encountered free-living leucocoprineaceous fungi (Vellinga 2004; Schultz et al. 2005; Vo, Mueller, and Mikheyev 2009). If the ancestral ant fed on them, the fungi could have benefited by being dispersed. All ants strain food through their infrabuccal pockets, a specialized organ in the mouth that filters out solid particles, including viable fungal mycelia and spores. The filtrate accumulates into "infrabuccal pellets," which ants expel one or more times a day in colony refuse piles or in other locations inside or in the vicinity of the nest (Bailey 1920; Letourneau 1998; Mueller et al. 2001; Little et al. 2003). Some fungi are known to be dispersed in this way, including other ant-associated fungi (Bailey 1920; Wheeler and Bailey 1920; Mueller et al. 2001; Blatrix et al. 2013; Mayer et al. 2018).

Lower-attine fungi occasionally produce structures that appear to be homologous with gongylidia, the previously mentioned nutritious swollen hyphal tips that are consistently produced in domesticated higher-attine fungi (Möller 1893; Urich 1895; Weber 1972, 1979; Masiulionis et al. 2014). Ancestral leucocoprineaceous fungi may have recruited ants as dispersal agents by providing gongylidia as food rewards (Mueller et al. 2001; Schultz et al. 2005), a strategy evolutionarily convergent with the recruitment of vertebrates (including humans) by plants using fruits and the recruitment of ants by plants using elaiosomes (nutritious seed appendages) (Lengyel et al. 2010; chapter 8, this volume). If leucocoprineaceous fungi are camp followers of leaf-litter-dwelling ants, then we might expect to find them more generally associated with colonies of extant leaf-litter-dwelling, non-fungus-growing close relatives of fungus-farming ants. Of note is that unidentified fungi have been reported to grow consistently on ant-constructed walls of accumulated detritus in the nests of species in the genera *Wasmannia* and *Blepharidatta*, both closely related to fungus-farming ants (Wheeler 1901; Diniz, Brandão, and Yamamoto 1998; Brandao et al. 2001; Rabeling, Verhaag, and Mueller 2006; Schultz and J. Sosa-Calvo, personal observation). Identifying these fungi will provide a first test of the camp-follower hypothesis for the origin of the fungus-farming ant agricultural symbiosis.

Domestication

Humans

In the introduction to this book, we define *domestication* as the genetic modification of one species by another in ways that benefit the modifying species (the farmer) but that would have reduced the fitness of the modified species (the domesticate) in its original niche. Note that this definition is neutral about genetic modifications in the farmer and that it is also neutral about the current fitness of the domesticate relative to its former fitness in the wild. It is possible that in the agricultural niche the domesticate has become more fit than, as equally fit as, or less fit than its undomesticated progenitor. Although they likely could not survive without humans, domesticated rice, wheat, and maize have become some of the most abundant plants on Earth (Collins 2015), and the worldwide population of cattle numbers over 1.4 billion (FAOSTAT 2020).

Domesticated plants possess suites of morphological traits that set them apart from their wild progenitors, some controlled by single loci, others multigenically controlled and producible via more than one developmental-genetic pathway. Depending on the plant (grass, legume, amaranth, cucurbit, and so on), this domestication syndrome includes increased seed size, decreased seed coat thickness, synchronous seed ripening, seed indehiscence, nondormant seeds, increased apical position of seeds on stalks, reduced stalk branching, and seed pod/spike/ear/panicle indehiscence (Smith 1998b; Bellwood 2005; Weiss, Kislev, and Hartmann 2006; Fuller 2007; chapter 10, this volume). Domestication traits in animals vary more widely across species, but include reduction in brain size (nearly universal), reduction in body size (common), reduction in female body size (common for managed herd animals), and reduction in tooth size (pigs) (Smith 1998b; Zeder 2012). Early evolving domestication traits are the products of unintentional artificial selection. For example, if preagricultural human cultivators harvested spikes from grasses when seeds were ripe, they would have missed seeds or spikes that had already fallen off (dehisced), and, if they sowed some of the harvested seeds the next year, those seeds would contain a higher frequency of indehiscent individuals. Repeated from year to year, such harvesting and sowing would result in an increasing frequency of the indehiscent trait over time, which is in fact what is observed in some well-preserved archaeological records.

Domestication is speciation. In the simplest case, what was formerly one population was subdivided into two populations, one of which, the wild progenitor, remained subject to the ancestral natural-selection regime whereas the other, cultivated population, was subjected to a human-mediated artificial-selection regime. Over time, the two populations diverged and a phenotypically and genetically distinct domesticated population was created. Whether one regards the domesticated population as a species, subspecies, race, variety, and so on, depends largely on one's species concept, but observed levels of genetic divergence separating human plant domesticates from their wild progenitors are typical of species-level differences (Dempewolf et al. 2012).

It has long been assumed that a high level of reproductive isolation (i.e., absence of gene flow) between the cultivar and its wild progenitor, coupled with strong artificial selection, were required—at least initially—to drive the fixation of domestication traits. In this intuitively simple scenario, a relatively small number of individual proto-domesticates was reproductively isolated from the wild population and interbred over multiple generations. Reproductive isolation could have been achieved by, for example, transporting the captive population beyond its ancestral range, cultivating self-pollinating species, or, in the case of an animal species, confining it in livestock pens. Undesirable phenotypes were (perhaps unintentionally) culled and desirable phenotypes retained until desirable traits became fixed in the domesticated population, a process that was thought to have proceeded relatively rapidly—that is, over the course of a few decades to a few hundred years. One problem with this scenario is that it entails a genetic bottleneck: depending on its size, genetic variability in the initial captive population would have been significantly reduced relative to that in the wild population. As a result, the population would have suffered from all of the problems associated with inbreeding depression (Barton and Charlesworth 1998), which would have affected its ability to adaptively respond to artificial selection for domestication traits and to natural selection for traits not directly associated with domestication, as in the case of transport by humans to a new and different environment.

At the other end of the gene-flow spectrum, the observed overlap in geographic distributions of some domesticates with their wild progenitors suggests that domestication may have occurred wholly or partly in sympatry. In this scenario, cultivars were subjected to artificial selection by humans while remaining in genetic contact with their progenitor populations, which, in contrast to the genetic-bottleneck scenario described above, would result in a domesticate with all or most of the background genetic variability present in the progenitor population. The problem with this scenario, which is shared with all sympatric-speciation scenarios (Bolnick and Fitzpatrick 2007), is that it requires genetic divergence to take place in the presence of unobstructed gene flow between the natural-selection and artificial-selection populations. Even if artificial selection on the captive population was strong, if the captive population size was small compared with the wild population size, then introgression from the parent population would be expected to have repeatedly diluted divergence, especially at domestication-trait loci where alternative alleles were favored in the captive versus the wild populations.

As in most biological systems, the population-level processes underlying domestication were far more complex than these contrasting allopatry/sympatry domestication scenarios in that they incorporated elements of both. Although the evolutionary histories of human domesticates differ in detail, the emerging convergent pattern, based on archaeological and archaeogenomic data, is that in most cases domestication took place over protracted time periods measured in many hundreds or in thousands of years—thus, for example >1,000 years for wheat and barley (Tanno and Willcox 2006), >4,000 years for maize (geographically divergent domestication lineages) (Kistler et al. 2018), ~3,000 years for rice (two Asian domestications) (Fuller 2007; Sweeney and McCouch 2007; Fuller et al. 2009; McCouch et al. 2012; chapter 10, this volume), and 1,000–2,000 years for pearl millet (Manning et al. 2011). In addition to long time spans, domestication also took place over large geographic distances and across a fragmented population-genetic landscape. At any given point in time, multiple differently sized artificial-selection populations were in greater or lesser degrees of genetic contact with, or completely isolated from, each other and from multiple differently sized natural-selection populations. Over the same time spans humans transported domesticates to new localities, traded seeds, and exerted artificial selection, which, on average, was weakly applied across many loci (McCouch et al. 2012; Allaby, Smith, and Kistler 2018; Burgarella et al. 2018; Kistler et al. 2018). Although localized genetic bottlenecks probably did repeatedly occur in this temporally and spatially fragmented landscape, there is little evidence that bottlenecks played a significant role in the long-term evolution of most domesticates; instead, most domesticated crops have retained levels of genetic diversity comparable to those present in their wild progenitors (Allaby, Fuller, and Brown 2008; Purugganan and Fuller 2009; Allaby 2010; McCouch et al. 2012; Allaby, Ware, and Kistler 2019). This pattern holds even for self-pollinating crops such as wheat and barley, which in theory could have been domesticated in time spans measured in decades rather than centuries or millennia (Hillman and Davies 1990; Honne and Heun 2009). There is evidence that, in some domesticates (e.g., rice, the common bean), artificial selection has increased the frequency of domestication alleles in wild populations; in other words, rather than wild populations diluting the genetic profiles of artificial-selection populations, human-mediated artificial selection relentlessly (even if weakly) applied over thousands of years has instead altered the genetic profiles of wild populations (Kaplan 1981; Papa et al. 2005; McCouch et al. 2012).

In general, the evolutionary histories of domesticated animals conform to the same general pattern as that of plants, including lack of bottlenecks, large overall levels of gene flow between artificial-selection and natural-selection populations, and long histories of population fragmentation and intermittent contact through time. This has been complicated by frequent deliberate outcrossing with wild populations in more than a few species (Zeder 2012; Marshall et al. 2014), which has proven advantageous for maintaining desirable traits, and by multiple species-specific pathways to animal domestication (Zeder 2012).

Ants

As discussed above, early humans coevolved with plants and animals in human-constructed niches for many tens of thousands of years prior to the advent of agriculture. This resulted in genetic change in the various species, including humans. It is probable that lower-attine ants likewise diffusely coevolved with leucocoprineaceous fungi, albeit over far longer time spans measured in tens of millions of years, because (1) a comparison of ant and fungal phylogenies suggests that fungus-farming ants co-diversified with their fungal cultivars (figure 14.2) and (2) the previously discussed utilization of ants as dispersal agents by leucocoprineaceous fungi also suggests a prolonged coevolutionary interaction. With regard to our definition of domestication, the scant available evidence suggests that lower-attine fungi are undomesticated, attine coral fungi may be domesticated, and attine yeast fungi are undomesticated (but see below). In contrast, multiple lines of evidence unequivocally indicate that the higher-attine fungi are domesticated.

Lower-Attine Fungal Cultivars

The lower-attine fungal cultivars, consisting of dozens of species in the Leucocoprineae, are thought to be undomesticated and capable of living freely outside the ant-fungus symbiosis without any reduction in fitness. As previously mentioned, although they are typically vertically transmitted across ant generations and clonally propagated when associated with ants, cultivated lower-attine fungal populations are thought to be genetically continuous with free-living, sexually reproducing conspecific populations. This is because, over evolutionary time spans measured in, for example, hundreds of thousands of ant-colony generations (Mehdiabadi et al. 2012), lower-attine fungal cultivars frequently "escape" from the symbiosis into the wild, where they sexually reproduce and are frequently recruited from the wild by ant workers or by foundress queens that have lost their resident cultivars. Data in support of these assumptions include (1) the collection of free-living fungi that are, based on internal transcribed spacer (ITS) "DNA-barcoding" sequences, conspecific with ant-cultivated lower-attine fungi (Mueller, Rehner, and Schultz 1998; Vo, Mueller, and Mikheyev 2009) and (2) the fungal phylogeny, which indicates that lower-attine fungi belong to two phylogenetically disjunct clades that also contain fungi not known to be cultivated by ants (Schultz et al. 2015). Although the lower-attine leucocoprineaceous fungi are undomesticated, it is plausible that, due to millions of years of diffuse coevolution, at least some species are genetically modified to take advantage of their frequent associations with ants. The only available lower-attine fungal genome indicates that the fungal cultivar of the ant *Cyphomyrmex costatus* has more carbohydrate-degrading enzyme genes than three other fungal species in the order Agaricales for which there are comparable data. This genetic difference is consistent with the role of the fungal cultivar in degrading plant material (insect frass, flower parts) gathered by the ants for their fungus garden (Nygaard et al. 2016). Based on available evidence, the lower-attine cultivars might currently best be regarded as highly but diffusely coevolved attine-ant commensals, likely possessing numerous genetic modifications for prolonged associations with ants during which they are protected, increased in biomass, and dispersed, but not at the expense of being able to live independently from ants when necessary or advantageous.

Attine Coral-Fungus Cultivars

Genomic data for the attine coral-fungus cultivars are currently unavailable, but, based on reasonably good recent sampling, many close relatives but no free-living conspecifics of the coral-fungus cultivars are known (Dentinger et al. 2009; Leal-Dutra et al. 2020), suggesting that they may be domesticated. Although also in the order Agaricales, the pterulaceous cultivars (family Pterulaceae) are not closely related to the leucocoprineaceous cultivars (family Agaricaceae) grown by all other attine ants (figure 14.2), so the origin of coral-fungus agriculture necessarily involved an evolutionary leap in which an ancestral *Apterostigma* ant species transferred its fungus-growing behavioral repertoire to a species in a previously uncultivated fungal family. Although the application of previously acquired background knowledge to the cultivation of new, previously uncultivated species is a common theme in human agriculture, this is the only known example from ant agriculture.

Attine Yeast Cultivars

As mentioned previously, based on the occurrence of free-living conspecifics (Mueller, Rehner, and Schultz 1998; Vo, Mueller, and Mikheyev 2009), the attine yeast fungi grow in typical mycelial form while living in the leaf litter apart from ants but in a unicellular yeast-like form when associated with ants (figure 14.1C). Genetic continuity between cultivated and wild populations suggests that the yeast cultivars are equivalently fit whether living with ants or growing in the wild (i.e., that they are undomesticated), but the occurrence of yeast-phase growth suggests genetic modification. Alternatively, yeast-phase growth could be induced entirely by the ants and the yeast cultivars could be genetically unmodified. As previously pointed out, however, unicellular yeast-phase growth is otherwise unknown in the order Agaricales, and the *Cyphomyrmex* yeast-cultivating ants are never found associated with—and are not known to be able to induce yeast-phase growth in—leucocoprineaceous species other than the yeast cultivars. Both the yeast cultivars and the ants that grow them belong to compact clades descended from a lower fungal and a lower-attine ant ancestor, respectively (figure 14.2) (Schultz and Brady 2008; Branstetter et al. 2017); this is consistent with ancestral species-to-species coevolution followed by an obligate association between fungal and ant descendants. Of note is that ambrosia-beetle-associated fungi in the phylum Ascomycota are also dimorphic (or "pleomorphic"), expressing yeast-like growth when growing in beetle mycangia or when actively tended by beetles but growing as mycelium in the absence of beetles (Batra and Michie 1963; Beaver 1989; Blackwell 2017; chapter 7, this volume). It thus remains possible that such fungi may be facultatively dimorphic, expressing one of two alternate phenotypes depending on their current environment. If so, this challenges our definition of domestication because it suggests a genetically modified cultivar that is equivalently fit in artificial-selection and natural-selection environments.

Higher-Attine Fungal Cultivars

The clearest case of domestication in attine-ant agriculture is the origin of the higher-attine fungi, which are never found apart from their ant farmers and which are genetically modified for life with ants. Unlike lower-attine fungi, higher-attine fungi are polyploid, they consistently produce gongylidia, and they express significantly different enzyme profiles, most associated with the degradation of fresh plant material. Higher-attine fungi sometimes produce sporocarps (mushrooms), such as on the surfaces of abandoned ant nests, but it is not known whether the spores are viable and therefore whether sexual recombination occurs (Kooij et al. 2015). When the ancestral higher-attine fungus was domesticated, a metabolic rate transition occurred. Unlike lower-attine colonies, which have metabolic rates lower than hunter-gatherer ant colonies, higher-attine ant colonies have significantly higher metabolic rates than hunter-gatherer colonies, even though the ratio of ant to fungal biomass remains the same as in lower-attine colonies. This difference may be due to the higher-energy chemical work performed by the fungus in order to produce a higher-value nutritional resource and/or to the requirement of the fungus to process a greater amount of substrate biomass with a greater amount of waste (Shik et al. 2014).

As mentioned, higher-attine fungi are polyploid (Kooij et al. 2015). This suggests that the putative benefits of polyploidy in human domesticates may also obtain in the domesticated higher-attine fungi, especially in *Leucoagaricus gongylophorus*, in which genetic variation significantly exceeds diploid levels, including (1) genome buffering (i.e., mitigation against the effects of deleterious mutant alleles), (2) increased allelic diversity and heterozygosity, and (3) generation of novel phenotypic variation (Udall and Wendel 2006). Kooij et al. (2015) suggest that, as in many human domesticates, such mechanisms may be responsible for the higher productivity of the higher-attine fungi, including their ability to degrade secondary plant compounds in the fresh vegetation substrates on which they grow.

Although gongylidia or gongylidia-like structures occur sporadically in lower-attine fungi (Masiulionis et al. 2014), these swollen hyphal tips are consistently expressed in higher-attine fungi and are preferentially harvested by the ants for food. Beyond nutrition, gongylidia serve to concentrate fungal enzymes (pectinases, proteases, and laccases) in the guts of garden-tending ants, which distribute them via their fecal droplets (Martin and Martin 1970; Schiøtt et al. 2010; De Fine Licht et al. 2014; Kooij et al. 2014). It can thus be said that the fungus uses the ants as vectors to transfer its digestive enzymes from the most luxuriant, gongylidia-rich middle layer of the garden to the most recently added and inoculated substrate in the upper layer of the garden. This allows for the rapid degradation of pectin, which makes starch and proteins accessible, and for the detoxification of toxic plant phenolic compounds (De Fine Licht et al. 2010; Schiøtt et al. 2010; Moller et al. 2011; Aylward et al. 2013; Grell et al. 2013; Kooij et al. 2014; Somera et al. 2015; Worsley et al. 2018). Although the *Leucoagaricus gongylophorus* cultivar genome encodes 145 lignocellulase enzymes (Aylward et al. 2013), these are generally expressed only in the older and least productive lower layers of the gardens, which are constantly being dismantled and discarded in the waste dumps. Higher-attine agriculture—particularly leafcutter agriculture—thus depends on considerable waste of substrate. It takes advantage of the increased levels of protein in fresh plant material by foregoing the digestion of recalcitrant polysaccharides and failing to utilize less accessible proteins (De Fine Licht et al. 2010, 2014; Schiøtt et al. 2010; Moller et al. 2011; Grell et al. 2013; Kooij et al. 2014; Worsley et al. 2018).

Climate Change

The desertification of the Sahara ~5000 years ago (Hély et al. 2009) and the resulting decreased access to free-living populations of food plants are hypothesized to have triggered the domestication of African rice, pearl millet, and possibly yams, the latter descended from a wet-forest-dwelling species (McKey et al. 2012; Burgarella et al. 2018; Cubry et al. 2018; Scarcelli et al. 2019). Climate change toward drier conditions has also been suggested as a possible cause of the domestication of higher-attine fungi and the origin of higher-attine agriculture. Higher-attine ants likely arose around 30 Mya (figure 14.2) in a seasonally dry habitat in South America following the Terminal Eocene Event, a period of global cooling that began 35 Mya, when drier habitats, including grasslands, greatly expanded globally (Prothero 1994; Graham 2011). As hypothesized by Branstetter et al. (2017), as fungus-farming ants dispersed out of wet forests and adapted to expanding grasslands and other seasonally dry habitats, they would have carried their wet-forest-adapted lower-attine fungal cultivars along with them into a geographic mosaic of populations in which gene flow between fungal cultivars and their wet-forest-dwelling wild progenitor populations would vary from strong to sporadic to nonexistent. Albeit occurring over a presumably vastly longer time span, these conditions are comparable to those described previously for the origins of many human domesticates: protracted periods of varying gene flow across a spatial and temporal mosaic of natural-selection and artificial-selection populations, the latter under generally weak but persistent selection across a large number of loci, ultimately producing new species dependent upon the farmers for survival. Some human domesticates can still outcross with their wild progenitors in spite of significant reproductive isolation and depressed hybrid fitness (Dempewolf et al. 2012; McKey et al. 2012), but after tens of millions of years, higher-attine fungi have become obligate symbionts lacking any genetic connection with extant free-living progenitors, if such still exist. Perhaps not surprisingly, higher-attine fungi have also continued to speciate, but because species appear to be separated by differing ploidy levels (Kooij et al. 2015), it is possible that subsequent evolution in this group has been driven at least in some cases by polyploidization events.

Who Domesticated Whom?

Over hundreds of thousands of years, humans evolved to accommodate an expanding diet (Mathieson et al. 2015). Over the past ten thousand years, humans have continued to evolve in order to accommodate diets containing lactose and wheat, to resist diseases acquired from their animal domesticates, and to detect and/or tolerate a variety of plant secondary compounds (Jackson 1991; Johns 1996; Diamond 1997; chapter 12, this volume). According to our definition, none of these modifications qualify as the reciprocal domestication of humans by their domesticates. In contrast, it can credibly be argued that, rather than the ants domesticating the fungi, it was the fungi that initially domesticated the fungus-farming ants. Perhaps most persuasively, all attine ants are obligate fungus farmers that cannot survive without their fungi, whereas at least the lower-attine and yeast cultivars are able to survive without the ants. More like domesticates than domesticators, all fungus-farming ants have lost the ability to synthesize the amino acid arginine, which they obtain from their cultivated fungi, and they express high levels of a modified chitinase in their labial gland fluids in order to digest fungal chitin. Based on analyses of seven attine-ant genomes, fungus-farming ants have experienced rates of structural gene rearrangement (i.e., loss of synteny) far more rapid than in any animal group of comparable age. Early in fungus-farming ant evolution many gene families contracted. In the ancestor of the leaf-cutting ant genus *Atta*, however, major gene families expanded, producing 129 novel genes with no clear homologies to known genes (Nygaard et al. 2016).

Agriculture

Humans

In this book we define *agriculture* as large-scale cultivation upon which the farmers have become economically (i.e., obligately) dependent. Although the definition does not require the cultivated species to be domesticated, domestication preceded agriculture in the histories of all or most human agricultural origins (Rindos et al. 1980; Fuller et al. 2014; chapter 10, this volume). The rise of agriculture allowed Holocene human societies to accumulate surplus food supplies, which in turn gave rise, in a positive feedback loop, to increasing population growth and expanding settlements (Smith 1998b; Tudge 1998; Fuller et al. 2014). This then led to increasingly complex social systems, increasingly refined division of labor (chapter 2, this volume), and, ultimately, the emergence of world civilizations. The economists Gowdy and Krall (2014, 179) argue that postagricultural, global human society is the result of a major evolutionary transition that is rare in nature, which they call "ultrasociality"; it is characterized by "superorganisms exhibiting an unparalleled degree of division of labor and an economic organization centered around surplus production."

Relevant to the subject of this book, the other examples they cite include fungus-farming termites and fungus-farming ants. The origins of human agriculture are consistently associated with major landscape modifications such as slash-and-burn clearing of forests (Neolithic Europe, New Guinea, China, Mesoamerica), excavation of irrigation ditches (northeastern Peru, New Guinea), raised fields, and anthropogenic "black soils" (Amazonia and eastern South America) (Denevan 2001; Fuller et al. 2014; McKey and Rostain 2014; Lombardo et al. 2020). With regard to nutrition, early agriculture was not necessarily superior to hunting-gathering, cultivation, and mixed food-acquisition strategies. Although human biological responses to agriculture varied considerably across agricultural origins (Pinhasi and Stock 2011; chapter 12, this volume), in at least some cases early agriculturalists were malnourished compared to hunter-gatherers due to overdependence on a limited range of foods, especially carbohydrates. For this reason and because of the accelerated population expansion that accompanied agriculture, early agriculturalists also suffered from increased exposure to pathogens, including newly evolved pathogens and pathogens vectored by their managed/domesticated animals (Larsen 1995; Mummert et al. 2011; Cordain, Hickey, and Kim 2012; Berbesque et al. 2014; Dyble et al. 2019; Larsen et al. 2019; chapter 2, this volume).

Agroecosystems

With the exception of microorganisms grown axenically in the lab, agriculture does not consist only of a farmer and a cultivar; rather, it is a multipartite symbiosis. Agricultural ecosystems provide reliable niches for diseases of farmers, for diseases of cultivars, for diseases of diseases, for pests of stored products, and for "weeds" (i.e., for organisms able to coevolve with and adapt to agroecosystems in order to directly or indirectly exploit the resources supplied by the farmers to their cultivars such as cleared land, tilled soils, fertilizers, and water) (Fuller and Stevens 2017). In many cases weeds have been unintentionally artificially selected by farmers to mimic their domesticates (Harlan 1992; Spahillari et al. 1999; Senda, Hiraoka, and Tominaga 2006; Thomas, Archer, and Turley 2011; Howard, Archer, and Turley 2016; Fuller and Stevens 2017; chapter 10, this volume). In some systems (e.g., maize, sorghum, rye, squash) even the ancestral progenitor may function as a weed (Wilkes 1977; Wilson 1990; Harlan 1992). Humans have developed many methods for excluding or controlling unwanted agricultural symbionts, including, in low-technology agriculture, the use of fire, mixed cropping, the cultivation of plant and animal biocontrol agents, and weeding (Denevan 2001; Anderson and Wohlgemuth 2012), and, in modern intensive agriculture, the use of herbicides, pesticides, biocontrol agents, and genetic engineering.

Ants

Around 55–65 Mya, an ancestral hunting-gathering ant became physiologically dependent on cultivating undomesticated leucocoprineaceous fungi and gave rise to lower-attine agriculture. In so doing, it was able to escape competition with other hunting-gathering ants by becoming fungivorous. Instead of competing with other litter-dwelling ants for arthropod prey and plant carbohydrates, it competed instead with microbes for bits of organic detritus, including insect frass, seeds, flower parts, wood particles, and dead insects (Weber 1972; Murakami and Higashi 1997; Leal and Oliveira 2000; Ješovnik,

Chaul, and Schultz 2018; Ronque, Feitosa, and Oliveira 2019). In terms of productivity, lower-attine ant agriculture is not superior to hunting-gathering. Lower-attine ants generate similar yields of ant biomass per unit foraging effort as equivalently sized colonies of non-attine hunting-gathering ants (Turner 1974; Quinlan and Cherrett 1979; Bass and Cherrett 1995). Most notably, by directly measuring whole-colony (ants plus fungus) metabolism (i.e., mass-specific CO_2 emission rate), Shik et al. (2014) show that, relative to hunting-gathering ants, lower-attine fungus-farming ant colonies have significantly lower mass-specific metabolic rates and that a significant part of the colony biomass is shifted into fungal (rather than ant) tissue. They hypothesize that transferring biomass from ants into a fungal mutualist may generate energetic storage advantages, paralleling a frequently-cited benefit of early human agriculture (Hayden 1995; Diamond 1997; Piperno and Pearsall 1998).

Around 30 Mya, following the Terminal Eocene Event, a lower-attine ant ancestor domesticated a lower-attine fungus and gave rise to higher-attine agriculture. As mentioned above, this domestication event may have been driven by the increase in grasslands and other seasonally dry habitats in South America, which separated fungal cultivars from continuous genetic contact with their free-living, wet-forest-dwelling conspecifics. Although extant non-leaf-cutting, higher-attine ants forage for substrates similar to those taken by lower-attine ants, some species additionally cut tender shoots, leaflets, and flowers (Leal and Oliveira 2000; Mehdiabadi and Schultz 2009; Ješovnik, Chaul, and Schultz 2018; Ronque, Feitosa, and Oliveira 2019). Unlike lower-attine ants, the colony-level metabolisms of higher-attine ants are higher than those of hunting-gathering ants (Shik et al. 2014). However, the colony sizes (around 3,000 workers maximum) and levels of social complexity of non-leaf-cutting higher-attine ants overlap with those of some derived species of lower-attine ants (Mehdiabadi and Schultz 2009).

Around 20 Mya, a non-leaf-cutting higher-attine ant ancestor acquired the ability to cut and process living plant material as the primary fungus-garden substrate. In so doing, it ceased to compete with (mostly microbial) detritivores that feed on the same substrates as lower-attine fungi and instead began to compete with vertebrate and invertebrate herbivores. In addition to the key adaptation of utilizing fresh vegetation, extant leaf-cutting ants in the genera *Amoimyrmex*, *Acromyrmex*, and *Atta* differ from non-leaf-cutting higher-attine ants in their vastly larger colony sizes and polymorphic worker castes (Mehdiabadi and Schultz 2009). In these traits *Amoimyrmex* and *Acromyrmex* species may be regarded as "transitional" or less derived than *Atta* species. For example, *Amoimyrmex* colonies contain approximately 5,000 individuals and *Acromyrmex* colony sizes approach a maximum of ~270,000 individuals, whereas standing populations of *Atta* colonies can reach 8 million individuals (Fowler et al. 1986; Wetterer 1995; Wetterer, Gruner, and Lopez 1998; Ferguson-Gow et al. 2014; M. P. Cristiano, personal communication). Likewise, the degree of physical caste differentiation in *Amoimyrmex* and *Acromyrmex* is less extreme than in *Atta* (Wetterer 1999; Cristiano et al. 2020). Following nest founding, an *Amoimyrmex* or *Acromyrmex* queen, like almost all other foundress ant queens, must forage for substrate until her first brood of adult worker ants emerges to take over that task, a period of 40–60 days during which she is exposed to very high risks of predation and disease. Nest-founding *Atta* queens, in contrast, found their nests "claustrally" by excavating a chamber in the soil, sealing off the entrance tunnel, and nourishing both the fungus garden and the first brood of workers with trophic (nonviable) eggs generated by catabolized stored fat and wing muscle tissue (figure 14.1E) (Autuori 1956; Hölldobler and Wilson 2010).

Atta leaf-cutting ants are among the most highly eusocial of all organisms (Hölldobler and Wilson 2009, 2010) and are the obvious exemplars for comparisons with modern human industrial-scale agriculture. Leaf-cutter colonies, which Carroll and Janzen (1973) call "subdivided cows," are the ecological equivalents of large herbivorous mammals in terms of collective biomass, lifespan, and plant material harvested (Herz, Beyschlag, and Hölldobler 2007). Peregrine (chapter 2, this volume) concludes that increasingly refined division of labor is one of a number of important components of agriculture, and this is certainly the case for *Atta*. Edward O. Wilson (1980) identified 29 discrete tasks performed by *Atta* workers, the majority involved in the assembly-line processing of fresh vegetation and performed by workers of specific size categories. *Atta* workers are highly polymorphic, spanning a broad range of sizes in which the smallest, the gardener-nurses, weigh 200 times less than the largest, the defensive soldiers (figure 14.1F). Unlike all other attine ants, *Amoimyrmex*, *Acromyrmex*, and *Atta* leaf-cutting ants are multiply mated: during their mating flights queens mate with an average of three males, storing and nourishing their sperm for life (Fjerdingstad, Boomsma, and Thorén 1998; Boomsma, Fjerdingstad, and Frydenberg 1999; Fjerdingstad and Boomsma 2000; Murakami, Higashi, and Windsor 2000; Villesen et al. 2002; Wirth et al. 2003; M. P. Cristiano, personal communication; chapter 5, this volume). In addition to supplying the sperm necessary for the many tens of millions of female offspring (males are the products of unfertilized, haploid eggs) produced by an *Atta* queen during her ~15-year lifespan, multiple mating serves to increase the genetic variability of the work force and thus overall resistance to disease (Hughes and Boomsma 2004). The disadvantage of multiple mating is that it decreases the genetic relatedness of workers in the colony, potentially decreasing the binding force of kin selection and increasing conflicts of interest over reproductive offspring. In leaf-cutter ants, however, the potential for conflict of interest has been largely or entirely removed both in the realm of sperm competition (den Boer, Baer, and Boomsma 2010) and in the realm of worker-worker reproductive conflict (chapter 5, this volume). Primitively, non-fungus-farming ant workers are able to lay haploid eggs that develop into males able to carry the worker's genes into the next generation. Workers of most attine ants, including non-leaf-cutting higher-attine ants, retain this ability, which may be employed, for example, when the colony queen dies. Among the leaf-cutting ants, workers of at least some *Acromyrmex* species also retain the ability to lay haploid (male) eggs, whereas *Atta* workers are sterile (Dijkstra, Nash, and Boomsma 2005). In the rare, small subset of *Atta* workers that lay viable eggs, the eggs hatch into sexually dysfunctional "dwarf" males (Dijkstra and Boomsma 2006), which indicates that the ability to lay haploid eggs persists only as a vestigial trait. In the logic of kin selection, the highly eusocial *Atta* superorganism could not have evolved without a gigantic, polymorphic, genetically variable work force, which could not have evolved without multiple mating, which could not have evolved without the elimination of sperm and worker-worker conflicts over reproductive control. That all of these conditions are met in *Atta* species is the hard-wired ant equivalent of "increased social control," cited by Peregrine (chapter 2, this volume), as one of the consistent components of human agriculture.

Agroecosystems

Like human agroecosystems, leaf-cutter agriculture provides a vast, reliable niche for thousands of symbiotic organisms, including arthropods, mollusks, annelid worms, nematodes, and reptiles (Autuori 1942; Weber 1972; Waller and Moser 1990; Schultz and McGlynn 2000). In the beetles (Coleoptera) alone, 411 species belonging to 25 families have been recorded from leaf-cutter nests (Navarette-Heredia 2001). Some species of non-leaf-cutting higher- and lower-attine ants are attacked by "agro-predatory" ants in the distantly related genera *Megalomyrmex* (subfamily Myrmicinae) and *Gnamptogenys* (subamily Ectatomminae), which raid nests and consume both fungi and brood. In contrast, the semi-mutualistic *Megalomyrmex symmetochus* cohabits with non-leaf-cutting higher-attine ant hosts for many years as a social parasite, consuming resources but not at the expense of host reproduction and even protecting the host from *Gnamptogenys* raids (Adams et al. 2000b, 2013; Dijkstra and Boomsma 2003; chapter 8, this volume).

The gardens of leaf-cutter ants host an unknown but large number of bacterial species, including nitrogen-fixing bacteria (Pinto-Tomás et al. 2009; Aylward et al. 2012) and a cohort of three bacterial genera that consistently occurs in the gardens of fungus-farming ants, beetles, and termites (Aylward, Currie, and Suen 2012; Aylward et al. 2014; Barcoto et al. 2020). They also host hundreds of fungal species (Little and Currie 2007; Pagnocca et al. 2008; Rodrigues et al. 2008, 2013; Mendes et al. 2012; Pagnocca, Masiulionis, and Rodrigues 2012), including the consistently occurring mycoparasitic genera *Escovopsis* and *Escovopsioides* (Ascomycota: Hypocreales), which have co-diversified with attine-ant-associated fungi since the beginning of attine-ant agriculture (Currie et al. 2003; Gerardo, Mueller, and Currie 2006; Augustin et al. 2013; Meirelles 2015a; Meirelles et al. 2015b; Osti and Rodrigues 2018; chapter 11, this volume).

Conclusion

As pointed out by McGhee (chapter 1, this volume), the most likely reason for evolutionary convergence of human and insect agricultural symbioses is functional constraint: agricultural systems are bound to share similar elements because there is a limited number of ways to solve the problem of growing your own food. This chapter has summarized the many similarities as well as the obvious differences between human and fungus-farming ant agriculture. Here I will point out five broad similarities that are likely due to evolutionary convergence.

1. The ancestors of both human and ant agriculturalists lived in social groups that obtained nourishment through central-place hunting-gathering. They altered their local environments through niche construction/ecological engineering, creating reliable ecosystems for and coevolving with preadapted plants and animals. At some point they added incidental cultivation to their repertoires of mixed food acquisition. Central-place hunting-gathering is an ancestral trait in ants, most ants are premiere ecological engineers, and incidental fungal cultivation may be present in the close relatives of fungus-farming ants in the genera *Wasmannia* and *Blepharidatta.*

2. At some point a commitment to cultivation occurred in both human and ant agriculturalists. It occurred at the origin of fungus-farming ants, probably due to the reliability

of "stored" surplus in the form of fungus gardens rather than to superior productivity relative to hunting-gathering because fungus-farming ants continue to forage for garden substrate and their colonies have lower overall metabolic rates than those of equivalently sized hunting-gathering ant colonies. It is likely that commitment to agriculture occurred in human cultural evolution for similar reasons—namely, because of the environmental-buffering effect of stored surpluses and in spite of the longer-term positive feedback loop between increasing reliance on stored surpluses, population growth, and initial adverse effects on human health.

3. Human and ant cultivars became domesticated. Higher-attine fungi are genetically and phenotypically modified, polyploid, and more productive than lower-attine fungi. They have diversified into multiple species, and their wild ancestor is likely extinct. Higher-attine ant colonies have higher overall metabolic rates than those of equivalently sized hunting-gathering ant colonies. Most human domesticates are more productive than, and are essentially separate species from, their wild progenitors, most are polyploid, and some are arguably, along with their human farmers, the most successful species on planet Earth.

4. Both human and ant agriculture have produced large, complex societies with unprecedented levels of division of labor and social complexity. Colonies of leaf-cutting *Atta* species are among the most highly eusocial of all organisms, possessing a range of physical castes unparalleled among social insects and having evolved a suite of biological traits that together reduce conflicts of interest among colony members and bind them and their resident fungal clone into a cohesive superorganism (chapter 5, this volume). Hölldobler and Wilson (2010) refer to this extreme social cohesion as "civilization by instinct."

5. Human and ant agriculture have produced agroecosystems consisting of hundreds or thousands of species of animals, plants, fungi, and bacteria. In both systems, in addition to farmers and cultivars, large numbers of species have come to rely on the huge, predictable niches produced by agriculture, including many species that have coevolved with the farmers and cultivars. Because some of them feed on or compete with their cultivars, farmers have evolved methods for controlling these "parasites"; some of these methods involve promoting the growth of predators/pathogens of the parasites.

If these similarities are due to evolutionarily convergent solutions to shared ecological problems, then it is very likely that, via a process of reciprocal illumination, we can discover the set of shared evolutionary and ecological rules that have produced and that govern both human and ant agriculture (Schultz et al. 2005). A similarly comparative study of convergence in strictly human agricultural evolution is underway due to a renaissance in synthesizing and analyzing new knowledge from archaeology, genomics, archaeogenomics, and population genetics (Allaby, Fuller, and Brown 2008; Fuller et al. 2014; Larson et al. 2014; chapter 10, this volume). Perhaps knowledge gained in that endeavor can inform the search for similar patterns in the comparative population genetics and genomics of higher-attine fungi, lower-attine fungi, and free-living leucoprineaceous fungi. During their evolution from wild progenitors to human domesticates, our plant and animal symbionts traveled down fragmented temporal and spatial population-genetic pathways that, in spite of their complexity, produced convergent phenotypic traits and genomic architectures. Did similar patterns of population-genetic evolution produce ant-cultivated domesticated fungi? Although the vastly greater time spans involved could make answering this question challenging, there are many extant species of lower-attine and higher-attine fungi cultivated by many extant species of ants that provide dozens of comparative microcosms for exploring population-level ant-fungus agricultural evolution, including domestication, which may be in progress in a number of extant systems (Gerardo et al. 2004; Mehdiabadi, Hughes, and Mueller 2006; Mehdiabadi et al. 2012).

Boomsma (chapter 5, this volume) points out a fundamental difference between human and leaf-cutter agriculture: the latter requires a lifetime "matrimonial" commitment between the ant colony and its single clonal fungal cultivar in order to ameliorate potential conflicts of interest between them. Humans, in contrast, retain the ability to recruit new cultivar strains or species, whereas their cultivars, which are obligately dependent on them, are under the asymmetrical control of their human farmers. Although this may be true of human agricultural symbioses in principle, during the past century humans have become increasingly reliant on the large-scale cultivation of monocultures of genetically monotonous strains of a very few plant and animal species (Harris 2012; US Department of Agriculture 2020a, 2020b). This increasing dependency is so far associated with only a few known human genetic adaptations, none of which is fixed across all human populations, but that seems unlikely to remain the case. As we contemplate the long-term future of the human species, it might be useful to do so with a better understanding of the industrial-scale agriculture practiced by leaf-cutting ant species in the genus *Atta*. Given current concerns about preserving genetic diversity in our domesticates and about maintaining biodiversity in our agroecosystems, how is it that, possibly for as many as 20 million years, leaf-cutting ants have managed to sustainably propagate a single, possibly asexual, fungal species; manage an ever-present, ever-evolving microfungal crop disease; manage a diverse array of other macro- and micro-organismal parasites and commensals; and manage the trees and grasses on which they forage for substrates in their local environments? In spite of considerable research, the answers to these questions are not at all well understood and their implications for human agriculture, if any, are almost entirely unexplored.

Acknowledgments

I am grateful to the administration and staff of the Konrad Lorenz Institute for sponsoring and hosting the 28th Altenburg Workshop in Theoretical Biology, which initiated the process that ultimately culminated in this book. I am also grateful to all of the workshop participants for many stimulating presentations, discussions, and conversations, seminal contributions toward our goal of synthesizing ideas from multiple, disparate disciplines and examining agricultural symbiotic evolution from a novel perspective. I thank Rebecca Wilson for help with the figures and I appreciate the comments and suggestions of Peter Peregrine, Rick Gawne, and Dorian Fuller, which greatly improved the quality of this chapter. My research was supported by US National Science Foundation awards DEB 1654829 and DEB 1927224.

References

Adams, Rachelle M. M., Joanito Liberti, Anders A. Illum, Tappey H. Jones, David R. Nash, et al. 2013. "Chemically Armed Mercenary Ants Protect Fungus-Farming Societies." *Proceedings of the National Academy of Sciences* 110(39): 15752–15757. https://doi.org/10.1073/pnas.1311654110.

Adams, Rachelle M. M., Ulrich G. Mueller, Alisha K. Holloway, Abigail M. Green, and Joanie Narozniak. 2000a. "Garden Sharing and Garden Stealing in Fungus-Growing Ants." *Naturwissenschaften* 87(11): 491–493.

Adams, Rachelle M. M., Ulrich G Mueller, Ted R. Schultz, and Beth Norden. 2000b. "Agro-Predation: Usurpation of Attine Fungus Gardens by *Megalomyrmex* ants." *Naturwissenschaften* 87(12): 549–554.

Allaby, Robin G. 2010. "Integrating the Processes in the Evolutionary System of Domestication." *Journal of Experimental Botany* 61(4): 935–944. https://doi.org/10.1093/jxb/erp382.

Allaby, Robin G., Dorian Q. Fuller, and Terence A. Brown. 2008. "The Genetic Expectations of a Protracted Model for the Origins of Domesticated Crops." *Proceedings of the National Academy of Sciences* 105(37): 13982. https://doi.org/10.1073/pnas.0803780105.

Allaby, Robin G., Logan Kistler, Rafal M. Gutaker, Roselyn Ware, James L. Kitchen, et al. 2015. "Archaeogenomic Insights into the Adaptation of Plants to the Human Environment: Pushing Plant–Hominin Co-Evolution Back to the Pliocene." *Journal of Human Evolution* 79:150–157. https://doi.org/https://doi.org/10.1016/j.jhevol.2014.10.014.

Allaby, Robin G., Oliver Smith, and Logan Kistler. 2018. "Archaeogenomics and Crop Adaptation." In *Paleogenomics*, edited by Charlotte Lindqvist and Om P. Rajora, 189–203. Cham: Springer.

Allaby, Robin G., Roselyn L. Ware, and Logan Kistler. 2019. "A Re-Evaluation of the Domestication Bottleneck from Archaeogenomic Evidence." *Evolutionary Applications* 12(1): 29–37. https://doi.org/10.1111/eva.12680.

Ames, Oakes. 1939. *Economic Annuals and Human Culture*. Cambridge, MA: Botanical Museum of Harvard University.

Anderson, M. Kat, and Eric Wohlgemuth. 2012. "California Indian Proto-Agriculture: Its Characterization and Legacy." In *Biodiversity in Agriculture: Domestication, Evolution, and Sustainability*, edited by Paul Gepts, Thomas R. Famula, Robert L. Bettinger, Stephen B. Brush, Ardeshir B. Damania, et al., 190–224. New York: Cambridge University Press.

Augustin, Juliana O., Johannes Z. Groenewald, Robson J. Nascimento, Eduardo S. G. Mizubuti, Robert W. Barreto, et al. 2013. "Yet More 'Weeds' in the Garden: Fungal Novelties from Nests of Leaf-Cutting Ants." *PLoS ONE* 8(12): e82265. https://doi.org/10.1371/journal.pone.0082265.s007.

Autuori, M. 1942. "Contribuição para o conhecimento da saúva (*Atta* spp.-Hymenoptera-Formicidae). III. Escavação de um sauveiro (*Atta sexdens rubropilosa* Forel, 1908)." *Arquivos do Instituto Biologico* 13(14): 137–148.

Autuori, M. 1956. "La fondation des sociétés chez les fourmis champignonnistes du genre *Atta* (Hym. Formicidae)." In *L'instinct dans le Comportement des Animaux et de L'homme*, edited by Pierre-P. Grassé, 77–104. Paris: Masson et Cie.

Aylward, Frank O., Kristin E. Burnum, Jarrod J. Scott, Garret Suen, Susannah G. Tringe, et al. 2012. "Metagenomic and Metaproteomic Insights into Bacterial Communities in Leaf-Cutter Ant Fungus Gardens." *ISME Journal* 6:1688–1701. https://doi.org/10.1038/ismej.2012.10.

Aylward, Frank O., Kristin E. Burnum-Johnson, Susannah G. Tringe, Clotilde Teiling, Daniel M. Tremmel, et al. 2013. "*Leucoagaricus gongylophorus* Produces Diverse Enzymes for the Degradation of Recalcitrant Plant Polymers in Leaf-Cutter Ant Fungus Gardens." *Applied and Environmental Microbiology* 79(12): 3770–3778. https://doi.org/10.1128/AEM.03833-12.

Aylward, Frank O., Cameron R. Currie, and Garret Suen. 2012. "The Evolutionary Innovation of Nutritional Symbioses in Leaf-Cutter Ants." *Insects* 3(1): 41–61. https://doi.org/10.3390/insects3010041.

Aylward, Frank O., Garret Suen, Peter H. W. Biedermann, Aaron S. Adams, Jarrod J. Scott, et al. 2014. "Convergent Bacterial Microbiotas in the Fungal Agricultural Systems of Insects." *mBio* 5(6): e02077–e02014. https://doi.org/10.1128/mBio.02077-14.

Bailey, Irving W. 1920. "Some Relations between Ants and Fungi." *Ecology* 1:174–189.

Barcoto, Mariana O., Camila Carlos-Shanley, Huan Fan, Milene Ferro, Nilson S. Nagamoto, et al. 2020. "Fungus-Growing Insects Host a Distinctive Microbiota Apparently Adapted to the Fungiculture Environment." *Scientific Reports* 10(1): 12384. https://doi.org/10.1038/s41598-020-68448-7.

Barton, N., and B. Charlesworth. 1998. "Why Sex and Recombination?" *Science* 281 (5385): 1986–1990.

Bass, M., and J. M. Cherrett. 1995. "Fungal Hyphae as a Source of Nutrients for the Leaf-Cutting Ant *Atta sexdens*." *Physiological Entomology* 20(1): 1–6.

Batra, Lekh R., and Mary Downing Michie. 1963. "Pleomorphism in Some Ambrosia and Related Fungi." *Transactions of the Kansas Academy of Science (1903-)* 66(3): 470–481. https://doi.org/10.2307/3626545.

Beattie, Andrew J. 1985. *The Evolutionary Ecology of Ant-Plant Mutualisms*. London: Cambridge University Press.

Beaver, R. A. 1989. "Insect-Fungus Relationships in the Bark and Ambrosia Beetles." In *Insect-Fungus Interactions*, edited by N. Wilding, N. M. Collins, P. M. Hammond, and J. F. Webber. London: Academic Press.

Beeren, C. Von, M. M. Mair, and V. Witte. 2014. "Discovery of a Second Mushroom Harvesting Ant (Hymenoptera: Formicidae) in Malyan Tropical Rainforests." *Myrmecological News* 20:37–42.

Bellwood, Peter. 2005. *First Farmers: The Origins of Agricultural Societies*. Malden, MA: Blackwell Publishing.

Belt, T. 1874. *The Naturalist in Nicaragua*. London: John Murray.

Berbesque, J. Colette, W. Marlowe Frank, Peter Shaw, and Peter Thompson. 2014. "Hunter–Gatherers Have Less Famine than Agriculturalists." *Biology Letters* 10(1): 20130853. https://doi.org/10.1098/rsbl.2013.0853.

Blackwell, Meredith. 2017. "Yeasts in Insects and Other Invertebrates." In *Yeasts in Natural Ecosystems: Diversity*, edited by Pietro Buzzini, Marc A. Lachance and Andrey Yurkov, 397–433. Cham: Springer.

Blatrix, Rumsaïs, Sarah Debaud, Alex Salas-Lopez, Céline Born, Laure Benoit, et al. 2013. "Repeated Evolution of Fungal Cultivar Specificity in Independently Evolved Ant-Plant-Fungus Symbioses." *PLoS ONE* 8(7): e68101. https://doi.org/10.1371/journal.pone.0068101.

Blatrix, Rumsaïs, Champlain Djiéto-Lordon, Laurence Mondolot, Philippe La Fisca, Hermann Voglmayr, et al. 2012. "Plant-Ants Use Symbiotic Fungi as a Food Source: New Insight into the Nutritional Ecology of Ant–Plant Interactions." *Proceedings of the Royal Society B—Biological Sciences* 279(1744): 3940–3947. https://doi.org/10.1098/rspb.2012.1403.

Bliege Bird, R., D. W. Bird, B. F. Codding, C. H. Parker, and J. H. Jones. 2008. "The "Fire Stick Farming" Hypothesis: Australian Aboriginal Foraging Strategies, Biodiversity, and Anthropogenic Fire Mosaics." *Proceedings of the National Academy of Sciences* 105(39): 14796. https://doi.org/10.1073/pnas.0804757105.

Bolnick, Daniel I., and Benjamin M. Fitzpatrick. 2007. "Sympatric Speciation: Models and Empirical Evidence." *Annual Review of Ecology, Evolution, and Systematics* 38(1): 459–487. https://doi.org/10.1146/annurev.ecolsys.38.091206.095804.

Boomsma, Jacobus J., Else J. Fjerdingstad, and Jane Frydenberg. 1999. "Multiple Paternity, Relatedness and Genetic Diversity in *Acromyrmex* Leaf-Cutter Ants." *Proceedings of the Royal Society B—Biological Sciences* 266(1416): 249–254.

Borowiec, Marek L., Christian Rabeling, Seán G. Brady, Brian L. Fisher, Ted R. Schultz, and Philip S. Ward. 2019. "Compositional Heterogeneity and Outgroup Choice Influence the Internal Phylogeny of the Ants." *Molecular Phylogenetics and Evolution* 134: 111–121. http://www.sciencedirect.com/science/article/pii/S1055790317305808.

Brandão, C. Roberto F., Jorge L. M. Diniz, P. R. Silva, N. L. Albuquerque, and R. Silvestre. 2001. "The First Case of Intranidal Phragmosis in Ants: The Ergatoid Queen of *Blepharidatta conops* (Formicidae, Myrmicinae) Blocks the Entrance of the Brood Chamber." *Insectes Sociaux* 48(3): 251–258. https://doi.org/10.1007/PL00001774.

Branstetter, Michael G., Ana Ješovnik, Jeffrey Sosa-Calvo, Michael W. Lloyd, Brant C. Faircloth, et al. 2017. "Dry Habitats Were Crucibles of Domestication in the Evolution of Agriculture in Ants." *Proceedings of the Royal Society B—Biological Sciences* 284(1852): 20170095. https://doi.org/10.1098/rspb.2017.0095.

Brown, Jr., W. L. 2000. "Diversity of Ants." In *Ants: Standard Methods for Measuring and Monitoring Biodiversity*, edited by Donat Agosti, Jonathan D. Majer, Leeanne E. Alonso, and T. R. Schultz, 45–79. Washington, DC: Smithsonian Institution Press.

Burgarella, Concetta, Philippe Cubry, Ndjido A. Kane, Rajeev K. Varshney, Cedric Mariac, et al. 2018. "A Western Sahara Centre of Domestication Inferred from Pearl Millet Genomes." *Nature Ecology & Evolution* 2(9): 1377–1380. https://doi.org/10.1038/s41559-018-0643-y.

Carroll, C. Ronald, and Daniel H. Janzen. 1973. "Ecology of Foraging by Ants." *Annual Review of Ecology and Systematics* 4:231–257.

Cherrett, J. M. 1989. "Leaf-Cutting Ants." In *Ecosystems of the World 14B*, edited by H. Lieth and M. J. A. Werger, 473–488. Amsterdam, The Netherlands: Elsevier.

Collins, Nancy C. 2015. "Mutualisms and Ecosystem-Level Processes." In *Mutualism*, edited by Judith L. Bronstein, 221–238. Oxford: Oxford University Press.

Cordain, Loren, Matthew S. Hickey, and Kami Kim. 2012. "Malaria and Rickets Represent Selective Forces for the Convergent Evolution of Adult Lactase Persistence." In *Biodiversity in Agriculture: Domestication, Evolution, and Sustainability*, edited by Paul Gepts, Thomas R. Famula, Robert L. Bettinger, Stephen B. Brush, Ardeshir B. Damania, et al., 299–308. New York: Cambridge University Press.

Cristiano, Maykon Passos, Danon Clemes Cardoso, Vivian E. Sandoval-Gómez, and Flávia Carolina Simões-Gomes. 2020. "*Amoimyrmex* Cristiano, Cardoso & Sandoval, gen. nov. (Hymenoptera: Formicidae): A New Genus of Leaf-Cutting Ants Revealed by Multilocus Molecular Phylogenetic and Morphological Analyses." *Austral Entomology* 59(4): 643–676. https://doi.org/https://doi.org/10.1111/aen.12493.

Cubry, Philippe, Christine Tranchant-Dubreuil, Anne-Céline Thuillet, Cécile Monat, Marie-Noelle Ndjiondjop, et al. 2018. "The Rise and Fall of African Rice Cultivation Revealed by Analysis of 246 New Genomes." *Current Biology* 28(14): 2274–2282.

Currie, Cameron R., Ulrich G. Mueller, and David Malloch. 1999. "The Agricultural Pathology of Ant Fungus Gardens." *Proceedings of the National Academy of Sciences* 96:7998–8002.

Currie, Cameron R., Michael Poulsen, John Mendenhall, Jacobus J. Boomsma, and Johan Billen. 2006. "Coevolved Crypts and Exocrine Glands Support Mutualistic Bacteria in Fungus-Growing Ants." *Science* 311(5757): 81–83.

Currie, Cameron R., James A. Scott, Richard C. Summerbell, and David Malloch. 1999. "Fungus-Growing Ants Use Antibiotic-Producing Bacteria to Control Garden Parasites." *Nature* 398:701–704.

Currie, Cameron R., Bess Wong, Alison E. Stuart, Ted R. Schultz, Stephen A. Rehner, et al. 2003. "Ancient Tripartite Coevolution in the Attine Ant-Microbe Symbiosis." *Science* 299(5605): 386. https://doi.org/10.1126/science.1078155.

De Fine Licht, Henrik H., Jacobus J. Boomsma, and Anders Tunlid. 2014. "Symbiotic Adaptations in the Fungal Cultivar of Leaf-Cutting Ants." *Nature Communications* 5:1–10. https://doi.org/doi:10.1038/ncomms6675.

De Fine Licht, Henrik H., Morten Schiøtt, Ulrich G. Mueller, and Jacobus J. Boomsma. 2010. "Evolutionary Transitions in Enzyme Activity of Ant Fungus Gardens." *Evolution* 64(7): 2055–2069. https://doi.org/10.1111/j.1558-5646.2010.00948.x.

Della Lucia, Terezinha M. C., ed. 2011. *Formigas-Cortadeiras: da Bioecologia ao Manejo*. Viçosa, Minas Gerais, Brazil: Editoria Universidade Federal de Viçosa.

Dempewolf, Hannes, Kathryn A. Hodgins, Sonja E. Rummell, Norman C. Ellstrand, and Loren H. Rieseberg. 2012. "Reproductive Isolation during Domestication." *Plant Cell* 24(7): 2710–2717. https://doi.org/10.1105/tpc.112.100115.

Den Boer, Susanne P. A., Boris Baer, and Jacobus J. Boomsma. 2010. "Seminal Fluid Mediates Ejaculate Competition in Social Insects." *Science* 327(5972): 1506. https://doi.org/10.1126/science.1184709.

Denevan, William M. 2001. *Cultivated Landscapes of Native Amazonia and the Andes*. Edited by Gordon Clark, Andrew Goudie, and Ceri Peach. Oxford: Oxford University Press.

Dentinger, B. T. M., D. J. Lodge, A. B. Munkacsi, D. E. Desjardin, and D. J. McLaughlin. 2009. "Phylogenetic Placement of an Unusual Coral Mushroom Challenges the Classic Hypothesis of Strict Coevolution in the *Apterostigma pilosum* Group Ant-Fungus Mutualism." *Evolution* 63(8): 2172–2178.

Diamond, Jared. 1997. *Guns, Germs, and Steel: The Fates of Human Societies*. New York: Norton.

Diamond, Jared. 2002. "Evolution, Consequences and Future of Plant and Animal Domestication." *Nature* 418(6898): 700–707. https://doi.org/10.1038/nature01019.

Diamond, Jared. 2012. "The Local Origins of Domestication." In *Biodiversity in Agriculture: Domestication, Evolution, and Sustainability*, edited by Paul Gepts, Thomas R. Famula, Robert L. Bettinger, Stephen B. Brush, Ardeshir B. Damania, et al., 9–17. New York: Cambridge University Press.

Dijkstra, Michiel B., and Jacobus J. Boomsma. 2003. "*Gnamptogenys hartmani* Wheeler (Ponerinae: Ectatommini): An Agro-Predator of *Trachymyrmex* and *Sericomyrmex* Fungus-Growing Ants." *Naturwissenschaften* 90(12): 568–571. https://doi.org/10.1007/s00114-003-0478-4.

Dijkstra, Michiel B., and Jacobus J. Boomsma. 2006. "Are Workers of *Atta* Leafcutter Ants Capable of Reproduction?" *Insectes Sociaux* 53(2): 136–140. https://doi.org/10.1007/s00040-005-0848-3.

Dijkstra, Michiel B., David R. Nash, and Jacobus J. Boomsma. 2005. "Self-Restraint and Sterility in Workers of *Acromyrmex* and *Atta* Leafcutter Ants." *Insectes Sociaux* 52(1): 67–76. https://doi.org/10.1007/s00040-004-0775-8.

Diniz, J. L. M., C. R. F. Brandão, and C. I. Yamamoto. 1998. "Biology of *Blepharidatta* Ants, the Sister Group of the Attini: A Possible Origin of Fungus-Ants Symbiosis." *Naturwissenschaften* 85:270–274.

Dyble, Mark, Jack Thorley, Abigail E. Page, Daniel Smith, and Andrea Bamberg Migliano. 2019. "Engagement in Agricultural Work Is Associated with Reduced Leisure Time among Agta Hunter-Gatherers." *Nature Human Behaviour* 3(8): 792–796. https://doi.org/10.1038/s41562-019-0614-6.

FAOSTAT. 2020. Food and Agriculture Organization of the United Nations Statistical Database. Rome, Italy.

Ferguson-Gow, Henry, Seirian Sumner, Andrew F. G. Bourke, and Kate E. Jones. 2014. "Colony Size Predicts Division of Labour in Attine Ants." *Proceedings of the Royal Society B— Biological Sciences* 281(1793): 20141411. https://doi.org/10.1111/j.1469-185X.2010.00156.x.

Fjerdingstad, Else J., and Jacobus J. Boomsma. 2000. "Queen Mating Frequency and Relatedness in Young *Atta sexdens* Colonies." *Insectes Sociaux* 47(4): 354–356.

Fjerdingstad, Else J., Jacobus J. Boomsma, and Peter Thorén. 1998. "Multiple Paternity in the Leafcutter Ant *Atta colombica*—a Microsatellite DNA Study." *Heredity* 80(1): 118–126. https://doi.org/10.1038/sj.hdy.6882470.

Folgarait, Patricia J. 1998. "Ant Biodiversity and Its Relationship to Ecosystem Functioning: A Review." *Biodiversity & Conservation* 7(9): 1221–1244. https://doi.org/10.1023/A:1008891901953.

Fowler, H. G., V. Pereira-da-Silva, L. C. Forti, and N. B. Saes. 1986. "Population Dynamics of Leaf-Cutting Ants: A Brief Review." In *Fire Ants and Leaf-Cutting Ants*, edited by C. S. Lofgren and R. K. Vander Meer, 123–145. Boulder, CO: Westview Press.

Fuller, Dorian Q. 2007. "Contrasting Patterns in Crop Domestication and Domestication Rates: Recent Archaeobotanical Insights from the Old World." *Annals of Botany* 100:903–924.

Fuller, Dorian Q., Tim Denham, Manuel Arroyo-Kalin, Leilani Lucas, Chris J. Stevens, et al. 2014. "Convergent Evolution and Parallelism in Plant Domestication Revealed by an Expanding Archaeological Record." *Proceedings of the National Academy of Sciences* 111(17): 6147. https://doi.org/10.1073/pnas.1308937110.

Fuller, Dorian Q., Ling Qin, Yunfei Zheng, Zhijun Zhao, Xugao Chen, et al. 2009. "The Domestication Process and Domestication Rate in Rice: Spikelet Bases from the Lower Yangtze." *Science* 323(5921): 1607. https://doi.org/10.1126/science.1166605.

Fuller, Dorian Q., and Chris J. Stevens. 2017. "Open for Competition: Domesticates, Parasitic Domesticoids and the Agricultural Niche." *Archaeology International* 20(3): 110–121. http://doi.org/10.5334/ai.359.

Gerardo, Nicole M., Ulrich G. Mueller, and Cameron R. Currie. 2006. "Complex Host-Pathogen Coevolution in the *Apterostigma* Fungus-Growing Ant-Microbe Symbiosis." *BioMed Central Evolutionary Biology* 6:88.

Gerardo, Nicole M., Ulrich G. Mueller, Shauna L. Price, and Cameron R. Currie. 2004. "Exploiting a Mutualism: Parasite Specialization on Cultivars within the Fungus-Growing Ant Symbiosis." *Proceedings of the Royal Society B—Biological Sciences* 271(1550). https://doi.org/10.1098/rspb.2004.2792.

Gotwald, William H., Jr. 1995. *Army Ants: The Biology of Social Predation*. Ithaca, NY: Cornell University Press.

Gowdy, John, and Lisi Krall. 2014. "Agriculture as a Major Evolutionary Transition to Human Ultrasociality." *Journal of Bioeconomics* 16(2): 179–202. https://doi.org/10.1007/s10818-013-9156-6.

Graham, Alan. 2011. *A Natural History of the New World: The Ecology and Evolution of Plants in the Americas*. Chicago: University of Chicago Press.

Green, Abigail M., Ulrich G. Mueller, and Rachelle M. M. Adams. 2002. "Extensive Exchange of Fungal Cultivars between Sympatric Species of Fungus-Growing Ants." *Molecular Ecology* 11(2): 191–195.

Grell, Morten N., Tore Linde, Sanne Nygaard, Kåre L. Nielsen, Jacobus J. Boomsma, et al. 2013. "The Fungal Symbiont of *Acromyrmex* Leaf-Cutting Ants Expresses the Full Spectrum of Genes to Degrade Cellulose and Other Plant Cell Wall Polysaccharides." *BMC Genomics* 14(1): 928. https://doi.org/10.1186/1471-2164-14-928.

Harlan, J. R. 1992. *Crops and Man*. Madison, WI: American Society of Agronomy.

Harris, David R. 1989. "An Evolutionary Continuum of People-Plant Interaction." In *Foraging and Farming: The Evolution of Plant Exploitation*, edited by David R. Harris and Gordon C. Hillman, 11–26. London: Routledge.

Harris, David R. 2012. "Evolution of Agrosystems: Biodiversity, Origins, and Differential Development." In *Biodiversity in Agriculture: Domestication, Evolution, and Sustainability*, edited by Paul Gepts, Thomas R. Famula, Robert L. Bettinger, Stephen B. Brush, Ardeshir B. Damania, et al., 21–56. New York: Cambridge University Press.

Hayden, B. 1995. "A New Overview of Domestication." In *Last Hunters—First Farmers*, edited by T. D. Price and A. B. Gebauer, 273–299. Santa Fe, NM: School of American Research Press.

Hély, Christelle, Pascale Braconnot, Julie Watrin, and Weipeng Zheng. 2009. "Climate and Vegetation: Simulating the African Humid Period." *Comptes Rendus Geoscience* 341(8): 671–688. https://doi.org/10.1016/j.crte.2009.07.002.

Henry, Amanda G., Alison S. Brooks, and Dolores R. Piperno. 2011. "Microfossils in Calculus Demonstrate Consumption of Plants and Cooked Foods in Neanderthal Diets (Shanidar III, Iraq; Spy I and II, Belgium)." *Proceedings of the National Academy of Sciences* 108(2): 486. https://doi.org/10.1073/pnas.1016868108.

Henry, Amanda G., Alison S. Brooks, and Dolores R. Piperno. 2014. "Plant Foods and the Dietary Ecology of Neanderthals and Early Modern Humans." *Journal of Human Evolution* 69:44–54. https://doi.org/10.1016/j.jhevol.2013.12.014.

Herz, Hubert, Wolfram Beyschlag, and Berthold Hölldobler. 2007. "Herbivory Rate of Leaf-Cutting Ants in a Tropical Moist Forest in Panama at the Population and Ecosystem Scales." *Biotropica* 39(4): 482–488. https://doi.org/10.1111/j.1744-7429.2007.00284.x.

Hillman, Gordon C. 1989. "Late Paleolithic Plant Foods from Wadi Kubbaniya in Upper Egypt: Dietary Diversity, Infant Weaning, and Seasonality in a Riverine Environment." In *Foraging and Farming: The Evolution of Plant Exploitation*, edited by David R. Harris and Gordon C. Hillman, 207–239. London: Routledge.

Hillman, Gordon C., and M. Stuart Davies. 1990. "Measured Domestication Rates in Wild Wheats and Barley under Primitive Cultivation, and Their Archaeological Implications." *Journal of World Prehistory* 4(2): 157–222. https://doi.org/10.1007/BF00974763.

Hillman, Gordon C., and Michéle M. Wollstonecroft. 2014. "Dietary Diversity: Our Species-Specific Dietary Adaptation." In *Archaeology of African Plant Use*, 37–49. Walnut Creek, CA: Left Coast Press/UCL Institute of Archaeology Publications.

Hölldobler, Bert, and Edward O. Wilson. 1990. *The Ants*. Cambridge, MA: Belknap Press.

Hölldobler, Bert, and Edward O. Wilson. 2009. *The Superorganism: The Beauty, Elegance, and Strangeness of Insect Societies*. New York: W. W. Norton.

Hölldobler, Bert, and Edward O. Wilson. 2010. *The Leafcutter Ants: Civilization by Instinct*. New York: W. W. Norton.

Honne, Bjørn Ivar, and Manfred Heun. 2009. "On the Domestication Genetics of Self-Fertilizing Plants." *Vegetation History and Archaeobotany* 18(3): 269–272. https://doi.org/10.1007/s00334-009-0213-y.

Howard, Thomas, Jayne Elisabeth Archer, and Richard Marggraf Turley. 2016. "Remembering Darnel, a Forgotten Plant of Literary, Religious, and Evolutionary Significance." *Journal of Ethnobiology* 36(1): 29–44. https://doi.org/10.2993/0278-0771-36.1.29.

Howe, Jack, Morten Schiøtt, and Jacobus J. Boomsma. 2018. "Horizontal Partner Exchange Does Not Preclude Stable Mutualism in Fungus-Growing Ants." *Behavioral Ecology* 30(2): 372–382. https://doi.org/10.1093/beheco/ary176.

Hughes, William, and Jacobus Boomsma. 2004. "Genetic Diversity and Disease Resistance in Leaf-Cutting Ant Societies." *Evolution* 58(6): 1251–1260.

Jackson, Fatimah Linda Collier. 1991. "Secondary Compounds in Plants (Allelochemicals) as Promoters of Human Biological Variability." *Annual Review of Anthropology* 20(1): 505–544. https://doi.org/10.1146/annurev.an.20.100191.002445.

Janzen, Daniel. 1995. "Who Survived the Cretaceous?" *Science* 268:785.

Ješovnik, Ana, Júlio Chaul, and Ted Schultz. 2018. "Natural History and Nest Architecture of the Fungus-Farming Ant Genus *Sericomyrmex* (Hymenoptera: Formicidae)." *Myrmecological News* 26:65–80.

Johns, Timothy. 1996. *The Origins of Human Diet and Medicine*. Tucson: University of Arizona Press.

Jolivet, Pierre. 1996. *Ants and Plants: An Example of Coevolution*. Leiden, The Netherlands: Backhuys.

Jouquet, Pascal, Jens Dauber, Jan Lagerlöf, Patrick Lavelle, and Michel Lepage. 2006. "Soil Invertebrates as Ecosystem Engineers: Intended and Accidental Effects on Soil and Feedback Loops." *Applied Soil Ecology* 32(2): 153–164. https://doi.org/https://doi.org/10.1016/j.apsoil.2005.07.004.

Kaiho, Kunio, Naga Oshima, Kouji Adachi, Yukimasa Adachi, Takuya Mizukami, et al. 2016. "Global Climate Change Driven by Soot at the K-Pg Boundary as the Cause of the Mass Extinction." *Scientific Reports* 6(1): 28427. https://doi.org/10.1038/srep28427.

Kaplan, Lawrence. 1981. "What Is the Origin of the Common Bean?" *Economic Botany* 35(2): 240–254.

Kermarrec, A., M. Decharme, and G. Febvay. 1986. "Leaf-Cutting Ant Symbiotic Fungi: A Synthesis of Recent Research." In *Fire Ants and Leaf-Cutting Ants*, edited by C. S. Lofgren and R. K. Vander Meer, 231–246. Boulder, CO: Westview Press.

Kirkbride, Joseph H. Jr. 1993. *Biosystematic Monograph of the Genus Cucumis (Cucurbitaceae): Botanical Identification of Cucumbers and Melons*. Boone, NC: Parkway Publishers.

Kistler, Logan, S. Yoshi Maezumi, Jonas Gregorio De Souza, Natalia A. S. Przelomska, Flaviane Malaquias Costa, et al. 2018. "Multiproxy Evidence Highlights a Complex Evolutionary Legacy of Maize in South America." *Science* 362(6420): 1309–1313. https://doi.org/10.1126/science.aav0207.

Kooij, Pepijn W., Duur K. Aanen, Morten Schiøtt, and Jacobus J. Boomsma. 2015. "Evolutionarily Advanced Ant Farmers Rear Polyploid Fungal Crops." *Journal of Evolutionary Biology* 28(11): 1911–1924. https://doi.org/10.1111/jeb.12718.

Kooij, Pepijn W., Adelina Rogowska-Wrzesinska, Daniel Hoffmann, Peter Roepstorff, Jacobus J. Boomsma, et al. 2014. "*Leucoagaricus gongylophorus* Uses Leaf-Cutting Ants to Vector Proteolytic Enzymes towards New Plant Substrate." *ISME Journal* 8: 1032. https://doi.org/10.1038/ismej.2013.23.1

Larsen, Clark Spencer. 1995. "Biological Changes in Human Populations with Agriculture." *Annual Review of Anthropology* 24(1): 185–213. https://doi.org/10.1146/annurev.an.24.100195.001153.

Larsen, Clark Spencer, Christopher J. Knüsel, Scott D. Haddow, Marin A. Pilloud, Marco Milella, et al. 2019. "Bioarchaeology of Neolithic Çatalhöyük Reveals Fundamental Transitions in Health, Mobility, and Lifestyle in Early Farmers." *Proceedings of the National Academy of Sciences* 116(26): 12615. https://doi.org/10.1073/pnas.1904345116.

Larson, Greger, Dolores R. Piperno, Robin G. Allaby, Michael D. Purugganan, Leif Andersson, et al. 2014. "Current Perspectives and the Future of Domestication Studies." *Proceedings of the National Academy of Sciences* 111(17): 6139. https://doi.org/10.1073/pnas.1323964111.

Leal, Inara R., and Paulo S. Oliveira. 2000. "Foraging Ecology of Attine Ants in a Neotropical Savanna: Seasonal Use of fungal Substrate in the Cerrado Vegetation of Brazil." *Insectes Sociaux* 47(4): 376–382. https://doi.org/10.1007/PL00001734.

Leal-Dutra, Caio A., Gareth W. Griffith, Maria Alice Neves, David J. McLaughlin, Esther G. McLaughlin, et al. 2020. "Reclassification of Pterulaceae Corner (Basidiomycota: Agaricales): Introducing the Ant-Associated Genus *Myrmecopterula* gen. nov., *Phaeopterula* Henn. and the Corticioid Radulomycetaceae fam. nov." *IMA Fungus* 11(1): 2. https://doi.org/10.1186/s43008-019-0022-6.

Lengyel, Szabolcs, Aaron D. Gove, Andrew M. Latimer, Jonathan D. Majer, and Robert R. Dunn. 2010. "Convergent Evolution of Seed Dispersal by Ants, and Phylogeny and Biogeography in Flowering Plants: A Global Survey." *Perspectives in Plant Ecology, Evolution and Systematics* 12(1): 43–55. https://doi.org/https://doi.org/10.1016/j.ppees.2009.08.001.

Letourneau, Deborah K. 1998. "Ants, Stem-Borers, and Fungal Pathogens: Experimental Tests of a Fitness Advantage in *Piper* Ant-Plants." *Ecology* 79(2): 593–603. https://doi.org/10.1890/0012-9658(1998)079[0593:asbafp]2.0.co;2.

Li, Hongjie, Jeffrey Sosa-Calvo, Heidi A. Horn, Mônica T. Pupo, Jon Clardy, et al. 2018. "Convergent Evolution of Complex Structures for Ant–Bacterial Defensive Symbiosis in Fungus-Farming Ants." *Proceedings of the National Academy of Sciences* 115(42): 10720. http://www.pnas.org/content/115/42/10720.abstract.

Little, Ainslie E. F., and Cameron R. Currie. 2007. "Symbiotic Complexity: Discovery of a Fifth Symbiont in the Attine Ant-Microbe Symbiosis." *Biology Letters* 3(5): 501–504. https://doi.org/10.1098/rsbl.2007.0253.

Little, Ainslie E. F., Takahiro Murakami, Ulrich G. Mueller, and Cameron R. Currie. 2003. "The Infrabuccal Pellet Piles of Fungus-Growing Ants." *Naturwissenschaften* 90:558–562.

Lombardo, Umberto, José Iriarte, Lautaro Hilbert, Javier Ruiz-Pérez, José M. Capriles, et al. 2020. "Early Holocene Crop Cultivation and Landscape Modification in Amazonia." *Nature* 581(7807): 190–193. https://doi.org/10.1038/s41586-020-2162-7.

Manning, Katie, Ruth Pelling, Tom Higham, Jean-Luc Schwenniger, and Dorian Q. Fuller. 2011. "4500-Year-Old Domesticated Pearl Millet (Pennisetum glaucum) from the Tilemsi Valley, Mali: New Insights into an Alternative Cereal Domestication Pathway." *Journal of Archaeological Science* 38(2): 312–322. https://doi.org/https://doi.org/10.1016/j.jas.2010.09.007.

Marshall, Fiona B., Keith Dobney, Tim Denham, and José M. Capriles. 2014. "Evaluating the Roles of Directed Breeding and Gene Flow in Animal Domestication." *Proceedings of the National Academy of Sciences* 111(17): 6153. https://doi.org/10.1073/pnas.1312984110.

Martin, Joan Stadler, and Michael M. Martin. 1970. "The Presence of Protease Activity in the Rectal Fluid of Attine Ants." *Journal of Insect Physiology* 16(2): 227–232. https://doi.org/10.1016/0022-1910(70)90164-2.

Masiulionis, Virginia E., Christian Rabeling, Henrik H. De Fine Licht, Ted R. Schultz, Maurício Bacci, et al. 2014. "A Brazilian Population of the Asexual Fungus-Growing Ant *Mycocepurus smithii* (Formicidae, Myrmicinae, Attini) Cultivates Fungal Symbionts with Gongylidia-Like Structures." *PLoS ONE* 9(8): e103800. https://doi.org/10.1371/journal.pone.0103800.g003.

Mathieson, Iain, Iosif Lazaridis, Nadin Rohland, Swapan Mallick, Nick Patterson, et al. 2015. "Genome-Wide Patterns of Selection in 230 Ancient Eurasians." *Nature* 528(7583): 499–503. https://doi.org/10.1038/nature16152.

Mayer, Veronika E., Maximilian Nepel, Rumsais Blatrix, Felix B. Oberhauser, Konrad Fiedler, et al. 2018. "Transmission of Fungal Partners to Incipient Cecropia-Tree Ant Colonies." *PLoS ONE* 13(2): e0192207. https://doi.org/10.1371/journal.pone.0192207.

McCouch, Susan R., Michael J. Kovach, Megan Sweeney, Hui Jiang, and Mande Semon. 2012. "The Dynamics of Rice Domestication: A Balance between Gene Flow and Genetic Isolation." In *Biodiversity in Agriculture: Domestication, Evolution, and Sustainability*, edited by Paul Gepts, Thomas R. Famula, Robert L. Bettinger, Stephen B. Brush, Ardeshir B. Damania, et al., 311–329. New York: Cambridge University Press.

McKey, Doyle B., Marianne Elias, Benoit Pujol, and A. Duputié. 2012. "Ecological Approaches to Crop Domestication." In *Biodiversity in Agriculture: Domestication, Evolution, and Sustainability*, edited by Paul Gepts, Thomas R. Famula, Robert L. Bettinger, Stephen B. Brush, Ardeshir B. Damania, et al., 377–406. New York: Cambridge University Press.

McKey, Doyle B., and S. Rostain. 2014. "Farming Technology in Amazonia." In *Encyclopaedia of the History of Science, Technology, and Medicine in Non-Western Cultures*, edited by H. Selin, 114. Dordrecht, The Netherlands: Springer.

Mehdiabadi, Natasha J., Benjamin Hughes, and Ulrich G. Mueller. 2006. "Cooperation, Conflict, and Coevolution in the Attine Ant-Fungus Symbiosis." *Behavioral Ecology* 17(2): 291–296. https://doi.org/10.1093/beheco/arj028.

Mehdiabadi, Natasha J., Ulrich G. Mueller, Seán G. Brady, Anna G. Himler, and Ted R. Schultz. 2012. "Symbiont Fidelity and the Origin of Species in Fungus-Growing Ants." *Nature Communications* 3:840. https://doi.org/DOI10.1038/Ncomms1844.

Mehdiabadi, Natasha J., and Ted R. Schultz. 2009. "Natural History and Phylogeny of the Fungus-Farming Ants (Hymenoptera: Formicidae: Myrmicinae: Attini)." *Myrmecological News* 13:37–55.

Meirelles, Lucas A., Quimi V. Montoya, Scott E. Solomon, and Andre Rodrigues. 2015a. "New Light on the Systematics of Fungi Associated with Attine Ant Gardens and the Description of *Escovopsis kreiselii* sp. nov." *PLoS ONE* 10(1): e0112067. https://doi.org/10.1371/journal.pone.0112067.

Meirelles, Lucas A., Scott E. Solomon, Mauricio Bacci, April M. Wright, Ulrich G. Mueller, et al. 2015b. "Shared *Escovopsis* Parasites between Leaf-Cutting and Non-Leaf-Cutting Ants in the Higher Attine Fungus-Growing Ant Symbiosis." *Royal Society Open Science* 2(9): 150257. https://doi.org/10.1098/rsos.150257.

Mendes, Thais D., Andre Rodrigues, Ifeloju Dayo-Owoyemi, Fernando A. L. Marson, and Fernando C. Pagnocca. 2012. "Generation of Nutrients and Detoxification: Possible Roles of Yeasts in Leaf-Cutting Ant Nests." *Insects* 3(4): 228–245. http://www.mdpi.com/2075-4450/3/1/228/.

Mercader, Julio. 2009. "Mozambican Grass Seed Consumption during the Middle Stone Age." *Science* 326(5960): 1680. https://doi.org/10.1126/science.1173966.

Meyer, Sebastian T., Meike Neubauer, Emma J. Sayer, Inara R. Leal, Marcelo Tabarelli, et al. 2013. "Leaf-Cutting Ants as Ecosystem Engineers: Topsoil and Litter Perturbations around Atta Cephalotes Nests Reduce Nutrient Availability." *Ecological Entomology* 38(5): 497–504. https://doi.org/10.1111/een.12043.

Miles, Philip G, and Shu-Ting Chang. 2004. *Mushrooms: Cultivation, Nutritional Value, Medicinal Effect, and Environmental Impact.* Boca Raton, FL: CRC Press.

Möller, Alfred. 1893. "Die Pilzgärten einiger Südamerikansicher Ameisen." *Botanische Mitheilungen aus den Tropen* 6:1–127.

Moller, Isabel E., Henrik H. De Fine Licht, Jesper Harholt, William G. T. Willats, and Jacobus J. Boomsma. 2011. "The Dynamics of Plant Cell-Wall Polysaccharide Decomposition in Leaf-Cutting Ant Fungus Gardens." *PLoS ONE* 6(3): e17506. https://doi.org/10.1371/journal.pone.0017506.

Mueller, Ulrich G., Nicole M. Gerardo, Duur K. Aanen, Diana L. Six, and Ted R. Schultz. 2005. "The Evolution of Agriculture in Insects." *Annual Review of Ecology, Evolution, and Systematics* 36:563–595.

Mueller, Ulrich G., Melissa R. Kardish, Heather D. Ishak, April M. Wright, Scott E. Solomon, et al.2018. "Phylogenetic Patterns of Ant–Fungus Associations Indicate That Farming Strategies, Not Only a Superior Fungal Cultivar, Explain the Ecological Success of Leafcutter Ants." *Molecular Ecology* 27(10): 2414–2434. https://doi.org/10.1111/mec.14588.

Mueller, Ulrich G., Stephen A. Rehner, and Ted R. Schultz. 1998. "The Evolution of Agriculture in Ants." *Science* 281(5385): 2034–2038.

Mueller, Ulrich G., Ted R. Schultz, Cameron R. Currie, Rachelle M. M. Adams, and David Malloch. 2001. "The Origin of the Attine Ant-Fungus Mutualism." *Quarterly Review of Biology* 76(2): 169–197.

Mummert, Amanda, Emily Esche, Joshua Robinson, and George J. Armelagos. 2011. "Stature and Robusticity during the Agricultural Transition: Evidence from the Bioarchaeological Record." *Economics & Human Biology* 9(3): 284–301. https://doi.org/10.1016/j.ehb.2011.03.004.

Murakami, Takahiro, and Seigo Higashi. 1997. "Social Organization in Two Primitive Attine Ants, *Cyphomyrmex rimosus* and *Myrmicocrypta ednaella*, with Reference to Their Fungus Substrates and Food Sources." *Journal of Ethology* 15:17–25.

Murakami, Takahiro, Seigo Higashi, and Donald M. Windsor. 2000. "Mating Frequency, Colony Size, Polyethism and Sex Ratio in Fungus-Growing Ants (Attini)." *Behavioral Ecology and Socibiology* 48(4): 276–284.

Navarette-Heredia, J. L. 2001. "Beetles Associated with *Atta* and *Acromyrmex* Ants (Hymenoptera: Formicidae: Attini)." *Transactions of the American Entomological Society* 127(3): 381–429.

North, R. D., C. W. Jackson, and P. E. Howse. 1997. "Evolutionary Aspects of Ant-Fungus Interactions in Leaf-Cutting Ants." *Trends in Ecology & Evolution* 12(10): 386–389.

Nygaard, Sanne, Haofu Hu, Cai Li, Morten Schiøtt, Zhensheng Chen, et al. 2016. "Reciprocal Genomic Evolution in the Ant–Fungus Agricultural Symbiosis." *Nature Communications* 7:12233. https://doi.org/10.1038/ncomms12233.

Osti, Julio Flavio, and Andre Rodrigues. 2018. "*Escovopsioides* as a Fungal Antagonist of the Fungus Cultivated by Leafcutter Ants." *BMC Microbiology* 18(1): 130. https://doi.org/10.1186/s12866-018-1265-x.

Pagnocca, Fernando C., Virginia E. Masiulionis, and Andre Rodrigues. 2012. "Specialized Fungal Parasites and Opportunistic Fungi in Gardens of Attine Ants." *Psyche: A Journal of Entomology* 2012:1–9. https://doi.org/10.1155/2012/905109.

Pagnocca, Fernando C., Andre Rodrigues, Nilson S. Nagamoto, and Mauricio Bacci. 2008. "Yeasts and Filamentous Fungi Carried by the Gynes of Leaf-Cutting Ants." *Antonie Van Leeuwenhoek International Journal of General and Molecular Microbiology* 94(4): 517–526. https://doi.org/10.1007/s10482-008-9268-5.

Papa, R., J. Acosta, A. Delgado-Salinas, and P. Gepts. 2005. "A Genome-Wide Analysis of Differentiation between Wild and Domesticated *Phaseolus vulgaris* from Mesoamerica." *Theoretical and Applied Genetics* 111(6): 1147–1158. https://doi.org/10.1007/s00122-005-0045-9.

Paterson, Andrew H., WenQian Kong, Robyn M. Johnston, Pheonah Nabukalu, Guohong Wu, et al. 2020. "The Evolution of an Invasive Plant, *Sorghum halepense* L. ('Johnsongrass')." *Frontiers in Genetics* 11:317. https://www.frontiersin.org/article/10.3389/fgene.2020.00317.

Petraglia, Michael. 2017. "Hominins on the Move: An Assessment of Anthropogenic Shaping of Environments in the Palaeolithic." In *Human Dispersal and Species Movement: From Prehistory to the Present*, edited by Michael Petraglia, Nicole Boivin, and Rémy Crassard, 90–118. Cambridge: Cambridge University Press.

Pinhasi, Ron, and Jay T. Stock, eds. 2011. *Human Bioarchaeology of the Transition to Agriculture*. West Sussex, UK: Wiley.

Pinto-Tomás, Adrián A., Mark A. Anderson, Garret Suen, David M. Stevenson, Fiona S. T. Chu, et al. 2009. "Symbiotic Nitrogen Fixation in the Fungus Gardens of Leaf-Cutter Ants." *Science* 326(5956): 1120. https://doi.org/10.1126/science.1173036.

Piperno, Dolores R., and Deborah M. Pearsall. 1998. *The Origins of Agriculture in the Lowland Neotropics*. San Diego, CA: Academic Press.

Prothero, Donald R. 1994. "The Late Eocene-Oligocene Extinctions." *Annual Review of Earth and Planetary Sciences* 22(1): 145–165. https://doi.org/10.1146/annurev.ea.22.050194.001045.

Purugganan, Michael D., and Dorian Q. Fuller. 2009. "The Nature of Selection during Plant Domestication." *Nature* 457(7231): 843–848. https://doi.org/10.1038/nature07895.

Quinlan, R. J., and J. M. Cherrett. 1979. "The Role of Fungus in the Diet of the Leaf-Cutting Ant *Atta cephalotes* (L.)." *Ecological Entomology* 4(2): 151–160.

Rabeling, C., M. Verhaag, and U. G. Mueller. 2006. "Behavioral Ecology and Natural History of *Blepharidatta brasiliensis* (Formicidae, Blepharidattini)." *Insectes Sociaux* 53(3): 300–306.

Rico-Gray, Victor, and Paulo S. Oliveira. 2007. *The Ecology and Evolution of Ant-Plant Interactions*. Chicago: University of Chicago Press.

Ridley, P., P. E. Howse, and C. W. Jackson. 1996. "Control of the Behaviour of Leaf-Cutting Ants by Their 'Symbiotic' Fungus." *Experientia* 52(6): 631–635. https://doi.org/10.1007/BF01969745.

Rindos, David. 1984. *The Origins of Agriculture: An Evolutionary Perspective*. New York: Academic Press.

Rindos, David. 1989. "Darwinism and Its Role in the Explanation of Domestication." In *Foraging and Farming: The Evolution of Plant Exploitation*, edited by David R. Harris and Gordon C. Hillman, 27–41. London: Routledge.

Rindos, David, Homer Aschmann, Peter Bellwood, Lynn Ceci, Mark Cohen, et al. 1980. "Symbiosis, Instability, and the Origins and Spread of Agriculture: A New Model [and Comments and Reply]." *Current Anthropology* 21(6): 751–772. http://www.jstor.org/stable/2742515.

Roberts, Patrick, Chris Hunt, Manuel Arroyo-Kalin, Damian Evans, and Nicole Boivin. 2017. "The Deep Human Prehistory of Global Tropical Forests and Its Relevance for Modern Conservation." *Nature Plants* 3(8): 17093. https://doi.org/10.1038/nplants.2017.93.

Rodgers, D. M. 2008. *Debugging the Link between Social Theory and Social Insects*. Baton Rouge: Louisiana State University Press.

Rodrigues, Andre, Mauricio Bacci, Jr., Ulrich G Mueller, A. Ortiz, and Fernando C. Pagnocca. 2008. "Microfungal 'Weeds' in the Leafcutter Ant Symbiosis." *Microbial Ecology* 56(4): 604–614.

Rodrigues, Andre, Michel R. Z. Passarini, Milene Ferro, Nilson S. Nagamoto, Luiz C. Forti, et al. 2013. "Fungal Communities in the Garden Chamber Soils of Leaf-Cutting Ants." *Journal of Basic Microbiology* 54(11): 1186–1196. http://doi.wiley.com/10.1002/jobm.201200458.

Ronque, Mariane U. V., Rodrigo M. Feitosa, and Paulo S. Oliveira. 2019. "Natural History and Ecology of Fungus-Farming Ants: A Field Study in Atlantic Rainforest." *Insectes Sociaux* 66(3): 375–387. https://doi.org/10.1007/s00040-019-00695-y.

Sanders, Dirk, and F. J. Frank van Veen. 2011. "Ecosystem Engineering and Predation: The Multi-Trophic Impact of Two Ant Species." *Journal of Animal Ecology* 80(3): 569–576. https://doi.org/10.1111/j.1365-2656.2010.01796.x.

Scarcelli, Nora, Philippe Cubry, Roland Akakpo, Anne-Céline Thuillet, Jude Obidiegwu, et al. 2019. "Yam Genomics Supports West Africa as a Major Cradle of Crop Domestication." *Science Advances* 5(5): eaaw1947. https://doi.org/10.1126/sciadv.aaw1947.

Schiøtt, Morten, Adelina Rogowska-Wrzesinska, Peter Roepstorff, and Jacobus J. Boomsma. 2010. "Leaf-Cutting Ant Fungi Produce Cell Wall Degrading Pectinase Complexes Reminiscent of Phytopathogenic Fungi." *BMC Biology* 8(1): 156. https://doi.org/10.1186/1741-7007-8-156.

Schultz, Ted R., and Seán G. Brady. 2008. "Major Evolutionary Transitions in Ant Agriculture." *Proceedings of the National Academy of Sciences* 105(14): 5435–5440.

Schultz, Ted R., and Terry P. McGlynn. 2000. "The Interactions of Ants with Other Organisms." In *Ants: Standard Methods for Measuring and Monitoring Biodiversity*, edited by D. Agosti, J. D. Majer, L. E. Alonso and T. R. Schultz, 35–44. Washington, DC: Smithsonian Institution Press.

Schultz, Ted R., Ulrich G. Mueller, Cameron R. Currie, and Stephen A. Rehner. 2005. "Reciprocal Illumination: A Comparison of Agriculture in Humans and in Fungus–Growing Ants." In *Insect–Fungal Associations: Ecology and Evolution*, edited by Fernando E. Vega and Meredith Blackwell, 149–190. New York: Oxford University Press.

Schultz, Ted R., Jeffrey Sosa-Calvo, Seán G. Brady, Cauê T. Lopes, Ulrich G. Mueller, et al. 2015. "The Most Relictual Fungus-Farming Ant Species Cultivates the Most Recently Evolved and Highly Domesticated Fungal Symbiont Species." *American Naturalist* 185(5): 693–703. https://doi.org/10.1086/680501.

Seal, Jon N., and U. G. Mueller. 2014. "Instability of Novel Ant-Fungal Associations Constrains Horizontal Exchange of Fungal Symbionts." *Evolutionary Ecology* 28(1): 157–176. https://doi.org/10.1007/s10682-013-9665-8.

Senda, Takayuki, Yuhi Hiraoka, and Tohru Tominaga. 2006. "Inheritance of Seed Shattering in *Lolium temulentum* and *L. persicum* hybrids." *Genetic Resources and Crop Evolution* 53(3): 449–451. https://doi.org/10.1007/s10722-005-6096-6.

Shik, Jonathan Z., Juan C. Santos, Jon N. Seal, Adam Kay, Ulrich G. Mueller, et al. 2014. "Metabolism and the Rise of Fungus Cultivation by Ants." *American Naturalist* 184(3): 364–373. https://doi.org/10.1086/677296.

Smith, Bruce D. 1998a. "Between Foraging and Farming." *Science* 279(5357): 1651. https://doi.org/10.1126/science.279.5357.1651.

Smith, Bruce D. 1998b. *The Emergence of Agriculture*. New York: Scientific American Library.

Smith, Bruce D. 2001a. "Low-Level Food Production." *Journal of Archaeological Research* 9(1): 1–43. https://doi.org/10.1023/A:1009436110049.

Smith, Bruce D. 2001b. "The Transition to Food Production." In *Archaeology at the Millennium: A Sourcebook*, edited by G. M. Feinmann and T. D. Price, 199–229. New York: Kluwer Academic.

Smith, Bruce D. 2011a. "A Cultural Niche Construction Theory of Initial Domestication." *Biological Theory* 6(3): 260–271. https://doi.org/10.1007/s13752-012-0028-4.

Smith, Bruce D. 2011b. "General Patterns of Niche Construction and the Management of 'Wild' Plant and Animal Resources by Small-Scale Pre-Industrial Societies." *Philosophical Transactions of the Royal Society B—Biological Sciences* 366(1566): 836–848. https://doi.org/10.1098/rstb.2010.0253.

Snir, Ainit, Dani Nadel, Iris Groman-Yaroslavski, Yoel Melamed, Marcelo Sternberg, et al. 2015. "The Origin of Cultivation and Proto-Weeds, Long before Neolithic Farming." *PLoS ONE* 10(7): e0131422. https://doi.org/10.1371/journal.pone.0131422.

Solomon, Scott E., Christian Rabeling, Jeffrey Sosa-Calvo, Cauê T. Lopes, André Rodrigues, et al. 2019. "The Molecular Phylogenetics of Trachymyrmex Forel Ants and Their Fungal Cultivars Provide Insights into the Origin and Coevolutionary History of 'Higher-Attine' Ant Agriculture." *Systematic Entomology* 44(4): 939–956. https://doi.org/10.1111/syen.12370.

Somera, Alexandre F., Adriel M. Lima, Álvaro J. dos Santos-Neto, Fernando M. Lanças, and Maurício Bacci. 2015. "Leaf-Cutter Ant Fungus Gardens Are Biphasic Mixed Microbial Bioreactors That Convert Plant Biomass to Polyols with Biotechnological Applications." *Applied and Environmental Microbiology* 81(13): 4525–4535. https://doi.org/10.1128/aem.00046-15.

Spahillari, Merita, Karl Hammer, Thomas Gladis, and Axel Diederichsen. 1999. "Weeds as Part of Agrobiodiversity." *Outlook on Agriculture* 28-(4): 227–232. https://doi.org/10.1177/003072709902800405.

Sweeney, Megan, and Susan McCouch. 2007. "The Complex History of the Domestication of Rice." *Annals of Botany* 100(5): 951–957. https://doi.org/10.1093/aob/mcm128.

Tanno, Ken-ichi, and George Willcox. 2006. "How Fast Was Wild Wheat Domesticated?" *Science* 311(5769): 1886. https://doi.org/10.1126/science.1124635.

Thomas, Howard, Jayne E. Archer, and Richard M. Turley. 2011. "Evolution, Physiology and Phytochemistry of the Psychotoxic Arable Mimic Weed Darnel (*Lolium temulentum* L.)." *Progress in Botany* 72:73–104.

Tobin, J E. 1994. "Ants as Primary Consumers: Diet and Abundance in the Formicidae." In *Nourishment and Evolution in Insect Societies*, edited by J. H. Hunt and C. A. Nalepa, 279–307. Boulder, CO: Westview Press.

Tudge, Colin. 1998. *Neanderthals, Bandits and Farmers: How Agriculture Really Began*. New Haven, CT: Yale University Press.

Turner, J. A. 1974. "The Bioenergetics of Leaf-Cutting Ants in Trinidad." MS thesis, University of the West Indies.

Udall, Joshua A., and Jonathan F. Wendel. 2006. "Polyploidy and Crop Improvement." *Crop Science* 46(Supp. 1): S-3. https://doi.org/10.2135/cropsci2006.07.0489tpg.

Urich, F. W. 1895. "Notes on Some Fungus-Growing Ants of Trinidad." *Journal of the Trinidad Field Naturalists' Club* 2(7): 175–182.

US Department of Agriculture. 2020a. *Livestock and Poultry: World Markets and Trade*. Washington, DC: Foreign Agricultural Service.

US Department of Agriculture. 2020b. *World Agricultural Production 5–20*. Washington, DC: Foreign Agricultural Service.

Vandermeer, John, and Ivette Perfecto. 2007. "The Diverse Faces of Ecosystem Engineers in Agroecosystems." In *Ecosystem Engineers: Plants to Protists*, edited by Kim Cuddington, James E. Byers, William G. Wilson, and Alan Hastings, 367–385. London: Academic Press.

Vellinga, Else C. 2004. "Ecology and Distribution of Lepiotaceous Fungi (Agaricaceae): A Review." *Nova Hedwigia* 78(3): 273–299. https://doi.org/10.1127/0029-5035/2004/0078-0273.

Villesen, Palle, Takahiro Murakami, Ted R. Schultz, and Jacobus J. Boomsma. 2002. "Identifying the Transition between Single and Multiple Mating of Queens in Fungus-Growing Ants." *Proceedings of the Royal Society B—Biological Sciences* 269(1500): 1541–1548.

Vo, Tanya L., Ulrich G. Mueller, and Alexander S. Mikheyev. 2009. "Free-Living Fungal Symbionts (Lepiotaceae) of Fungus-Growing Ants (Attini: Formicidae)." *Mycologia* 101(2): 206–210.

Wadley, Lyn, Lucinda Backwell, Francesco d'Errico, and Christine Sievers. 2020. "Cooked Starchy Rhizomes in Africa 170 Thousand Years Ago." *Science* 367(6473): 87. https://doi.org/10.1126/science.aaz5926.

Waller, D. A., and J. C. Moser. 1990. "Invertebrate Enemies and Nest Associates of the Leaf-Cutting Ant *Atta texana* (Buckley) (Formicidae, Attini)." In *Applied Myrmecology: A World Perspective*, edited by R. K. Vander Meer, K. Jaffe, and A. Cedeno, 255–273. Boulder, CO: Westview Press.

Ward, Philip S. 2014. "The Phylogeny and Evolution of Ants." *Annual Review of Ecology, Evolution, and Systematics* 45(1): 23–43. https://doi.org/10.1146/annurev-ecolsys-120213-091824.

Weber, Neal A. 1972. *Gardening Ants: The Attines*. Philadelphia: American Philosophical Society.

Weber, Neal A. 1979. "Fungus-Culturing by Ants." In *Insect-Fungus Symbiosis: Nutrition, Mutualism, and Commensalism*, edited by Lekh R. Batra, 77–116. Montclair, NJ: Allanheld, Osmun.

Weiss, Ehud, Mordechai E. Kislev, and Anat Hartmann. 2006. "Autonomous Cultivation before Domestication." *Science* 312(5780): 1608. https://doi.org/10.1126/science.1127235.

Wetterer, James K. 1995. "Forager Size and Ecology of *Acromyrmex coronatus* and Other Leaf-Cutting Ants in Costa Rica." *Oecologia* 104(4): 409–415. http://www.jstor.org/stable/4221127.

Wetterer, James K. 1999. "The Ecology and Evolution of Worker Size-Distribution in Leaf-Cutting Ants (Hymenoptera: Formicidae)." *Sociobiology* 34(1): 119–144.

Wetterer, James K., Daniel Gruner, and Jorgé Lopez. 1998. "Foraging and Nesting Ecology of Acromyrmex octospinosus (Hymenoptera: Formicidae) in a Costa Rican Tropical Dry Forest." *Florida Entomologist* 81(1): 61–67. http://www.jstor.org/stable/3495996.

Wheeler, William M. 1901. "Notices biologiques sur les fourmis Mexicaines." *Annales de la Société Entomologique de Belgique* 45:199–205.

Wheeler, William M., and Irving W. Bailey. 1920. "The Feeding Habits of Pseudomyrmine and Other Ants." *Transactions of the American Philosophical Society New Series* 22(4): 235–279. http://www.jstor.org/stable/1005485.

Wilkes, H. G. 1977. "Hybridization of Maize and Teosinte, in Mexico and Guatemala and the Improvement of Maize." *Economic Botany* 31(3): 254–293. . http://www.jstor.org/stable/4253844.

Wilson, Edward O. 1980. "Caste and Division of Labor in Leaf-Cutter Ants (Hymenoptera: Formicidae: *Atta*): I. The Overall Pattern in *A. sexdens*." *Behavioral Ecology and Sociobiology* 7(2): 143–156. http://www.jstor.org/stable/4599319.

Wilson, Hugh D. 1990. "Gene Flow in Squash Species." *Bioscience* 40(6): 449–455. https://doi.org/10.2307/1311392.

Wirth, R., H. Herz, R. J. Ryel, W. Beyschlag, and Bert Hölldobler. 2003. *Herbivory of Leaf-Cutting Ants: A Case Study on Atta colombica in the Tropical Rainforest of Panama*, vol. 164, *Ecological Studies* series, edited by I. T. Baldwin, M. M. Caldwell, G. Heldmaier, O. L. Lange, H. A. Mooney, et al. New York: Springer.

Witte, Volker, and Ulrich Maschwitz. 2008. "Mushroom Harvesting Ants in the Tropical Rain Forest." *Die Naturwissenschaften* 95(11): 1049–1054. https://doi.org/10.1007/s00114-008-0421-9.

Worsley, Sarah F., Tabitha M. Innocent, Daniel Heine, J. Colin Murrell, Douglas W. Yu, et al. 2018. "Symbiotic Partnerships and Their Chemical Interactions in the Leafcutter Ants (Hymenoptera: Formicidae)." *Myrmecological News* 27:59–74.

Zeder, Melinda A. 2012. "Pathways to Animal Domestication." In *Biodiversity in Agriculture: Domestication, Evolution, and Sustainability*, edited by Paul Gepts, Thomas R. Famula, Robert L. Bettinger, Stephen B. Brush, Ardeshir B. Damania, et al., 227–259. New York: Cambridge University Press.

Zeder, Melinda A. 2018. "Why Evolutionary Biology Needs Anthropology: Evaluating Core Assumptions of the Extended Evolutionary Synthesis." *Evolutionary Anthropology: Issues, News, and Reviews* 27(6): 267–284. https://doi.org/10.1002/evan.21747.

Contributors

Duur K. Aanen
Wageningen University

Niels P. R. Anten
Wageningen University

Peter H. W. Biedermann
University of Freiburg

Jacobus J. Boomsma
University of Copenhagen

Laura T. Buck
Liverpool John Moores University

Guillaume Chomicki
University of Sheffield

Tim Denham
Australian National University

R. Ford Denison
University of Minnesota

Dorian Q. Fuller
University College London

Richard Gawne
Tufts University

Nicole M. Gerardo
Emory University

Thomas C. Harrington
Iowa State University

Ana Ješovnik
Smithsonian Institution

Judith Korb
University of Freiburg

Chase G. Mayers
Cornell University

George R. McGhee
Rutgers University

Kenneth Z. McKenna
University of California San Diego

Lumila P. Menéndez
Konrad Lorenz Institute

Peter N. Peregrine
Lawrence University

Ted R. Schultz
Smithsonian Institution

Index

Aanen, D. K., 3, 4, 10, 11, 12, 15, 26, 33, 44, 49, 50, 61–72, 95, 98, 99, 177, 295
Aardvark pumpkin. See *Cucumis humifructus*
Acacia, 146–147, 162
Acorn, 232
Acromyrmex, 16, 75, 76, 78, 79, 80, 81, 216, 299, 300
Acropyga, 145
Acropyga goeldii, 147
Actinobacteria, 77
Actinomycetes, 286
Actinopterygii, 8, 13, 27
Adaptive landscapes, 18–27
Aechmea mertensii, 163
Afalou site, Morocco, 230
AFC. *See* Ambrosia *Fusarium* clade
Africa, 108, 146, 150, 170, 187, 193–194, 245
  Algeria, 230
  Egypt, 194
  Ethiopia, 195
  Ivory Coast, 90–91, 97–98
  Maghreb, 226–227, 237. *See also* Africa: Sub-Saharan Africa
  Morocco, 230
  Nile Valley, 234
  North Africa, 235
  Northeastern Africa, 229
  Northwestern Africa, 230, 241
  Nubia, 226–227, 229–234, 237–238, 243–244, 246
  Savannah, 89–93, 147
  Southern Africa, 74, 98–99, 226, 240, 245
  Sub-Saharan Africa, 32, 187, 189, 225–226
  Togo, 95
  Tunis, 230
  Western Africa, 32, 90, 94, 98, 188
African people, 240–241, 245
African rice, 296
Afroraffaelea, 113
Afroraffaelea ambrosiae, 108, 118, 123
Agaricaceae, 151, 283, 294
Agaricales, 42, 151, 283, 287, 294, 295
Agaricus bisporus, 283
Agriculture
  ant-plant, 148–149, 161–174
  as a form of symbiosis, 282
  commensal organisms of, 182, 298, 301–302
  comparative analyses of, 7–46, 168–171
  definition of, 3, 31, 143, 161, 210, 282, 297
  diseases of, 16, 35, 52, 55, 77, 100–101, 151, 153, 167–168, 172, 209–223, 286, 299, 303 (*see also* Pests, pest control; Weeds, weeding)
  division of labor and, 44–45, 90, 117, 152, 153, 239, 297, 300, 302
  eukaryotic pathogens of, 209–223
  evolutionary convergence on, types of, 17–18, 24–27
  firestick, 288
  human genetic evolution and, 245, 297, 303
  landscape modifications and, 186, 199, 298
  "material-goods" (*see* Agriculture: non-nutritional)
  morphological changes and (humans), 225–256
  mycangial facilitation of (ambrosia beetles), 128–130
  non-nutritional, 37, 150, 199, 282
  origins of (human), 1, 11, 32, 129, 170, 178–180, 182–183, 191–192, 199, 225–227, 287–289
  origins of (non-human), 1, 7–46, 77, 118, 168–171, 177–178, 286, 289–291, 301
  pathogens of (*see* Agriculture: diseases of; Pests, pest control; Weeds, weeding)
  proto-agriculture (ants), 144
  task specialization and, (*see* Agriculture: division of labor and)
Agroecosystems, 49–51, 177, 182, 282, 288, 298, 301, 302
Agro-predation, 152, 301
AGs. *See* Ant gardens
Alfalfa. See *Medicago sativa*
Algaculture, 11, 26
Algae, 10–14, 62, 70, 161
Allium, 191
Allo-convergent evolution, 25
Allomyrmococcini, 145
Allopatric speciation, 292
Alocasia, 193
Alpaca, 282
Amaranthus, 188
Amasa, 108
Ambriosiophilus spp., 16

Ambrosia beetles, 11–12, 15–16, 18, 20–24, 26–27, 42, 57, 80–82, 105–142, 167, 169–170, 295
auxiliary fungi, 115–117
definition of, 105
epidermal gland cytology, 112
gut transmission of fungi, 108–109, 120, 123, 129
primary ambrosia fungi, 42, 105–109
sociality, 116–117
subsociality, 117
Ambrosia *Fusarium* clade (AFC), 118, 119
Ambrosiella, 108, 114, 117, 124, 128, 129
Ambrosiella remansi, 108
Ambrosiella xylebori, 106
Ambrosiodmus, 16, 108, 122, 123
Ambrosiophilus, 16, 108, 122, 123
Ambrosiophilus atratus, 11
Amino acids, 111, 144, 151, 165, 240, 263, 297. *See also* Protein
Amoimyrmex, 299
Amorphophallus abyssinicus, 195
Amorphophallus, 193
Amygdalus persicus, 197
Ancistrotermes, 95, 98
Ancistrotermes cavithorax, 96
Angelis, K., 17
Anisandrus, 108, 112, 122, 124, 128
Annelida, 8–9, 11–14, 18, 20–21, 23–24, 27, 301
Ant agriculture, 11, 16–24, 33–34, 53, 143–159, 161–174, 281–331. *See also* Fungus-farming ants; Higher-attine ant agriculture; Lower-attine ant agriculture; Plant-farming ants
Anten, N. P. R., 44, 61–72
Ant gardens (AGs), 148, 149, 161–174
Ant-Homoptera associations, 144–147, 149–150, 152, 200, 290
Antibiotics, 16, 63, 66, 70, 149, 151, 153, 210, 286. *See also* Herbicides
Ant-Lepidoptera associations, 144, 146
Ant-plant associations, 146–149, 161–174
Ants, 1–2, 15–16, 18, 21, 23–24, 26, 33–34, 42, 44, 49, 51–54, 57, 63–64, 73–85, 92–93, 97–98, 106, 116, 118, 129, 143–159, 161–174, 209–210, 213, 215, 218, 225, 272–273, 281–313
Aphidae, 144–145, 147, 150, 152, 200, 290
Aphids. *See* Aphidae
Apocynaceae, 148, 162
Apterostigma, 216, 283, 294
Apterostigma collare, 147
Apterostigma dentigerum, 284
Apterostigma megacephala, 285
Arable habitat (defined), 178, 181–182
Araceae, 163, 191, 195
Archaeobotany, 185, 193–197, 288
Archaeogenomics, 292, 302
Arikara, 242
Arisaema schimperianum, 195
Armelagos, G. J., 244
Army ants, 92–94, 145, 290
Arrowroot. See *Tacca leontopetaloides*
Artificial selection, 11–13, 16, 55–56, 61–63, 73, 83, 130, 168, 191, 200, 257–259, 262, 270, 288, 291–293, 295–296
Asclepiadaceae, 148
Asia, 16, 49, 108, 146, 211, 245, 292
Anatolia, 227, 244
Central Asia, 190
China, 32, 184–185, 187, 189–190, 197, 226, 298
East Asia, 187–188, 193, 197, 235
India, 182–183, 188, 190–191, 193, 197, 225–226, 267
Japan, 197, 226–227, 229, 231–232, 237–238, 242–244, 246
South Asia, 187–188, 226, 231–232, 244
Southeast Asia, 145, 148, 162–164, 170, 193, 195
Southern China, 183, 287
Southwest Asia, 181, 183–187
West Asia, 188–189, 197
Yangtze basin, China, 186, 225
Yellow River basin, 225
Asian people, 245
Atkinson, T. H., 110
Atta, 16, 44, 75, 78, 79, 80, 216, 299, 300, 303
Atta cephalotes, 284
Atta colombica, 74
Atta texana, 11
Attina, 151, 282
Attini, 106, 282
Auerbach, B. M, 233, 234
Australasia, 148
Australia, 147, 193, 195, 288
Austroplatypus, 107
Austroplatypus incompertus, 112, 117, 120, 122
Avena abyssinica, 188
Avena byzantina, 188, 189
Avena chinensis, 188
Avena sativa, 188
Avena strigosa, 189
Avirulence genes, 217–218
Avocado, 197
Azteca, 146–148, 151, 162

Bacteria, 15–16, 51, 53–57, 62–66, 77, 116, 151, 153, 161, 168, 171, 210, 213, 218–219, 282, 286–287, 301–302
Baiocis, 107
Banana (*Musa* spp.), 194, 195, 200, 211
Bark beetles, 27, 105–106, 109, 112, 115–116, 118–121, 127
Barley. See *Hordeum vulgare*
Barteria fistulosa, 147
Bar-Yosef, O., 241, 244
Basidiomycota, 108, 118–120, 151
Batesian mimicry, 258
Batra, L. R., 110, 113
Beans, 42, 183, 212, 293. See also *Fabales*
Beaver, R. A., 110
Beaverium, 16
Beccarian bodies, 147
Beeson, C. F. C., 109
Béguelin, M., 234
Belt, T., 281
Beltian bodies, 147
Berger, B., 109
Bernal, V., 244
Biedermann, P. H. W., 105–142
Bijma, P., 68
Bilateria, 7–8, 17–18, 25–27

Black soils. See *Terra preta*
Black yeasts. *See* Chaetothyriales (black yeasts)
Blaeuer, M. W., 234
Blepharidatta, 290, 301
Blumeria graminis, 211
Blumeria spp., 210
Bolton, J. J., 14
Boomsma, J. J., 73–85, 98, 99, 295, 303
Bothrosternus, 107, 127
Botrytis cinerea, 210, 211
Bottle gourd, 32, 150, 282
Brachiaria deflexa, 188
Branch, G. M., 13
Branstetter, M. G., 296
Brassica, 212
Broadbean. See *Vicia faba*
Brown, A. C., 14
Buck, L. T., 225–256
Buckwheat. See *Fagopyrum*
Buia, M. C., 14
Bullia, 26
Bullia digitalis, 11–12, 13–14, 20–21, 23–24, 26
Burnett, B. A., 234
Burying beetles. *See* Silphidae

CAM. *See* Crassulacean-acid metabolism
Camels, 282
Camp-following organisms, 191–192, 199, 288–291
Camponotus, 148, 162
Camponotus femoratus, 163
Camptocerus, 108
Canarium, 197
Candela, M., 243
Cannabis, 282
Cao, C., 56
Carassius auratus, 261
Carassius sp., 261
Carbohydrates, 51, 66, 89, 217–218, 238–240, 273, 289–290, 294, 298
Carchesiopygus, 107
"Cargo-cult science," 50
Caribbean, 194
Carlson, D. S., 236, 244
Carroll, C. R., 300
Carrot. See *Daucus carota*
Carton nests, 147–150, 162–164, 168
Cassava. *See* Manioc
Cassava manihot, 191
Cats, 182, 288
Cattle, 12, 49, 61, 67, 153, 259, 267–268, 270, 272, 291, 300. *See also* Livestock
Cecropia, 147
Cenocephalus, 107
Centanni, M., 243
Central America, 162, 194, 281. *See also* Mesoamerica
Ceratocystidaceae, 112–113, 118–119, 124–126
Cercopidae, 144
Cereal crops, 178, 181–183, 185–191, 195–196, 199–200, 211–212, 238, 240
Chaetastus, 107
Chaetothyriales (black yeasts), 53, 147, 150–151
Character-character competition, 263
Chen, X., 289
Chenopodium, 183, 188, 288
Chicken, 68, 70, 260
Chickpea. See *Cicer arietinum*
Chilca Canyon, Peru, 194
Childe, G., 225
Cholodkovsky, N, 109
Chomicki, G., 148, 161–174
Cicadellidae, 144
Cicer arietinum, 183, 187
Cladosporium myrmecophilum, 150
Claustral colony founding, 78, 80, 299
Claw, K. G., 245
Climate change, 153–154, 296–297
Clonal propagation, 76–77, 82–83, 129–130, 191–192, 195, 200, 303. *See also* Monoculture; Vegetative propagation
Cnesinus, 107, 127
Cnestus, 108, 123
Cnestus mutilatus, 112, 124, 128
CNP. *See* Comoé National Park
Coccidae, 144
Codonanthe calcarata, 163
Coevolution, 98, 116, 128, 153, 164, 177, 182, 190, 201, 215, 219, 282, 286, 287, 293–295
Coevolutionary arms race, 190, 215
Coffee, 282
Coleoptera, 8, 10, 27, 105, 109, 301
Colletotrichum spp., 211
Colocasia esculenta, 191–194, 200
Commensalism, 115, 128, 143, 146, 152, 182, 199, 213–214, 282, 294, 303. *See also* Agriculture: commensal organisms of; Symbiosis
defined, 282
Common vetch. See *Vicia sativa*
Comoé National Park, Ivory Coast, Africa (CNP), 90–92, 94, 98
Consolandi, C., 243
Cónsole, G. M., 240
Convergent evolution, general principles of, 7, 17, 24–25
Coptodryas, 108, 123
Coptonotus, 121
Coral fungi. *See* Pterulaceae
Coral-fungus agriculture (ants), 283–285, 293–294
Corn. *See* Poales; *Zea mays*
Corthylina, 124
Corthylini, 107, 117, 124–125
Corthylus, 107, 111, 122, 124–125, 128
Corthylus punctatissimus, 111
Corthylus zulmae, 128
Costaroplatus, 107
Cotton, 68, 150, 199, 282
Cows. *See* Cattle; Livestock
Cranium, human, 229–231
Crassulacean-acid metabolism (CAM), 171
Crematogaster, 146, 148, 162
Crematogaster levior, 163
Crop rotation, 35, 52, 210–213, 219
Crop-weed complexes, 189
Crossotarsus niponicus, 107, 121, 122
Cryptoxyleborus, 108, 123
Cucumber, 288
Cucumis humifructus, 288
Cucurbita pepo, 260, 288

Cultivation
ambrosia beetles, 105–142
ants, 143–159, 161–174, 281–313
definition of, 2–3, 143
humans, 177–208
pre-domestication (ants), 144
pre-domestication (humans), 177, 183, 186, 199
termites, 89–104
Curculionidae, 10, 105, 109, 182, 288
Curcurbitales, 42. *See also* Squash
Cyclorhipidion, 108, 122
Cylindropalpus, 107
Cynodon, 195
Cyperus rotundatus, 195
Cyphomyrmex, 283, 284, 295
Cyphomyrmex costatus, 294
Cystoseria, 14
Cytoplasmic genes, 51. *See also* Mitochondria

Daceton armigerum, 144
Dactylipalpus, 107, 127
Damselfish. See *Stegastes nigricans*
Darwin, C., 7, 49, 73, 182, 191, 259
da Silva, F. M., 14
Daucus carota, 191
Debets, A. J. M., 99
de Fine Licht, H. H., 99
Degenerative joint disease, human, 232, 234
Dendroctonus, 119, 120
Denham, T., 177–208
Denison, R. F., 49–59, 68
Dentition, human. *See* Teeth, human
Deuterostomia, 8, 11, 17–18, 20–21
Developmental constraints, 17–18, 27, 257, 259–262, 275
Developmental plasticity, 192–193, 227, 233, 236, 241–245, 247, 269, 291
Devil's gardens, 81, 149
Deyrup, M. A., 110
Diacamma, 148
Diapus, 107
Diapus quinquespinatus, 121, 122
Diatrypella japonica, 108, 118, 128
Dickson Mounds, Illinois, 232–234
Digitaria exilis, 188
Digitaria iburua, 188
Dimorphism, fungi, 112–113, 295; *See also* Developmental plasticity; Sexual dimorphism, human
Dinoplatypus, 107
Dioscorea, 191, 193. *See also* Yams (*Dioscorea* spp.)
Dioscorea alata, 191, 193
Dioscorea bulbifera, 193
Dioscorea esculenta, 193
Directional selection, 185, 197, 227, 236, 241, 243–245, 247–248, 270
Dischidia, 162
Dischidia nummularia, 163
Diuncus, 124
Division of labor. *See* Agriculture: division of labor and
DNA, 52, 236, 294
Dogs, 182, 282, 288
Dolichoderinae, 144, 164, 169
Dolichoderus cuspidatus, 145, 146, 152
Doliopygus, 107, 121
Domatia, 82–83, 146–151, 161–166, 168–169, 171
Domestication
as speciation, 291
de-domestication, 189
definition of, 2, 31, 61, 83, 143–144, 177, 198, 291
development and, 257–280
of fungi by ambrosia beetles, 105–106, 118, 128–129
of fungi by termites, 89–104
of landscape. *See* Niche construction
of microbes, 210
of symbionts by ants, 143–159, 161–174, 293–296
of symbionts by humans, 1–2, 31–32, 129–130, 177–208, 291–293
pathways to, animal, 199, 291
pathway to, cereal, 178, 181–191
pathway to, long-lived perennials, 181, 196–198
pathway to, vegetative, 181, 191–195
population genetics of, 291–293
sociobiology of, 61–72
syndrome, 178, 181, 183, 185–186, 190, 291
traits, 32, 177–178, 181–183, 185, 191, 193, 195–196, 199, 258, 260–261, 291–293
Dominy, N. J., 245
Donald, C., 55
Donoghue, P. C. J., 17
Dos Reis, M., 17
dos Santos, M., 99
Doubledaya bucculenta, 109
Driesch, H., 273
Drosophila, 270
Drosophila melanogaster, 269
Dryocoetes, 120
Duroia hirsuta, 149

Eccoptopterus, 108, 123
Ecosystem engineering. *See* Niche construction
Ecosystem organization, natural selection and, 50–52
EFNs. *See* Extrafloral nectaries (EFNs)
Eggleton, P., 98
Einkorn wheat, 184, 187
Elaiosomes, 148, 290
Elephants, 147, 282
Elias, R., 56
Enamel hypoplasia, 32, 246
Ensete ventricosum, 195
Epiplatypus, 107
Erotylidae, 109
Escovopsioides, 286, 301
Escovopsis, 53, 54, 56, 57, 77, 210, 213, 215, 216, 217, 218, 286, 301
Eshed, V., 243
Ethiopian banana. See *Ensete ventricosum*
Eugomontia sacculata, 11, 14
Eupagiocerus, 107, 127
Euplatypus, 107
Eurasian people, 245
Europe, 145, 150, 187, 190, 194, 241, 298
Balkan Peninsula, 227, 231
Danube Gorges, 234
Greece, 232
Southeastern Europe, 226, 234

Western Europe, 248
Yugoslavia, 232, 237
European people, 245
Euwallacea, 108, 122, 123
Evo-devo, 227
Ewing, G. H., 242
Excoffier, L., 245
Extrafloral nectaries (EFNs), 82, 146, 148–149, 162. *See also* Post-anthetic nectaries

Fabales, 42
Fagopyrum, 188
Faircloth, B. C., 296
Feralization, 189–190, 195, 200. *See also* Domestication: de-domestication
Fertile Crescent, 31–32
Fertilization (of crops), 10, 12, 13–16, 18, 22, 36, 143, 148, 152–153, 164–166, 171–172, 186, 287, 298
Fiegler, H., 245
Fig wasps, 51, 64
Fiji, 81, 148–149, 163–164, 168–172
Fitness, 2, 31, 44–45, 49, 51–55, 57, 61, 63, 65–67, 80, 83, 96–97, 99–101, 128, 130, 143, 216, 257–258, 261–262, 268–269, 272, 291, 294, 296. *See also* Inclusive fitness; Kin selection
Flavodon ambrosius, 11, 16
Flavodon subulatus, 108, 118, 119, 123, 129
Flax. See *Linum usitatissimum*
Fleisch, S., 238, 244
Fodder crops, 188–190
Foll, M., 245
Forager, 1, 31, 196, 229–230, 232–233, 237–239, 243–244, 289. *See also* Hunter-gatherer
Formica francoeuri, 147
Formica rufa, 144
Formicidae, 10, 153
Formicinae, 144, 200
Foxtail millet. See *Setaria italica*
Fragaria, 260
Francke-Grosmann, H., 109, 110, 112
Frank, S. A., 64, 76
Froneman, P. W., 13
Fruit trees. *See* Domestication: pathway to, of long-lived perennials (by humans)
Fukase, H., 244
Fulgoridae, 144
Fuller, D. Q., 177–208, 289
Functional constraints, 17, 18, 27, 45, 269, 301
Fungus-farming ants, 2, 16, 49, 63, 73–85, 98, 118, 129, 147, 149–153, 161, 167–169, 170, 210, 281–313
  agricultural systems of, 283–286
  crop diseases of, 209–223, 286
  queen mating of, 74–76, 300
  structural gene rearrangement of, 297
Fungus-growing termites, 1, 15–16, 18, 23–24, 26, 33–34, 52, 57, 63–64, 73–75, 77–82, 89–104, 106, 116, 118, 129, 162, 167–170, 177–178, 210, 225, 298, 301
  community ecology of, 94–96
  growth rates, limiting factor of, 89
  interspecific competition, 94
  intraspecific competition, mound distribution and, 92
  isotope signatures between species, 96
  Macrotermes bellicosus, niche expansion and contraction of, 90–93
  niche differentiation through fungal symbionts, 93–96
Furniss, M. M., 110
Fusarium, 108, 113, 119, 123
Fusarium graminearum, 211
Fusarium oxysporum, 210, 211

Gallus gallus, 260
Gambi, M. C., 14
Ganyocerus papuanus, 122
Gardner, A., 53
Gawne, R., 1–4, 245, 257–280
Genetic bottleneck, 54, 63, 99–100, 128–129, 213, 292–293
Genetic determinism, 273–274
Genetic drift, 130, 226, 244
Genetic engineering, 11–13, 16, 171, 298
Genyocerus, 107
Geosmithia, 118
Geosmithia cnesini, 107
Geosmithia eupagioceri, 107
Geosmithia microcorthyli, 107
Gerardo, N. M., 10, 49, 50, 177, 209–223
Gesneriaceae, 148, 163
GH. *See* Growth hormone
Ginger. See *Zingiber officinale*
Ginter, J. K., 245
Gnamptogenys, 301
Gnamptogenys hartmani, 152
Gnathotrichus, 107, 117, 124, 125
Goldfish. See *Carassius auratus*
Gongylidia, 151, 272, 283, 290, 295–296
Gonzalez, P. N., 244
González-José, R., 242
Gowdy, J., 297
Gracilization, 228, 230, 235, 239, 241, 244, 246–247
Grafen, A., 53
Grain crops, 55, 178, 182, 185–188, 190, 195, 199, 236
Grapes, 197, 211
Grazer, 13, 15, 25–27
Greene, D. L., 242
Green Revolution, 55, 68
Groman-Yaroslavski, I., 288
Group selection, 49, 52–54, 56, 61, 66–67, 69
Growth hormone (GH), 240
Guldberg-Froslev, T., 98

Hadrodemius, 108
Haematopinus suis, 63, 65, 70
Hammond, J., 259
Hammond's law, 257–280
  defined, 262–267
Harlan, J. R., 185
Harrington, T. C., 105–142
Harris, S. A., 14
Hartig, T., 109
Hata, K., 12
Hayes, W. C., 239
Herbicides, 16, 167, 181, 201, 286, 298. *See also* Antibiotics
Hernández, M., 242

Heteroborips, 108, 123
HG economy. *See* Hunter-gatherer
Higher-attine ant agriculture, 16, 34, 151–152, 199, 272, 283–286, 290, 293, 295–297, 299, 300–302
Hoekstra, R. F., 99
Hölldobler, B., 302
Holmes, M. A., 241
Holocene, 187, 193–194, 287, 297
Homoptera, 144–146, 149, 152. *See also* Ant-Homoptera associations
Homo sapiens, 11, 27, 177, 286, 287
Honeydew, 144–146, 149–150, 152, 290
Hordeum vulgare, 182–183, 185–186, 189, 288, 292–293
Horizontal symbiont transmission, 64, 66, 78, 80–82, 90, 96–100, 116, 118, 129, 162, 169–170, 201, 284–285
Horntail. *See* Xiphydriidae
Horses, 282
House mouse. *See* Mice; *Mus musculus; Mus musculus domesticus*
Howe, J., 78, 79
Hubbard, H. G., 117, 119
Hulcr, J., 110, 121, 122
Human technology, analysis of convergent evolution of, 7
Hunter-gatherer, 11, 25–27, 228, 233–234, 289–290, 295, 298. *See also* Forager
Hydnophytum, 162
Hymenoptera, 8, 10, 27, 97, 106
Hypothenemus curtipennis, 107, 127

Iberomaurisian Complex, 230
ILPs. *See* Insulin-like peptides
Inbreeding depression, 69, 292
Inclusive fitness, 57, 63, 75, 258. *See also* Kin selection
Indocryphalus, 108, 122, 125
Indocryphalus pubipennis, 108, 126
Insulin-like peptides (ILPs), 269, 270
Intermediate economies, 177, 186, 194, 289
Internal transcribed spacer (ITS), 294
Inuit, 242, 245
Ipomoea batatas, 191, 193–194, 200
Ips, 120
Iran, 226
Iraq, 31, 226
Irish potato famine, 100, 213
Iso-convergent evolution, 25–26
Iterative evolution, 26
ITS. *See* Internal transcribed spacer

Janzen, D. H., 300
Japanese chestnuts, 197
Jennings, P., 55
Ješovnik, A., 44, 143–159, 296
Johnsongrass, 289
Jomon, 229, 232, 243–244
Jurado, S. B., 240
Jute, 199

Kaarikia abrahamsonii, 108, 118, 125
Kaifu, Y., 242
Kaneko, T., 112
Kato, M., 12
Kava, 282
Kay, A., 299
Kerstes, N. A. G., 99
Kiers, E. T., 49
Kilmer, W. L., 56
King Solomon, 154
Kin selection, 44, 52–53, 62–66, 129, 300. *See also* Inclusive fitness
Kodo millet. See *Paspalum scrobiculatum*
Kooij, P. W., 295
Korb, J., 89–104
Krall, L., 297
KT boundary, 153
Kubanek, J., 15
Kudzu. See *Pueraria montana*
Kuk Swamp, New Guinea, 193–194

Landscape domestication. *See* Niche construction
Languriinae, 109
Larsen, C. S., 234, 239
Lasius flavus, 145
Lasius fuliginosus, 145, 147, 150, 152, 153
Lateral symbiont transmission. *See* Horizontal symbiont transmission
Leaf-cutting ants, 1, 11–12, 15–16, 18, 20–21, 23–24, 33–34, 44, 74–75, 77, 79, 116, 151–153, 212, 216, 281, 283–286, 296–297, 299–303. See also *Acromyrmex*; *Amoimyrmex*; *Atta*
Lebanon, 31
Lecanopteris, 162
Lee, A. S., 245
Leeks, A., 99
Leguminosae, 50–51, 63–64, 66, 183, 185, 195, 199–200, 211, 291
Lens culinaris, 183–184, 187
Lentil. See *Lens culinaris*
Leonardoxa africana, 150
Lepiota, 283
Leucoagaricus, 283
Leucoagaricus gongylophorus, 11, 16, 80, 286, 295, 296
Leucocoprineae, 273, 283, 290, 293, 294–295, 298, 302
Leucocoprinus, 283
Levant, 225–227, 229–232, 235, 237–238, 241, 246, 248
Levin, B. R., 56
Levinton, J. S., 14
Lieberman, D. E., 242
Lifetime commitment, 73–85, 100, 117, 303. *See also* Symbiont fidelity
Limpet snail (*Scutellastra* spp.), 11–14, 20–21, 23–24, 26
Linseed, 211, 217
Linum usitatissimum, 183, 211, 217
Lipids, 111–112, 147, 263, 290
Littoraria, 22, 26
Littoraria irrorata, 11, 14, 15
Littorinidae, 8–9, 14, 27
Livestock, 1, 61, 66, 69–70, 73, 80, 188, 200, 257, 262, 292. *See also* Cattle; Horses; Oxen; Pigs
Lizard beetles. See *Doubledaya bucculenta*; Languriinae
Lloyd, M. W., 296

Lower-attine ant agriculture, 151, 290, 293–294, 298–300
Lucanidae, 109
Lycaena xanthoides, 147
Lycaenidae, 144, 146
Lycett, S. J., 7
Lymexylidae, 109

Macaranga, 146, 147
Machua. See *Tropaeolum tuberosum*
Macrotermes annandalei, 96
Macrotermes bellicosus, 64, 80, 90, 91, 92, 93, 94, 95, 97, 98, 100
Macrotermes carbonarius, 96
Macrotermes gilvus, 96
Macrotermes muelleri, 96
Macrotermes natalensis, 74, 98, 99
Macrotermes subhyalinus, 95, 98
Macrotermitinae, 10–12, 15, 18, 20–21, 23–24, 26, 74, 89–104, 106, 167–170
Magnaporthe, 213
Magnaporthe oryzae, 212, 217
Maize. *See* Poales; *Zea mays*
Mandible, human, 227–229, 235, 237–238, 240, 242–244, 246
Mandible, insect, 145, 281
Mango, 197
Manihot, 289
Manihot esculenta, 191
Manioc, 191, 193, 289. See also *Cassava manihot; Manihot; Manihot esculenta*
Martínez-Abadías, N., 242
Masticatory-functional hypothesis (MFH), 236–237, 242
May, H., 243
Mayers, C. G., 105–142
Mayr, E., 236
Mazella, L., 14
McGhee, G. R., 7–30, 45, 301
McKenna, K. Z., 245, 257–280
McQuaid, C. D., 13
Mealybugs. *See* Pseudococcidae
Mecopelmus, 121
Medicago sativa, 188, 189–190
Medicago truncatula, 51
Mediterranean, 190, 194, 197, 231
Mediterranean seagrass. See *Posidonia oceanica*
Megalomyrmex symmetochus, 152, 301
Megaplatypus, 107
Meiklejohn, C., 244
Melamed, Y., 288
Melampsora lini, 211, 217
Melons, 178, 184, 288
Membracidae, 144
Menéndez, L. P., 225–256
Meredithiella, 125
Mesoamerica, 32–34, 188, 197, 287, 298. *See also* Central America
Mesolithic, 229–232, 242
Mesoplatypus, 107
Mesopotamia, 33–34
MFH. *See* Masticatory-functional hypothesis
Mice, 64, 182, 288. See also *Mus musculus*; *Mus musculus domesticus*
Microascales, 107–108, 112, 117–119
Microbial consortia. *See* Microbiomes
Microbiomes, 54, 56–57, 116, 171, 243, 286
Microcorthylus, 107, 124
Microperus, 108, 123
Microtermes, 64
Microtermes lepidus, 95
Microtermes subhyalinus, 95
Microtermes toumodiensis, 96
Millet, 185, 188, 190, 213, 292, 296. See also *Brachiaria deflexa; Digitaria exilis; Digitaria iburua; Paspalum scrobiculatum; Pennisetum glaucum*
Mitochondria, 61, 63–64, 66, 76, 82. *See also* Cytoplasmic genes
Mitosomac, 107
Mixed food-acquisition strategies. *See* Intermediate economies
Mollusca, 8–9, 13, 20–21, 27, 45, 301
Monarthrum, 107, 117, 122
Monarthrum bicallosum, 107, 122, 124
Monoculture, 50, 64, 70, 101, 149, 151, 153, 169–171, 191, 201, 210, 212, 219, 303. *See also* Clonal propagation; Vegetative propagation
Monogamy. *See* Lifetime commitment
Morphospace, 259, 262
Mouse. *See* Mice
Mueller, U. G., 10, 49, 50, 57, 177, 299
Muir, W. M., 67, 68
Mullerian bodies, 147
Musa, 194. *See also* Banana
Mus musculus, 182
Mus musculus domesticus, 182
Mutualism, 2, 10–16, 61–64, 69–70, 73, 76–82, 96–97, 105, 109, 114–115, 126, 143–147, 150, 161–164, 178, 189, 200, 213–214, 282, 286, 299, 301. *See also* Symbiosis
  defined, 282
Mycangia
  coxal mycangia, 124
  cranial sphere mycangia, 121
  dual preoral pocket mycangia, 121–123
  elytral notch mycangia, 123
  fin-type pleural mycangia, 126
  genal brush mycangia, 127
  horseshoe pleural cavity mycangia, 125–126
  mesonotal pouch mycangia, 123–124
  mycangium cycle, 113–115
  nonglandular mycangia, 121, 127
  pleural cavity mycangia, 125–126
  pro-mesonotal mycangia, 123
  pronotal mycangia, 120–121, 126
  prothoracic tube mycangia, 124–125
  pseudomycangia, 127
  setose patch mycangia, 126–127
  single oral pocket mycangia, 124
  understudied mycangia, 127–128
Mycetophylax asper, 284
Mycosphaerella, 11, 15
Mycosphaerella graminicola, 211
Myoplatypus, 107
Myrmecodia, 147, 162, 169
Myrmecodia tuberosa, 163

Myrmecopterula, 283
Myrmecotrophy, 148
Myrmelachista schumanni, 149
Myrmicinae, 10, 106, 144, 282, 301

Nadel, D., 288
Nakashima, T., 120
Nassaney, M. S., 234
Nassariidae, 9, 14, 26
Natufian, 229
Natural selection, 2, 7, 18–19, 49–54, 55, 57, 73, 77, 81, 130, 185, 241, 258–259, 262, 272, 281, 291–293, 295–296
Nectaries. *See* Extrafloral nectaries (EFNs); Post-anthetic nectaries
Neogastropoda, 8, 9, 27
Neolithic, 34, 185, 187, 198, 229–233, 241, 244, 248, 298
Neolithic revolution, 32, 225
Neotrachyostus, 107
Neotropics, 146, 148, 151–152, 162–164, 215
Nereididae, 9, 14
Neutral theory of ecology, 258
Newell, S. Y., 14, 15, 22, 26
New Guinea, 32, 163–164, 169–170, 191, 193, 200, 225–226, 298, 301
Niche construction, 73, 90, 93, 100, 144, 199, 225, 282, 286–291, 293, 298, 301. *See also* Agriculture: landscape modifications and
Nitrogen, 13–14, 51, 64, 66, 89, 95, 111, 147, 166, 171–172, 248, 301–302
Nomadic or migrating herdsman, 145, 152
North America, 32, 183, 188, 194–195, 231, 234–235, 239–240
  Arkansas, 233
  California, 147, 227, 232
  Eastern North America, 225–226
  Georgia, 227, 231–234, 239
  Illinois, 226–227, 232–234
  Illinois Valley, 233
  Louisiana, 233
  Mississippi Valley, 32, 227, 232–234
  Ohio, 226–227, 248
  Ohio Valley, 227, 231–232, 234
  South Dakota, 242
  Southern North America, 233
  Western and southern North America, 226
Notoplatypus, 107
Nuclear genes, 51–52
Nunberg, M., 109, 126, 127

Oats, 188–190, 216. See also *Avena abyssinica*; *Avena byzantina*; *Avena chinensis*; *Avena sativa*; *Avena strigosa*
Oca. See *Oxalis tuberosa*
Odontomachus, 148
Odontotermes, 96, 98, 99
Odontotermes formosanus, 11
Onion. See *Allium*
On the Origin of Species, 7, 259, 275
Oomycetes, 209–212, 217
Ophiostoma novo-ulmi, 113
Ophiostomatales, 107–108, 112–113, 118–120
Ophiostoma ulmi, 113
Ortiz-Santiago, V., 15
Oryza sativa, 49–50, 52, 55–56, 61, 178, 181–186, 189, 212–213, 215–217, 232, 235, 289, 291–293
Osteoarthritis, human, 234
Oti Keran National Park (Togo), 95
Oxalis tuberosa, 194
Oxen, 282
Oxoplatypus, 107
Oxoplatypus quadridentatus, 122
Oyhenart, E. E., 240

Pachycondyla, 148
Palau, 195
Paleolithic, 34, 230, 241–242, 287, 289
Paleotropics, 146, 148
Papua New Guinea. *See* New Guinea
Parabiosis, 162
Parasitic domesticoids, 177, 181–182, 190, 195
Parasitism, 2, 10, 12–13, 15–16, 18, 96, 115–116, 143–144, 146, 152–153, 161, 177, 181–182, 195, 209–210, 282, 286, 301–303. *See also* Agriculture: diseases of; Parasitic domesticoids; Pests, pest control; Symbiosis; Weeds, weeding
Paspalum scrobiculatum, 188, 190
Pastoralist, 228
Patella [=*Scutellastra*] *longicosta*, 13
Patellogastropoda, 8–9, 27
Pathogens (non-agricultural), 22, 38, 53–54, 101, 113, 147, 298, 302. *See also* Agriculture: diseases of; Pests, pest control; Weeds, weeding
Pea. See *Pisum sativum*
Peach. See *Amygdalus persicus*
Pearl millet. See *Pennisetum glaucum*
Pei, A., 245
Peled, N., 243
Peng, S., 56
Pennisetum glaucum, 185, 292, 296
Peregrine, P. N., 1–4, 31–46, 300
Pereioplatypus, 107
Perez, S. I., 244
Periwinkle snails (*Littoraria* spp.), 11–15, 20–21, 23–24, 26
Peroplatypus, 107
Perry, G. H., 245
Pests, pest control, 10–13, 15–16, 22, 35, 42, 52, 54, 55, 68–69, 143, 151, 153, 167, 172, 201, 283, 287, 298. *See also* Agriculture: diseases of; Weeds, weeding
Petalomyrmex-Leonardoxa system, 151
Petalomyrmex phylax, 150
Phaerosphaeria, 11, 15
Phakopsora pachyrhizi, 210, 211
Pheidole, 148
Pheidole bicornis, 162
Phialophoropsis, 108, 126
Philidris, 81, 147
Philidris cordata, 162
Philidris nagasau, 148, 153, 163–172
Philodendron sp., 163
Phloeoborus, 107, 122, 127
Phloeoborus cf. *niger*, 111
Phoenix dactylifera, 198
Photosynthesis, 50, 52, 67, 153, 171, 282
Phytophthora infestans, 211, 213, 216

Pigs, 264–267, 291
Pinhasi, R., 243, 244
Piper, 162
Pisum sativum, 183–184
Pityoborus, 127
Plagányi, E. E., 13
Plant-farming ants, 146–149, 161–174
Platynereis, 26
Platynereis dumerilii, 11, 14
Platyphysus, 107
Platypodinae, 105, 107, 109, 117–118, 120–121, 124, 127–128
Platypodini, 121, 130
Platypus, 107, 120
Platypus cylindrus, 122
Pleomorphism. *See* Developmental plasticity; Dimorphism, fungi; Sexual dimorphism, human
Poales, 42. See also *Zea mays*
Pokhojaev, A., 243
Polychaeta, 8–9, 11–12, 14, 20–21, 23–24, 26–27
Polymorphism. *See* Developmental plasticity; Dimorphism, fungi; Sexual dimorphism, human
Polyneoptera, 8, 10, 27
Polynesia, 194–195, 200. *See also* Fiji
Polyploidy, 79, 151, 283, 295, 297, 302
Polysiphonia sp. 1, 11, 12
Porodic hyperostosis, 32
Posidonia oceanica, 14
Post-anthetic nectaries, 164–165. *See also* Extrafloral nectaries (EFNs)
Potato. See *Solanum tuberosum*
Pre-domestication cultivation. *See* Cultivation: pre-domestication (humans); Intermediate economies
Premnobius, 108
Premnobius cavipennis, 111, 118, 122, 123
Protein, 1, 75–76, 89, 112, 144, 147, 152, 239–240, 245–246, 270–271, 273, 290, 296. *See also* Amino acids
Proto-agriculture. *See* Agriculture: proto-agriculture (ants)
Protohylastes, 121
Protoplatypus, 121
Protostomia, 9, 12–13, 15, 17–18, 20–21
Pseudacanthotermes militaris, 95, 96
Pseudo-cereals, 188
Pseudococcidae, 145–147
Pseudomyrmex, 146
Pseudomyrmex peperi, 147
Pseudomyrmex triplarinus, 149
Pseudonocardia, 16, 53, 286
Psyllidae, 144
Pterocyclon bicallosum, 124
Pterulaceae, 151, 170, 283, 294
Pucciarelli, H. M., 230, 240, 241
Puccinia, 210, 211
Pueraria lobata, 195
Pueraria montana, 195
Pumpkin. See *Cucumis humifructus; Cucurbita pepo*
Pyricularia graminis-tritici, 211, 213, 216
Pyricularia oryzae, 212, 213, 215, 217
Pyricularia spp., 213

Qin, L., 289
Quantitative traits, 217, 269
Quseir al-Qadim (Egypt), 194

Raffaelea, 111, 113, 117, 118, 119, 121, 122, 123, 125, 128
Raffaelea cf. *canadensis*, 108
Raffaelea scolytidis, 107
Raffaelea sensu lato, 108
Raffaelea sulphurea, 107, 108
Raj, S. M., 245
Ralfsia verrucosa, 11, 13
Ramirez-Rozzi, F., 230, 241, 242
Rampelli, S., 243
Rat. See *Rattus norvegicus*
Rattus norvegicus, 270, 271
Reaction norms, 269–272
Red clover. See *Trifolium pratense* var. *sativa*
Red jungle fowl. See *Gallus gallus*
Reindeer, 52
Remansus mutabilis, 108
Retro-convergent evolution, 25, 27
Rhizobia, 51, 57, 63–64, 66
Rhizoctonia solani, 210, 212
Riccillo, F. L., 240
Rice. See *Oryza sativa*
Riodinidae, 144, 146
Roeper, R. A., 117
Rose, J. C., 234
Rosendahl, S., 98
Rouland-Lefevre, C., 98
Rubiaceae, 148–149, 162–163
Ruderal, 181, 189–190
Rudinsky, J. A., 114, 116
Ruff, C. B., 234, 239, 241
Rye. See *Secale cereale*

Saccharum officinarum, 194, 195, 200
Sachs, J. L., 57
Sahel. *See* Africa: Sub-Saharan Africa
Saint-Hilaire, G., 257
Santos, J. C., 299
Sarcopterygii, 8, 27
Sardi, M., 242
Sardi, M. L., 230, 234, 241
Sarig, R., 243
Sauer, C. O., 200
Sauromatum nubicum, 195
Schedl, W., 110, 124
Schedlarius, 121
Schiøtt, M., 78, 79, 295
Schlamp, F., 245
Schneider, I., 114, 116
Schnorr, S. L., 243
Schultz, T. R., 1–4, 10, 44, 49, 50, 143–159, 177, 281–313
Scolytinae, 10, 105, 107–108, 118–119, 125–126
Scolytini, 127
Scolytodes culcitatus, 107, 127
Scolytodes unipunctatus, 107, 127, 128
Scolytoplatypodini, 108, 126
Scolytoplatypus, 108–109, 122
Scolytotarsus, 121
Scolytus ventralis, 120

Scutellastra chapmani, 13
Scutellastra cochlear, 13
Scutellastra flexuosa, 13
Scutellastra laticostra, 13
Scutellastra longicosta, 11, 13
Scutellastra mexicana, 13
Scutellastridae, 9, 13, 26
Seal, J. N., 299
Sebright, J., 259
Secale cereale, 188, 190, 298
Segetal, 181, 189–190
Sella-Tunis, T., 243
Sericomyrmex amabilis, 152
Setaria italica, 185
Sexual dimorphism, human, 229–230, 232, 234–235, 239, 244, 246
Sexual selection, 36, 258
Shade avoidance syndrome, 68
Shaw, P., 243
Shik, J. Z., 299
Ship-timber beetles. *See* Lymexylidae
Sieg, R. D., 15
Silkworm, 282
Sillen, A., 241, 244
Silliman, B. R., 14, 15, 22, 26
Silphidae, 109
Siricidae, 109
Sitophilus granarius, 182
Sitophilus oryzae, 182
Six, D. L., 10, 49, 50, 110, 177
Skeleton, human, 225–256
long bones, 227–228, 233
postcranial, 232–235
skull, 227–228, 236–237, 240–243, 246–247
Slime molds, 210
Smith, P., 241, 244
Snails, 171, 210, 257. See also *Bullia*; *Bullia digitalis; Littoraria*; *Littoraria irrorata*; Littorinidae; *Platynereis*; *Platynereis dumerili; Scutellastra chapmani*; *Scutellastra cochlear*; *Scutellastra flexuosa*; *Scutellastra laticostra*; *Scutellastra longicosta*; *Scutellastra mexicana*; Scutellastridae
Snir, A., 288
Social control, correlated with agriculture, 41, 43–45, 300
Solanum tuberosum, 42, 100, 191, 193–194, 211, 213, 216
Sorghum. See *Sorghum bicolor*
Sorghum bicolor, 182, 183, 189, 287, 289, 298
Sosa-Calvo, J., 296
South America, 188, 194, 213, 230, 232, 235, 239–240, 242, 244, 281–283, 290, 296, 299
Amazon, 81, 149, 170, 288, 298
Andes, 32
Andes, Argentinean, 227
Andes, Central, 33, 194, 226
Andes, Chilean, 227
Andes, Ecuadorian, 227
Andes, Peruvian, 194, 226–227
Andes, southern, 230, 234, 237
Argentina, 230, 232, 234
Brazil, 147, 213
Casma Valley, Peru, 194
Chilca Canyon, Peru, 194
Chile, 232
Eastern South America, 298
Ecuador, 194, 231–232
Paloma, Peru, 231–232, 234
Peruvian Andes, 194, 226
Western South America, 225
Soybeans, 51, 178, 184, 211
Spartina alternifora, 15
Spathidicerus, 107
Sperm competition, 75–76, 81, 300
Spiralia, 9, 12–13, 20–21
Sporothrix schenkii, 113
Spurway, H., 272
Squamellaria, 81, 148, 149, 163, 164, 165, 166, 168
Squamellaria imberbis, 165, 171
Squamellaria major, 163
Squamellaria thekii, 163
Squamellaria wilsonii, 165, 171
Squash, 42, 288, 298
Stegastes nigricans, 8, 11–13, 18, 20–21, 23–24, 26, 70, 161
Stelinski, L. L., 110, 121, 122
Sternberg, M., 288
Stock, J. T., 245
Stomaphis quercus, 145
Strawberry. See *Fragraria*
Streptomyces, 282, 286
"Subdivided cows" (leaf-cutter ant colonies), 44, 151–152, 300
Sucking louse. See *Haematopinus suis*
Sueus, 108, 128
Sugarcane. See *Saccharum officinarum*
Superorganisms, 1, 75, 77, 151, 297, 300, 302
Sustainable agriculture, 49–50, 153–154, 166–167, 172, 303
Suwa, G., 244
Sweet potato. See *Ipomoea batatas*
Swenson, W., 56
Symbiont fidelity, 119, 215, 285. *See also* Lifetime commitment
Symbiosis, 10, 12, 51, 66, 73–74, 78–79, 83, 89–90, 93, 99–100, 105–106, 116, 120, 126, 128–130, 144, 148–151, 162–165, 181, 198, 281–282, 286, 291, 294, 298. *See also* Agriculture: as a form of symbiosis; Agriculture: commensal organisms of; Agriculture: diseases of; Commensalism; Mutualism; Parabiosis; Parasitic domesticoids; Parasitism; Pests, pest control; Weeds, weeding
defined, 282
Sympatric speciation, 292
Syria, 31

Tacca leontopetaloides, 195
Taforalt site, Algeria, 230
Taro. See *Colocasia esculenta*
Task specialization. *See* Agriculture: division of labor and
Taurodemus, 108
Taveuni. *See* Fiji
Teeth, human, 32, 227, 231–232, 235, 237–238, 242, 244, 246–247
Telford, M. J., 17
Teloplatypus, 107

Temporomandibular joint (TMJ), 229
Terminal Eocene Event, 296
Termites, 75, 79, 290. *See also* Fungus-growing termites; Macrotermitinae
Termitomyces, 11, 15, 16, 78, 79, 81, 89, 90, 97, 99, 100
Terra preta, 288, 298
Tetraponera aethiops, 147
Thawornwattana, Y., 17
TMJ. *See* Temporomandibular joint
Tobacco, 282
TOCs. *See* Tragedies of the commons
Tomato, 211, 216, 288
Tomicus, 120
Toshionella, 108, 126
Trachymyrmex septentrionalis, 284
Trachyostus, 107
Tragedies of the commons (TOCs), 61, 67–70, 75–76
Trapa natans, 178, 183
Treptoplatypus, 107
Trifolium pratense var. *sativa*, 188
Triozastus, 107
Triticum dicoccon, 183
Triticum monococcum, 183
Triticum timopheevi, 183
Tropaeolum tuberosum, 194
Trypodendron, 108, 112, 113, 117, 122, 125, 126
Trypodendron betulae, 111
Tudge, C., 289
Turkey, 31

Ule, Ernst, 162
Ulluco. See *Ullucus tuberosus*
Ullucus tuberosus, 194
Ultrasociality, 297
Ulva, 14
Unconscious selection, 61, 182–183, 191–192, 200, 213, 288
Ustilago maydis, 212
Ustilago spp., 210

Vachellia, 162
Van Gerven, D. P., 236, 244
Vanua Levu. *See* Fiji
Vavilov, N. I., 188
Vegecultural pathway to domestication. *See* Domestication: pathway to, vegetative
Vegetative propagation, 181, 189, 191–197, 199–201, 272–273, 289. *See also* Clonal propagation; Domestication: pathway to, vegetative; Monoculture
Vertical symbiont transmission, 63–64, 78–82, 90, 96–100, 108, 128–129, 150–151, 169–170, 210, 283, 294
Vicia faba, 183
Vicia sativa, 188–190
Virk, P., 56

Wasmannia, 290, 301
Watanabe, K., 12
Water chestnuts. See *Trapa natans*
Weeds, weeding, 10, 12, 13–16, 18, 22, 35, 42, 44, 53, 63, 70, 114, 116–117, 147, 151–153, 161, 167–168, 178–182, 187–191, 195, 199, 201, 213, 286, 288–290, 298. *See also* Agriculture: diseases of; Pests, pest control
Weevils. *See* Curculionidae; *Sitophilus granarius; Sitophilus oryzae*
Weismann, A., 257
Wells, J. C., 245
West, S. A., 49, 99
Wheat, 55, 182–183, 185–187, 211, 213, 216, 288, 291–293, 297. See also *Triticum dicoccon*; *Triticum monococcum*; *Triticum timopheevi*
Wheeler, W. M., 75, 77
Whelk snail. See *Bullia digitalis*
Willey, D., 15
Wilson, D. S., 56
Wilson, E. O., 300, 302
Wolfe, K., 15
Wolff's law, 237–239, 242, 247
Wolfgangiella, 108
Woltereck, R., 269
Woo, J. Y., 110
Woodin, S. A., 14
Wood wasp. *See* Siricidae
Wright, S., 18

Xing, D., 56
Xiphydriidae, 109
Xyleborini, 10, 108, 117, 122–123, 167, 169–170
Xyleborinus, 108, 117, 122
Xyleborinus attenuatus, 123
Xyleborinus gracilis, 123
Xyleborinus saxesenii, 111, 123
Xyleborinus schreineri, 123
Xyleborinus sentosus, 123
Xyleborus, 108, 117, 122
Xylosandrus, 108, 111, 114, 117, 122
Xylosandrus amputatus, 111
Xylosandrus compactus, 106
Xylosandrus germanus, 111, 112
Xyloterini, 108, 117, 122, 125–126
Xyloterinus politus, 108, 111, 122, 125

Yams (*Dioscorea* spp.), 191, 193, 289, 296
Yayoi, 229, 232
Yeast agriculture, fungus-farming ants, 283–285, 293–295, 297
Yeasts, 116, 282. *See also* Chaetothyriales (black yeasts); Yeast agriculture, fungus-farming ants; Yeasts, "yeast-like" growth, ambrosia beetles
Yeasts, "yeast-like" growth, ambrosia beetles, 113, 115, 123
Y'Edynak, G., 238, 244
Younger Dryas, 287
Yuan, W., 56

Zea mays, 32, 42, 55, 68, 182, 185, 211–212, 240, 261, 291–292, 298
Zhao, Z., 289
Zheng, Y., 289
Zingiber officinale, 191
Zoonotic diseases, 219
Zupo, V., 14
Zymoseptoria tritici, 211